BASIC IDEAS AND CONCEPTS IN NUCLEAR PHYSICS
THIRD EDITION

T0289667

Series in Fundamental and Applied Nuclear Physics

Other books in the series

Accelerator Driven Subcritical Reactors
H Nifenecker, O Meplan and S David

Nuclear Dynamics in the Nucleonic Regime
D Durand, E Suraud and B Tamain

Statistical Models for Nuclear Decay
A J Cole

Nuclear Methods in Science and Technology
Y M Tsipenyuk

Nuclear Decay Modes
D N Poenaru

SERIES IN FUNDAMENTAL AND APPLIED NUCLEAR PHYSICS

Series Editors
R R Betts and W Greiner

BASIC IDEAS AND CONCEPTS IN NUCLEAR PHYSICS

AN INTRODUCTORY APPROACH
THIRD EDITION

K Heyde

*Department of Subatomic and Radiation Physics,
Universiteit Gent, Belgium*

INSTITUTE OF PHYSICS PUBLISHING
BRISTOL AND PHILADELPHIA

British Library Cataloguing-in-Publication Data

A catalogue record for this book is available from the British Library.

ISBN 0 7503 0980 6

Library of Congress Cataloging-in-Publication Data are available

First edition 1994
Second edition 1999
Third edition 2004

Front cover image: ISOLDE collaboration/CERN, CERN Courier May 2004 p 26, © CERN Courier.

Commissioning Editor: John Navas
Production Editor: Simon Laurenson
Production Control: Leah Fielding
Cover Design: Victoria Le Billon
Marketing: Nicola Newey

Published by Institute of Physics Publishing, wholly owned by The Institute of Physics, London

Institute of Physics Publishing, Dirac House, Temple Back, Bristol BS1 6BE, UK

US Office: Institute of Physics Publishing, The Public Ledger Building, Suite 929, 150 South Independence Mall West, Philadelphia, PA 19106, USA

Typeset in LaTeX 2_ε by Text 2 Text Limited, Torquay, Devon
Printed in the UK by MPG Books Ltd, Bodmin, Cornwall

For Daisy, Jan and Mieke

Contents

Preface to the third edition

Since the preface to the second edition was written, back in September 1998, nuclear physics has been confronted with a large number of new and often ground-breaking results all through the field of nuclear physics as discussed in the present book. Therefore, but also because copies of the former edition were running out, working on a new and fully updated edition became mandatory. Even though the overall structure has not been changed, a number of the boxes have become obsolete and have thus either been removed or modified in a serious way. The number of highlights in the period 1999–2003 has been so large that a good number of new boxes have been inserted, indicating the rapidly changing structure of a lively field of physics. Of course, this has the drawback that a number of topics do not always 'age' very well. In this sense, the new edition tries to convey part of the dynamics in the field of nuclear structure, in particular over the period between the appearance of the second edition and the time of writing this new preface.

First of all, some new problems have been added to the existing list and I am particularly grateful to E Jacobs, who has taught the course on Subatomic Physics at the University of Gent in recent years, for supplying hints on these extra exercises. They have served as examination questions over a number of years and, are thus, well tested, feasible and form an essential part of the book in order to acquire a good knowledge of the basic ideas and concepts.

In part A, most changes in the general text are not very extensive, except in the chapter on β-decay and new results introduced by modifying boxes, deleting some and introducing new ones. In chapter 1, a new box has been introduced with the most recent results on electric charge and magnetic density distributions of the proton and neutron and the box on super-heavy elements has, of course, been updated. Box 1f has been deleted and the figures, related to the largest detectors and accelerators, have likewise been updated. In chapter 2, on radioactivity, the section on exotic decay modes has been updated and recent examples of proton radioactivity have been included. It is in part B, in chapter 3 on β-decay that there has been a substantial addition when discussing the role of the neutrino. Sections on inverse β-decay and double β-decay have been updated. In the section on the neutrino mass, however, we discuss in some detail (new section 5.4.5) the issue of neutrino oscillations. Here we have tried, albeit in a very concise way, to summarize the ground-breaking results on (a) the solar neutrino problem and its solution, (b) the consequences derived from the study of atmospheric neutrinos and (c) the results on earth-based neutrino experiments. Extensive references to the recent literature are given, a number of keynote figures are included and a new box on the vanishing of neutrinos is included. In part C, no major changes have been made, except for an extended discussion on two-neutron separation energies within the liquid-drop model, discussed in chapter 7.

Since part D is the part in which the relation between 'basic ideas and concepts' in nuclear physics, presented in parts A B and C that lends itself to a general course and the more advanced topics in nuclear structure research is made, it is obvious that a large number of changes have been included.

In chapter 11, we have updated the present text in the discussion of state-of-the-art large-scale shell-model calculations (text and new figures) and have also replaced box 11a with more recent results. We have added a part on level densities in the nuclear shell model and brought in a new section, describing how a detailed study of the spectral properties of the shell-model many-body system can teach us about symmetries or randomness that may be present in the nuclear interactions.

We have brought in a new chapter 12, describing our present highly increased knowledge on very light systems (nuclei up to mass $A \approx 10$). Here, both a number of the general properties of bound systems in nuclear stability on a more phenomenological level as well as exciting results derived from *ab initio* methods have been included. This chapter is well illustrated with typical figures.

Before entering, with the discussion in chapters 13 and 14, the domain of nuclear collective motion, we present a transition section indicating the present stage in describing atomic nuclei as obtained through the discussion presented in chapters 7–12.

In chapter 13, we have updated the section on the Interacting Boson Model description of collective low-lying excitations by incorporating the most recent results. In this chapter, we have also included a discussion on shape coexistence and phase transitions, a theme that forms an intensive topic of research, both experimentally and theoretically and we highlight this with a box on triple shape coexistence in ^{186}Pb. In chapter 14, at the end, we have updated and extended the discussion on gamma-arrays and presented the strong activity in building detector systems with a highly increased sensitivity.

Chapter 15, in which we discuss nuclei far from stability, has been extended at a number of places. We have inserted a more detailed discussion on possible modifications of the nuclear mean field as well as on the experimental evidence for such changes in particular for light nuclei. In particular, section 16.3 which discusses radioactive beams (RIB) has been reworked in the light of the world-wide efforts for exploring exotic nuclei. It is, in particular, the box on RIB facilities and projects that has been totally rewritten in the light of the huge efforts in this domain since the appearance of the second edition. For most facilities and projects, reference to the appropriate web page is given, precisely because in this discipline, during the time between finishing this editorial and the appearance of this new edition, new decisions may have been taken on RIB facilities. It is suggested that the reader consults these web pages for updates in the field of RIB.

Chapter 16, in which an excursion is made into the field beyond nucleon degrees of freedom, important modifications, updating and additions have been implemented. The old boxes 15b and 15c have been left out and a new box discussing the very recent observation of pentaquark configurations introduced. The boxes on the spin structure of the nucleon and the first hints for quark–gluon plasma formation, as derived from the very recent Relativistic Heavy Ion Collider (RHIC) experiments, have been greatly modified (highlighted with new figures). The text on the Continuous Electron Beam Accelerator Facility (CEBAF) (now renamed as JLab) and on the quark–gluon phase of matter has been brought up-to-date so as to enable this chapter to be a starting point for further reading and study.

Finally, chapter 17 has been modified in a number of places but the main philosophy remains in that new developments in accelerator methods, detectors and analysing methods as well as increased computer capabilities for modelling the nuclear many-body system form the essential ingredients for progression in the field. The past five years in nuclear physics research since the appearance of the second edition is a vivid illustration of this point.

I hope that the many changes with respect to the former two editions indicate the very dynamic nature of the nuclear physics community. There will be omissions and my selection of new boxes definitely contains bias. In this respect, I am open to any comments that readers and students may have. I am also very grateful to all my students and colleagues who have used the book and pointed out ways for improvement. I have benefitted a lot from the valuable remarks I received from people who have used the book in their teaching of nuclear physics. I would like to thank in particular, N Jachowicz for a critical reading of the new sections on neutrino physics and I am grateful to P Van Duppen for checking my 'update' on RIB facilities and projects. He helped me towards completeness but for all remaining errors and omissions, I am to blame. I am, moreover, particularly grateful to the new group of young PhD students for creating an exciting atmosphere that has influenced the chapters on the nuclear shell model and on collective models. I would also like to thank the CERN-ISOLDE group for its hospitality during the phase of the 'final touch', CERN, the FWO-Vlaanderen for financial support in various phases during this reworking of the book and definitely my own university, the University of Gent (UGent) for strong overall support in my research projects.

Finally, I should thank R Verspille for the great care he took in modifying figures, preparing new figures up to the highest standards of lay-out and precision. His artistic touch can be seen on almost every page of the book. I am also particularly grateful to J Navas for initiating this third edition of the book and to the whole group at the IOP for taking my many pieces of text, figures, new references and perfectly constructing this new edition. Last but not least, I have to thank my wife, Daisy, for support during the whole process which is more than just a simple writing task. Without her presence, this new edition would never have been finished in time and I dedicate this new version to her and the kids Jan and Mieke.

Kris Heyde
CERN—February 2004
Gent—April 2004

Preface to the second edition

The first edition of this textbook was used by a number of colleagues in their introductory courses on nuclear physics and I received very valuable comments, suggesting topics to be added and others to be deleted, pointing out errors to be corrected and making various suggestions for improvement. I therefore decided the time had come to work on a revised and updated edition.

In this new edition, the basic structure remains the same. Extensive discussions of the various basic elements, essential to an intensive introductory course on nuclear physics, are interspersed with the highlights of recent developments in the very lively field of basic research in subatomic physics. I have taken more care to accentuate the unity of this field: nuclear physics is not an isolated subject but brings in a large number of elements from different scientific domains, ranging from particle physics to astrophysics, from fundamental quantum mechanics to technological developments.

The addition of a set of problems had been promised in the first edition and a number of colleagues and students have asked for this over the past few years. I apologize for the fact these have still been in Dutch until now. The problems (collected after parts A, B and C) allow students to test themselves by solving them as an integral part of mastering the text. Most of the problems have served as examination questions during the time I have been teaching the course. The problems have not proved to be intractable, as the students in Gent usually got good scores.

In part A, most of the modifications in this edition are to the material presented in the boxes. The heaviest element, artificially made in laboratory conditions, is now $Z = 112$ and this has been modified accordingly. In part B, in addition to a number of minor changes, the box on the 17 keV neutrino and its possible existence has been removed now it has been discovered that this was an experimental artefact. No major modifications have been made to part C.

Part D is the most extensively revised section. A number of recent developments in nuclear physics have been incorporated, often in detail, enabling me to retain the title 'Recent Developments'.

In chapter 11, in the discussion on the nuclear shell model, a full section has been added about the new approach to treating the nuclear many-body problem using shell-model Monte Carlo methods.

When discussing nuclear collective motion in chapter 12, recent extensions to the interacting boson model have been incorporated.

The most recent results on reaching out towards very high-spin states and exploring nuclear shapes of extreme deformation (superdeformation and hyperdeformation) are given in chapter 13.

A new chapter 14 has been added which concentrates on the intensive efforts to reach out from the valley of stability towards the edges of stability. With the title 'Nuclear physics at the extremes of stability: weakly bound quantum systems and exotic nuclei', we enter a field that has progressed in major leaps during the last few years. Besides the physics underlying atomic nuclei far from stability, the many technical efforts to reach into this still unknown region of 'exotica' are addressed. Chapter 14 contains two boxes: the first on the discovery of the heaviest $N = Z$ doubly-closed shell nucleus, ^{100}Sn, and the second on the present status of radioactive ion beam facilities (currently active, in the building stage or planned worldwide).

Chapter 15 (the old chapter 14) has been substantially revised. Two new boxes have been added: 'What is the nucleon spin made of?' and 'The quark–gluon plasma: first hints seen?'. The box on the biggest Van de Graaff accelerator at that time has been deleted.

The final chapter (now chapter 16) has also been considerably modified, with the aim of showing how the many facets of nuclear physics can be united in a very neat framework. I point out the importance of technical developments in particle accelerators, detector systems and computer facilities as an essential means for discovering new phenomena and in trying to reveal the basic structures that govern the nuclear many-body system.

I hope that the second edition is a serious improvement on the first in many respects: errors have been corrected, the most recent results have been added, the reference list has been enlarged and updated and the problem sections, needed for teaching, have been added.

The index to the book has been fully revised and I thank Phil Elliott for his useful suggestions.

I would like to thank all my students and colleagues who used the book in their nuclear physics courses: I benefited a lot from their valuable remarks and suggestions. In particular, I would like to thank E Jacobs (who is currently teaching the course at Gent) and R Bijker (University of Mexico) for their very conscientious checking and for pointing out a number of errors that I had not noticed. I would particularly like to thank R F Casten, W Nazarewicz and P Van Duppen for critically reading chapter 14, for many suggestions and for helping to make the chapter readable, precise and up-to-date.

I am grateful to the CERN-ISOLDE group for its hospitality during the final phase in the production of this book, to CERN and the FWO (Fund for Scientific Research-Flanders) for their financial support and to the University of Gent (RUG) for having made the 'on-leave' to CERN possible.

Finally, I must thank R Verspille for the great care he took in modifying figures and preparing new figures and artwork, and D dutré-Lootens and L Schepens for their diligent typing of several versions of the manuscript and for solving a number of TeX problems.

Kris Heyde
September 1998

Acknowledgments to the first edition

The present book project grew out of a course taught over the past 10 years at the University of Gent aiming at introducing various concepts that appear in nuclear physics. Over the years, the original text has evolved through many contacts with the students who, by encouraging more and clearer discussions, have modified the form and content in almost every chapter. I have been trying to bridge the gap, by the addition of the various boxed items, between the main text of the course and present-day work and research in nuclear physics. One of the aims was also of emphasizing the various existing connections with other domains of physics, in particular with the higher energy particle physics and astrophysics fields. An actual problem set has not been incorporated as yet: the exams set over many years form a good test and those for parts A, B and C can be obtained by contacting the author directly.

I am most grateful to the series editors R Betts, W Greiner and W D Hamilton for their time in reading through the manuscript and for their various suggestions to improve the text. Also, the suggestion to extend the original scope of the nuclear physics course by the addition of part D and thus to bring the major concepts and basic ideas of nuclear physics in contact with present-day views on how the nucleus can be described as an interacting many-nucleon system is partly due to the series editors.

I am much indebted to my colleagues at the Institute of Nuclear Physics and the Institute for Theoretical Physics at the University of Gent who have contributed, maybe unintentionally, to the present text in an important way. More specifically, I am indebted to the past and present nuclear theory group members, in alphabetical order: C De Coster, J Jolie, L Machenil, J Moreau, S Rombouts, J Ryckebusch, M Vanderhaeghen, V Van Der Sluys, P Van Isacker, J Van Maldeghem, D Van Neck, H Vincx, M Waroquier and G Wenes in particular relating to the various subjects of part D. I would also like to thank the many experimentalist, both in Gent and elsewhere, who through informal discussions have made many suggestions to relate the various concepts and ideas of nuclear physics to the many observables that allow a detailed probing of the atomic nucleus.

The author and Institute of Physics Publishing have attempted to trace the copyright holders of all the figures, tables and articles reproduced in this

xxiv   *Acknowledgments to the first edition*

publication and would like to thank the many authors, editors and publishers for their much appreciated cooperation. We would like to apologize to those few copyright holders whose permission to publish in the present form could not be obtained.

Introduction

On first coming into contact with the basics of nuclear physics, it is a good idea to obtain a feeling for the range of energies, densities, temperatures and forces that are acting on the level of the atomic nucleus. In figure I.1, we introduce an energy scale placing the nucleus relative to solid state chemistry scales, the atomic energy scale and, higher in energy, the scale of masses for the elementary particles. In the nucleus, the lower energy processes can come down to 1 keV, the energy distance between certain excited states in odd-mass nuclei and x-ray or electron conversion processes, and go up to 100 MeV, the energy needed to induce collisions between heavy nuclei. In figure I.2 the density scale is shown. This points towards the extreme density of atomic nuclei compared to more ordinary objects such as most solid materials. Even densities in most celestial objects (regular stars) are much lower. Only in certain types of stars—neutron stars that can be compared to huge atomic nuclei (see chapter 7)—do analogous densities show up. The forces at work and the different strength scales, as well as ranges on which they act and the specific aspects in physics where they dominate, are presented in figure I.3. It is clear that it is mainly the strong force between nucleons or, at a deeper level, the strong force between the nucleon constituents

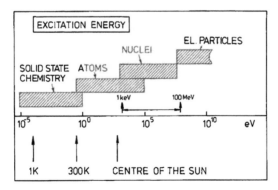

Figure I.1. Typical range of excitation energies spanning from the solid state phase towards elementary particles. In addition, a few related temperatures are indicated.

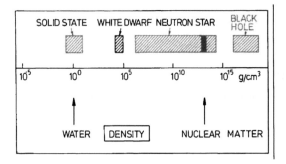

Figure I.2. Typical range of densities spanning the interval from the solid state phase into more exotic situations like a black hole.

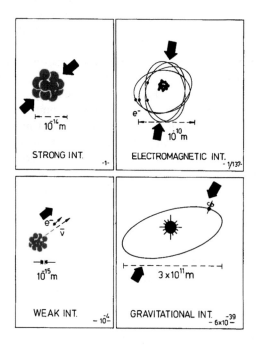

Figure I.3. Schematic illustration of the very different distance scales over which the four basic interactions act. A typical illustration for those four interactions is given at the same time. Relative interaction strengths are also shown.

(quarks) that determines the binding of atomic nuclei. Electromagnetic effects cannot be ignored in determining the nuclear stability since a number of protons occur in a small region of space. The weak force, responsible for beta-decay processes, also cannot be neglected.

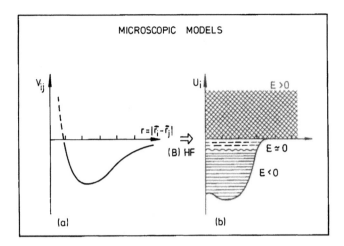

Figure I.4. Illustration of how the typical form of the nucleon–nucleon two-body interaction $V_{ij}(|\vec{r}_i - \vec{r}_j|)$ (*a*) connects to the nuclear average one-body field (*b*) making use of (Brueckner)–Hartree–Fock theory. The region of strongly bound $(E < 0)$ levels near the Fermi energy $(E \simeq 0)$ as well as the region of unbound $(E > 0)$ particle motion is indicated on the one-body field $U_i(|\vec{r}|)$.

In attempting a description of bound nuclei (a collection of A strongly interacting nucleons) in terms of the nucleon–nucleon interaction and of processes where nuclear states decay via the emission of particles or electromagnetic radiation, one has to make constant use of the quantum mechanical apparatus that governs both the bound $(E < 0)$ and unbound $(E > 0)$ nuclear regime. Even though the n–n interaction, with a short range attractive part and repulsive core part (figure I.4), would not immediately suggest a large mean-free path in the nuclear medium, a quite regular average field becomes manifest. It is the connection between the non-relativisitic A-nucleon interacting Hamiltonian

$$\hat{\mathcal{H}} = \sum_{i=1}^{A} \frac{\vec{p}_i^2}{2m_i} + \sum_{i<j=1}^{A} V(\vec{r}_i, \vec{r}_j), \tag{I.1}$$

and the one-body plus residual interaction Hamiltonian

$$\hat{\mathcal{H}} = \sum_{i=1}^{A} \left(\frac{\vec{p}_i^2}{2m_i} + U(\vec{r}_i) \right) + \hat{\mathcal{H}}_{\text{res}}, \tag{I.2}$$

that is one of the tasks in understanding bound nuclear structure physics. If, as in many cases, the residual interactions $\hat{\mathcal{H}}_{\text{res}}$ can be left out initially, an independent-particle nucleon motion in the nucleus shows up and is quite well verified experimentally.

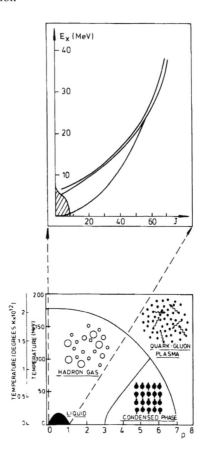

Figure I.5. Part of a nuclear phase diagram (schematic). In the upper part possible configurations in the excitation energy (E_x) and angular momentum (J) are indicated, at normal nuclear density. In the lower part a much longer interval of the nuclear phase diagram, containing the upper part, is shown. Here nuclear temperature ($T \lesssim 200$ MeV) and nuclear density ($\rho \lesssim 8\rho_0$) represent the variables. Various possible regions—hadron gas, liquid, condensed phase and quark–gluon plasma—are also presented (adapted from Greiner and Stöcker 1985).

Concerning decay processes, where transitions between initial and final states occur, time-dependent perturbation theory will be the appropriate technique for calculating decay rates. We shall illustrate this, in particular for the α-, β- and γ-decay processes, showing the very similar aspects in the three main decay processes that spontaneously occur in standard nuclear physics. At the same time, we shall highlight the different time-scales and characteristics distinguishing α-decay (strong interaction process via almost stationary states), β-decay (weak decay creating electrons (positrons) and neutrinos (or antineutrinos)) and γ-decay (via electromagnetic interaction).

Of course, nuclear physics is a field that interconnects very much to adjacent fields such as elementary particle physics (at the higher energy end), astrophysics (via nuclear transmutation processes) and solid state physics (via the nuclear hyperfine field interactions). Various connections will be highlighted at the appropriate place.

We shall not concentrate on reaction processes in detail and will mainly keep to the nuclear excitation region below $E_x \simeq 8\text{--}10$ MeV.

This presents only a rather small portion of the nuclear system (figure I.5) but this domain is already very rich in being able to offer a first contact with nuclear physics in an introductory course requiring a knowledge of standard, non-relativistic quantum mechanics.

The text is devoted to a typical two-semester period with one lecture a week. Optional parts are included that expand on recent developments in nuclear physics ('boxes' of text and figures) and extensive references to recent literature are given so as to make this text, at the same time, a topical introduction to the very alive and rapidly developing field of nuclear physics.

Kris Heyde
1 June 1994

PART A

KNOWING THE NUCLEUS:
THE NUCLEAR CONSTITUENTS AND
CHARACTERISTICS

Chapter 1

Nuclear global properties

1.1 Introduction and outline

In this chapter, we shall discuss the specific characteristics of the atomic nucleus that make it a unique laboratory where different forces and particles meet. Depending on the probe we use to 'view' the nucleus different aspects become observable. Using probes (e^-, p, $\pi^\pm$, ...) with an energy such that the quantum mechanical wavelength $\lambda = h/p$ is of the order of the nucleus, global aspects do show up such that collective and surface effects can be studied. At shorter wavelengths, the A-nucleon system containing Z protons and N neutrons becomes evident. It is this and the above 'picture' that will mainly be of use in the present discussion. Using even shorter wavelengths, the mesonic degrees and excited nucleon configurations (Δ, ...) become observable. At the extreme high-energy side, the internal structure of the nucleons shows up in the dynamics of an interacting quark–gluon system (figure 1.1).

Besides more standard characteristics such as mass, binding energy, nuclear extension and radii, nuclear angular momentum and nuclear moments, we shall try to illustrate these properties using up-to-date research results that point towards the still quite fast evolving subject of nuclear physics. We also discuss some of the more important ways the nucleus can interact with external fields and particles: hyperfine interactions and nuclear transmutations in reactions.

1.2 Nuclear mass table

Nuclei, consisting of a bound collection of Z protons and N neutrons (A nucleons) can be represented in a diagrammatic way using Z and N as axes in the plane. This plane is mainly filled along or near to the diagonal $N = Z$ line with equal number of protons and neutrons. Only a relatively small number of nuclei form stable nuclei, stable against any emission of particles or other transmutations. For heavy elements, denoted as $^A_Z X_N$, with $A \gtrsim 100$, a neutron excess over the proton number shows up along the line where most stable nuclei

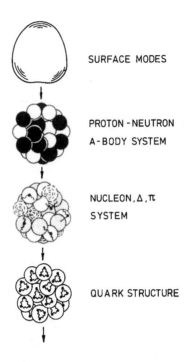

SURFACE MODES

PROTON - NEUTRON
A - BODY SYSTEM

NUCLEON, Δ, π
SYSTEM

QUARK STRUCTURE

Figure 1.1. Different dimensions (energy scales) for observing the atomic nucleus. From top to bottom, increasing resolving power (shorter wavelengths) is used to see nuclear surface modes, the A-body proton–neutron system, the more exotic nucleon, isobar, mesonic system and, at the lowest level, the quark system interacting via gluon exchange.

are situated and which is illustrated in figure 1.3 as the grey and dark zone. Around these stable nuclei, a large zone of unstable nuclei shows up: these nuclei will transform the excess of neutrons in protons or excess of protons in neutrons through β-decay. These processes are written as

$$^{A}_{Z}X_N \rightarrow {}^{A}_{Z+1}Y_{N-1} + e^- + \bar{\nu}_e,$$

$$^{A}_{Z}X_N \rightarrow {}^{A}_{Z-1}Y_{N+1} + e^+ + \nu_e,$$

$$^{A}_{Z}X_N + e^- \rightarrow {}^{A}_{Z-1}Y_{N+1} + \nu_e,$$

for β^-, β^+ and electron capture, respectively. (See chapter 5 for more detailed discussions.) In some cases, other, larger particles such as α-particles (atomic nucleus of a ^{4}He atom) or even higher mass systems can be emitted. More particularly, it is spontaneous α-decay and fission of the heavy nuclei that makes the region of stable nuclei end somewhat above uranium. Still, large numbers of radioactive nuclei have been artificially made in laboratory conditions using various types of accelerators. Before giving some more details on the heaviest elements (in Z) observed and synthesized at present we give an excerpt of the nuclear system of nuclei in the region of very light nuclei (figures 1.2(a), (b)) using the official chart of nuclides. These mass charts give a wealth of information such as explained in figure 1.2(b). In this mass chart excerpt (figure 1.2(a)), one can see how far from stability one can go: elements like $^{8}_{2}\text{He}_{6}$, $^{11}_{3}\text{Li}_{8}$, ... have been synthesized at various accelerator, isotope separator labs like GANIL (Caen) in France, CERN in Switzerland using the Isolde separator facility and, most spectacularly, at GSI, Darmstadt. In a separate box (Box 1a) we illustrate the heaviest elements and their decay pattern as observed.

In this division of nuclei, one calls isotopes nuclei with fixed proton number Z and changing neutron number, i.e. the even–even Sn nuclei forms a very long series of stable nuclei (figure 1.3). Analogously, one has isotones, with fixed N and isobars (fixed A, changing Z and N). The reason for the particular way nuclei are distributed in the (N, Z) plane, is that the nuclear strong binding force maximizes the binding energy for a given number of nucleons A. This will be studied in detail in chapter 7, when discussing the liquid drop and nuclear shell model.

1.3   Nuclear binding, nuclear masses

As pointed out in the introductory section, the nuclear strong interaction acts on a very short distance scale, i.e. the n–n interaction becomes very weak beyond nucleon separations of 3–4 fm.

The non-relativistic A-nucleon Hamiltonian dictating nuclear binding was given in equation (I.1) and accounts for a non-negligible 'condensation' energy when building the nucleus from its A-constituent nucleons put initially at very large distances (see figure 1.4). Generally speaking, the solution of this A-body strongly interacting system is highly complicated and experimental data can give an interesting insight in the bound nucleus. Naively speaking, we expect $A(A - 1)/2$ bonds and, if each bond between two nucleons amounts to a fairly constant value E_2, we expect for the nuclear binding energy per nucleon

$$BE(^{A}_{Z}\text{X}_N)/A \propto E_2(A - 1)/2, \qquad (1.1)$$

or, an expression that increases with A. The data are completely at variance with this two-body interaction picture and points to an average value for $BE(^{A}_{Z}\text{X}_N)/A \simeq 8$ MeV over the whole mass region. The above data therefore imply at least two important facets of the n–n interaction in a nucleus:

(i)   nuclear, charge independence,
(ii) saturation of the strong interaction.

The above picture, pointing out that the least bound nucleon in a nucleus is bound by $\simeq 8$ MeV, independent of the number of nucleons, also implies an independent particle picture where nucleons move in an average potential

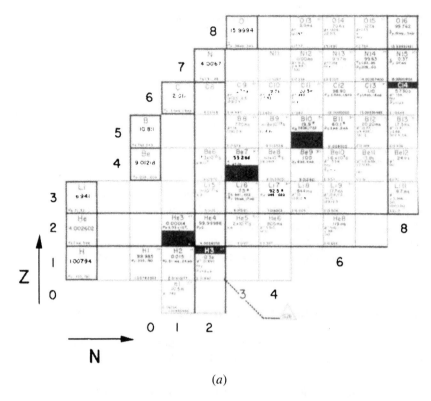

(*a*)

Figure 1.2. (*a*) Sections of the nuclear mass chart for light nuclei. (*b*) Excerpt from the Chart of Nuclides for very light nuclei. This diagram shows stable as well as artifical radioactive nuclei. Legend to discriminate between the many possible forms of nuclei and their various decay modes, as well as the typical displacements caused by nuclear processes. (Taken from *Chart of Nuclides*, 13th edition, General Electric, 1984.)

(figure 1.5) In section 1.4, we shall learn more about the precise structure of the average potential and thus of the nuclear mass and charge densities in this potential.

The binding energy of a given nucleus $^A_Z X_N$ is now given by

$$BE(^A_Z X_N) = Z \cdot M_p c^2 + N \cdot M_n c^2 - M'(^A_Z X_N)c^2, \qquad (1.2)$$

where M_p, M_n denote the proton and neutron mass, respectively and $M'(^A_Z X_N)$ is the actual nuclear mass. The above quantity is the nuclear binding energy. A total, atomic binding energy can be given as

$$BE(^A_Z X_N; \text{atom}) = Z \cdot M_{1_H} \cdot c^2 + N \cdot M_n c^2 - M(^A_Z X_N; \text{atom})c^2. \qquad (1.3)$$

where M_{1_H} is the mass of the hydrogen atom. If relative variations of the order of eV are neglected, nucleon and atomic binding energies are equal (give a proof of

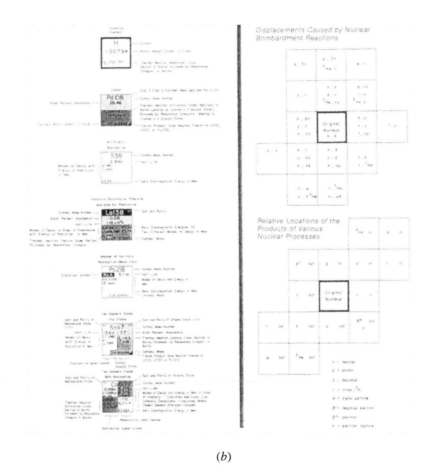

(b)

Figure 1.2. (Continued.)

this statement). In general, we shall for the remaining part of this text, denote the nuclear mass as $M'(^A_ZX_N)$ and the atomic mass as $M(^A_ZX_N)$.

Atomic (or nuclear) masses, denoted as amu or m.u. corresponds to 1/12 of the mass of the atom ^{12}C. Its value is

$$1.660566 \times 10^{-27} \text{ kg} = 931.5016 \pm 0.0026 \text{ MeV}/c^2.$$

In table 1.1 we give a number of important masses in units of amu and MeV. It is interesting to compare rest energies of the nucleon, its excited states and e.g. the rest energy of a light nucleus such as ^{58}Fe. The nucleon excited states are very close and cannot be resolved in figure 1.6. A magnified spectrum, comparing the spectrum of ^{58}Fe with the nucleon exited spectrum is shown in figure 1.7, where a difference in scale of $\times 10^3$ is very clear.

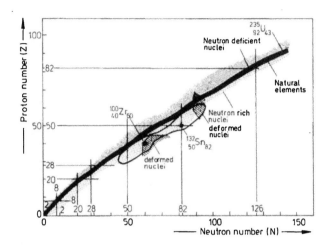

Figure 1.3. Chart of known nuclei in which stable nuclei (natural elements showing up in nature), neutron-rich and neutron-deficient nuclei are presented. Magic (closed shell) nuclei occur where the horizontal and vertical lines intersect. A few regions of deformed nuclei are also shown as well as a few key nuclei: ^{100}Zr, ^{132}Sn, ^{235}U.

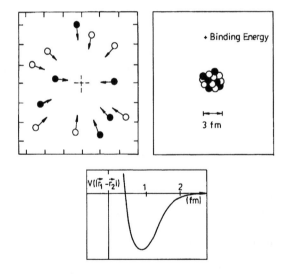

Figure 1.4. Representation of the condensation process where free nucleons (protons and neutrons), under the influence of the two-body, charge-independent interaction $V(|\vec{r}_i - \vec{r}_j|)$, form a bound nucleus at a separation of a few fermi and release a corresponding amount of binding (condensation) energy.

Before leaving this subject, it is interesting to note that, even though the average binding energy amounts to $\simeq 8$ MeV, there is a specific variation in $BE(^A_Z X_N)/A$, as a function of A. The maximal binding energy per nucleon is situated near mass $A = 56$–62^1, light and very heavy nuclei are containing less bound nucleons. Thus, the source of energy production in fusion of light nuclei or fission of very heavy nuclei can be a source of energy. They are at the basis of fusion and fission bombs and (reactors), respectively, even though fusion reactors are not yet coming into practical use.

The most tightly bound nucleus

Richard Shurtleff and Edward Derringh
Department of Physics, Wentworth Institute of Technology, Boston, Massachusetts 02115

(Received 1 March 1988; accepted for publication 5 October 1988)

In many textbooks,[1–3] we are told that ^{56}Fe is the nuclide with the greatest binding energy per nucleon, and therefore is the most stable nucleus, the heaviest that can be formed by fusion in normal stars.

But we calculate the binding energy per nucleon BE/A, for a nucleus of mass number A, by the usual formula,

$$BE/A = (1/A)(Zm_H + Nm_n - M_{atom})c^2, \qquad (1)$$

where m_H is the hydrogen atomic mass and m_n is the neutron mass, for the nuclides ^{56}Fe and ^{62}Ni (both are stable) using data from Wapstra and Audi.[4] The results are 8.790 MeV/nucleon for ^{56}Fe and 8.795 MeV/nucleon for ^{62}Ni. The difference,

$$(0.005 \text{ MeV/nucleon})(\approx 60 \text{ nucleons}) = 300 \text{ keV}, \quad (2)$$

is much too large to be accounted for as the binding energy of the two extra electrons in ^{62}Ni over the 26 electrons in ^{56}Fe.

^{56}Fe is readily produced in old stars as the end product of the silicon-burning series of reactions.[5] How, then, do we explain the relative cosmic deficiency of ^{62}Ni compared with ^{56}Fe? In order to be abundant, it is not enough that ^{62}Ni be the most stable nucleus. To be formed by charged-particle fusion (the energy source in normal stars), a reaction must be available to bridge the gap from ^{56}Fe to ^{62}Ni.

To accomplish this with a single fusion requires a nuclide with $Z = 2$, $A = 6$. But no such stable nuclide exists. The other possibility is two sequential fusions with ^{3}H, producing first ^{59}Co then ^{62}Ni. However, the ^{3}H nucleus is unstable and is not expected to be present in old stars synthesizing heavy elements. We are aware that there are element-generating processes other than charged-particle fusion, such as processes involving neutron capture, which could generate nickel. However, these processes apparently do not occur in normal stars, but rather in supernovas and post-supernova phases, which we do not address.

We conclude that ^{56}Fe is the end product of normal stellar fusion not because it is the most tightly bound nucleus, which it is not, but that it is in close, but unbridgeable, proximity to ^{62}Ni, which is the most tightly bound nucleus.

[1] Arthur Beiser, *Concepts of Modern Physics* (McGraw-Hill, New York, 1987), 4th ed , p. 421
[2] Frank Shu, *The Physical Universe* (University Science Books, Mill Valley, CA, 1982), 1st ed., pp. 116–117.
[3] Donald D. Clayton, *Principles of Stellar Evolution and Nucleosynthesis* (McGraw-Hill, New York, 1968), p. 518.
[4] A. H. Wapstra and G. Audi, Nucl. Phys. A **432**, 1 (1985).
[5] William K. Rose, *Astrophysics* (Holt, Rinehart and Winston, New York, 1973), p. 186.

---

[1] It is often stated that ^{56}Fe is the most tightly bound nucleus—this is not correct since ^{62}Ni is more bound by a difference of 0.005 MeV/nucleon or, for $\simeq 60$ nucleons, with an amount of 300 keV. For more details, see Shurtleff and Derringh's article reproduced below.

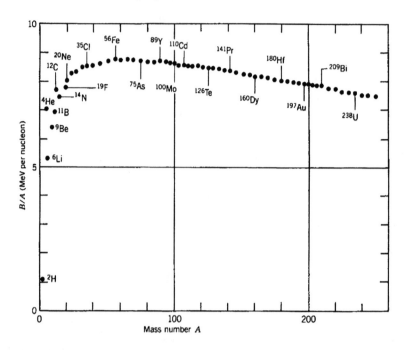

Figure 1.5. The binding energy per nucleon B/A as a function of the nuclear mass number A. (Taken from Krane, *Introductory Nuclear Physics* © 1987 John Wiley & Sons. Reprinted by permission.)

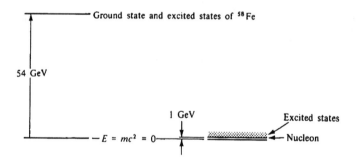

Figure 1.6. Total rest energy of the states in ^{58}Fe (typical atomic nucleus) and of the nucleon and its excited states. On the scale, the excited states in ^{58}Fe are so close to the ground state that they cannot be observed without magnification. This view is shown in figure 1.7. (Taken from Frauenfelder and Henley (1991) *Subatomic Physics* © 1974. Reprinted by permission of Prentice-Hall, Englewood Cliffs, NJ.)

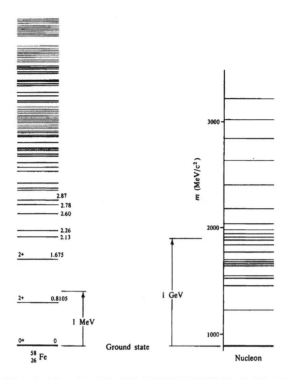

Figure 1.7. Ground state and excited states in ^{58}Fe and of the nucleon. The region above the ground state in ^{58}Fe in figure 1.7 has been exploded by a factor 10^4. The spectrum of the nucleon in figure 1.6 has been expanded by a factor 25. (Taken from Frauenfelder and Henley (1991) *Subatomic Physics* © 1974. Reprinted by permission of Prentice-Hall, Englewood Cliffs, NJ.)

Table 1.1. Some important masses given in units amu and MeV respectively.

	amu	MeV
^{12}C/12	1	931.5016
1 MeV	1.073535×10^{-3}	1
Electron	5.485580×10^{-4}	0.511003
Neutron	1.008665	939.5731
Proton	1.007276	938.2796
Deuterium atom	2.0141014	1876.14
Helium atom	4.002600	3728.44

1.4   Nuclear extension: densities and radii

The discussion in section 1.3 indicated unambiguous evidence for saturation in the nuclear strong force amongst nucleons in an atomic nucleus. Under these assumptions of saturation and charge independence each nucleon occupies an almost equal size within the nucleus. Calling r_0 an elementary radius for a nucleon in the nucleus, a most naive estimate gives for the nuclear volume

$$V = \tfrac{4}{3}\pi r_0^3 A, \tag{1.4}$$

or

$$R = r_0 A^{1/3}. \tag{1.5}$$

This relation describes the variation of the nuclear radius, with a value of $r_0 \cong$ 1.2 fm when deducing a 'charge' radius, and a value of $r_0 \cong 1.4$ fm for the full 'matter' radius.

The experimental access to obtain information on nuclear radii comes from scattering particles (e^-, p, $\pi^{\pm}$, ...) off the atomic nucleus with appropriate energy to map out the nuclear charge and/or matter distributions. A corresponding typical profile is a Fermi or Woods–Saxon shape, described by the expression

$$\rho(r) = \frac{\rho_0}{1 + e^{(r-R_0)/a}}, \tag{1.6}$$

with ρ_0 the central density. R_0 is then the radius at half density and a describes the diffuseness of the nuclear surface.

Electron scattering off nuclei is, for example, one of the most appropriate methods to deduce radii. The cross-sections over many decades have been measured in e.g. ^{208}Pb (see figure 1.8) and give detailed information on the nuclear density distribution $\rho_c(r)$ as is discussed in Box 1b. We also point out the present day level of understanding of the variation in charge and matter density distributions for many nuclei. A comparison between recent, high-quality data and Hartree–Fock calculations for charge and mass densities are presented in figures 1.9 giving an impressive agreement between experiment and theory.

Here, some details should be presented relating to the quantum mechanical expression of these densities. In taking collective, nuclear models (liquid drop, ...) a smooth distribution $\rho_c(\vec{r})$, $\rho_{mass}(\vec{r})$ can be given (figure 1.10). In a more microscopic approach, the densities result from the occupied orbitals in the nucleus. Using a shell-model description where orbitals are characterized by quantum numbers $\alpha \equiv n_a, l_a, j_a, m_a$ (radial, orbital, total spin, magnetic quantum number) the density can be written as (figure 1.10).

$$\rho_{mass}(\vec{r}) = \sum_{k=1}^{A} |\varphi_{\alpha_k}(\vec{r})|^2, \tag{1.7}$$

where α_k denotes the quantum numbers of all occupied ($k = 1, \ldots, A$) nucleons. Using an A-nucleon product wavefunction to characterize the nucleus in an

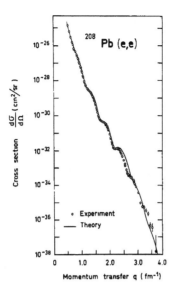

Figure 1.8. Typical cross-section obtained in electron elastic scattering off ^{208}Pb as a function of momentum transfer. The full line is a theoretical prediction. (Taken from Frois 1987).

independent-particle model (neglecting the Pauli principle for a while)

$$\psi(\vec{r}_1, \ldots, \vec{r}_A) = \prod_{k=1}^{A} \varphi_{\alpha_k}(\vec{r}_k), \qquad (1.8)$$

the density should appear as the expectation value of the density 'operator' $\hat{\rho}_{\text{mass}}(\vec{r})$, or

$$\rho_{\text{mass}}(\vec{r}) = \int \psi^*(\vec{r}_1, \ldots \vec{r}_A) \hat{\rho}_{\text{mass}}(\vec{r}) \psi(\vec{r}_1, \ldots \vec{r}_A) d\vec{r}_1 \ldots d\vec{r}_A. \qquad (1.9)$$

From this, an expression for $\hat{\rho}_{\text{mass}}(\vec{r})$ is derived as

$$\rho_{\text{mass}}(\vec{r}) = \sum_{k=1}^{A} \delta(\vec{r} - \vec{r}_k), \qquad (1.10)$$

as can be easily verified. The above expression for the density operator (a similar one can be discussed for the charge density) shall be used later on.

As a final comment, one can obtain a simple estimate for the nuclear matter density by calculating the ratio

$$\rho = \frac{M}{V} = \frac{1.66 \times 10^{-27} \text{ A kg}}{1.15 \times 10^{-44} \text{ A m}^3} = 1.44 \times 10^{17} \text{ kg m}^{-3},$$

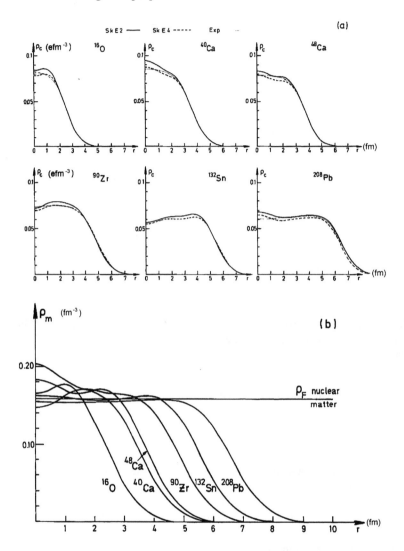

Figure 1.9. (*a*) Charge density distributions $\rho_c(r)$ for the doubly-magic nuclei ^{16}O, ^{40}Ca, ^{48}Ca, ^{90}Zr, ^{132}Sn and ^{208}Pb. The theoretical curves correspond to various forms of effective nucleon–nucleon forces, called Skyrme forces and are compared with the experimental data points (units are ρ_c (efm^{-3}) and r (fm)). (*b*) Nuclear matter density distributions ρ_m (fm^{-3}) for the magic nuclei. (Taken from Waroquier 1987.)

and is independent of A.   This density is (see the introductory chapter) approximately 10^{14} times normal matter density and expresses the highly packed density of nucleons.

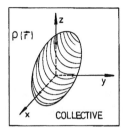 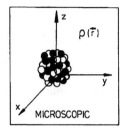

Figure 1.10. Nuclear density distributions $\rho(\vec{r})$. Both a purely collective distribution (left-hand part) and a microscopic description, incorporating both proton and neutron variables (right-hand part) are illustrated.

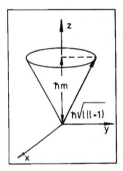 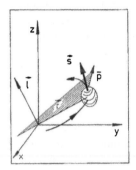

Figure 1.11. Angular momentum ($\hat{\ell}$) connected to the orbital motion of a nucleon (characterized by radius vector $\vec{r}$ and linear momentum $\vec{p}$). The intrinsic angular momentum (spin $\hat{s}$) is also indicated. On the left-hand side, the semiclassical picture of angular momentum in quantum mechanics is illustrated and is characterized by the length ($\hbar[\ell(\ell+1)]^{1/2}$) and projection ($\hbar m$).

1.5 Angular momentum in the nucleus

Protons and neutrons move in an average field and so cause orbital angular momentum to build up. Besides, nucleons, as fermions with intrinsic spin $\hbar/2$, will add up to a total angular momentum of the whole nucleus. The addition can be done correctly using angular momentum techniques, in a first stage combining orbital and intrinsic angular momentum to nucleon total angular momentum and later adding individual 'spin' (used as an abbreviation to angular momentum) to the total nuclear spin I (figure 1.11).

Briefly collecting the main features of angular momentum quantum mechanics, one has the orbital eigenfunctions (spherical harmonics) $Y_\ell^m(\hat{r})$ with eigenvalue properties

$$\hat{\ell}^2 Y_\ell^{m_\ell}(\hat{r}) = \hbar^2 \ell(\ell+1) Y_\ell^{m_\ell}(\hat{r})$$
$$\hat{\ell}_z Y_\ell^{m_\ell}(\hat{r}) = \hbar m_\ell Y_\ell^{m_\ell}(\hat{r}). \qquad (1.11)$$

Here, $\hat{r}$ denotes the angular coordinates $\hat{r} \equiv (\theta, \varphi)$. Similarly, for the intrinsic spin properties, eigenvectors can be obtained with properties (for protons and neutrons)

$$\hat{s}^2 \chi_{1/2}^{m_s}(s) = \hbar^2 \cdot \tfrac{3}{4} \cdot \chi_{1/2}^{m_s}(s)$$
$$\hat{s}_z \chi_{1/2}^{m_s}(s) = \hbar m_s \chi_{1/2}^{m_s}(s), \qquad (1.12)$$

where $m_s = \pm 1/2$ and the argument s just indicates that the eigenvectors relate to intrinsic spin. A precise realization using for $\hat{s}^2$, $\hat{s}_z$ and $\chi_{1/2}^{m_s} 2 \times 2$ matrices and 2-row column vectors, respectively, can be found in quantum mechanics texts.

Now, total 'spin' $\hat{j}$ is constructed as the operator sum

$$\hat{j} = \hat{\ell} + \hat{s}, \qquad (1.13)$$

which gives rise to a total 'spin' operator for which $\hat{j}^2$, $\hat{j}_z$ commute and also commute with $\hat{\ell}^2, \hat{s}^2$. The precise construction of the single-particle wavefunctions, that are eigenfunctions of $\hat{\ell}^2$, $\hat{s}^2$ and also of $\hat{j}^2$, $\hat{j}_z$ needs angular momentum coupling techniques and results in wavefunctions characterized by the quantum numbers $(\ell, \tfrac{1}{2})j, m$ with $j = \ell \pm 1/2$ and is denoted as

$$\psi(\ell\tfrac{1}{2}, jm) = [Y_\ell \otimes \chi_{1/2}]_j^{(m)}, \qquad (1.14)$$

in vector-coupled notation (see quantum mechanics). In a similar way one can go on to construct the total spin operator of the whole nucleus

$$\hat{J} = \sum_{i=1}^{A} \hat{j}_i, \qquad (1.15)$$

where still $\hat{J}^2$, $\hat{J}_z$ will constitute correct spin operators. These operators still commute with the individual operators $\hat{j}_1^2, \hat{j}_2^2, \ldots, \hat{j}_A^2$ but no longer with the $\hat{j}_{i,z}$ operators. Also, extra internal momenta will be needed to correctly couple spins.

This looks like a very difficult job. Many nuclei can in first approximation be treated as a collection of largely independent nucleons moving in a spherical, average field. Shells j can contain $(2j+1)$ particles that constitute a fully coupled shell with all m-states $-j \le m \le j$ occupied thus forming a $J = 0$, $M = 0$ state. The only remaining 'valence' nucleons will determine the actual nuclear 'spin' J. As a consequence of the above arguments and the fact that the short-range nucleon–nucleon interaction favours pairing nucleons into angular momentum 0^+ coupled pairs, one has that:

- even–even nuclei have $J = 0$ in the ground state
- odd-mass nuclei will have a half-integer spin J since j itself is always half-integer
- odd–odd nuclei have integer spin J in the ground state, resulting from combining the last odd-proton spin with the last odd-neutron spin, i.e.

$$\hat{J} = \hat{j}_{\mathrm{p}} + \hat{j}_{\mathrm{n}}. \tag{1.16}$$

For deformed nuclei (nuclei with a non-spherical mass and charge density distribution) some complications arise that shall not be discussed in the present text.

1.6 Nuclear moments

Since in the nucleus, protons (having an elementary charge +e) and neutrons are both moving, charge, mass and current densities result. We shall give some attention to the magnetic dipole and electric quadrupole moment, two moments that are particularly well measured over many nuclei in different mass regions.

1.6.1 Dipole magnetic moment

With a particle having orbital angular momentum, a current and thus a magnetic moment vector $\vec{\mu}$ can be associated. In the more simple case of a circular, orbital motion (classical), one has

$$\vec{\ell} = \vec{r} \times \vec{p}, \tag{1.17}$$

and

$$|\vec{\ell}| = rmv.$$

For the magnetic moment one has

$$\vec{\mu} = \pi r^2 \cdot i \vec{1}, \tag{1.18}$$

(with $\vec{1}$ a unit vector, vertical to the circular motion, in the rotation sense going with a positive current). For a proton (or electron) one has, in magnitude

$$|\vec{\mu}| = \pi r^2 \frac{ev}{2\pi r} = \frac{e}{2m} |\vec{\ell}|, \tag{1.19}$$

and derives (for the circular motion still)

$$\vec{\mu}_\ell = \frac{e}{2m} \vec{\ell} \qquad \left(\frac{e}{2mc} \text{ in Gaussian units} \right). \tag{1.20}$$

Moving to a quantum mechanical description of orbital motion and thus of the magnetic moment description, one has the relation between operators

$$\hat{\mu}_\ell = \frac{e}{2m} \hat{\ell}, \tag{1.21}$$

and

$$\hat{\mu}_{\ell,z} = \frac{e}{2m}\hat{\ell}_z. \tag{1.22}$$

The eigenvalue of the orbital, magnetic dipole operator, acting on the orbital eigenfunctions $Y_\ell^{m_\ell}$ then becomes

$$\hat{\mu}_{\ell,z} Y_\ell^{m_\ell}(\hat{r}) = \frac{e}{2m}\hat{\ell}_z Y_\ell^{m_\ell}(\hat{r})$$
$$= \frac{e\hbar}{2m} m_\ell Y_\ell^{m_\ell}(\hat{r}). \tag{1.23}$$

If we call the unit $e\hbar/2m$ the nuclear (if m is the nucleon mass) or Bohr (for electrons) magneton, then one has for the eigenvalue $\mu_N(\mu_B)$

$$\mu_{\ell,z} = m_\ell \mu_N. \tag{1.24}$$

For the intrinsic spin, an analoguous procedure can be used. Here, however, the mechanism that generates the spin is not known and classic models are doomed to fail. Only the Dirac equation has given a correct description of intrinsic spin and of its origin. The picture one would make, as in figure 1.12, is clearly not correct and we still need to introduce a proportionality factor, called gyromagnetic ratio g_s. for intrinsic spin $\hbar/2$ fermions. One obtains

$$\mu_{s,z} = g_s \mu_N m_s, \tag{1.25}$$

as eigenvalue, for the $\hat{\mu}_{s,z}$ operator acting on the spin $\chi_{1/2}^{m_s}(s)$ eigenvector. For the electron this g_s factor turns out to be almost -2 and at the original time of introducing intrinsic $\hbar/2$ spin electrons this factor (in 1926) was not understood and had to be taken from experiment. In 1928 Dirac gave a natural explanation for this fact using the now famous Dirac equation. For a Dirac point electron this should be exact but small deviations given by

$$a = \frac{|g|-2}{2}, \tag{1.26}$$

were detected, giving the result

$$a_{e^-}^{\exp} = 0.001159658(4). \tag{1.27}$$

Detailed calculations in QED (quantum electrodynamics) and the present value give

$$a_{e^-}^{th} = \frac{1}{2}\left(\frac{\alpha}{\pi}\right) - 0.328479\left(\frac{\alpha}{\pi}\right)^2 + 1.29\left(\frac{\alpha}{\pi}\right)^3, \tag{1.28}$$

with $\alpha = e^2/\hbar c$, and the difference $(a_{e^-}^{th} - a_{e^-}^{\exp})/a^{th} = (2 \pm 5) \times 10^{-6}$, which means 1 part in 10^5 (for a nice overview, see Crane (1968) and lower part of figure 1.12).

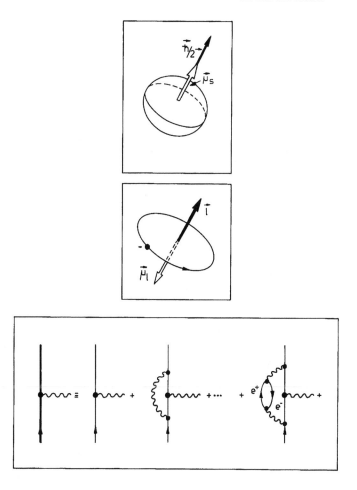

Figure 1.12. In the upper part, the relationships between the intrinsic $(\vec{\mu}_s)$ and orbital $(\vec{\mu}_\ell)$ magnetic moments and the corresponding angular moment vectors $(\vec{\hbar}/2$ and $\vec{\ell}$, respectively) are indicated. Thereby gyromagnetic factors are defined. In the lower part, modifications to the single-electron g-factor are illustrated. The physical electron g-factor is not just a pure Dirac particle. The presence of virtual photons, e^+e^- creation and more complicated processes modify these free electron properties and are illustrated. (Taken from Frauenfelder and Henley (1991) *Subatomic Physics* © 1974. Reprinted by permission of Prentice-Hall, Englewood Cliffs, NJ.)

   This argumentation can also be carried out for the intrinsic spin motion of the single proton and neutron, and results in non-integer values for both the proton and the neutron, i.e. g_s(proton) $= 5.5855$ and g_s(neutron) $= -3.8263$. The fact is that, even for the neutron with zero charge, an intrinsic, non-vanishing moment

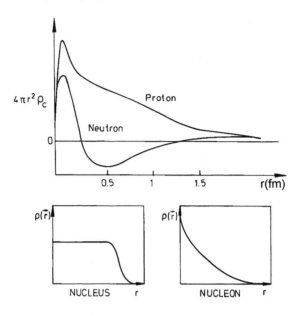

Figure 1.13. Charge distributions of nucleons deduced from the analyses of elastic electron scattering off protons (hydrogen target) and off neutrons (from a deuterium target). In the lower parts, the typical difference between a nuclear and a nucleon density distribution are presented.

shows up and points towards an internal charge structure for both the neutron and proton that is not just a simple distribution (see Box 1c).

From electron high-energy scattering off nucleons (see section 1.4) a charge form factor can be obtained (see results in figure 1.13 for the charge density distributions $\rho_{charge}(r)$ for proton and neutron). As a conclusion one obtains that:

- Nucleons are not point particles and do not exhibit a well-defined surface in contrast with the total nucleus, as shown in the illustration. Still higher energy scattering at SLAC (Perkins 1987) showed that the scattering process very much resembled that of scattering on points inside the proton. The nature of these point scatterers and their relation to observed and anticipated particles was coined by Feynman as 'partons' and attempts have been made to relate these to the quark structure of nucleons (see Box 1d).

One can now combine moments to obtain the total nuclear magnetic dipole moment and obtain:

$$\mu_{J,z} = g_J \mu_N m_J, \tag{1.29}$$

with g_J the nuclear gyromagnetic ratio. Here too, the addition rules for angular momentum can be used to construct (i) a full nucleon g-factor after combining

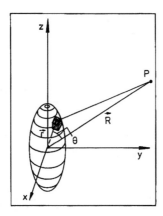

Figure 1.14. Coordinate system for the evaluation of the potential generated at the point $P(\vec{r})$ and caused by a continuous charge distribution $\rho_c(\vec{r})$. Here we consider, an axially symmetric distribution along the z-axis.

orbital and intrinsic spin and (ii) the total nuclear dipole magnetic moment. We give, as an informative result, the g-factor for free nucleons (combining $\hat{\ell}$ and $\hat{s}$ to the total spin $\hat{j}$) as

$$g = g_\ell \pm \frac{1}{2\ell + 1}(g_s - g_\ell), \tag{1.30}$$

where the upper sign applies for the $j = \ell + \frac{1}{2}$ and lower sign for the $j = \ell - \frac{1}{2}$ orientation. Moreover, these g-factors apply to free 'nucleons'. When nucleons move inside a nuclear medium the remaining nucleons modify this free g-value into 'effective' g-factors. This aspect is closely related to typical shell-model structure aspects which shall not be discussed here.

1.6.2 Electric moments—electric quadrupole moment

If the nuclear charge is distributed according to a smooth function $\rho(\vec{r})$, then an analysis in multipole moments can be made. These moments are quite important in determining e.g. the potential in a point P at large distance $\vec{R}$, compared to the nuclear charge extension. The potential $\Phi(\vec{R})$ can be calculated as (figure 1.14)

$$\Phi(\vec{R}) = \frac{1}{4\pi\epsilon_0} \int_{\text{Vol}} \frac{\rho(\vec{r})}{|\vec{R} - \vec{r}|} \, d\vec{r}, \tag{1.31}$$

which, for small values of $|\vec{r}/\vec{R}|$ can be expanded in a series for (r/R), and gives as a result

$$\Phi(\vec{R}) = \frac{1}{4\pi\epsilon_0}\frac{q}{R} + \frac{1}{4\pi\epsilon_0} \int \frac{\rho(\vec{r})r\cos\theta \, d\vec{r}}{R^2}$$

$$+ \frac{1}{2} \cdot \frac{1}{4\pi\epsilon_0} \int \frac{\rho(\vec{r})(3\cos^2\theta - 1)r^2 \, d\vec{r}}{R^3} + \cdots. \qquad (1.32)$$

This expression can be rewritten by noting that

$$r\cos\theta = \vec{r} \cdot \vec{R}/R = \sum_i x_i X_i / R, \qquad (1.33)$$

(with x_i for $i = 1, 2, 3$ corresponding to x, y, z, respectively). In the above expression, q is the total charge $\int_{\text{Vol}} \rho(\vec{r}) \, d\vec{r}$. Using the Cartesian expansion we obtain

$$\Phi(\vec{R}) = \frac{1}{4\pi\epsilon_0} q/R + \sum_i \frac{p_i}{4\pi\epsilon_0} \frac{X_i}{R^3} + \sum_{ij} \frac{1}{2} \frac{1}{4\pi\epsilon_0} \frac{Q_{ij}}{R^5} X_i X_j + \cdots, \qquad (1.34)$$

with

$$p_i = \int \rho(\vec{r}) x_i \, d\vec{r},$$

$$Q_{ij} = \int \rho(\vec{r})(3x_i x_j - r^2 \delta_{ij}) \, d\vec{r}, \qquad (1.35)$$

the dipole components and quadrupole tensor, respectively. The quadrupole tensor Q_{ij} can be expressed via its nine components in matrix form (with vanishing diagonal sum)

$$Q \Rightarrow \begin{pmatrix} 3x^2 - r^2 & 3xy & 3xz \\ 3xy & 3y^2 - r^2 & 3yz \\ 3xz & 3yz & 3z^2 - r^2 \end{pmatrix}. \qquad (1.36)$$

Transforming to diagonal form, one can find a new coordinate system in which the non-diagonal terms vanish and one gets the new quadrupole tensor

$$Q \Rightarrow \begin{pmatrix} 3\bar{x}^2 - r^2 & & \\ & 3\bar{y}^2 - r^2 & \\ & & 3\bar{z}^2 - r^2 \end{pmatrix}. \qquad (1.37)$$

The quantity $Q_{\bar{z}\bar{z}} = \int \rho(\vec{r})(3\bar{z}^2 - r^2) \, d\vec{r}$ is also denoted as the quadrupole moment of the charge distribution, relative to the axis system $(\bar{x}, \bar{y}, \bar{z})$. For a quantum mechanical system where the charge (or mass) density is given as the modulus squared of the wavefunction $\psi_J^M(\vec{r}_i)$, one obtains, the quadrupole moment as the expectation value of the operator $\sum_i (3z_i^2 - r_i^2)$, or $\sum_i \sqrt{(16\pi/5)} r_i^2 Y_2^0(\hat{r}_i)$ and results in

$$Q(J, M) = \int \psi_J^{*M}(\vec{r}_i) \sum_i (3z_i^2 - r_i^2) \psi_J^M(\vec{r}_i) \, d\vec{r}_i, \qquad (1.38)$$

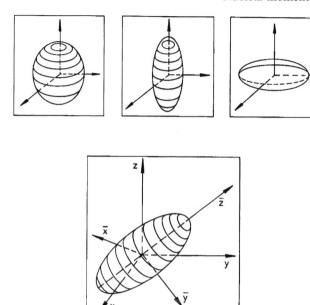

Figure 1.15. In the upper part the various density distributions that give rise to a vanishing, positive and negative quadrupole moment, respectively. These situations correspond to a spherical, prolate and oblate shape, respectively. In the lower part, the orientation of an axially symmetric nuclear density distribution $\rho(\vec{r})$ relative to the body-fixed axis system $(\bar{x}, \bar{y}, \bar{z})$ and to a laboratory-fixed axis system x, y, z is presented. This situation is used to relate the intrinsic to the laboratory (or spectroscopic) quadrupole moment as discussed in the text.

for a microscopic description of the nuclear wavefunction or,

$$Q(J, M) = \int \psi_J^{*M}(\vec{r})(3z^2 - r^2)\psi_M^J(\vec{r})\, d\vec{r}, \qquad (1.39)$$

in a collective model description where the wavefunction depends on a single coordinate $\vec{r}$ describing the collective system with no internal structure.

Still keeping somewhat to the classical picture (see equation (1.39)), one has a vanishing quadrupole moment $Q = 0$ for a spherical distribution $\rho(\vec{r})$, a positive value $Q > 0$ for a prolate (cigar-like shape) distribution and a negative value $Q < 0$ for an oblate distribution (discus-shape) (figure 1.15).

For the rest of the discussion we shall restrict to cylindrical symmetric $\rho(\vec{r})$ distributions since this approximation also is quite often encountered when discussing actual, deformed nuclear charge and mass quadrupole distributions. In this latter case, an interesting relation exists between the quadrupole moment in a body-fixed axis system $(1, 2, 3)$ (which we denote with $\bar{x}, \bar{y}, \bar{z}$ coordinates) and

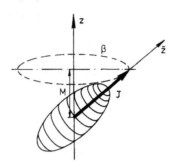

Figure 1.16. The spectroscopic (or laboratory) quadrupole moment can be obtained using a semiclassical procedure of averaging the intrinsic quadrupole moment (defined relative to the $\bar{z}$-axis) over the precession of J about the laboratory z-axis. The intrinsic system is tilted out of the laboratory system over an angle β.

in a fixed (x, y, z) laboratory axis system. If β is the polar angle (angle between the z and $\bar{z}$ axes), one can derive (figure 1.16).

$$Q_{\text{lab}} = \tfrac{1}{2}(3\cos^2\beta - 1)Q_{\text{intr}}, \tag{1.40}$$

where Q_{lab} and Q_{intr} mean the quadrupole moment in the laboratory axis system $\int \rho(\vec{r})(3z^2 - r^2)\,d\vec{r}$ and in the body-fixed axis system $\int \rho(\vec{r})(3\bar{z}^2 - r^2)\,d\vec{r}$, respectively. For angles with $\cos\beta = \pm 1/\sqrt{3}$, the laboratory quadrupole moment will vanish even though Q_{intr} differs from zero, indicating that Q_{intr} is carrying the most basic information on deformed distributions.

1.7 Hyperfine interactions

The hyperfine interaction between a nucleus (made-up of a collection of A nucleons) and its surroundings via electromagnetic fields generated by the atomic and molecular electrons is an interesting probe in order to determine some of the above nuclear moments.

The precise derivation of the electromagnetic coupling Hamiltonian (in describing non-relativistic systems) will be discussed in chapter 6. We briefly give the main results here.

The Hamiltonian describing the nucleons (described by charges e_i and currents $\vec{j}_i$) in interaction with external fields, described by the scalar and vector potentials $\Phi(\vec{r})$ and $\vec{A}(\vec{r})$, is obtained via the 'minimal electromagnetic coupling' substitution $\vec{p}_i \rightarrow \vec{p}_i - e_i\vec{A}$, or

$$\hat{H} = \sum_{i=1}^{A} \frac{1}{2m_i}(p_i - e_i\vec{A})^2 + \sum_{i=1}^{Z} e_i\Phi. \tag{1.41}$$

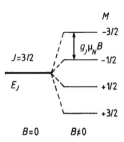

Figure 1.17. Zeeman splitting of the energy levels of a quantum-mechanical system characterized by angular momentum $J = \frac{3}{2}$ and a g-factor g_J in an external field B. The $\vec{B}$ field is oriented along the positive z-axis with $g_J > 0$.

This Hamiltonian can be rewritten as

$$\hat{H} = \sum_{i=1}^{A} \frac{p_i^2}{2m_i} - \sum_{i=1}^{A} \frac{\vec{p_i}}{m_i} e_i \cdot \vec{A} + \sum_{i=1}^{A} \frac{e_i^2}{2m_i} \vec{A}^2 + \sum_{i=1}^{Z} e_i \Phi. \tag{1.42}$$

In neglecting the term, quadratic in $\vec{A}$ at present, we obtain

$$\hat{H} = \hat{H}_0 + \hat{H}_{\text{e.m.}}(\text{coupling}), \tag{1.43}$$

with

$$\hat{H}_{\text{e.m.}}(\text{coupling}) = - \sum_{i=1}^{A} \frac{e_i \vec{p_i} \cdot \vec{A}}{m_i} + \sum_{i=1}^{Z} e_i \Phi. \tag{1.44}$$

For a continuous charge and current distribution this results in

$$\hat{H}_{\text{e.m.}}(\text{coupling}) = - \int \vec{j} \cdot \vec{A} \, d\vec{r} + \int \rho \Phi \, d\vec{r}. \tag{1.45}$$

Using this coupling Hamiltonian, we shall in particular describe the magnetic dipole moment (via Zeeman splitting) and the electric quadrupole moment interacting with external fields.

With $\hat{H}_0$ describing the nuclear Hamiltonian with energy eigenvalues E_J and corresponding wavefunctions ψ_J^M, there remains in general an M-degeneracy for the substates since

$$\hat{H}_0 \psi_J^M = E_J \psi_J^M. \tag{1.46}$$

Including the electromagnetic coupling Hamiltonian, the degeneracy will be lifted, and depending on the interaction $\hat{H}_{\text{e.m.}}$ (coupling) used, specific splitting of the energy levels results. For the Zeeman splitting, one has (figure 1.17)

$$\hat{H}_{\text{hyperfine}} = -\hat{\vec{\mu}} \cdot \vec{B}, \tag{1.47}$$

and, relating $\hat{\vec{\mu}}$ to the nuclear total spin, via the g-factor as follows

$$\hat{\vec{\mu}} = g_J \mu_N \frac{1}{\hbar} \hat{\vec{J}}, \tag{1.48}$$

the expectation value of the hyperfine perturbing Hamiltonian becomes

$$\int \psi_J^{M*} \hat{H}_{\text{hyperfine}} \psi_J^M \, d\vec{r}. \tag{1.49}$$

For a magnetic induction, oriented along the z-axis (quantization axis) $\hat{\vec{J}} \cdot \vec{B}$ becomes $\hat{J}_z B$ and the interaction energy reduces to

$$-g_J \mu_N M B. \tag{1.50}$$

Here now, the $2J + 1$ substates are linearly split via the magnetic interaction and measurements of these splittings not only determine the number of states (and thus J) but also g_J when the induction B is known. The above method will be discussed in some detail for the electric quadrupole interaction too, for axially symmetric systems. We first discuss the general, classical interaction energy for a charge distribution $\rho(\vec{r})$ with an external field $\Phi(\vec{r})$ and secondly derive the quantum mechanical effects through the degeneracy splitting. The classical interaction energy reads

$$E_{\text{int}} = \int_{\text{vol}} \rho(\vec{r}) \Phi(\vec{r}) \, d\vec{r}, \tag{1.51}$$

integrating over the nuclear, charge distribution $\rho(\vec{r})$ volume, and $\Phi(\vec{r})$ denotes the potential field generated by electrons of the atomic or molecular environment. In view of the distance scale relating the charges generating $\Phi(\vec{r})$ (the charge density $\rho_{e^-}(\vec{r})$) and the atomic nucleus volume, $\Phi(\vec{r})$ will in general be almost constant or varying by a small amount over the volume only, allowing for a Taylor expansion into

$$\Phi(\vec{r}) = \Phi(\vec{r})_0 + \sum_i \left(\frac{\partial \Phi}{\partial x_i} \right)_0 x_i + \frac{1}{2} \sum_{i,j} \left(\frac{\partial^2 \Phi}{\partial x_i \partial x_j} \right)_0 x_i x_j + \cdots. \tag{1.52}$$

The corresponding energy then separates as follows into

$$E_{\text{int}} = \Phi(\vec{r})_0 \cdot q + \sum_i \left(\frac{\partial \Phi}{\partial x_i} \right)_0 p_i + \frac{1}{2} \sum_{i,j} \left(\frac{\partial^2 \Phi}{\partial x_i \partial x_j} \right)_0 Q'_{ij} + \cdots,$$

with

$$Q'_{i,j} = \int \rho(\vec{r}) x_i x_j \, d\vec{r}. \tag{1.53}$$

Then a monopole, a dipole, a quadrupole, ... interaction energy results where the monopole term will induce no degeneracy splitting, the dipole term gives the Stark splitting, etc. In what follows, we shall concentrate on the quadrupole term only (q). By choosing the coordinate system (x, y, z) such that the non-diagonal terms in $(\partial^2\Phi/\partial x_i \partial x_j)_0$, $(i \neq j)$ vanish, we obtain for the interaction energy, the expression

$$E^q_{int} = \frac{1}{2}\sum_i \left(\frac{\partial^2\Phi}{\partial x_i^2}\right)_0 \int \rho(\vec{r})x_i^2 \, d\vec{r}. \tag{1.54}$$

Adding, and subtracting a monopole-like term, this expression can be rewritten as

$$E^q_{int} = \frac{1}{6}\sum_i \left(\frac{\partial^2\Phi}{\partial x_i^2}\right)_0 \int \rho(\vec{r})(3x_i^2 - r^2) \, d\vec{r}$$

$$+ \frac{1}{6}\sum_i \left(\frac{\partial^2\Phi}{\partial x_i^2}\right)_0 \int \rho(\vec{r})r^2 \, d\vec{r}. \tag{1.55}$$

Two situations can now be distinguished:

- In situations where electrons at the origin are present, i.e. s-electrons, and these electrons determine the external potential field $\Phi(\vec{r})$, which subsequently becomes spherically symmetric, the first term disappears and we obtain (since $\Delta\Phi = -\rho/\epsilon_0$)

$$E^{q(m)}_{int} = -\frac{\rho_{el}(0)}{2\epsilon_0} \int \rho(\vec{r})r^2 \, d\vec{r}, \tag{1.56}$$

as the interaction energy. This represents a 'monopole' shift and all levels (independent of M) receive a shift in energy, expressed by $E^{q(m)}_{int}$, a shift which is proportional to both the electron density at the origin $\rho_{el}(0)$ and the mean-square radius describing the nucleon charge distribution $\rho(\vec{r})$. Measurements of $E^{q(m)}_{int}$ leads to information about the nuclear charge radius.
- In situations where a vanishing electron density at the origin occurs, $\Delta\Phi = 0$ and only the first term contributes. In this case the particular term can be rewritten as

$$E^q_{int} = \frac{1}{6}\left(\frac{\partial^2\Phi}{\partial x^2}\right)_0 \int \rho(\vec{r})(3x^2 - r^2) \, d\vec{r}$$

$$+ \frac{1}{6}\left(\frac{\partial^2\Phi}{\partial y^2}\right)_0 \int \rho(\vec{r})(3y^2 - r^2) \, d\vec{r}$$

$$+ \frac{1}{6}\left(\frac{\partial^2\Phi}{\partial z^2}\right)_0 \int \rho(\vec{r})(3z^2 - r^2) \, d\vec{r}. \tag{1.57}$$

In cases where the cylindrical symmetry condition for the external potential

$$\left(\frac{\partial^2 \Phi}{\partial x^2}\right)_0 = \left(\frac{\partial^2 \Phi}{\partial y^2}\right)_0 = -\frac{1}{2}\left(\frac{\partial^2 \Phi}{\partial z^2}\right)_0,$$

holds, the quadrupole interaction energy becomes

$$E_{\text{int}}^q = \frac{1}{4}\left(\frac{\partial^2 \Phi}{\partial z^2}\right)_0 \int \rho(\vec{r})(3z^2 - (x^2 + y^2 + z^2))\, d\vec{r}, \tag{1.58}$$

or

$$E_{\text{int}}^q = \frac{1}{4}\left(\frac{\partial^2 \Phi}{\partial z^2}\right)_0 \sqrt{\frac{16\pi}{5}} \int \rho(\vec{r})r^2 Y_2^0 \, d\vec{r}. \tag{1.59}$$

The classical expression can now be obtained, by replacing the density $\rho(\vec{r})$ by the modulus squared of the nuclear wavefunction

$$\langle E_{\text{int}}^q \rangle = \frac{1}{4}\left(\frac{\partial^2 \Phi}{\partial z^2}\right)_0 \int \psi_J^{M*}(\vec{r})(3z^2 - r^2)\psi_J^M(\vec{r})\, d\vec{r}, \tag{1.60}$$

and now rewriting $(3z^2 - r^2)$ as $r^2(3\cos^2\theta - 1)$ or $r^2\sqrt{(16\pi/5)}Y_2^0(\hat{r})$, the quadrupole interaction energy becomes

$$\langle E_{\text{int}}^q \rangle = \frac{1}{4}\left(\frac{\partial^2 \Phi}{\partial z^2}\right)_0 \left\langle \sqrt{\frac{16\pi}{5}}r^2 Y_2^0(\hat{r}) \right\rangle_{J,M}. \tag{1.61}$$

One can now also use a semi-quantum-mechanical argument to relate the laboratory quadrupole moment to the intrinsic quadrupole moment, using equation (1.40), with β the angle between the laboratory z-axis and intrinsic $\bar{z}$-axis and where the quantum-mechanical labels J and M are replaced in terms of the tilting angle β in the classical vector model for angular momentum $\vec{J}$ (vector with length $\hbar\sqrt{J(J+1)}$ and projection $\hbar M$) (figure 1.16). Thereby

$$\langle E_{\text{int}}^q \rangle = \frac{1}{8}\left(\frac{\partial^2 \Phi}{\partial z^2}\right)_0 \langle(3\cos^2\beta - 1)\rangle Q_{\text{intr}}$$

$$= \frac{1}{8}\left(\frac{\partial^2 \Phi}{\partial z^2}\right)_0 \frac{3M^2 - J(J+1)}{J(J+1)} Q_{\text{intr}}. \tag{1.62}$$

Finally, one obtains the result that the interaction energy is quadratic in the projection quantum number M and, because of the quadrupole interaction, breaks the degeneracy level J into $J + \frac{1}{2}$ doubly-degenerate levels. We point out that a correct quantum-mechanical description, using equation (1.61) where the expectation value is

$$\frac{1}{4}\left(\frac{\partial^2 \Phi}{\partial z^2}\right)_0 \langle JM|(16\pi/5)^{1/2}r^2Y_2^0(\hat{r})|JM\rangle, \tag{1.63}$$

has to be evaluated, making use of the Wigner–Eckart theorem (Heyde 1991). Thus, one obtains

$$\frac{1}{4}\left(\frac{\partial^2 \Phi}{\partial z^2}\right)_0 \frac{\langle JM, 20|JM\rangle}{\sqrt{2J+1}} \langle J \| (16\pi/5)^{1/2}r^2Y_2(\hat{r}) \| J\rangle, \tag{1.64}$$

with a separation into a Clebsch–Gordan coupling coefficient and a reduced ($\langle \| \ldots \| \rangle$) matrix element. Putting the explicit value of the Clebsch–Gordan coefficient, one has

$$\frac{1}{4}\left(\frac{\partial^2 \Phi}{\partial z^2}\right)_0 \sqrt{\frac{16\pi}{5}} \cdot \frac{3M^2 - J(J+1)}{(J(J+1)(2J-1)(2J+1)(2J+3))^{1/2}} \langle J\|r^2Y_2\|J\rangle. \tag{1.65}$$

For $M = J$, one obtains the result (maximal interaction)

$$\frac{1}{4}\left(\frac{\partial^2 \Phi}{\partial z^2}\right)_0 \sqrt{\frac{16\pi}{5}} \cdot \sqrt{\frac{J(2J-1)}{(J+1)(2J+1)(2J+3)}} \langle J\|r^2Y_2\|J\rangle, \tag{1.66}$$

which vanishes for $J = 0$ and $J = \frac{1}{2}$.

Since typical values of the intrinsic quadrupole moment are 5×10^{-24} cm^2 (5 barn), one needs the fields that atoms experience in solids to get high enough field gradients in order to give observable splittings. Atomic energies are of the order of eV and atomic dimensions of the order of 10^{-8} cm. So, typical field gradients are of the order of $|\partial \Phi/\partial z| \cong 10^8$ V cm^{-1} or $|\partial^2 \Phi/\partial z^2| \simeq 10^{16}$ V cm^{-2}, and splittings of the order of $\langle E_{\text{int}}^q \rangle \propto e(\partial^2 \Phi/\partial z^2)_0 \cdot Q_{\text{intr}} \propto 5 \times 10^{-8}$ eV can result. A typical splitting pattern for the case of ^{57}Fe with a spin value of $J^\pi = \frac{3}{2}^-$ in its first excited state is shown in figure 1.18. Resonant absorption from the $J^\pi = \frac{1}{2}^-$ ground state can now lead to the absorption pattern and from the measurement of a splitting to a nuclear quadrupole moment whenever $(\partial^2 \Phi/\partial z^2)_0$ is known. On the other hand, starting from a known quadrupole moments, an internal field gradient $(\partial^2 \Phi/\partial z^2)_0$ can be determined. In the other part of the figure, the magnetic hyperfine or Zeeman splitting of ^{57}Fe is also indicated. A discussion on one-particle nucleon quadrupole moments and its relation to nuclear shapes is presented in Box 1e.

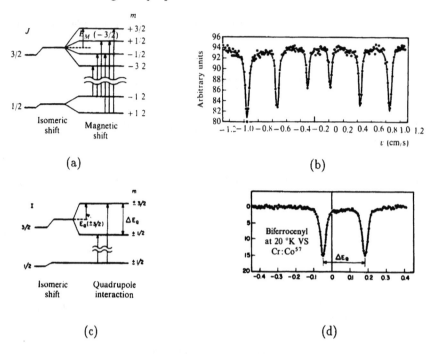

Figure 1.18. Typical hyperfine interactions. (*a*) Removal of the *magnetic* degeneracy resulting from Zeeman splitting. Magnetic moments in the $\frac{3}{2}^{-}$ and $\frac{1}{2}^{-}$ levels have opposite signs. (*b*) The hyperfine absorption structure in a sample of FeF_3. (*c*) Removal of the *quadrupole* degeneracy. Here, only the $\frac{3}{2}^{-}$ substates are split in energy for the same sample as discussed above. (*d*) The observed splitting using Mössbauer spectroscopy. (Taken from Valentin 1981.)

1.8   Nuclear reactions

In analysing nuclear reactions, nuclear transmutations may be written

$$_{Z}^{A}X_N + {}_{Z'}^{A'}a_{N'} \rightarrow {}_{Z''}^{A''}Y_{N''} + {}_{Z'''}^{A'''}b_{N'''}, \qquad (1.67)$$

where we denote the lighter projectiles and fragments with a and b, and the target and final nucleus with X and Y. In the above transformations, a number of conservation laws have to be fulfilled.

1.8.1   Elementary kinematics and conservation laws

These may be listed as follows.

(i)   Conservation of linear momentum, $\sum \vec{p}_i = \sum \vec{p}_f$,

(ii) Conservation of total angular momentum i.e.

$$\sum \vec{J}_i + \vec{J}_{\text{rel},i} = \sum \vec{J}_f + \vec{J}_{\text{rel},f}, \tag{1.68}$$

where $\vec{J}_i$, $\vec{J}_f$ denote the angular momenta in the initial and final nuclei and $\vec{J}_{\text{rel},i}$, $\vec{J}_{\text{rel},f}$ denote the relative angular momenta in the entrance (X, a) and final (Y, b) channels.

(iii) Conservation of proton (charge) and neutron number is not a strict conservation law. Under general conditions, one has conservation of *charge* and *conservation* of *nucleon or baryon* (strongly interacting particles) number.

(iv) Conservation of parity, π, such that

$$\pi_X \cdot \pi_a \cdot \pi_{(X,a)} = \pi_Y \cdot \pi_b \cdot \pi_{(Y,b)}, \tag{1.69}$$

where the parities of the initial and final nuclei and projectiles (incoming, outgoing) are considered.

(v) Conservation of total energy, which becomes

$$T_X + M'_X \cdot c^2 + T_a + M'_a \cdot c^2 = T_Y + M'_Y \cdot c^2 + T_b + M'_b \cdot c^2, \tag{1.70}$$

with T_i the kinetic energy and $M'_i \cdot c^2$ the mass–energy. In the non-relativistic situation, the kinetic energy $T_i = \frac{1}{2} M'_i v_i^2$. One defines the Q-value of a given reaction as

$$Q = \left(\sum M'_i c^2 - \sum M'_f c^2 \right)$$
$$= (M'_X + M'_a - M'_Y - M'_b)c^2, \tag{1.71}$$

which can be rewritten using the kinetic energies as

$$Q = \sum T_f - \sum T_i$$
$$= T_Y + T_b - T_X - T_a. \tag{1.72}$$

The kinematic relations, relating the linear momenta, in the reaction process can be drawn in either the laboratory frame of reference ('lab' system) or in the system where the centre-of-mass is at rest ('COM' system). We show these for the reaction given above in figure 1.19 in shorthand notation X(a, b)Y. A relation between scattering properties in the two systems (in particular relating the scattering angles θ_{lab} and θ_{COM}) is easily obtained, resulting in the expression (see also figure 1.20)

$$\tan \theta_{\text{lab}} = \frac{\sin \theta_{\text{COM}}}{\cos \theta_{\text{COM}} + (v_{\text{COM}}/v'_b)}. \tag{1.73}$$

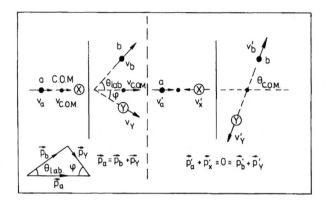

Figure 1.19. Kinematic relations, for the reaction as discussed in the text, i.e. X(a, b)Y, between the laboratory (lab) and the centre-of-mass (COM) system. Indices with accent (′) refer to the COM system and v_{COM} is the centre-of-mass velocity. (Taken from Mayer-Kuckuk 1979.)

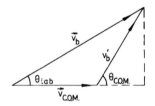

Figure 1.20. Relation between the scattering angle θ_{lab} and θ_{COM}. An explanation of the various kinematic quantities is given in the text.

The various quantities are denoted in the figure. Since one has $M_a' \cdot \vec{v}_a' = -M_X' \cdot \vec{v}_X'$ and also $\vec{v}_{COM} = -\vec{v}_X'$, one can derive the relation $\vec{v}_{COM} = (M_a'/M_X')\vec{v}_a'$. The angle relation can be rewritten since

$$v_{COM}/v_b' = \frac{M_a' \cdot v_a'}{M_X' \cdot v_b'}.$$

For elastic processes, $v_a' = v_b'$ and then the approximate relation

$$\tan \theta_{lab} = \frac{\sin \theta_{COM}}{\cos \theta_{COM} + (M_a'/M_x')}$$
$$\simeq \tan \theta_{COM}(M_a' \ll M_X'). \qquad (1.74)$$

In the Q-value expressions, the various quantities are lab system quantities for kinetic energies. Since Q, however, can also be expressed in terms of the

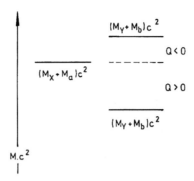

Figure 1.21. Mass relations for the nuclear reaction $X(a, b)Y$ for both the exothermic ($Q > 0$) and endothermic ($Q < 0$) situations.

nuclear masses, the Q-value is independent of the nuclear coordinate system used (see figure 1.21).

From the expressions, giving conservation of linear momentum, one finds a relation between the incoming and outgoing kinetic energies T_a, T_b eliminating the angle-dependence φ and the kinetic energy T_Y since it is most often very difficult to measure the kinetic energy T_Y. The two relations are

$$\sqrt{2M_a'T_a} = \sqrt{2M_Y'.T_Y}\cos\varphi + \sqrt{2M_b' \cdot T_b}\cos\theta$$
$$\sqrt{2M_b' \cdot T_b}\sin\theta = \sqrt{2M_Y' \cdot T_Y}\sin\varphi, \tag{1.75}$$

and, eliminating T_Y and the angle φ, one obtains the relation

$$Q = T_b\left(1 + \frac{M_b'}{M_Y'}\right) - T_a\left(1 - \frac{M_a'}{M_Y'}\right)$$
$$- \frac{2\sqrt{M_a'M_b'T_a}\sqrt{T_b}}{M_Y'} \cdot \cos\theta. \tag{1.76}$$

So, it is possible to determine an unknown Q-value, starting from the kinetic energies T_a, T_b and the angle θ. It is interesting to use the above expression in order to determine T_b as a function of the incoming energy T_a and the angle θ. Using the above equation, as a quadratic equation in $\sqrt{T_b}$, we find

$$\sqrt{T_b} = r \pm \sqrt{r^2 + s},$$

with

$$r = \frac{\sqrt{M_a'M_b'T_a}}{M_b' + M_Y'}\cos\theta,$$

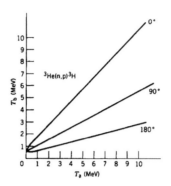

Figure 1.22. Relation between the kinetic energy for incoming (T_a) and outgoing particle (T_b) as illustrated for the reaction ^{3}He(n, p) ^{3}H. (Taken from Krane, *Introductory Nuclear Physics* © 1987 John Wiley & Sons. Reprinted by permission.)

$$s = \frac{M'_Y \cdot Q + T_a(M'_Y - M'_a)}{M'_b + M'_Y}. \tag{1.77}$$

Recall that we are working in the non-relativistic limit. In discussing the result for $\sqrt{T_b}$ we have to distinguish between exothermic ($Q > 0$) and endothermic ($Q < 0$) reactions.

(i) $Q > 0$.
    In this case, as long as $M'_a < M'_Y$ or, if the projectile is much lighter than the final nucleus, one always finds a solution for T_b as a positive quantity. Because of the specific angular dependence on $\cos\theta$, the smallest value of T_b will appear for $\theta = 180°$. Even when $T_a \to 0$, a positive value of T_b results and $T_b \simeq Q \cdot M'_Y/(M'_b + M'_Y)$. We illustrate the above case, for the reaction ^{3}He(p, n) ^{3}H in figure 1.22.

(ii) $Q < 0$.
    In the case when we choose $T_a \to 0$, $r \to 0$ and now s becomes negative so that no solution with positive T_b value can result. This means that for each angle θ, there exists a minimal value $T_a^{th}(\theta)$ below which no reaction is possible. This value is lowest at $\theta = 0°$ and is called the threshold energy. The value is obtained by putting $r^2 + s = 0$, with a result

$$T_a^{th} = -Q \cdot \frac{M'_b + M'_Y}{M'_b + M'_Y - M'_a}. \tag{1.78}$$

It is clear that $T_a^{th} > |Q|$ since in the reaction process of a with X, some kinetic energy is inevitably lost because of linear momentum conservation. Just near to the threshold value, the reaction products Y and b appear with

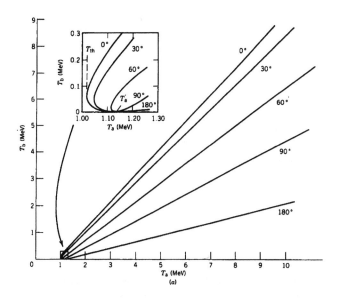

Figure 1.23. Relation between kinetic energy of incoming (T_a) and outgoing (T_b) particles for the reaction ^{3}H(p, n) ^{3}He. The insert shows the region of double-valued behaviour near $T_a \simeq 1.0$ MeV. (Taken from Krane, *Introductory Nuclear Physics* © 1987 John Wiley & Sons. Reprinted by permission.)

an energy in the lab system

$$T_b = r^2 = T_a^{th} \frac{M_a' M_b'}{(M_b' + M_Y')^2}, \qquad (1.79)$$

even though the energy in the COM system is vanishingly small. Increasing T_a somewhat above the threshold energy then for $\theta = 0°$, *two* positive solutions for T_b and a double-valued behaviour results. Two groups of particles b can be observed with different, discrete energies. We illustrate the relation T_b against T_a for the endothermic reaction ^{3}H(p, n) ^{3}He, (figure 1.23).

The above result can most clearly be understood, starting from the diagram in figure 1.24 where the COM velocity v_{COM} and the velocity v_b' of the reaction product for different emission directions in the COM system are shown.

At the angle θ_{lab}, two velocities v_b' result, one resulting from the addition of v_b and v_{COM}, one from subtracting. No particles b can be emitted outside of the cone indicated by the dashed lines. The angle becomes 90° whenever $v_b' = v_{COM}$. Then, in each direction only one energy value T_b is obtained.

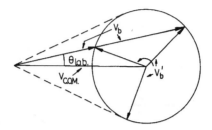

Figure 1.24. Relation between the velocity vectors of the outgoing particles v_b and v'_b for a nuclear reaction X(a, b)Y with negative Q value. (Taken from Mayer-Kuckuk 1979.)

The condition for having a single, positive value of T_b is

$$T_a = -Q \frac{M'_Y}{M'_Y - M'_a}. \qquad (1.80)$$

So, the expression (1.77), for T_b, indeed reveals many details on the nuclear reaction kinematics, as illustrated in the above figures for some specific reactions.

After a reaction, the final nucleus Y may not always be in its ground state. The Q-value can be used as before but now, one has to use the values $M'^*_Y \cdot c^2$ and T^*_Y, corresponding to the excited nucleus Y. This means that to every excited state in the final nucleus Y, with internal excitation energy $E_x(Y)$, a unique Q-value

$$Q_x = Q - E_x(Y), \qquad (1.81)$$

occurs. The energy spectrum of the emitted particles T_b at a certain angle θ_{lab}, shows that for every level in the final nucleus Y, a related T_b value, from equation (1.77) will result. Conversely, from the energy spectrum of particles b, one can deduce information on the excited states in the final nucleus.

A typical layout for the experimental conditions is illustrated below (figures 1.25 and 1.26)

1.8.2   A tutorial in nuclear reaction theory

In a nuclear reaction $a + X \rightarrow \cdots$, the nucleus X seen by the incoming particle looks like a region with a certain nuclear potential, which even contains an imaginary part. Thereby both scattering and absorption can result from the potential

$$U(r) = V(r) + iW(r). \qquad (1.82)$$

The potentials $V(r)$ and $W(r)$ can, in a microscopic approach, be related to the basic nucleon–nucleon interactions. This process of calculating the 'optical' potential is a difficult one.

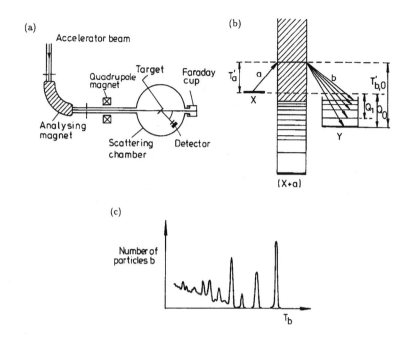

Figure 1.25. Outline of a typical nuclear reaction experiment. (*a*) The experimental layout and set-up of the various elements from accelerator to detector. (*b*) The energy relations (COM energies T_a', T_b') for the nuclear reaction X(a, b)Y. (*c*) The corresponding spectrum of detected particles of type b as a function of the energy of these particles (after Mayer-Kuckuk 1979).

The incoming particle can undergo diffraction by the nucleus without loss of energy (elastic processes). If the energy of the incoming particle T_a is large enough, such that $\lambda_a \ll R$ (R is the nuclear radius) and there is a negligible absorption component ($W(r) \simeq 0$), the nucleus shows up as a 'black sphere' and a typical diffraction pattern results (figure 1.27(*a*)). This is also discussed in Box 1b and is illustrated here for ^{4}He scattering by ^{24}Mg. In cases where T_a is small, and with negligible absorption, standing 'waves' can be formed within the nuclear interior and thus, a large cross-section results, for certain values of R (and thus of A) with respect to the value of T_a. Such states are called, 'resonance' states. The giant resonances are typical examples of such excited states in which large cross-sections occur (figure 1.28).

It is possible that the light projectile *a* enters in the nucleus thereby exciting a single nucleon from an occupied orbital into an 'unbound' ($E > 0$) state, with enough energy to leave the nucleus (see figure 1.27(*b*)). In these cases, we speak of 'direct' nuclear reactions. It could also be such that the incoming projectile *a* loses so much energy that it cannot leave the nucleus within a short time interval.

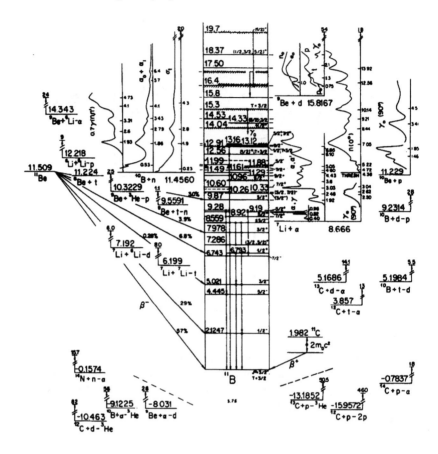

Figure 1.26. Realistic energy level scheme for ^{11}B with all possible reaction channels. (Taken from Ajzenberg-Selove 1975.)

This process can lead to the formation of a 'compound' nuclear state. Only with a statistical concentration of enough energy into a single nucleon or a cluster of nucleons can a nucleon (or a cluster of nucleons), leave the nucleus. The spectrum of emitted particles in such a process closely follows a Maxwellian distribution for the velocity of the outgoing particles with an almost isotropic angular emission distribution.

In contrast, direct nuclear reactions give rise to very specific angular distributions which are characteristic of the energy, and angular momenta of initial and final nucleus. Between the above two extremes (direct reactions—compound nucleus formation), a whole range of pre-equilibrium emission processes can occur. In these reactions a nucleon is emitted only after a number of 'collisions' have taken place, but before full thermalization has taken place.

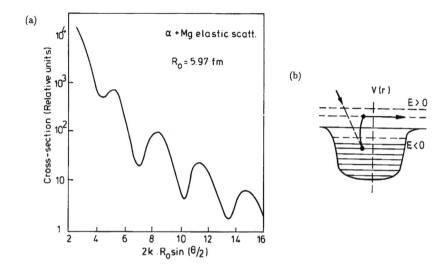

Figure 1.27. (*a*) Illustration of the reaction cross-section for elastic scattering of alpha particles on ^{24}Mg. The various diffraction minima are indicated as a function of the momentum transfer. (Taken from Blair *et al* 1960). (*b*) Schematic illustration of a typical direct reaction process (for explanation see the text.)

The whole series of possible reaction processes is illustrated in figure 1.29 in an intuitive way.

We return briefly to the direct nuclear reaction mechanism by which a reaction a + X → Y + b can be described as follows. We can describe both the entrance and exit channels in terms of an optical potential $U_{a,X}(\vec{r})$ and $U_{b,Y}(\vec{r})$. The Schrödinger equations describing the scattering process become:

$$-\frac{\hbar^2}{2m_a}\Delta\chi_{a,X} + U_{a,X}(\vec{r})\chi_{a,X} = E_{a,X}\chi_{a,X}$$

$$-\frac{\hbar^2}{2m_b}\Delta\chi_{b,Y} + U_{b,Y}(\vec{r})\chi_{b,Y} = E_{b,Y}\chi_{b,Y}. \tag{1.83}$$

Here, $m_a(m_b)$ is the reduced mass of projectile (ejectile) relative to the nucleus X(Y). In a direct reaction, a direct 'coupling' occurs between the two unperturbed channels (a, X) and (b, Y) described by the distorted waves $\chi_{a,X}$ and $\chi_{b,Y}$. The transition probability for entering by the channel $\chi_{a,X}$ at given energy $E_{a,X}$ and leaving the reaction region by the outgoing channel $\chi_{b,Y}$ is then giving using lowest-order perturbation theory by the matrix element (figure 1.30)

$$\int \chi^*_{a,X}(\vec{r})H_{\text{int}}(\vec{r})\chi_{b,Y}(\vec{r})\,d\vec{r}. \tag{1.84}$$

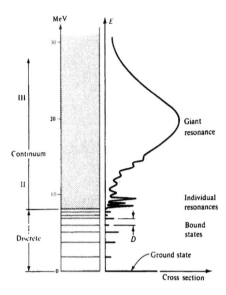

Figure 1.28. Typical regions of an excited nucleus where we make a schematic separation into three different regimes: I—region of bound states (discrete states); II—continuum with discrete resonances; and III—statistical region with many overlapping resonances. The cross-section function is idealized in the present case. (Taken from Frauenfelder and Henley (1991) *Subatomic Physics* © 1974. Reprinted by permission of Prentice-Hall, Englewood Cliffs, NJ.)

The description of the process is, in particular, rather straightforward if a zero-range interaction $H_{\rm int}(\vec{r})$ is used to describe the nuclear interaction in the interior of the nucleus. The method is called DWBA (distorted wave Born approximation).

A more detailed but still quite transparent article on direct reactions is that by Satchler (1978).

1.8.3 Types of nuclear reactions

The domain of nuclear reactions is very extensive and, as discussed before, ranges from the simple cases of elastic scattering and one-step direct processes to the much slower, compound nuclear reaction formation. In table 1.2, we give a schematic overview depending on the type and energy of the incoming particle and on the target nucleus (X), characterized by its nuclear mass A. In this table, though, only a limited class of reactions is presented. In the present, short paragraph, we shall illustrate some typical reactions that are of recent interest, bridging the gap between 'classical' methods and the more advanced 'high-energy' types of experiments.

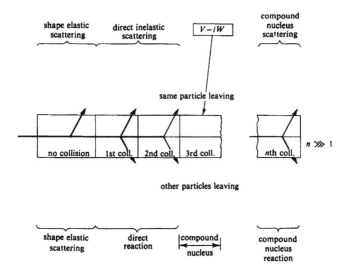

Figure 1.29. Evolution of a nuclear reaction according to increasing complexity in the states formed. In the first stage, shape-elastic scattering is observed. The other processes excite more and more individual particles in the nucleus, eventually ending up at a fully compound nuclear state. (Taken from Weisskopf 1961.)

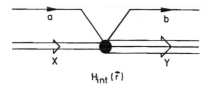

Figure 1.30. Direct reaction process where the nucleus X takes up part of the light fragment a, releasing the fragment b and the final nucleus Y. The time scale is too short for equilibrium to occur and a compound nuclear state to form.

(i)   The possible, natural decay processes can also be brought into the class of reaction processes with the conditions: no incoming particle a and $Q > 0$. We list them in the following sequence.

- α-decay:

$$^{A}_{Z}X_N \rightarrow \, ^{A-4}_{Z-2}Y_{N-2} + \, ^{4}_{2}\text{He}_2.$$

- β-decay:

$$^{A}_{Z}X_N \rightarrow \, ^{A}_{Z-1}Y_{N+1} + e^{+} + \nu_e \qquad (\text{p} \rightarrow \text{n-type}),$$

$$^{A}_{Z}X_N \rightarrow \, ^{A}_{Z+1}Y_{N-1} + e^{-} + \bar{\nu}_e \qquad (\text{n} \rightarrow \text{p-type}),$$

Table 1.2. Table of nuclear reactions listing reaction products for a variety of exprimental conditions. (Taken from Segré *Nuclei and Particles* 2nd edn © 1982 Addison-Wesley Publishing Company. Reprinted by permission.)

	Intermediate nuclei (30 > A > 90)					Heavy nuclei (A > 90)		
	Incident particle							
Energy of incident particle	n	p	α	d	n	p	α	d
Low, 0–1 keV	n(el.) γ (res.)	No appreciable reaction	No appreciable reaction	No appreciable reaction	γ n(el.) (res.)	No appreciable reaction	No appreciable reaction	No appreciable reaction
Intermediate, 1–500 keV	n(el.) γ (res.)	n γ α (res.)	n γ p (res.)	p n	n(el.) γ (res.)	Very small reaction cross section	Very small reaction cross section	Very small reaction cross section
High, 0.5–10 MeV	n(el.) n(inel.) p α (res. for lower energies)	n p(inel.) α (res. for lower energies)	n p α(inel.) (res. for lower energies)	p n pn $2n$	n(el.) n(inel.) p γ	n p(inel.) γ	n p γ	p n pn $2n$
Very high, 10–50 MeV	$2n$ n(inel.) n(el.) p np $2p$ α Three or more particles	$2n$ n p(inel.) np $2p$ α Three or more particles	$2n$ n p np $2p$ α(inel.) Three or more particles	p $2n$ pn $3n$ d(inel.) tritons Three or more particles	$2n$ n(inel.) n(el.) p pn $2p$ α Three or more particles	$2n$ n p(inel.) np $2p$ α Three or more particles	$2n$ n p np $\cdot 2p$ α(inel.) Three or more particles	p $2n$ np $3n$ d(inel.) tritons Three or more particles

$$^{A}_{Z}X_N + e^- \rightarrow {}^{A}_{Z-1}Y_{N+1} + \nu_e \qquad (e^- - \text{capture}).$$

- γ-decay:

$$^{A}_{Z}X^*_N \rightarrow {}^{A}_{Z}X_N + h\nu.$$

Here, the nucleus could also decay via pair $(e^- + e^+)$ creation (if $E \gtrsim 1.022$ MeV) or e^- internal conversion, in particular for $0^+ \rightarrow 0^+$ decay processes.

- Nuclear fission

$$^{A}_{Z}X_N \rightarrow {}^{A_1}_{Z_1}Y_{N_1} + {}^{A_2}_{Z_2}U_{N_2} + x \cdot n.$$

- Particle emission (near the drip-line) with $Q > 0$, in particular proton and neutron emission.

Various other, even exotic radioactive decay mechanisms do show up which are not discussed in detail in the present text (see also Box 1f).

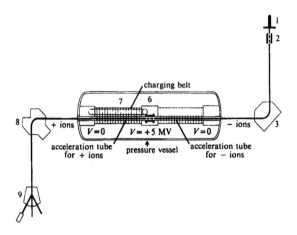

Figure 1.31. Two-state tandem accelerator. 1—source of positive ions. 2—canal for adding electrons. 3—negative ions are pre-accelerated to 80 keV and injected into the Van de Graaff accelerator where they receive a potential energy of 5 MeV. 6—the ions are stripped of electrons and become positively charged by passing through a gas at low pressure. 7—the positive ions are accelerated by a further 5 MeV so that their total potential energy is 10 MeV. The kinetic energy of the emerging ions depends on their charge states in the two stages of acceleration, e.g. if single charged in each stage their kinetic energy will be 10 MeV. 8—deflecting and analysing magnet. 9—switching magnet. (High-Voltage Engineering Corp., Burlington, MA.)

(ii) As outlined before, and presented in table 1.2, a large variety of nuclear, induced reactions can be carried out. Most of the easy processes use accelerated, charged particle beams with projectiles ranging from e^-, e^+ particles up to highly stripped heavy ions. In the accelerating process, depending on the particle being accelerated and the energy one is interested in, various accelerator techniques are in use.

One of the oldest uses a static potential energy difference realized in the Van de Graaff accelerator, mainly used for light, charged particles. Various new versions, such as the two-stage tandem were developed, and is explained in figure 1.31.

Linear accelerators have been used mainly to obtain high-energy e^- beams. At the high-energy accelerator centres (CERN, SLAC, ...) linear accelerators have been used to pre-accelerate protons before injecting them in circular accelerators. Even circular, e^- accelerators have been constructed like LEP at CERN. By constructing with a very large radius and thus, a small curvature, rather low losses due to radiation are present. Also, special 'low-loss' radiofrequency cavities have been built.

Many accelerators make use of circular systems: cyclotron accelerators are very often used in nuclear structure research (see the SATURNE set up

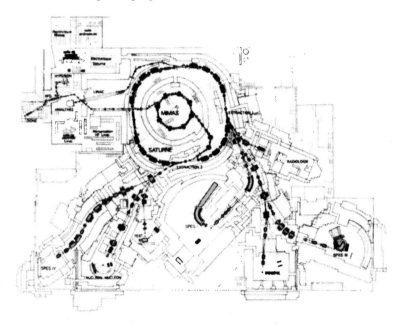

Figure 1.32. Typical view of an accelerator laboratory. In the present case, the SATURNE accelerator with the various beam lines and experiments is shown. (Taken from Chamouard and Durand 1990.)

Figure 1.33. View of the alternating-gradient synchrotron (AGS) at the Brookhaven National Laboratory showing a few of the beam bending magnetic sections (reprinted with permission of Brookhaven National Laboratory).

Figure 1.34. Aerial view of the CERN accelerator complex and its surrounding region. Three rings are shown: PS (small), SPS (middle) and LHC (large). (Taken from CERN photo archive CERN-SI-910 5065, with permission.)

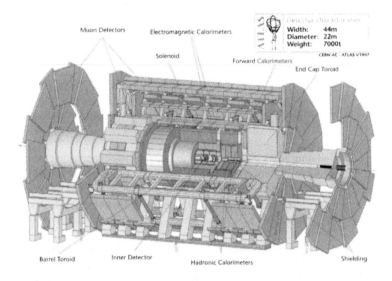

Figure 1.35. Layout of the ATLAS detector. (Taken from CERN photo archive CERN-DI-980 3026, with permission.)

at Saclay where projectiles ranging from protons to heavy ions have been used figure 1.32). At the high- and very-high-energy side, starting from the early AGS (Alternating-gradient synchrotron) at the Brookhaven National

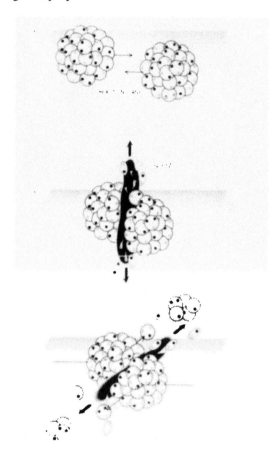

Figure 1.36. Colliding heavy nuclei are used to study the nuclear equation of state. Here, two Au nuclei collide slightly off-centre (1). Matter is squeezed out at right angles to the reaction plane (2). The remaining parts of the two nuclei then bounce off each other (3). (Taken from Gutbrod and Stöcker 1991. Reprinted with permission of *Scientific American.*)

Laboratory (figure 1.33), over the LEP e^-, e^+ facility at CERN, the Tevatron at Fermilab, the LHC (Large Hadron Collider), presently under construction at CERN, is planned to become operational in 2007 and will then represent the highest energy accelerator project.

The scale of the AGS (Alternating Gradient Synchrotron) and the CERN-LHC installations are illustrated on the following pages and, thus, give an impression of the huge scale of such projects (see figure 1.34 to appreciate the scale of the LHC).

The accelerator mechanisms, projectiles used and various recent possibilities are discussed in e.g. *Nuclei and Particles*, second edition (Segré 1982).

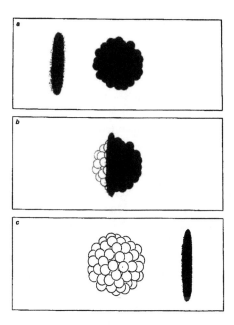

Figure 1.37. Relativistic contraction is illustrated in an ultra-high-energy collision between two uranium nuclei. (*a*) A 1 TeV uranium nucleus, having a velocity of 99.999% of that of light, appears as a disk, with a contraction predicted by the theory of special relativity. It encounters the target-nucleus for about 10^{-22} s; a time much too short to reach equilibrium so the projectile passes through (*c*). In this process, the temperatures reached may create conditions similar to the early Universe, shortly after the big bang (*b*). (Taken from Harris and Rasmussen 1983. Reprinted with permission of *Scientific American*)

In conjunction with the accelerator processes, the detectors needed to measure the appropriate particles (energy, angular distributions, multiplicities, ...), can sometimes be huge, as in the high-energy experiments (ATLAS, CMS, ALICE to name just a few of the detectors presently under construction at CERN) (see also figure 1.35 for the typical scale of such detectors).

Reactions that are attracting much attention at present, in recent projects and in planned experiments, are heavy ion reactions. These will eventually be done using heavy ions moving at relativistic speeds. At Brookhaven National Laboratory (BNL), the Relativistic Heavy-Ion Collider (RHIC for short), has been built in order to study heavy ion collisions carried out with the heaviest stable elements in order to try to create the quark–gluon phase of matter, that has been anticipated for some time and for which circumstantial evidence has already been found through the SPS program at CERN (see chapter 16 for more details on the physics issues at stake).

> **Box 1a. The heaviest artificial elements in nature: up to $Z = 112$
> and beyond**

Where are the heaviest nuclei made artificially under laboratory conditions situated? The hunt for superheavies or answering the question about the maximum charge and mass that a nucleus may acquire makes an interesting saga. On a classical basis, the surface tension and the Coulomb repulsion are the essential players and this shows that elements with $Z > 104$ would fission immediately. Quantal shell-effects can bring in the necessary stabilizing energy to make the formation of these superheavy nuclei possible. At present, efforts at the GSI, Dubna and LBL have resulted in the unambigous existence of elements $Z = 110, 111$ and 112 (Hofmann and Münzenberg 2000). Experimental research programs have also been started at GANIL (France) and RIKEN (Japan).

The 'Gesellschaft für Schwerionenforschung' (GSI for short) in Darmstadt succeeded in 1982 in fusing atomic nuclei to form the element with $Z = 109$ and a few years later to form the element with $Z = 108$. In 1994 successful synthesis of elements with $Z = 110$ and 111 was carried out. These elements indeed have very short lifetimes, of the order of a few milliseconds (ms), but could be identified uniquely by their unambiguous decay chain. The group of Armbruster and Münzenberg, followed by Hofmann, in a momentous series of experiments making heavier and heavier artificial elements, in February 1996, formed the heaviest element now known with $Z = 112$.

The group of physicists at the GSI, together with scientists from a number of laboratories with long-standing experience in fusing lighter elements in increasingly super-heavy elements (Dubna (Russia), Bratislava (Slovakia) and Jyväskylä (Finland)) made this reaction possible.

The particular fusion reaction that was used consisted of accelerating ^{70}Zn nuclei with the UNILAC at GSI and colliding them with an enriched target

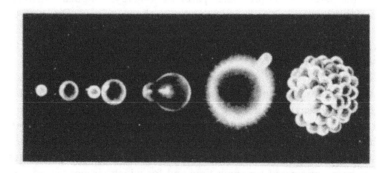

Figure 1a.1. Birth of an element: a Zn nucleus (left) hits a Pb nucleus, merges, cools by ejecting a neutron and ends as an element with $Z = 112$. (Adapted from Taubes 1982.)

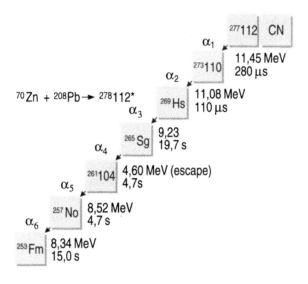

Figure 1a.2. Decay chain with α-decays characterizing the decay of the element with charge $Z = 112$. (Taken from GSI-Nachreichten, 1996a, with permission.) See also Armbruster and Münzenberg (1989) and Hofmann (1996).

consisting of ^{208}Pb. If an accelerated nucleus reacts with a target nucleus fusion can happen but this process is rare: only about one collision in 10^{18} produces the element with $Z = 112$. Even with a successful fusion between the projectile and the target nucleus, the 'compound' nucleus formed is still heated and must be 'cooled' very quickly otherwise it would fall apart in a short time. If a single neutron is emitted, this cooling process just dissipates enough energy to form a single nucleus of the element with $Z = 112$ which can subsequently be detected. The formation is shown in figure 1a.1 (an artistic view) and in figure 1a.2 in which the precise decay chain following the formation of element $Z = 112$ is given. Both the alpha-decay kinetic energies and half-lives for the disintegration (see chapter 4) are given at the various steps.

   In addition to the discovery of the, at present, heaviest element with a charge of $Z = 112$, the nuclear chemistry of elements up to $Z = 108$ has also been studied. A dedicated analysis, discussed by Düllmann in *Nature* (Dullman 2002), has investigated the chemistry of the element Hassium ($Z = 108$) and the methods used there may pave the way towards understanding still heavier elements and their position in the Mendeleyv table. Moreover, an element with charge number $Z = 110$ has recently been credited, through IUPAP, as the element Darmstadtium after its discovery at the GSI in Darmstadt. Exciting new results on superheavy nuclei have been published recently by the Dubna group. The synthesis of element $Z = 114$ by complete fusion of ^{242,244}Pu with ^{48}Ca and, recently the discovery of element $Z = 116$ in the reaction ^{248}Cm plus ^{48}Ca with a correlated decay sequence leading to the previously known 114 chain has been announced (Oganesian 2000) (cross sections of the order of ~ 1 pb).

The search for these very elements has been very competitive with the announcement in 1999, of an element with $Z = 118$ at LBL in high-energy Pb–Kr interactions (Ninov 1999). Follow-up experiments at the LBL itself and at other places have since failed to confirm these first results. Subsequently, after re-analyses, the claim has been retracted from the earlier publication in *Physical Review Letters* in 1999 (Ninov 2002).

Still heavier and more neutron-rich nuclei are expected to see a spherical shell closure develop with an accompanying strong energy stabilization. Theoretical studies give indications for a 'doubly-magic' nucleus at $^{310}_{126}X_{184}$. This proton shell closure is quite different from the earlier suggestions that $Z = 114$ was a closed-shell configuration. Hartree–Fock studies that lead to these results are consistent with other self-consistent non-relativistic and relativistic mean field calculations, indicating the same neutron shell closure at $N = 184$ but with the proton shell closure at $Z = 120$ or 126 (Nazarewicz 1999).

Note added in proof. Very recent experimental evidence for element $Z = 115$ has been published by Oganessian *et al* 2004 using the reaction ^{243}Am + ^{48}Ca. This element then breaks into the element $Z = 113$.

Box 1b. Electron scattering: nuclear form factors

In describing the scattering of electrons from a nucleus, we can start from lowest-order perturbation theory and represent an incoming particle by a plane wave with momentum $\vec{k}_i$ and an outgoing particle by plane wave with momentum $\vec{k}_f$. The tranisition matrix element for scattering is given by

$$W(\vec{k}_i, \vec{k}_f) = \frac{2\pi}{\hbar}|M(\vec{k}_i, \vec{k}_f)|^2 \rho(E), \ (s^{-1}), \tag{1b.1}$$

where

$$M(\vec{k}_i, \vec{k}_f) = \frac{1}{V}\int \exp[i(\vec{k}_i - \vec{k}_f) \cdot \vec{r}]V(r)\, d\vec{r}. \tag{1b.2}$$

Starting from the knowledge of $W(\vec{k}_i, \vec{k}_f)$ and the incoming current density $\sim j$, one can evaluate the differential cross-section for scattering of the incoming particles by an angle θ within the angular element $d\Omega$ as

$$\frac{d\sigma}{d\Omega} = W_\theta(\vec{k}_i, \vec{k}_f)/j, \tag{1b.3}$$

or

$$\frac{d\sigma}{d\Omega} = \frac{m^2}{4\pi^3 \hbar^4}|M(\vec{k}_i, \vec{k}_f)|^2. \tag{1b.4}$$

This is the Born approximation equation which can be rewritten in terms of the scattering angle θ since we have (see figure 1b.1)

$$\vec{q} = \hbar(\vec{k}_i - \vec{k}_f). \tag{1b.5}$$

We then obtain the expression

$$\frac{d\sigma}{d\Omega} = |f(\theta)|^2, \tag{1b.6}$$

with

$$f(\theta) = \frac{m}{2\pi\hbar^2}\int V(r)\exp\left(\frac{i}{\hbar}\vec{q} \cdot \vec{r}\right)d\vec{r}. \tag{1b.7}$$

For a central potential $V(r)$, the above integral can be performed more easily if we choose the z-axis along the direction of the $\vec{q}$-axis. So, we have

$$\vec{q} \cdot \vec{r} = qr\cos\theta',$$
$$d\vec{r} = r^2 \sin\theta'\, dr\, d\theta'\, d\phi, \tag{1b.8}$$

and the integral becomes

$$\int \exp\left(\frac{i}{\hbar}\vec{q} \cdot \vec{r}\right)d\vec{r} = \int \exp\left(\frac{i}{\hbar}qr\cos\theta'\right)r^2 \sin\theta'\, dr\, d\theta'\, d\phi'. \tag{1b.9}$$

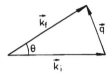

Figure 1b.1. Linear momentum diagram representing initial ($\vec{k}_i$), final ($\vec{k}_f$) and transferred ($\vec{q}$) vectors in electromagnetic interactions.

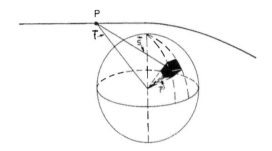

Figure 1b.2. Coordinates $\vec{r}$, $\vec{s}$ and $\vec{t}$ in the scattering process (see text).

Using the substitution $z = (i/\hbar)qr\cos\theta'$, this integral can quite easily be rewritten in the form

$$\int \exp\left(\frac{i}{\hbar}\vec{q} \cdot \vec{r}\right) d\vec{r} = \int_0^\infty \frac{\sin(qr/\hbar)}{(qr/\hbar)} 4\pi r^2 \, dr, \qquad (1b.10)$$

and for the scattering amplitude $f(\theta)$ we obtain the result

$$f(\theta) = \frac{2m}{\hbar q} \int_0^\infty V(r) r \sin(qr/\hbar) \cdot dr. \qquad (1b.11)$$

We now discuss three applications: (1) scattering by the Coulomb potential $1/r$; (2) scattering off a nucleus with a charge density $\rho(r)$ (spherical nucleus); (3) scattering by a square well potential.

(i) *Coulomb (Rutherford) scattering.* Using the potential $V(r) = Ze^2/r$, we immediately obtain

$$\begin{aligned}
f(\theta) &= \frac{2mZe^2}{\hbar q} \int_0^\infty \sin(qr/\hbar) \, dr \\
&= \frac{2mZe^2}{\hbar q} \lim_{a\to 0} \int_0^\infty \sin(qr/\hbar) e^{-r/a} \, dr \\
&= \frac{2mZe^2}{q^2},
\end{aligned} \qquad (1b.12)$$

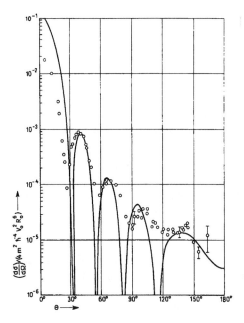

Figure 1b.3. Angular distribution for elastic scattering on the square-well potential, as discussed in the text, obtained using the Born approximation with plane waves (see equation (1.b.20)) and with $kR_0 = 8.35$. The data correspond to 14.5 MeV neutron scattering on Pb. (Taken from Mayer-Kuckuk 1979.)

and thus

$$\frac{d\sigma}{d\Omega} = |f(\theta)|^2 = 4m^2 e^4 Z^2 \cdot \frac{1}{q^4}, \qquad (1b.13)$$

which is indeed, the Rutherford scattering differential cross-section.

(ii) *Electron scattering off a nucleus.* In figure 1b.2 we show the scattering process where $\vec{r}$ is the radial coordinate from the center of the nucleus, $\vec{t}$ is the coordinate of the moving electron measured from the centre of the nucleus with $\vec{t} = \vec{r} + \vec{s}$. The volume element $Ze\rho(r)\,d\vec{r}$ contributes to the potential felt by the electron as

$$\Delta V(t) = \frac{Ze^2}{s}\rho(r)\,d\vec{r}. \qquad (1b.14)$$

Now, the scattering amplitude $f(\theta)$ becomes

$$f(\theta) = \frac{mZe^2}{2\pi\hbar^2} \iint \frac{\rho(r)}{s} \exp\left[\frac{i}{\hbar}\vec{q}\cdot(\vec{r}+\vec{s})\right] d\vec{r}\,d\vec{s}$$

$$= \int \rho(r) \exp\left(\frac{i}{\hbar}\vec{q}\cdot\vec{r}\right) d\vec{r} \cdot \frac{mZe^2}{2\pi\hbar^2} \int \frac{1}{s} \exp\left(\frac{i}{\hbar}\vec{q}\cdot\vec{s}\right) d\vec{s}$$

$$= \int_0^\infty \underbrace{\rho(r)\frac{\sin(qr/\hbar)}{(qr/\hbar)}4\pi r^2\,dr}_{F^2(q)} \cdot \underbrace{\frac{2mZe^2}{\hbar q}\int_0^\infty \sin(qs/\hbar)ds}_{(d\sigma/d\Omega)_{\text{Ruth}}}.$$

(1b.15)

The final result for $d\sigma/d\Omega$ is modified via the finite density distribution $\rho(r)$ and the correction factor is called the charge nuclear form factor $F^2(q)$ and contains the information about the nuclear charge density distribution $\rho(r)$. Thus one has

$$\frac{d\sigma}{d\Omega} = \left(\frac{d\sigma}{d\Omega}\right)_{\text{Ruth}} \cdot F^2(q).$$

(1b.16)

In comparing a typical electron scattering cross-section, as derived here, to the realistic situation for ^{208}Pb (figure 1.8), the measured cross-section, as a function of the momentum transfer $\vec{q}$, can be used to extract the $\rho(r)$ value in ^{208}Pb.

(iii) *Scattering by a square-well potential.* Using a potential which resembles closely a nuclear potential (compared to the always present Coulomb part) with $V(r) = -V_0$ for $r < R_0$ and $V(r) = 0$ for $r \geq R_0$ (where R_0 is the nuclear radius) we obtain for $f(\theta)$

$$f(\theta) = \frac{-2mV_0}{\hbar q}\int_0^{R_0} r\sin(qr/\hbar)\cdot dr.$$

(1b.17)

The integral becomes

$$(\hbar^2/q^2)[\sin(qr/\hbar) - (qr/\hbar)\cos(qr/\hbar)],$$

(1b.18)

and the cross-section is

$$\frac{d\sigma}{d\Omega} = 4m^2\frac{V_0^2\hbar^2}{q^6}[\sin(qR_0/\hbar) - (qR_0/\hbar)\cdot\cos(qR_0/\hbar)]^2.$$

(1b.19)

From figure 1b.1, we obtain the result that $q = 2k\hbar\sin(\theta/2)$ with $k = |\vec{k}_i| = |\vec{k}_f|$ for elastic scattering. Using this relation between q and the scattering angle, we can express $d\sigma/d\Omega$, for given k and R_0, as a function of the scattering angle θ and obtain

$$\frac{d\sigma}{d\Omega} = \frac{4m^2V_0^2R_0^6}{\hbar^4}\frac{[\sin(2kR_0\sin\theta/2) - 2kR_0\sin\theta/2\cos(2kR_0\sin\theta/2)]^2}{(2kR_0\sin\theta/2)^6}.$$

(1b.20)

This scattering cross-section, which is valid at high incident energies, results in a typical diffraction pattern. A large contribution occurs only for forward scattering if $qR_0 < 1$, or $\sin\theta/2 < 1/2kR_0$. In figure 1b.3, we compare the result using equation 1b.20, with $kR_0 = 8.35$, with data for 14.5 MeV neutron scattering on Pb. Even though the fit is not good at the diffraction minima, the overall behaviour in reproducing the data is satisfactory.

Box 1c. Proton and neutron charge distributions: status in 2004

The internal distribution of electric charge and magnetism inside the proton (charge e) and the neutron (charge 0) are of fundamental importance in understanding nucleons and nuclear structure. Electron scattering off these basic nuclear constituents makes up for the ideal probe in order to obtain a detailed view of the internal structure (a neutron is not just a global neutral particle but, through the internal quark content, will inevitably show a specific radial charge density distribution).

With the advent of high duty-factor polarized electron beam facilities, experiments using recoil polarimeters, polarized ^{3}He targets and polarized deuterium targets, have yielded very precise data on the neutron electric form factor (see (Madey *et al* 2003, Arrington 2003) and references therein). The Jefferson Laboratory (JLab for short) has played a pioneering role over the last couple of years in this respect.

A very detailed analysis using the best available data (emphasizing these data originate from recoil or target polarization experiments) has been carried out recently by Kelly (Kelly 2002) (see this reference for technical details on data selection and analysis methods).

In figure 1c.1, the proton charge and magnetization distributions are given. What should be noted is the softer charge distribution compared to the magnetic one for the proton. These resulting densities are quite similar to Gaussian density distributions that can be expected starting from the quark picture and, at the same

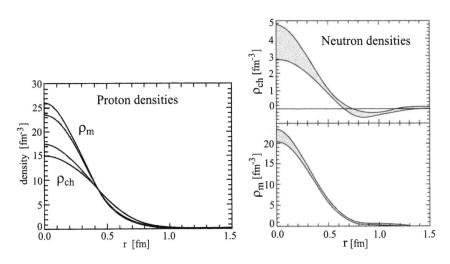

Figure 1c.1. Proton and neutron electric charge and magnetic density distributions. (Reprinted figures with permission from Kelly J J 2002 *Phys. Rev.* C **66** 065203 © 2002 by the American Physical Society.)

time, more realistic than the exponential density distributions that are derived using a dipole form factor and non-relativistic methods to deduce the density distributions from the data.

The neutron charge and magnetization distributions are also given in figure 1c.1. What is striking is that the magnetization distribution resembles very closely the corresponding proton distribution. Since scattering on neutrons normally carries the larger error, the neutron charge distribution is not that precisely fixed. Nonetheless, one notices that the interior charge density is balanced by a negative charge density, situated at the neutron surface region, thereby making up for the integral vanishing of the total charge of the neutron.

Box 1d. Observing the structure in the nucleon

Nobel Prize 1990

The most prestigious award in physics went this year to Jerome I. Friedman and Henry W. Kendall, both of the Massachusetts Institute of Technology (MIT), and Richard E. Taylor of Stanford 'for their pioneering investigations concerning deep inelastic scattering of electrons on protons and bound neutrons, which have been of essential importance for the development of the quark model in particle physics'.

Their experiments, carried out from 1967 at the then new two-mile linac at the Stanford Linear Accelerator Center (SLAC) showed that deep inside the proton there are hard grains, initially called 'partons' by Feynman and later identified with the quarks, mathematical quirks which since 1964 had been known to play an important role in understanding the observed variety of subatomic particles.

These initial forays into high energy electron scattering discovered that a surprisingly large number of the electrons are severely deflected inside the proton targets. Just as the classic alpha particle results of Rutherford earlier this century showed that the atom is largely

1990 Physics Nobel Prize people – left to right: Jerome I. Friedman and Henry W. Kendall of MIT, and Richard Taylor of Stanford

(Photos Keystone)

empty space with a compact nucleus at its centre, so the SLAC-MIT experiments showed that hard scattering centres lurked deep inside the proton.

Writing on the wall for this year's Nobel Prize was the award last year of the Wolfgang Panofsky Prize (sponsored by the Division of Particles and Fields of the American Physical Society) to the same trio for their leadership in the first deep-inelastic electron scattering experiments to explore the deep interior of nuclear particles.

In the early 1960s when construction of the SLAC linac was getting underway, the 1990 Nobel trio, who had first met in the 1950s as young researchers at Stanford's High Energy Physics Laboratory, came together in a collaboration preparing the detectors and experimental areas to exploit the new high energy electron beams

The outcome was described in an article (October 1987, page 9)

by Michael Riordan, subsequently a member of the experimental collaboration and now SLAC's Science Information Officer, for the 20th anniversary of SLAC's electron beams.

In October 1967, MIT and SLAC physicists started shaking down their new 20 GeV spectrometer; by mid-December they were logging electron-proton scattering in the so-called deep inelastic region where the electrons probed deep inside the protons. The huge excess of scattered electrons they encountered there · about ten times the expected rate · was later interpreted as evidence for pointlike, fractionally charged objects inside the proton.

The quarks we take for granted today were at best 'mathematical' entities in 1967 · if one allowed them any true existence at all. The majority of physicists did not. Their failure to turn up in a large number of intentional searches had convinced most of us that Murray Gell-Mann's whimsical entities could not possibly be 'real' particles in the usual sense, just as he had insisted from the very first.

58   *Nuclear global properties*

Jerome Friedman, Henry Kendall, Richard Taylor and the other MIT-SLAC physicists were not looking for quarks that year. SLAC Experiment 4B had originally been designed to study the electroproduction of resonances. But the proddings of a young SLAC theorist, James Bjorken, who had been working in current algebra (then an esoteric field none of the experimenters really understood), helped convince them to make additional measurements in the deep inelastic region, too.

Over the next six years, as first the 20 GeV spectrometer and then its 8 GeV counterpart swung out to larger angles and cycled up and down in momentum, mapping out this deep inelastic region in excruciating detail, the new quark parton picture of a nucleon's innards gradually took a firmer and firmer hold upon the particle physics community. These two massive spectrometers were our principal 'eyes' into the new realm, by far the best ones we had until more powerful muon and neutrino beams became

available at Fermilab and CERN. They were our Geiger and Marsden, reporting back to Rutherford the detailed patterns of ricocheting projectiles. Through their magnetic lenses we 'observed' quarks for the very first time, hard 'pits' inside hadrons.

These two goliaths stood resolutely at the front as a scientific revolution erupted all about them during the late 1960s and early 1970s. The harbingers of a new age in particle physics, they helped pioneer the previously radical idea that leptons, weakly interacting particles, of all things, could be used to plumb the mysteries of the strong force. Who would have guessed, in 1967, that such spindly particles would eventually ferret out their more robust cousins, the quarks? Nobody, except perhaps Bjorken - and he wasn't too sure himself.

(The saga is recounted in every detail in Riordan's book 'The Hunting of the Quark', published by Simon and Schuster.)

1990 Physics Nobel Prize apparatus – the big spectrometers at End Station A of the Stanford Linear Accelerator Center (SLAC) in 1967: right, the 8 GeV spectrometer and left, its 20 GeV and (extreme left) 1.6 GeV counterparts.

(Reprinted from *CERN Courier* 1990 (December) 1, with permission from CERN.)

Box 1e. One-particle quadrupole moment

Consider the motion of a point-like particle with charge q rotating in a circle with radius R outside a spherical core. Neglecting interaction effects between the particle and the core, we obtain for the intrinsic quadrupole moment Q_{intr}

$$Q_{\text{intr}} = q(3z'^2 - R^2) = -qR^2. \tag{1e.1}$$

Using the semiclassical arguments discussed before, one obtains for the possible values $Q(J, M)$, the expression

$$Q(J, M) = \frac{3M^2 - J(J+1)}{2J(J+1)} Q_{\text{intr}}. \tag{1e.2}$$

Thus, the quadrupole operator, $\hat{Q}$, has $J + 1/2$ distinct eigenvalues, corresponding to the value of $M(\pm M)$, when evaluating its expectation value by a semiclassical method. By convention, $Q(J, J)$ is called the quadrupole *moment* of the corresponding quantum state and is denoted by Q. For the above point-like single-particle motion one obtains

$$Q \equiv Q(J, J) = \frac{3J^2 - J(J+1)}{2J(J+1)} \cdot Q_{\text{intr}}, \tag{1e.3}$$

or

$$Q = -q\frac{2J - 1}{2J + 2} \cdot R^2, \tag{1e.4}$$

for $J \neq 0$.

In a more precise way, if a radial wavefunction $R_{n\ell}(r)$ is associated with the single-particle motion, one has to replace R^2 by the average value $\langle r^2 \rangle = \int_0^\infty R_{n\ell}(r) r^2 R_{n\ell}(r) r^2 \, dr$ and obtain the results

$$Q = -q \cdot \frac{2J - 1}{2J + 2} \langle r^2 \rangle_{n\ell} \qquad J \neq 0, \tag{1e.5}$$

$$Q = 0 \qquad J = 0, 1/2. \tag{1e.6}$$

This model can be used for a doubly-magic nucleus with an outer additional nucleon outside or a single nucleon missing from a closed shell. One can evaluate the average value of Q if one simplifies the radial wavefunction to a constant value within the nuclear radius $R = r_0 A^{1/3}$. One then obtains $\langle r^2 \rangle = \frac{3}{5}R^2$, and the results

$$Q = -\frac{3}{5}eR^2\frac{2J - 1}{2J + 2} \qquad \text{for an extra particle } J \neq 0, \neq \frac{1}{2}, \tag{1e.7}$$

$$Q = \frac{3}{5}eR^2\frac{2J - 1}{2J + 2} \qquad \text{for an extra 'hole' } J \neq 0, \neq \frac{1}{2}. \tag{1e.8}$$

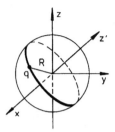

Figure 1e.1. Coordinates used to evaluate the intrinsic quadrupole moment of a single nucleon with charge q in a circular orbit with radius R.

Here, we have assumed that the extra particle is a proton with charge $+e$ and that a proton hole has charge $-e$. The corresponding distribution is represented in figure 1e.1. A measure of nuclear deformation, independent of the size of the nucleus is obtained from the value of the reduced quadrupole moments, Q/ZR^2, where R is the radius of the sphere. One observes in figure 1e.2 that near to closed shells, quadrupole moments agree well with a single particle (or hole) outside an inert, spherical core. In some regions though, very large quadrupole moments occur. Rainwater (1950) suggested that motion of the single particle deforms the whole nucleus and that the observed quadrupole moment results from a collective distortion of many orbits. This collective model has been developed by Bohr and Mottelson since 1952. Using this model and the concept of nuclear deformation, landscapes have been mapped (see figures 1e.2 and 1e.3). These dramatically show the strongly pronounced spherical shapes associated with closed shells at certain N and/or Z values. The measure of deviation from a spherical shape can be given using various parametrizations of the shape, e.g. the ellipsoid with axial symmetry, as in the Nilsson model (Nilsson 1955).

Moreover, a nucleus can change its shape when rotating. It has been shown, quite recently, that important modifications to the nuclear shape show up at high rotational frequencies (Simpson *et al* 1984).

The predicted shape variations are shown in figure 1e.4 where the nucleus is depicted as rotating in the plane of the figure. At low spin, the nucleus has a prolate shape relative to the symmetry axis, denoted by S. At intermediate spins the rotation forces some matter into the rotating plane, but now the axis S is no longer a symmetry axis and the nucleus becomes triaxial. At high spins, the matter is distributed in the plane of rotation, but again in a regular pattern. Now, the symmetry axis coincides with the rotational axis and the nucleus flattens into an 'oblate' shape. It is this latter set of shapes that was observed in [158]Er by Simpson *et al* (1984).

In order to understand the experimental signature for such shape changes, one must consider the concept of 'rotational frequency' for a microscopic object. The nuclear, excited states are strictly characterized by the angular momentum

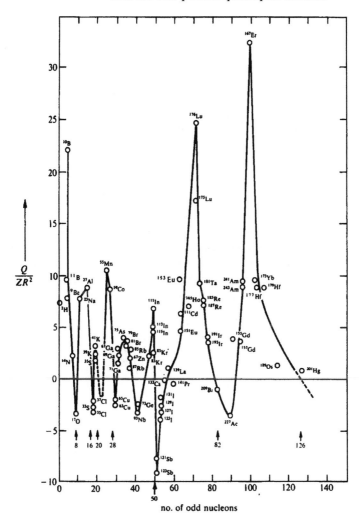

Figure 1e.2. The reduced nuclear quadrupole moments (with the quadrupole moments in units 10^{-26} cm^2) as a function of the number of odd nucleons. The quantity Q/ZR^2 gives a measure of the nuclear quadrupole deformation. (Taken from Segré *Nuclei and Particles* 2nd edn © 1982 Addison-Wesley Publishing Company. Reprinted by permission.)

quantum number J, rather than ω, which is a classical quantity. A rotating nucleus gives rise to a rotating, deformed electric field which can emit electromagnetic radiation. The rotational frequency of the nucleus can then be deduced from the frequency of the emitted γ-radiation. If the concept of genuine rotation is meaningful in the nucleus, this quantity should vary in a smooth way as a function of the energy in ^{158}Er, relative to that of a classical rotor with the same

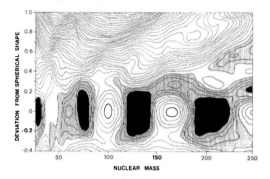

Figure 1e.3. A nuclear 'landscape' of shapes. The total energy as a function of nuclear mass and of the deviation from the spherical shape, as a reference level, is illustrated in a contour plot. Calculations are from Ragnarsson and Nilsson. The lowest-lying areas (dark regions) are the 'valleys' corresponding to minimum energy and to the stable, spherical nuclei. The most stable regions correspond to the $(40, 40)$, $(50, 82)$ $(82, 126)$ (Z, N) combinations. The side-valleys (shaded regions) correspond to less stable prolate-shaped nuclei.

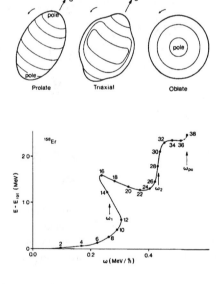

Figure 1e.4. Typical rare-earth shape changes as a function of increasing rotational frequency ω (see text in Box 1d). In the lower part, the energy of the nucleus ^{158}Er, relative to that of a classical rotor E_{rot} with the same angular momentum J and frequency ω. Angular momentum values are indicated along the curve and are in units of $\hbar$. (Taken from Rowley 1985. Reprinted with permission of *Nature* © 1985 MacMillan Magazines Ltd.)

spin and frequency. In figure 1e.4, one observes, however, some frequencies $\omega_1, \omega_2, \omega_{p0}, \ldots$ where the classical frequency almost stays constant whereas the energy increases rapidly. These points correspond to a breakdown of classical motion due to specific shell-model aspects, known as 'rotational alignment' by which nucleons tend to become localized in the plane of rotation and thus contribute to the triaxiality of the system.

Box 1f. An astrophysical application: alpha-capture reactions

---

Stars are ideal places for manufacturing elements such as carbon and oxygen, which are the basic constituents of planets and people. All those elements that have been formed, though, very much depend on some remarkable coincidences in the laws of physics, sometimes also called the 'antropic' principle.

About 75% of a star is hydrogen, about 25% is helium and a mere 1% or so consists of heavier elements: this proportion is an important clue that all heavier elements (C, O, ...) have been manufactured inside of the stars. This process of nuclear synthesis is quite well understood (see Fowler 1984), though a number of key difficulties remained for some time. It was known how the present abundances of hydrogen and helium and some other light elements such as deuterium and lithium were created in the early phase of the universe formation, starting from the 'standard' model. Even early studies in 1950 by Gamow indicated the difficulties of forming heavier elements by synthesis inside stars. But how do stars perform this 'trick'?

We know that the nucleus ^{4}He is extremely stable ($\simeq 28$ MeV of binding energy) and so, nuclei which form an α-like composition show an increased stability compared with nearby nuclei. Carbon (3 α particles) and oxygen (4 α particles) are particularly stable elements in that respect. Once these elements are formed in the right quantities, heavier element formation looks quite straightforward using the subsequent interaction with α-particles (in stars or in the laboratory).

A problem showed up, however, at the very first stages. Two ^{4}He nuclei can form ^{8}Be if they have the correct kinetic energy ($Q = -0.094$ MeV). The nucleus ^{8}Be, though, is extremely unstable, breaking apart again into two ^{4}He nuclei, within a lifetime of only $\simeq 10^{-17}$ s. So, how then can ^{12}C be formed which even needs an extra ^{4}He nucleus to merge with ^{8}Be to form a ^{12}C nucleus.

In 1952, Salpeter suggested that ^{12}C might be formed in a rapid two-step process (see figure 1f.1) with two ^{4}He nuclei, colliding to form a ^{8}Be nucleus which was then hit by a third ^{4}He nucleus in the very short time before desintegration took place. Hoyle, though, puzzled out that, even though the above process looked highly exotic and unprobable, the presence of a resonance level in ^{12}C might make the process feasible. This coincidence, proved later by Fowler, allows indeed the formation of ^{12}C and thus of all heavier elements. When two nuclei collide and stick together, the new nucleus that is formed carries the combined mass (minus the mass defect caused by the strong interaction forces) plus the combined, relative kinetic energy of motion. It now occurs that in the reaction

$$^8\text{Be} + {}^4\text{He} + T_{4_{\text{He}}} \rightarrow {}^{12}\text{C}^*,$$

a resonance in ^{12}C* is formed at 7.654 MeV excitation energy and a COM kinetic energy of 0.287 MeV is needed to come into 'resonance' with the excited state

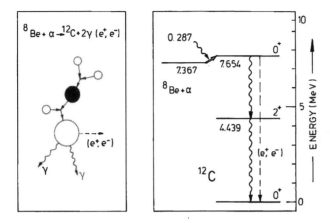

Figure 1f.1. The formation of ^{12}C: two ^{4}He nuclei collide to form ^{8}Be. Occasionally, a third ^{4}He nuclei can combine to form ^{12}C in an excited state. The pictorial description of this process is depicted in the left-hand part. The reaction energy balance is explained and illustrated in the right-hand side.

in ^{12}C. The temperature inside the stars is such that in He-burning, the thermal motion can supply this small energy excess.

The combined mass $M'(^8\text{Be})c^2 + M'(^4\text{He})c^2 = 7.367$ MeV, just about 4% less than the ^{12}C* excited state of 7.654 MeV. It is this coincidence which has allowed ^{12}C formation and that of heavier elements (figure 1f.2) (Gribben and Rees 1990 and Livio *et al* 1989).

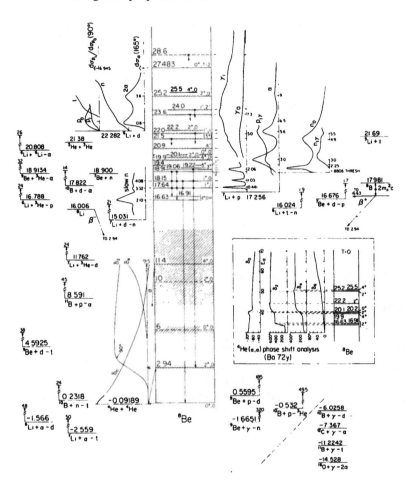

Figure 1f.2. The more detailed reactions leading to the possible formation of ^{8}Be (in the ground state or any of its excited states). (Taken from Ajzenberg-Selove 1960.)

Chapter 2

General nuclear radioactive decay properties and transmutations

2.1 General radioactive decay properties

Here, we shall start to discuss the decay of a given nucleus irrespective of the precise origin causing the transformation. This important problem and the related physics question on how to calculate the decay rate using time-dependent perturbation theory will be attacked in the following chapters on α-decay (strong interaction decay process), β-decay (decay caused by the weak interaction) and γ-decay (originating from the electromagnetic interaction). See figure 2.1.

The decay law of radioactive substances was first clearly formulated by Rutherford and Soddy as a result of their studies on the radioactivity of various substances (notably thorium, thorium-X and their emanations).

Disintegration constant, half-life and mean-life. A radioactive nucleus may be characterized by the rate at which it disintegrates, and the decay constant, half-life or mean-life are equivalent ways of characterizing this decay. The probability that a nucleus decays within a short time interval dt is $\lambda\,dt$. It is the quantity λ (T^{-1} is its dimension) which determines the rate of decay. This decay process is a statistical process and we cannot say anything particular for a given nucleus: we can only speak in terms of decay probabilities. Suppose we start with N_0 nuclei of a certain type at a hypothetical starting time $t_0 = 0$; then the number of nuclei disappearing can be given by the expression

$$dN = -\lambda N\,dt, \tag{2.1}$$

which is easily integrated (with initial conditions $N(t = 0) = N_0$) with the result

$$N(t) = N_0 e^{-\lambda t}. \tag{2.2}$$

The above law is the well-known and fundamental law of radioactive decay. Starting from this expression, a time can be defined $t = T$ (half-life) at which

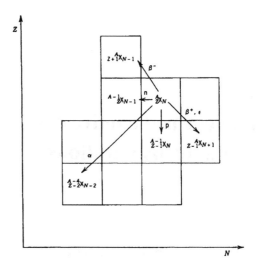

Figure 2.1. The initial nucleus $^{A}_{Z}X_{N}$ with the various final nuclei that can be reached via the indicated decay processes (taken from Krane 1987, *Introductory Nuclear Physics* © 1987 John Wiley & Sons. Reprinted by permission.)

the number of radioactive elements has decreased to half of its original value, i.e. $N_0/2$. This leads to the value

$$T = \frac{\ln 2}{\lambda} = \tau \ln 2 = 0.6931472\ldots\tau, \tag{2.3}$$

where $\tau (\equiv 1/\lambda)$ is a characteristic decay time which we show is equal to a mean lifetime for the decaying radioactive substance. The above exponential decay curve is illustrated in figure 2.2 where both the $e^{-\lambda t}$ as well as its presentation on a logarithmic scale are given.

Evaluating the average lifetime of the atoms, we known that at a certain time t, we still have $N(t) = N_0 e^{-\lambda t}$ radioactive nuclei. Of these $N(t)\lambda\,dt$ will decay between t and $t + dt$ and thus

$$\bar{t} = \frac{1}{N_0} \int_0^\infty \lambda t N_0 e^{-\lambda t}\,dt = \frac{1}{\lambda} = \tau. \tag{2.4}$$

In table 2.1, we show the half-life T for some of the longest-lived nuclei. These half-lives are of the order of the time scale describing slow transformations in the Universe.

For smaller values (ranging over the period years to seconds), the T value can be determined by measuring the 'activity' for the decaying radioactive substance. As a definition of the quantity 'activity', we specify the rate at which a certain radioactive species is generating decay products. In each decay ($\alpha, \beta, \gamma, \ldots$),

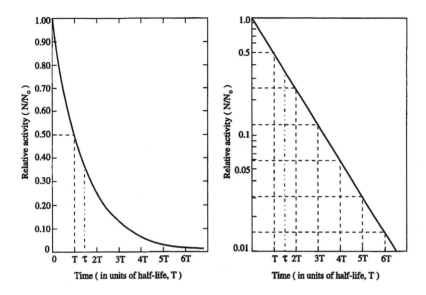

Figure 2.2. Radioactive decay curve (on both a linear and a log scale) as a function of the elapsed time (given in units of the half-life T). (Adapted from Kaplan, *Nuclear Physics* 2nd edn © 1962 Addison-Wesley Publishing Company. Reprinted by permission.)

Table 2.1. Some characteristics of the disintegration series of the heavy elements. (Taken from Krane 1987, *Introductory Nuclear Physics* © 1987 John Wiley & Sons. Reprinted by permission.)

| | | Final | Longest-lived member | |
| | | Nucleus | | Half-life |
Name of series	Type[a]	(Stable)	Nucleus	(y)
Thorium	$4n$	^{208}Pb	^{232}Th	1.41×10^{10}
Neptunium	$4n+1$	^{209}Bi	^{237}Np	2.14×10^{6}
Uranium	$4n+2$	^{206}Pb	^{238}U	4.47×10^{9}
Actinium	$4n+3$	^{207}Pb	^{235}U	7.04×10^{8}

[a]n is an integer.

some measurable objects are emitted that can be detected with a measuring apparatus, with the relation

$$dN_a = -dN = \lambda N(t)\, dt. \tag{2.5}$$

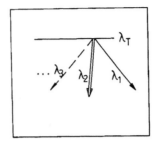

Figure 2.3. Schematic illustration of various, independent decay possibilities indicated via the decay constants $\lambda_1, \lambda_2, \lambda_3$, etc and the total level decay constant $\lambda_T = \sum_i \lambda_i$.

The rate of emission then reads

$$\frac{dN_a}{dt} = \lambda N(t) = \lambda N_0 e^{-\lambda t},$$

or

$$A = A_0 e^{-\lambda t}, \tag{2.6}$$

with A, the activity of the decaying substance which follows the same $e^{-\lambda t}$ decay law and thus leads to an easy way of determining experimentally the $\lambda(T)$ value by a graphical method.

We remind the reader here that activity is defined via $\lambda N(t)$ and not $dN(t)/dt$. For a simple, single decay process both quantities are identical. For more complex situations where a radioactive decay is only a certain part of a complex decay chain, both quantities are clearly different. One can even find situations (see section 2.2) where $dN(t)/dt = 0$ but $\lambda N(t) \neq 0$.

Some of the more often used units characterizing the activity are

$$1 \text{ Curie} = 1 \text{ Ci} = 3.7 \times 10^{10} \text{ disintegrations/s}$$
$$1 \text{ Rutherford} = 1 \text{ R} = 10^6 \text{ disintegrations/s}$$
$$1 \text{ Becquerel} = 1 \text{ Bq} = 1 \text{ disintegration/s}$$

To conclude this section, we point out that in many cases a nucleus can disintegrate via different decay channels: γ-decay in competition with β-decay. If, in general, different channels $\lambda_1, \lambda_2, \ldots$ exist, a total transition probability of $\lambda_T = \sum \lambda_i$ results (figure 2.3).

So, one obtains

$$N(t) = N_0 e^{-\lambda_T t} = N_0 e^{-(\sum_i \lambda_i)t}. \tag{2.7}$$

Branching ratios are thus expressed by $\lambda_1/\lambda_T, \lambda_2/\lambda_T, \ldots$ and partial activities read

$$A_i = \frac{\lambda_i}{\lambda_T} \cdot A_T,$$

with

$$A_T = A_0 e^{-\lambda_T t}. \tag{2.8}$$

2.2 Production and decay of radioactive elements

Suppose that a sample of stable nuclei is bombarded using a projectile that can induce transmutations at a given rate of Q atoms/s and so forms radioactive elements which decay again with a decay constant λ, an interesting situation results. In practice such methods are used for activation analyses, in particular to produce radioactive tracer elements (using neutron from reactors, or electron accelerators to induce $(e, p)(\gamma, p)(\gamma, n) \ldots$ reactions). The law, describing the change of the number of elements $dN(t)/dt$ is a balance between the formation rate Q and the decay activity $-\lambda N(t)$, or,

$$\frac{dN(t)}{dt} = Q - \lambda N. \tag{2.9}$$

This equation (for constant formation rate Q) can be rewritten as

$$\frac{d(Q - \lambda N)}{Q - \lambda N} = -\lambda \, dt. \tag{2.10}$$

Integrating this differential equation, with at the time at which the irradiation starts, $N(t = 0) = 0$ atoms, one obtains

$$Q - \lambda N(t) = (Q - \lambda N(t = 0))e^{-\lambda t},$$

or

$$N(t) = \frac{Q}{\lambda}(1 - e^{-\lambda t}). \tag{2.11}$$

Here, we observe a saturation curve, typical for these kinds of balancing situations where formation and decay compete. The corresponding results for the activity $A(t) \, (= \lambda N(t))$ and $dN(t)/dt$ are given in figure 2.4.

Here, it becomes clear that activity $\lambda N(t)$ and the quantity $dN(t)/dt$ are totally different: the activity grows to an almost constant value, approaching a saturation near $A(t) \simeq Q$ whereas $dN(t)/dt$ approaches the value of zero.

After irradiating during $t = 2T, 3T$ the number of radioactive elements formed is $\frac{3}{4}$ and $\frac{7}{8}$, respectively of the maximal number (when $t \to \infty$). So, in practice, irradiations should not last longer than 2 to 3 half-life periods.

2.3 General decay chains

In the more general situations, which also occur in actual decay chains, a long set of connections can show up. This is, in particular, the case for the actinide nuclei. In figure 2.5 we show the very long sequence starting at mass $A = 244$, eventually

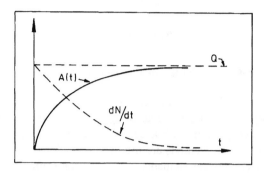

Figure 2.4. Schematic figure for a system where elements are formed at a rate of Q atoms/s; the activity $A(t) (\equiv \lambda N(t))$ and the derivative $dN(t)/dt$ are shown as functions of time t.

Table 2.2. Some natural radioactive isotopes

Isotope	T (y)
^{40}K	1.28×10^9
^{87}Rb	4.8×10^{10}
^{113}Cd	9.0×10^{15}
^{115}In	4.4×10^{14}
^{138}La	1.3×10^{11}
^{176}Lu	3.6×10^{10}
^{187}Re	5.0×10^{10}

decaying into the stable nucleus ^{208}Pb. Both α, β^+ (and electron capture) and β^- decay are present. Some half-life values in this series can be as long as 10^9 y. In table 2.2 we present some of the slowest decaying elements with ^{113}Cd ($T = 9 \times 10^{15}$ y) being the longest lived.

2.3.1 Mathematical formulation

The general decay chain

$$N_1 \xrightarrow{\lambda_1} N_2 \xrightarrow{\lambda_2} N_3 \to \cdots, \tag{2.12}$$

is characterized by the set of coupled, linear differential equations

$$\frac{dN_1}{dt} = -\lambda_1 N_1$$

$$\frac{dN_2}{dt} = \lambda_1 N_1 - \lambda_2 N_2$$

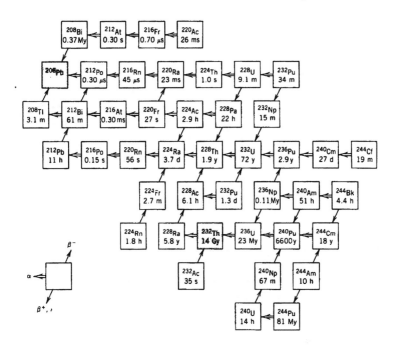

Figure 2.5. The thorium decay series. Some half-lives are indicated in My (10^6 y) and Gy (10^9 y). The shaded members are the longest-lived radioactive nuclei in the series as well as the stable end product. (Taken from Krane 1987, *Introductory Nuclear Physics* © 1987 John Wiley & Sons. Reprinted by permission.)

$$\frac{dN_3}{dt} = \lambda_2 N_2 - \lambda_3 N_3. \qquad (2.13)$$

$$\vdots$$

Using the theory of solving coupled, linear differential equations with constant coefficients λ_i, one can show that a general solution can be obtained as

$$N_1(t) = a_{11}e^{-\lambda_1 t}$$
$$N_2(t) = a_{21}e^{-\lambda_1 t} + a_{22}e^{-\lambda_2 t}$$
$$N_3(t) = a_{31}e^{-\lambda_1 t} + a_{32}e^{-\lambda_2 t} + a_{33}e^{-\lambda_3 t}. \qquad (2.14)$$

$$\vdots$$

with initial conditions

$$a_{11} = N_1(t = 0),$$
$$\sum_{i=1}^{k} a_{ki} = 0. \qquad (2.15)$$

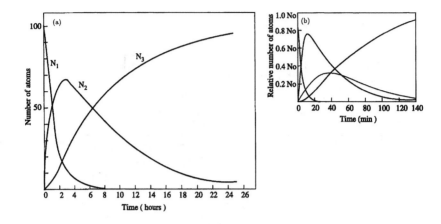

Figure 2.6. (*a*) Example of a radioactive series with three members: only the parent ($T = 1$ h) is present initially; the daughter nucleus has a half-life of 5 h and the third member is stable. (*b*) The decay of Radium A (Adapted from Kaplan, *Nuclear Physics* 2nd edn © 1962 Addison-Wesley Publishing Company. Reprinted by permission.)

One can obtain the coefficients $a_{k,i}$ by substituting the above general solution into the differential equation, leading to a set of algebraic equations in the unknown coefficients $a_{k,i}$. For the particular case of $N_2(t)$ we obtain

$$-\lambda_1 a_{21} e^{-\lambda_1 t} - \lambda_2 a_{22} e^{-\lambda_2 t} = \lambda_1 a_{11} e^{-\lambda_1 t} - \lambda_2 a_{21} e^{-\lambda_1 t} - \lambda_2 a_{22} e^{-\lambda_2 t},$$

or

$$a_{21} = \frac{\lambda_1}{\lambda_2 - \lambda_1} a_{11}, \qquad a_{22} = \frac{\lambda_1}{\lambda_1 - \lambda_2} a_{11}. \tag{2.16}$$

Carrying out this process, we obtain the general solution

$$a_{k,i} = \frac{\lambda_{k-1}}{\lambda_k - \lambda_i} a_{k-1,i}, \tag{2.17}$$

or

$$a_{k,i} = \frac{\lambda_1 \lambda_2 \ldots \lambda_{k-1}}{(\lambda_1 - \lambda_i)(\lambda_2 - \lambda_i) \ldots (\lambda_k - \lambda_i)} a_{11}. \tag{2.18}$$

The results for a radioactive series with $T = 1$ h, $T = 5$ h, $T = \infty$ (figure 2.6(*a*)) and secondly the decay of the radium A decay series where the RaA(^{218}Po) $\rightarrow$ RaB(^{214}Pb) $\rightarrow$ RaC(^{214}Bi) $\rightarrow$ RaD(^{210}Pb) (see Evans, p 251) decay chain occurs (figure 2.6(*b*)) are presented.

From the above radioactive decay equations, a rather general property shows up. One can write that

$$\frac{\mathrm{d}N_i}{\mathrm{d}t} = \lambda_{i-1} N_{i-1} - \lambda_i N_i, \tag{2.19}$$

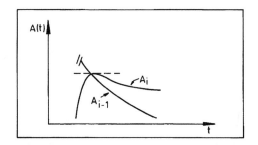

Figure 2.7. Illustration of the activity $A(t)$ in the decay of element A_i. Here, we schematically point out that at the time when the activities A_i and A_{i-1} become equal, the activity $A_i(t)$ becomes the extremum.

or

$$\lambda_i \frac{dN_i}{dt} = \lambda_i (A_{i-1} - A_i), \tag{2.20}$$

which can be rewritten as

$$\frac{dA_i}{dt} = \lambda_i (A_{i-1} - A_i). \tag{2.21}$$

This relation states that for a given term $A_i(t)$, the extremum (maximum) occurs when $A_{i-1} = A_i$ (figure 2.7).

2.3.2 Specific examples—radioactive equilibrium

If we consider the decay chain $N_1 \xrightarrow{\lambda_1} N_2 \xrightarrow{\lambda_2} N_3$; with N_3 a stable nucleus, according to the results of section 2.3.1, one obtains

$$N_1(t) = N_1^0 e^{-\lambda_1 t}$$

$$N_2(t) = \frac{\lambda_1}{\lambda_2 - \lambda_1} N_1^0 (e^{-\lambda_1 t} - e^{-\lambda_2 t})$$

$$N_3(t) = \frac{\lambda_1 \lambda_2}{\lambda_2 - \lambda_1} N_1^0 \left(\frac{1 - e^{-\lambda_1 t}}{\lambda_1} - \frac{1 - e^{-\lambda_2 t}}{\lambda_2} \right). \tag{2.22}$$

We study now, in particular, some special cases of this result

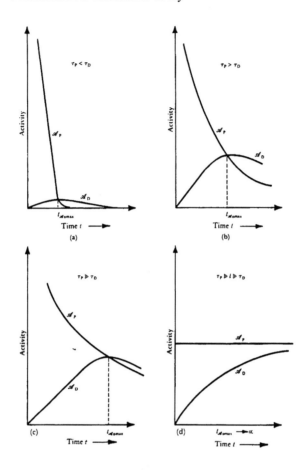

Figure 2.8. Activity curves for various parent–daughter relationships: (*a*) short-lived parent ($\tau_P < \tau_D$); (*b*) long-lived parent ($\tau_P > \tau_D$); (*c*) very long-lived parent ($\tau_P \gg \tau_D$); (*d*) almost stable, or constantly replenished parent ($\tau_P \gg t \gg \tau_D$). We also indicate the time $t_{A_D}(max)$ at which the daughter activity becomes maximal in the cases (*a*), (*b*) and (*c*)) (taken from Marmier and Sheldon, 1969)

(i) $\lambda_1 > \lambda_2$

In this case, and for a time $t \gg 1/\lambda_1$, one can neglect the first exponential and obtain

$$N_2(t) \cong \frac{\lambda_1}{\lambda_1 - \lambda_2} N_1^0 e^{-\lambda_2 t}, \qquad (2.23)$$

indicating that after some time $t \gg T(1)$; the elements N_2 will decay with there own decay constant λ_2 (see figure 2.8(*a*)).

(ii) $\lambda_1 < \lambda_2$

In this case, and for a time $t \gg 1/\lambda_2$ one can neglect the second exponential and obtain

$$N_2(t) \cong \frac{\lambda_1}{\lambda_2 - \lambda_1} N_1^0 e^{-\lambda_1 t}, \qquad (2.24)$$

indicating that for $t \gg T(2)$, the elements 2 will decay with the decay constant of the first element. This looks rather strange but can be better understood after some reflection: since the elements of type 1 decay very slowly, one can never observe the actual decay of element 2 (λ_2) but, on the contrary, one observes the element 2 decaying with the decay constant by which it is formed, i.e. λ_1 (figure 2.8(b)).

The above expression can be transformed into

$$A_2(t) \cong \frac{\lambda_2}{\lambda_2 - \lambda_1} A_1(t). \qquad (2.25)$$

(see figure 2.8(b)).

(iii) $\lambda_1 \ll \lambda_2$

In this situation, the activities $A_2(t) \simeq A_1(t)$ and a *transient equilibrium* results between the activities of elements 1 and 2 with $A_2/A_1 \simeq \lambda_2/(\lambda_2 - \lambda_1)$. The example of the case $T(1) = 8$ h and $T(2) = 0.8$ h is shown in figure 2.9. Here, we indicate the daughter activity ($A_2(t)$) growing into a purified parent fraction, the parent activity $A_1(t)$, the total activity of an initially pure parent fraction $A(t) = A_1(t) + A_2(t)$ as well as the decay activity of the freshly isolated daughter fraction and finally, the total daughter activity in the parent-plus-daughter fractions.

(iv) $\lambda_1 \simeq 0, \lambda_1 \ll \lambda_2$

In this case, we approach *secular equilibrium*. The expressions for $N_1(t), N_2(t)$ now become ($1/\lambda_1 \gg t \gg 1/\lambda_2$)

$$N_1(t) = N_1^0,$$
$$N_2(t) = \frac{\lambda_1}{\lambda_2} N_1^0 (1 - e^{-\lambda_2 t}), \qquad (2.26)$$

or

$$A_1(t) = A_1^0,$$
$$A_2(t) = A_1^0 (1 - e^{-\lambda_2 t}). \qquad (2.27)$$

We have a very small, but almost constant activity $A_1(t)$. The activity $A_2(t)$ reaches equilibrium at a time $t \gg T$, when $e^{-\lambda_2 t} \simeq 0$ and $A_2(t) \simeq A_1^0$ in the saturation (or equilibrium) regime (figure 2.8(d)). In the latter case, the condition of equilibrium (all $dN_i/dt = 0$) expresses the equality of *all* activities

$$A_1 = A_2 = A_3 = \cdots = A_i. \qquad (2.28)$$

These conditions cannot all be fulfilled, i.e. $dN_1/dt = 0$ would indicate a *stable* element (implying $\lambda_1 = 0$). It is now possible to achieve a state very

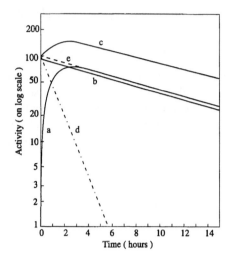

Figure 2.9. Transient equilibrium (*a*) Daughter activity growing in a freshly purified parent, (*b*) Activity of parent ($T = 8$ h), (*c*) Total activity of an initially pure parent fraction, (*d*) Decay of freshly isolated daughter fractions ($T = 0.8$ h), and (*e*) Total daughter activity in parent-plus-daughter fractions).

close to exact equilibrium, however, if the parent substance decays much more slowly than any of the other members of the chain then one reaches the condition

$$\frac{N_1}{T(1)} = \frac{N_2}{T(2)} = \cdots = \frac{N_n}{T(n)}. \tag{2.29}$$

It is only applicable when the material (containing *all* decay products) has been undisturbed in order for secular equilibrium to become established. Equilibrium can also be obtained (see section 2.2) in the formation of an element, which itself decays (formation activity Q, decay activity λN) since $\mathrm{d}N/\mathrm{d}t = 0$ if $Q = \lambda N$.

2.4   Radioactive dating methods

If, at the time $t = 0$, we have a collection of a large number N_0 of radioactive nuclei, then, after a time equal to T, we find a remaining fraction of $N_0/2$ nuclei. This indicates that, knowing the decay constant λ, the exponential decrease in activity of a sample can be used to determine the time interval. The difficulty in using this process occurs when we apply it to decays that occur over geological times ($\approx 10^9$ y) because in these situations, we do not measure the activity as a function of time. Instead, we use the relative number of parent and daughter nuclei at a time t_1 compared with the relative number at an earlier time t_0. In principle, this process is rather simple. Given the decay of a parent P to a daughter isotope D, we just count the number of atoms present $N_P(t_1)$ and $N_D(t_1)$. One has the relation

$$N_D(t_1) + N_P(t_1) = N_P(t_0),$$

and

$$N_P(t_1) = N_P(t_0)e^{-\lambda(t_1-t_0)}. \tag{2.30}$$

From this expression, it follows that

$$\Delta t \equiv t_1 - t_0 = \frac{1}{\lambda}\ln\left(1 + \frac{N_D(t_1)}{N_P(t_1)}\right). \tag{2.31}$$

So, given the decay constant λ and the ratio $N_D(t_1)/N_P(t_1)$ the age of the sample is immediately obtained with a precision determined by the knowledge of λ and the counting statistics with which we determine N_P and N_D.

We can now relax the condition (2.30) and permit daughter nuclei to be present at time $t = t_0$. These nuclei could have been formed from decay of parent nuclei at times prior to t_0 or from processes that formed the original daughter nuclei in an independent way. The means of formation of $N_D(t_0)$ is of no importance in the calculation given below. We have

$$N_D(t_1) + N_P(t_1) = N_D(t_0) + N_P(t_0), \tag{2.32}$$

where now, an extra unknown $N_D(t_0)$ shows up and we can no longer solve easily for $\Delta t \equiv t_1 - t_0$. If now, there is an isotope of the daughter nucleus, called D', present in the sample which is *neither* formed from decay of a long-lived parent *nor* is radioactive, we can again determine Δt. We call the population $N_{D'}$ and $N_{D'}(t_1) = N_{D'}(t_0)$. So we obtain

$$\frac{N_D(t_1)}{N_{D'}(t_1)} = \frac{N_P(t_1)}{N_{D'}(t_1)}(e^{\lambda(t_1-t_0)} - 1) + \frac{N_D(t_0)}{N_{D'}(t_0)}. \tag{2.33}$$

The ratios containing N_D and N_P at time t_1 can again be measured. Now, the above equation expresses a straight line $y = mx + b$ with slope $m = e^{\lambda(t_1-t_0)} - 1$ and intercept $N_D(t_0)/N_{D'}(t_0)$. Figure 2.10 is an example for the decay ^{87}Rb $\rightarrow$ ^{87}Sr ($T = 4.8 \times 10^{10}$ y) in which the comparison is made with the isotope ^{86}Sr of the daughter nucleus ^{87}Sr. We extract a time interval $\Delta t = 4.53 \times 10^9$ y and a really good linear fit is obtained. This fit indicates (i) no loss of parent nuclei and (ii) no loss of daughter nuclei during the decay processes.

For dating more recent samples of organic matter, the ^{14}C dating method is most often used. The CO_2 that is absorbed by organic matter consists almost entirely of stable ^{12}C (98.89%) with a small admixture of stable ^{13}C (1.11%). Radioactive ^{14}C is continuously formed in the upper atmosphere according to the (n, p) reaction

$$^{14}\text{N(n, p)} \, ^{14}\text{C,}$$

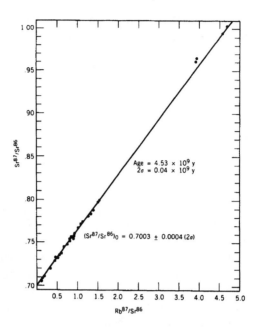

Figure 2.10. The Rb–Sr dating method, allowing for the presence of some initial ^{87}Sr. (Taken from Wetherhill 1975. Reproduced with permission from the *Annual Review of Nuclear Science* **25** © 1975 by Annual Reviews Inc.)

following cosmic ray neutron bombardment. Thus all living matter is slightly radioactive owing to its ^{14}C content. The ^{14}C decays via β^--decay back to ^{14}N ($T = 5730$ y). Since the production rate Q (^{14}C) by cosmic ray bombardment has been relatively constant for thousands of years, the carbon of living organic material has reached equilibrium with atmospheric carbon. Using the equilibrium condition

$$\frac{d}{dt} N(^{14}C) = 0,$$

or

$$Q(^{14}C) = \lambda(^{14}C) \cdot N(^{14}C), \qquad (2.34)$$

one derives that with about 1 atom of ^{14}C one has 10^{12} atoms of ^{12}C. Knowing the half-life of ^{14}C as 5730 y, each gramme of carbon has an activity of 15 decays/min. When an organism dies, equilibrium with atmospheric carbon ceases: it stops acquiring new ^{14}C and one now has

$$N(^{14}C, t_1) = N(^{14}C, t_0)e^{-\lambda(t_1-t_0)}. \qquad (2.35)$$

So, one can determine $\Delta t \equiv t_1 - t_0$ by measuring the specific activity (activity per gram) of its carbon content. The activity method breaks down for periods of the order of $\Delta t \gtrsim 10T$. Recent techniques, using accelerators and mass spectrometers, count the number of ^{14}C atoms directly and are much more powerful (see Box 2a).

A major ingredient is, of course, the constancy of Q for ^{14}C over the last 20 000 y. Comparisons using date determination of totally different origin give good confidence in the above method. During the last 100 years, the burning of fossil fuels has upset somewhat the atmospheric balance. During the 1950s–1960s, nuclear weapons have placed additional ^{14}C in the atmosphere leading to

$$Q(^{14}\text{C}) = Q(^{14}\text{C; cosmic ray}) + Q(^{14}\text{C; extra}), \qquad (2.36)$$

perhaps doubling the concentration over the equilibrium value for cosmic-ray production alone.

2.5 Exotic nuclear decay modes

Besides the typical and well-known nuclear transformations caused by α-decay (emission of the nucleus of the ^{4}He atom), β-decay (transformations of the type $n \rightarrow p + e^- + \bar{\nu}$; $p \rightarrow n + e^+ + \nu$, $p + e^- \rightarrow n + \nu, \ldots$) and γ-decay (emission of high-energy photons from nuclear excited states, decaying towards the ground state), a number of more exotic forms of nuclear radioactivity have been discovered, more recently (Rose and Jones 1984).

Since the discovery of the decay mode where ^{14}C nuclei are emitted, in particular in the transformation

$$^{223}\text{Ra} \rightarrow \ ^{209}\text{Pb} + \ ^{14}\text{C},$$

rapid progess has been made in the experimental observation of other ^{14}C, ^{24}Ne and ^{28}Mg decays (see Price 1989 and references therein). Very recently, the ^{14}C decay of ^{222}Ra was studied in much detail. In a 16 days counting period, 210 ^{14}C events were recorded. Almost all events proceed towards the ^{208}Pb 0^+ ground state with almost no feeding into the first excited state in ^{208}Pb. The total spectrum for the decay

$$^{222}\text{Ra} \rightarrow \ ^{208}\text{Pb} + \ ^{14}\text{C}$$

is presented in figure 2.11 (taken from Hussonnois *et al* (1991).

It is, in particular, the rarity of such new events, compared to the much more frequent α-decay mechanism (10^9 α-particles per ^{14}C nucleus in the ^{223}Ra decay), that prevented these decay modes of being found much earlier. The original Rose and Jones experiments took more than half a year of running. With better detection methods, however, some more examples were discovered as outlined before.

The calculation describing this heavy particle decay proceeds very much along the lines of α-decay penetration (tunnelling) through the Coulomb barrier around the nuclear, central potential field. A theoretical interpretation of the

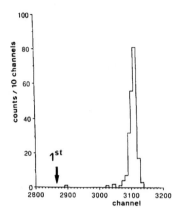

Figure 2.11. Total spectrum of the 210 ^{14}C events recorded from the 85 M Bq ^{230}U source from which the ^{222}Ra was extracted. The position of the ^{14}C group expected to feed the first excited state of ^{208}Pb is indicated by the arrow (taken from Hussonnois *et al* 1991).

peaks observed in the ^{14}C decay of ^{223}Ra has recently been tried by Sheline and Ragnarsson (1991).

When approaching very proton-rich nuclei, a most interesting phenomenon, first observed in 1970, is seen, namely proton radioactivity (Jackson 1970). At present, proton radioactivity has been mapped extensively near to the region where protons are just bound through the presence of the Coulomb barrier but when the Q-value for proton emission becomes positive. The decay originates, therefore, in a quasi-bound state that would otherwise be unbound if only the strong nuclear forces would be active. These particular nuclei eventually decay via the quantum tunnelling process of a proton through the Coulomb barrier. The last decade has seen an extensive exploration of this particular zone of the nuclear mass region and, this phenomenon has been observed in most odd-Z nuclei starting at I and continuing all the way up to Bi. A review of proton radioactivity was published in 1997 (Woods and Davids 1997) in which it is extensively studied. Later on, in 1999, a conference dedicated to the study of proton-emitting nuclei was held in Oak-Ridge, indicating the speed with which results became available.

More recently, evidence for ground-state two-proton radioactivity was observed in ^{45}Fe by teams at GSI, Darmstadt (Germany) and at GANIL (France) (Pfützner *et al* 2002, Giovanozzi *et al* 2002), almost 40 years after this form of radioactivity was first predicted. This same two-proton decay mode has been seen in ^{18}Ne, starting from an excited state (Gomez del Campo *et al* 2001) (see figure 2.12). Of course, sequential two-proton decay had been seen where the process passes through an intermediate state in the 'middle' nucleus between the

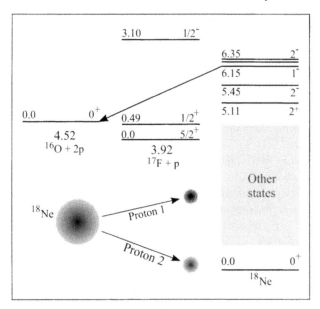

Figure 2.12. Two-proton decay starting from the nucleus ^{18}Ne leading to a final state in ^{16}O plus two protons as indicated pictorially in the lower part of the figure. (Reprinted figure with permission from Gomez del Campo *et al* 2001 *Phys. Rev. Lett.* **86** 43 © 2001 by the American Physical Society.)

initial decay and final observed nucleus. In the case of ^{18}Ne, such a sequential process is forbidden by kinematical conditions.

More exotic decay modes are even possible when approaching the limits of stability for the proton-rich nuclei. The reason is that the energy that is available when transforming a proton into a neutron, a positron and a neutrino (β^+-decay), for increasingly proton-rich nuclei, becomes so large that exotic decay modes come into reach. One of the most exotic ones is the discovery of a β^+-delayed three-proton emission from ^{31}Ar (Bazin *et al* 1992).

Box 2a. Dating the Shroud of Turin

The Shroud of Turin, which many people believe was used to wrap Christ's body, bears detailed front and back images of a man who appears to have suffered whipping and crucifixion. It was first displayed at Lirey (France) in the 1350s. After many journeys, the shroud was finally brought to Turin in 1578, where later, in 1694, it was placed in the Royal Chapel of the Turin Cathedral in a specifically designed shrine.

Photography of the shroud by Secondo Pia in 1898 indicated that the image resembled a photographic 'negative' and represents the first modern study to determine its origin. Subsequently, the shroud was made available for scientific examination, first in 1969 and 1973 by a committee[1] and then again in 1978 by the Shroud of Turin Research Project[2]. Even for the first investigation, there was a possibility of using radiocarbon dating to determine the age of the linen from which the shroud was woven. The size of the sample then required, however, was about $\approx$500 cm^2, which would have resulted in unacceptable damage, and it was not until the development in the 1970s of accelerator-mass-spectrometry techniques (AMS) together with small gas-counting methods (requiring only a few square centimetres) that radiocarbon dating of the shroud became a real possibility.

To confirm the feasibility of dating by these methods an intercomparison, involving four AMS and two small gas-counter radiocarbon laboratories and the dating of three known-age textile samples, was coordinated by the British Museum in 1983[3].

Following this intercomparison, a meeting was held in Turin over September–October 1986 at which seven radiocarbon laboratories recommended a protocol for dating the shroud. In October 1987, the offers from three AMS laboratories (Arizona, Oxford and Zürich) were selected.

The sampling of the shroud took place in the Sacristy at Turin Cathedral on the morning of 21 April 1988. Three samples, each $\approx$50 mg in weight were prepared from the shroud in well prepared and controlled conditions. At the same time, samples weighing 50 mg from two of the three controls were similarly packaged. The three containers, holding the shroud (sample 1) were then handed to representatives of each of the three laboratories together with a sample of the third control (sample 4), which was in the form of threads.

The laboratories were not told which container held the shroud sample. The three laboratories undertook not to compare results until after they had been transmitted to the British Museum. Also, at two laboratories (Oxford and Zürich),

---

[1] La S. Sindone-Ricerche e studi della Commissione di Esperti nominata dall'Arcivescovo di Torino, Cardinal M.Pellegrino, nel 1969, Suppl. Rivista Dioscesana Torinese (1976).
[2] Jumper E J *et al* 1984 *Arch Chemistry—III* (ed J B Lambert) (Washington, DC: American Chemical Society) pp 447–76
[3] Burleigh R, Leese M N and Tite M S 1986 *Radiocarbon* **28** 571.

Table 2a.1. A summary of the mean radiocarbon dates for the four samples.

Sample	1	2	3	4
Arizona	646±31	927±32	1995±46	722±43
Oxford	750±30	940±30	1980±35	755±30
Zurich	676±24	941±23	1940±30	685±34
Unweighted average	691±31	936±5	1972±16	721±20
Weighted average	689±16	937±16	1964±20	724±20
χ^2 (2df)	6.4	0.1	1.3	2.4
Significance level (%)	5	90	50	30

Dates are in years BP (years before 1950).

Radiocarbon dating of the Shroud of Turin

P. E. Damon[*], D. J. Donahue[†], B. H. Gore[*], A. L. Hatheway[*], A. J. T. Jull[*],
T. W. Linick[*], P. J. Sercel[*], L. J. Toolin[*], C. R. Bronk[‡], E. T. Hall[‡],
R. E. M. Hedges[‡], R. Housley[‡], I. A. Law[‡], C. Perry[‡], G. Bonani[§], S. Trumbore[‖*],
W. Woelfli[§], J. C. Ambers[¶], S. G. E. Bowman[¶], M. N. Leese[¶] & M. S. Tite[¶]

[*] Department of Geosciences, [†] Department of Physics, University of Arizona, Tucson, Arizona 85721, USA
[‡] Research Laboratory for Archaeology and History of Art, University of Oxford, Oxford, OX1 3QJ, UK
[§] Institut für Mittelenergiephysik, ETH-Hönggerberg, CH-8093 Zurich, Switzerland
[‖] Lamont-Doherty Geological Observatory, Columbia University, Palisades, New York 10964, USA
[¶] Research Laboratory, British Museum, London, WC1B 3DG, UK

Very small samples from the Shroud of Turin have been dated by accelerator mass spectrometry in laboratories at Arizona, Oxford and Zurich. As controls, three samples whose ages had been determined independently were also dated. The results provide conclusive evidence that the linen of the Shroud of Turin is mediaeval.

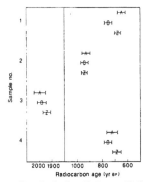

Mean radiocarbon dates, with ±1σ errors, of the Shroud of Turin and control samples, as supplied by the three laboratories (A, Arizona, O, Oxford; Z, Zurich) (See also Table 2) The shroud is sample 1, and the three controls are samples 2-4 Note the break in age scale Ages are given in yr BP (years before 1950) The age of the shroud is obtained as AD 1260-1390, with at least 95% confidence

Figure 2a.1. Taken from Damon *et al* 1989. Reprinted with permission of *Nature* © 1989 MacMillan Magazines Ltd.

after combustion to gas, the samples were recoded so that the staff making the measurements did not know the identity of the samples.

Details on measuring conditions have been discussed by Damon *et al* [4]. In the following table, we give a summary of the mean radiocarbon dates and assessment of interlaboratory scatter (table 2a.1).

We also illustrate the results of the sample measurements in figure 2a.1.

---

[4] Damon P E *et al* 1989 *Nature* **337** 611.

Box 2b. Chernobyl: a test-case in radioactive decay chains

In the Chernobyl reactor accident in April 1986, the radioactive material that was released came directly into the environment as a radioactive dust, propelled upward by intense heat and the rising plume of hot gases from the burning graphite. It would be difficult to design a better system for releasing the radioactivity so as to maximize its impact on public health. Another unfortunate circumstance was that the Soviets had not been using that reactor for producing bomb grade plutonium and therefore the fuel had been in the reactor accumulating radioactivity for over two years[5].

The most important radioactive releases were:

(i) Noble gases, radioactive isotopes of krypton and xenon which are relatively abundant among the fission products. Essentially all of these were released, but fortunately they do little harm because, when inhaled, they are promptly exhaled and so do not remain in the human body. They principally cause radiation exposure by external radiation from the surrounding air, and since most of their radiation is not very penetrating, their health effects are essentially negligible.

(ii) ^{131}I has an eight-day T value. Since it is highly volatile, it is readily released—at least 20% of the ^{131}I in the Chernobyl reactor was released into the environment. When taken into the human body by inhalation or by ingestion with food and drink, it is efficiently transferred to the thyroid gland where its radiation can cause thyroid nodules or thyroid cancers. These diseases represent a large fraction of all health effects predicted from nuclear accidents, but only a tiny fraction would be fatal.

(iii) ^{137}Cs with a 30-year T value and which decays with a 0.661 MeV gamma transition. About 13% of the ^{137}Cs at Chernobyl was released. It does harm by being deposited on the ground where its gamma radiation continues to expose those nearby for many years. It can also be picked up by plant roots and thereby get into the food chain which leads to exposure from within the body.

Here we discuss some results obtained by Uyttenhove[6] (2 June 1986) from measurements on the decay fission products signaling the Chernobyl accident. All results have been obtained using high-resolution gamma-ray spectroscopy. Air samples were taken by pumping 5 to 10 m^3 of air through a glass fibre filter. The filters were subsequently measured in a fixed geometry. In figure 2b.1, we show a typical spectrum, taken on 2 May at 12 noon with a 10000 litre of air sample. The measuring time was 10000 s. The main isotopes detected (with activity in Bq/m^3 and T value) were

[5] Cohen B L 1987 *Am. J. Phys.* **55** 1076.
[6] Uyttenhove J, Internal Report (2/6/1986).

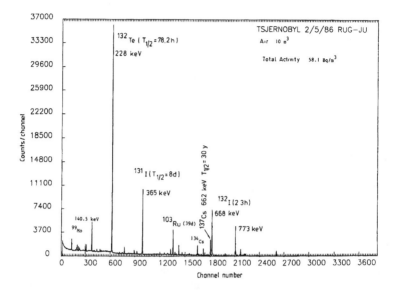

Figure 2b.1.

^{132}Te	(18; 78.2 h)	^{103}Ru	(4.5; 39.4 d)
^{132}I	(10.6; 2.3 h)	^{99}Mo	(1.4; 6.02 h)
^{131}I	(8.5; 8.04 d)	^{129}Te	(3.5; 33.6 d)
^{137}Cs	(4.3; 30.2 y)	^{140}Ba	(2.3; 12.8 d)
^{134}Cs	(2.1; 2.04 y)	^{140}La	(2.3; 40.2 h)
^{136}Cs	(0.6; 13.0 d)		

From these, one obtains a total activity of 58.1 Bq/m^3 (see figure 2b.2). All important fission fragments are present, indicating very serious damage on the fuel rods of the damaged reactor. The measuring technique only allows the measurement of those isotopes that adhere to dust particles. For ^{131}I, this seems to be only 30% of the total activity.

The highest activity was recorded on Friday, 2 May. In the evening, the activity dropped to about 23% and this was the same for *all* isotopes. On Saturday morning, 3 May, the activities stayed at about the same value. Then, the activity dropped rather quickly and at 22 h, it decreased to less than 2% of the maximal values (see figure 2b.3).

In figure 2b.4, we show part of a measurement for 120 g of milk. The iodine concentration decreases rather fast but the long-lived Cs isotopes (^{137}Cs in particular) are still very clearly observable.

To conclude, we have shown that using high-resolution gamma-ray spectroscopy, a rapid and unambiguous detection of fission products can monitor nuclear accidents, even over very long distances if enough activity is released.

ISOTOPE DISTRIBUTION

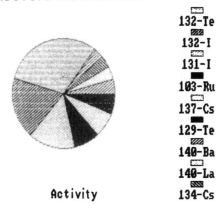

132-Te
132-I
131-I
103-Ru
137-Cs
129-Te
140-Ba
140-La
134-Cs

Activity

Figure 2b.2.

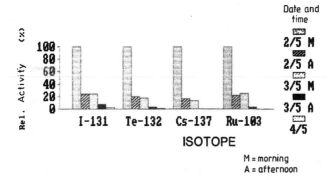

Figure 2b.3.

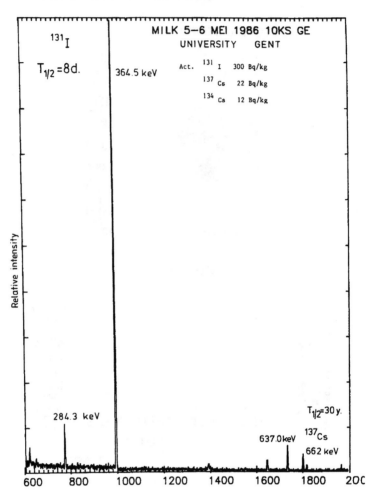

Figure 2b.4.

Problem set—Part A

1. For a single nucleon (proton, neutron), the total angular momentum is given by the sum of the orbital and the intrinsic angular momentum. Calculate the total magnetic dipole moment μ for both the parallel ($j = \ell + 1/2$) and anti-parallel ($j = \ell - 1/2$) spin-orbital orientation.

Hint: Because the magnetic dipole vector $\vec{\mu}$ is not oriented along the direction of the total angular momentum vector $\vec{j}$, a precession of $\vec{\mu}$ around $\vec{j}$ will result and, subsequently, the total magnetic dipole moment will be given by the expression

$$\frac{(\vec{\mu} \cdot \vec{j})\vec{j}}{\langle \vec{j}^2 \rangle}.$$

Furthermore, you can replace the expectation value of the angular momentum vector $\langle \vec{j}^2 \rangle$ by $j(j+1)$.

Use $g(p) = 5.58\mu_N$, $g(n) = -3.83\mu_N$, $g(p) = 1\mu_N$, $g(n) = 0\mu_N$.

2. Describe the quadrupole interaction energy for an axially symmetric nucleus with total angular momentum I and projection $M (-I \leq M \leq I)$ that is placed inside an external field with a field gradient that also exhibits axial symmetry around the z-axis. The nuclear distribution is put at an angle θ (the angle between the z-axis and the symmetry axis of the nucleur distribution). We give the relation $Q(z) = \frac{Q}{2}(3\cos^2\theta - 1)$. We give, furthermore, the following input data: $\partial^2\Phi/\partial z^2 = 10^{20}$ V m^{-2}, $Q = 5$ barn. Also draw, for the quantities as given before, the particular splitting of the magnetic substates for an angular momentum $I = 3/2$ caused by this quadrupole interaction.

3. Show that it is possible to derive a mean quadratic nuclear radius for an atomic nuclear charge distribution starting from the interaction energy of this atom placed in an external field, generated by the atomic s-electrons surrounding the atomic nucleus. Can we use this same method when considering the effect caused by p, d, etc, electrons?

Derive also the mean square radius for an atomic nucleus with Z protons, simplifying to a constant charge density inside the atomic nucleus.

4. In the nuclear reaction $a + X \rightarrow b + Y + Q$, the recoil energy of the nucleus Y cannot easily be determined, in general.

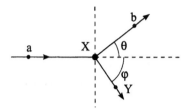

Figure PA.4.

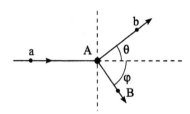

Figure PA.6.

Eliminate this recoil kinetic energy E_Y from the reaction equation for the Q-value, i.e. $Q = E_Y + E_b - E_a$ (the nucleus X is at rest in the laboratory coordinate system) making use of the conservation laws for momentum and energy (see figure PA.4). Discuss this result.

5.   As a consequence of the conservation of linear momentum in the laboratory coordinate axis system, the compound system C that is formed in the reaction $a + X \rightarrow C \rightarrow b + Y$ will move with a given velocity v_C. Show that the kinetic energy needed for the particle a to induce an endothermic reaction ($Q < 0$) has a threshold value which can be expressed as $E_{thr} = (-Q)(1 + m_a/m_X)$. Discuss this result.

6. In the nuclear reaction $a + A \rightarrow b + B$, in which the target nucleus A is initially at rest and the incoming particle a has kinetic energy E_a, the outgoing particle b and the recoiling nucleus B will move off with kinetic energies E_b and E_B, at angles described by θ and φ, respectively, with respect to the direction of the incoming particle a (see figure PA.6).

(a) Determine $\sqrt{E_b}$ as a function of the incoming energy E_a, the masses of the various particles and nuclei, the Q-value of the reaction and the scattering angle θ.
(b) Discuss the solutions for $\sqrt{E_b}$ for an exothermic reaction ($Q > 0$).
(c) Show, using the results from (a) and (b), that in the endothermic case ($Q < 0$) only one solution exists (and this for the angle $\theta = 0°$) starting at the threshold value $E_a > (-Q)(m_b + m_B)/(m_b + m_B - m_a)$.

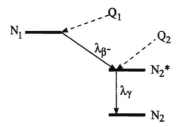

Figure PA.8.

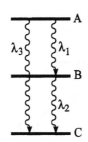

Figure PA.9.

7. In the nuclear reaction ^{27}Al(d, p) ^{28}Al, deuterons are used with a kinetic energy of 2.10 MeV. In this experiment (measuring in the laboratory system), the outgoing protons are detected at right angles with respect to the direction of the incoming deuterons. If protons are detected (with energy E_p) corresponding to ^{28}Al in its ground state and with energy E_p^* corresponding to ^{28}Al remaining in its excited state at 1.014 MeV, show that $|E_p - E_p^*| \neq 1.014$ MeV. Explain this difference.

Data given: d mass: 2.014 102 amu, p mass: 1.007 825 amu, ^{27}Al mass: 26.981 539 amu, ^{28}Al mass: 27.981 913 amu, 1 amu: 931.50 MeV). We use non-relativistic kinematics.

8. Within the nuclear fission process, atomic nuclei far from the region of beta-stability are formed with independent formation probabilities Q_1 and Q_2 (for species 1 and 2, respectively). These nuclei decay with decay probabilities of λ_{β^-} and λ_γ, respectively, to form stable nuclei (see figure PA.8).

Calculate N_1, N_2^* and N_2 as functions of time (with initial conditions of N_1, N_2^* and $N_2 = 0$ at $t = 0$).

Is there a possibility that $N_2^*(t)$ decays with a pure $\exp(-\lambda_{\beta^-}t)$ decay law?

9. Consider a nucleus with the following decay scheme (see figure PA.9). If, at time $t = 0$, all nuclei exist in the excited state A, calculate the occupation of the states B and C (as a function of time t). Study the particular cases in which one has

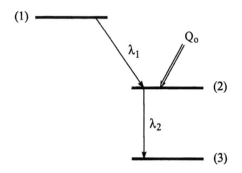

Figure PA.10.

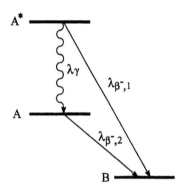

Figure PA.11.

(a) $\lambda_3 = 0$,
(b) $\lambda_1 = \lambda_2 = \lambda_3 = \lambda$,
(c) $\lambda_1 = \lambda_3 = \lambda/2; \lambda_2 = \lambda$.

10. In a radioactive decay process, nuclei of type 1 (with initial condition that $N_1(t = 0) = N_0$) decay with the decay constant λ_1 into nuclei of type 2. The nuclei of type 2, however, are formed in an independent way with formation probability of Q_0 atoms per second and then, subsequently decay into the stable form 3 with decay constant λ_2 (see figure PA.10).

Determine $N_1(t)$, $N_2(t)$ and $N_3(t)$ and discuss your results.

11. The atomic nucleus in its excited-state configuration, indicated by A*, can decay in two independent ways: once via gamma-decay with corresponding decay constant λ_γ into the ground-state configuration A and once via beta-decay into the nucleus B with decay constant $\lambda_{\beta^-,1}$ (see figure PA.11). After the gamma-decay, the nucleus A in its ground state is not stable and can on its way also decay into the nuclear form B with a beta-decay constant $\lambda_{\beta^-,2}$. If the initial conditions

at time $t = 0$ where $N_A = N_B = 0$ are imposed, calculate the time dependence of $N_A^*(t)$, $N_A(t)$ and $N_B(t)$. Make a graphical study of these populations.

Show that if the decay constant $\lambda_{\beta-,1} = 0$, the standard expression for a chain decay process $A^* \to A \to B$ is recovered.

12. We can discuss the radioactive decay chain

$$A \xrightarrow{\lambda_A} B \xrightarrow{\lambda_B} C \xrightarrow{\lambda_C} D \qquad \text{(stable nucleus),}$$

with $N_B = N_C = N_D = 0$ at time $t = 0$.

(a) Evaluate the total activity within the radioactive decay chain.
(b) How does this result change under the conditions $\lambda_A = 3\lambda$, $\lambda_B = 2\lambda$, $\lambda_C = \lambda$.
(c) Show that, in case (b), the total activity is independent of the initial conditions concerning the number of atoms of type A, B, C and D.
(d) Derive a general extension in the case with the decay chain

$$\cdots \to A_k \xrightarrow{k\cdot\lambda} \cdots \xrightarrow{2\lambda} A_1 \xrightarrow{\lambda} A_0.$$

13. At the starting time ($t = 0$) we have N_A^0 nuclei of type A. These nuclei decay into stable nuclei of type B with a decay constant λ_A. As soon as nuclei of type B are formed, we bring these back into the species A using an appropriate nuclear reaction, characterized by the formation probability λ_B.

Determine the occupation for nuclei of type A and B as a function of time and make a graphical study.

Study the special cases with (a) $\lambda_B = 0$, (b) $\lambda_A \ll \lambda_B$ and (c) $\lambda_A = \lambda_B$.

14. Show that in the decay chain $A \xrightarrow{\lambda_A} B \xrightarrow{\lambda_B} C \xrightarrow{\lambda_C} D$ (stable), in which the product $N(A)_0 \cdot \lambda_A$ represents the constant activity of a very long-lived radioactive source, the number of atoms of type D collected in the time interval $(0, t)$ is given by the expression

$$N(D) = N(A)_0 \lambda_A \left\{ t - \frac{1}{\lambda_B} - \frac{1}{\lambda_C} - \frac{\lambda_C}{(\lambda_B(\lambda_B - \lambda_C))} e^{-\lambda_B t} \right.$$
$$\left. + \frac{\lambda_B}{(\lambda_C(\lambda_B - \lambda_C))} e^{-\lambda_C t} \right\}.$$

How does this expression change if the decay chain stops at the element C.

15. A radioactive element B is being produced in the time interval Δt_0 with a formation rate of Q_0 (atoms per second). These nuclei of type B transform into elements of type C with decay constant λ_B, which, in their turn, decay into stable nuclei with decay constant λ_C.

(a) What happens to the activity of elements of type B and C at time t after the initial irradiation process forming the elements of type B has been stopped.

(b) If we have been performing the irradiation for a long enough time, such that an equilibrium has formed between the activities of elements B and C, show that the acitivity of element C, a time t after stopping the initial irradiation process, is given by the expression

$$A_C(t) = Q \frac{\lambda_C}{(\lambda_C - \lambda_B)} \left(e^{-\lambda_B t} - \frac{\lambda_B}{\lambda_C} e^{-\lambda_C t} \right).$$

16. The radioactive elements B are activated for a time t_0 with formation rate Q (atoms per second) and the decay chain then proceeds as B $\overset{\lambda_B}{\to}$ C $\overset{\lambda_C}{\to}$ D (stable).

(a) Find the activity of both the elements B and C, after the irradiation process is stopped.
(b) Show that the maximal activity of element C occurs at a time Δt_{max} after stopping the irradiation, which is given as

$$\Delta t_{max} = \frac{1}{\lambda_C - \lambda_B} \ln \frac{(1 - e^{-\lambda_C t_0})}{(1 - e^{-\lambda_B t_0})}.$$

(c) What is this maximal activity of element C?

17. The ratio between the number of daughter and parent nuclei in the decay chain P $\to$ D (with decay constant λ) is given at the time t_1, knowing that at the starting time t_0 all elements were of type P only.

(a) Calculate the time interval $\Delta t = t_1 - t_0$ as a function of this ratio.
(b) Show that this method, if λ is known, presents an appropriate dating method.
(c) Study the application in which we study the decay of ^{232}Th into the stable ^{208}Pb nuclei. In a piece of rock one observes the presence of 3.65 g of ^{232}Th and 0.75 g of ^{208}Pb. What is the age of the rock starting from the above Th/Pb ratio.

Given are: $\lambda(^{232}\text{Th})$ = 1.41 × 10^{10} year and Avogadro's number: 6.022 045 × 10^{23} mol^{-1}.

18. The production and decay (^{14}C $\to$ ^{14}N + β^- with half-life $T_{1/2}$ = 5730 year) of radioactive ^{14}C is in equilibrium in the earth's atmosphere. The equilibrium mass ratio of ^{14}C/^{12}C is 1.3 × 10^{-12} in organic material. Show that the decay rate of ^{14}C in 'living' organic material amounts to one disintegration per second for each 4 g of carbon.

Determine the age of a carbon species of 64 g in which two disintegrations per second are measured.

19. Within a given activation process, using a pulsating system, radioactive elements of type A are formed at a production rate of Q_0 atoms per second. These

nuclei A are unstable and decay into the elements B with a decay constant λ. If the process runs as follows: radiation in a time interval $\Delta t = T_{1/2}$, waiting time $\Delta t = T_{1/2}$, radiation time $\Delta t = T_{1/2}$, waiting time $\Delta t = T_{1/2}$, etc, determine the number of elements A formed straight after the nth irradiation period. What is the maximal number of nuclei of type A that can eventually be formed (or determine the limit for $n \rightarrow \infty$).

20. By the use of an intense neutron source and using the nuclear reaction $^{31}P(n, \gamma)\,^{32}P$, radioactive ^{32}P is produced at a production rate of $Q_0 = 10^7$ atoms per second.

Study the activity (as a function of time) if, at the start of the irradiation, an initial activity of $A_0 = 5 \times 10^6$ disintegrations per second of ^{32}P was already present. Make a graphical study of the various terms contributing to the total activity of ^{32}P (the half-life $T_{1/2}(^{32}P) = 14$ days). Also study the particular situation in which the initial activity A_0 is identical to the formation rate Q_0.

21. We consider the radioactive decay chain $X \rightarrow Y \rightarrow Z$ with decay constants λ_X and λ_Y respectively.

(a) Determine the time at which the daughter activity becomes maximal.
(b) Derive an approximate expression for this time t and for the corresponding maximal activity if we have $\lambda_X \simeq \lambda_Y$.
(c) In the case $\lambda_X \simeq 0$ (with $\lambda_X \ll \lambda_Y$) and with the initial condition that at time $t = 0$, all elements are of type X, determine the total activity in the decaying system.

22. We consider a decay chain $\cdots A_3 \xrightarrow{3\lambda} A_2 \xrightarrow{2\lambda} A_1 \xrightarrow{\lambda} A_0$, for which the decay constant decreases with a constant value of λ at each transformation step.

Determine the total activity of the decaying system.

23. Consider the decay in the following chain $X \rightarrow Y \rightarrow Z \rightarrow U \ldots$. Show that for a relatively short accumulation time t (short with respect to the various half-lives in the given decay chain), the number of atoms of type U is given by the expression $N(U) = N(X)_0 \lambda_X \lambda_Y \lambda_Z (t^3/3!)$.

PART B

NUCLEAR INTERACTIONS: STRONG, WEAK AND ELECTROMAGNETIC FORCES

Chapter 3

General methods

3.1 Time-dependent perturbation theory: a general method to study interaction properties

Having introduced the general properties of the atomic nucleus (chapter 1) and the study of nuclear decay characteristics, without concentrating on the precise origin of the decay processes however, we shall now discuss in more detail the various manifestations of transitions that happen in the nuclear A-body system.

The three basic interactions that are very important at the 'level' of the atomic nucleus and its constituents are the strong (hadronic), weak (beta-decay processes) and the electromagnetic (radiation processes) interactions. It is this last interaction, described by an interaction Hamiltonian $\hat{H}_{\text{int}}$ for characterizing systems non-relativistically which causes transitions between the various stationary states in the nucleus.

In figure 3.1, we illustrate a number of typical transitions caused by the various interactions: (a) the emission and absorption of a boson (the photon or a pion) by a fermion; the coupling constants are denoted by e and $f_{\pi NN^*}$, respectively. There also exists the possibility of changing a boson (photon) into another boson (vector ρ meson), as expressed in (b). The force between two nucleons is mediated by exchanging the charge carrier which, in this case is the pion, with coupling strength $f_{\pi NN}$ (c). Finally, also in (c), we present the basic diagram expressing the beta-decay process where protons (p), neutrons (n), neutrinos (ν) and electrons (e^-) couple with interaction strength G.

The above processes are the basic ingredients that indeed result in the observable transition processes that occur in the nucleus like α-decay or emission of particles within a cluster, the $\beta^{\mp}$ and electron capture processes and emission of gamma radiation. The general method to study the decay constant for these specific processes is time-dependent perturbation theory where transitions result from the specific interaction $\hat{H}_{\text{int}}$. One thus needs to evaluate (in lowest order, if allowed)

$$\lambda_{fi} = \frac{2\pi}{\hbar} |\langle \psi_f^{(0)} | \hat{H}_{\text{int}} | \psi_i^{(0)} \rangle|^2 \frac{dn}{dE}, \tag{3.1}$$

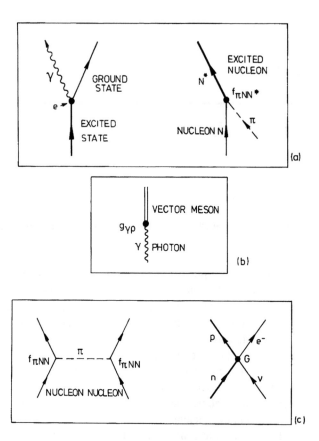

Figure 3.1. Various illustrations indicating basic interaction strengths. (*a*) Emission and absorption of a boson 'charge' carrier by a fermion. The coupling constants are denoted by e and $f_{\pi NN^*}$, respectively. (*b*) Transforming a boson into another boson with coupling strength $g_{\gamma\rho}$. (*c*) The force between two nucleons via the exchange of a meson (here, for instance, a pion) with coupling strength $f_{\pi NN}$ and the coupling strength G characterizing the weak interaction transformation processes. (Taken from Frauenfelder and Henley (1991) *Subatomic Physics* © 1974. Reprinted by permission of Prentice-Hall, Englewood Cliffs, NJ.)

where $\psi_i^{(0)}$, $\psi_f^{(0)}$ describe the stationary states and $\mathrm{d}n/\mathrm{d}E$ is the density of final states appropriate to the particular process and with conservation laws like energy conservation, momentum conservation, etc. These decay probabilities separate into two pieces: (i) a phase space term and (ii) the matrix element where the dynamics of the process enters via the initial and final state wavefunction and the interaction Hamiltonian.

In chapters 4, 5 and 6 we shall, by this method, study typical examples of the strong decay (α-decay), weak decay (β-decay) and electromagnetic decay (γ-decay), making use as much as possible of the common physics underlying the evaluation of the transition probabilities (λ-values). In each of the chapters we shall discuss in introductory sections the kinematics and phase space factors, before embarking on the more interesting implications of the interaction dynamics. We shall arrange the text such that these three chapters can be studied largely in an independent way.

Even though we shall illustrate the various basic interactions with the alpha-decay process, beta-decay and gamma-emission from excited states in atomic nuclei, we shall introduce in each chapter some of the most recent illustrative examples and point towards remaining problems.

It is clear that the approach, using non-relativistic perturbation theory, is restricted to low- and medium energy processes. More generally, one should use quantum field theory treating the matter and radiation fields (photon, pion, quark, gluon, etc) on the same footing and construct the various quantum mechanical amplitudes that lead from an initial to a given final state. The language of Feynmann- diagrams has been instrumental in depicting such calculations in a most illustrative and appealing way (see texts on relativistic quantum mechanics, QED and quantum field theory).

3.2 Time-dependent perturbation theory: facing the dynamics of the three basic interactions and phase space

As shown before and extensively discussed in quantum mechanics, the transition rate from initial rate i to the final state f becomes

$$\lambda_{fi} = \frac{2\pi}{\hbar} |\langle \psi_f^{(0)} | \hat{H}_{\text{int}} | \psi_i^{(0)} \rangle|^2 \frac{dn}{dE}, \tag{3.2}$$

in lowest order. The matrix element itself depends on the very specific interaction one is studying and the wavefunctions describing the initial and final nuclear wavefunctions.

Phase space dn/dE takes into account the *number* of final states and conforms with energy conservation, momentum conservation, etc knowing that we have the basic quantum mechanical uncertainty relation $\Delta x \cdot \Delta p_x \geq \hbar$ which gives the minimal element in phase space to which one can localize a particle. In a one-dimensional problem, with a particle constrained over a length L and moving with momentum in the interval $(0, p)$, the number of states of a freely moving particle is (figure 3.2)

$$n = \frac{L \cdot p}{2\pi \hbar}. \tag{3.3}$$

This equation can be verified solving the one-dimensional quantum mechanical problem for a particle constrained in the interval $(0, L)$ and counting the number

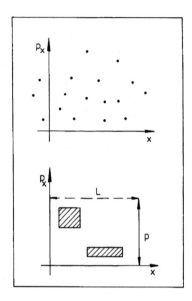

Figure 3.2. Classical and quantum mechanical one-dimensional phase space. In the classical case, the state is presented by a point, in the quantum mechanical case, on the other hand, a state has to be described by a phase space volume $p_x \cdot L = h$.

of momentum states up to the value p (figure 3.3). For a general, three-dimensional problem one has

$$n = \frac{1}{(2\pi\hbar)^3} \int d^3x \int d^3p,\qquad(3.4)$$

and the corresponding density of states reads

$$
\begin{aligned}
\frac{dn}{dE} &= \frac{V}{(2\pi\hbar)^3} \frac{d}{dE} \int d^3p \\
&= \frac{V}{2\pi^2\hbar^3} p^2 \frac{dp}{dE} \\
&= \frac{V}{2\pi^2c^2\hbar^3} pE,
\end{aligned}
\qquad(3.5)
$$

using the relation $E^2 = p^2c^2 + m_0^2c^4$.

In the case of two particles, moving fully independently, one has

$$n_2 = \frac{V^2}{(2\pi\hbar)^6} \int d^3p_1 \int d^3p_2,\qquad(3.6)$$

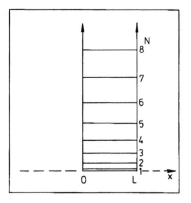

Figure 3.3. Energy spectrum for a particle moving inside a one-dimensional potential in the interval $(0, L)$, denoted by the quantum number N.

with its generalization to N independent particles

$$n_N = \frac{V^N}{(2\pi\hbar)^{3N}} \int d^3 p_1 \dots \int d^3 p_N. \tag{3.7}$$

In many cases constraints on conservation of linear momentum are present e.g. two particles with momentum such that (in the COM system) $\vec{p}_1 + \vec{p}_2 = 0$. In such situations the *total* number of final states is determined by just *one* of the particles, say 1. The number of states is identical to equation (3.4) or

$$n_2' = \frac{V}{(2\pi\hbar)^3} \int d^3 p_1, \tag{3.8}$$

but the density of final states will differ from equation (3.5) since E now denotes: the total energy of the *two* particles. Thus one obtains

$$\frac{dn_2'}{dE} = \frac{V}{(2\pi\hbar)^3} \frac{d}{dE} \int d^3 p_1, \tag{3.9}$$

with

$$dE = dE_1 + dE_2$$
$$= \left(\frac{p_1}{E_1} dp_1 + \frac{p_2}{E_2} dp_2 \right) \cdot c^2. \tag{3.10}$$

Because of momentum conservation, $p_1^2 = p_2^2$ and $p_1 dp_1 = p_2 dp_2$, so the differential dE becomes

$$dE = p_1 \frac{(E_1 + E_2)}{E_1 E_2} c^2 dp_1, \tag{3.11}$$

and the corresponding density for the two-body (correlated) decay becomes finally

$$\frac{dn_2'}{dE} = \frac{V}{2\pi^2 c^2 \hbar^3} \frac{E_1 E_2}{(E_1 + E_2)}. \tag{3.12}$$

This method can also be extended to many particles, but constraining

$$\vec{p}_1 + \vec{p}_2 + \cdots \vec{p}_N = 0, \tag{3.13}$$

leading to $(N - 1)$-independent particles in the final state.

We shall need the above phase space factors, in particular when studying beta-decay processes where one has a recoiling nucleus and the emission of an electron (positron) and an antineutrino (neutrino).

Chapter 4

Alpha-decay: the strong interaction at work

4.1 Kinematics of alpha-decay: alpha particle energy

Alpha radioactivity has been known for a long time in heavy nuclei. One of the main reasons for this decay mode to appear as a spontaneous process is the large binding energy of the alpha-particle (28.3 MeV).

The collection of data on α-decay energy shows a rather abrupt change, indicating a new region of α-emitters, at $Z = 82$ ($A = 212$) presenting an additional example of the stability of shells in the atomic nucleus. These features will be discussed later on in chapter 7 (see figure 4.1). In plotting the information on α-binding energy, i.e.

$$-Q_\alpha(^A_Z X_N) = (M'(^{A-4}_{Z-2} X_{N-2}) + M'(^4_2 He_2) - M'(^A_Z X_N))c^2$$
$$= BE(A, Z) - (BE(A - 4, Z - 2) + 28.3 \text{ MeV}), \quad (4.1)$$

one knows that when this quantity becomes negative, the α-particle will no longer be a bound particle and will be emitted spontaneously from the initial nucleus $^A_Z X_N$. The results are shown as dots in figure 4.2 with the first few negative values occurring at $A \simeq 150$.

In deriving some of the more important quantities on α-decay, we can write up (according to the discussion in chapter 1 on nuclear reactions) the Q-value equation for a decaying parent nucleus

$$M'_p c^2 = M'_D c^2 + M'_\alpha c^2 + T_\alpha + T_D, \quad (4.2)$$

where T_α(T_D) denote the kinetic energy of the outgoing α-particle (recoiling daughter nucleus). The masses denoted with a prime are nuclear masses and can be converted into atomic masses by adding $Z \cdot m_e c^2$ (and neglecting very small electron binding energy differences) resulting in

$$M_p c^2 = M_D c^2 + M_\alpha c^2 + T_\alpha + T_D. \quad (4.3)$$

107

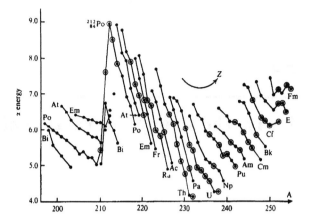

Figure 4.1. The α-decay energy in the mass region $200 < A < 260$. The lines connect the various isotopic chains. The large and rapid variation around $A = 212$ indicates evidence for nuclear shell structure due to the presence of closed shells in the nucleus ${}^{208}_{82}\text{Pb}_{126}$ (taken from Valentin 1981).

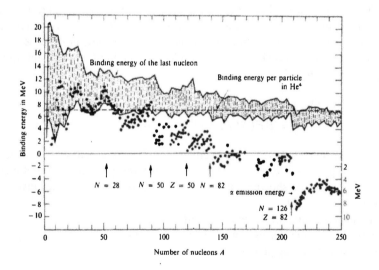

Figure 4.2. Alpha binding energy in a large number of nuclei (dots). Only from about $A \simeq 150$, does α-decay becomes a spontaneous process. The shaded zone is enclosed by the minimum and maximum separation energies. The shells are indicated by arrows at $N = 28$, $N = 50$, $Z = 50$, $N = 82$ and $Z = 82$, $N = 126$ (taken from Valentin 1981).

The Q_α value now reads

$$Q_\alpha = M_P c^2 - (M_D c^2 + M_\alpha c^2)$$

$$= T_\alpha + T_D. \tag{4.4}$$

In the emission process (a non-relativistic situation is considered which is well fulfilled in most cases), the α-particle and the recoiling daughter nucleus have the constraint

$$|\vec{p}_\alpha| = |\vec{p}_D|. \tag{4.5}$$

The Q_α value can then be rewritten as

$$\begin{aligned}
Q_\alpha &= \tfrac{1}{2} M_\alpha v_\alpha^2 + \tfrac{1}{2} M_D v_D^2 \\
&= \tfrac{1}{2} M_\alpha^{\text{reduced}} v_{\text{rel}}^2 (\alpha, D) \\
&= \tfrac{1}{2} M_\alpha v_\alpha^2 \left(\frac{M_D + M_\alpha}{M_D} \right),
\end{aligned}$$

or

$$Q_\alpha = T_\alpha \left(\frac{M_D + M_\alpha}{M_D} \right) \simeq T_\alpha \frac{A}{A - 4}. \tag{4.6}$$

4.2   Approximating the dynamics of the alpha-decay process

In treating the process of the emission of an α-particle through the Coulomb barrier surrounding the atomic nucleus, the strong attractive forces holding the nucleons together can be represented by an effective, strongly attractive nuclear potential. One immediately runs into a problem. How could a particle escape through the high Coulomb barrier?

This paradox (presented in figure 4.3) was first solved by Gamow (1928) and Gurney and Condon (1928), and represented the first triumph for quantum theory in its application to the atomic nucleus.

In the present section, we indicate the method one should use, in principle, and subsequently discuss various approximations that lead to a more tractable problem.

(i)   *Time-dependent approach.*
In describing α-decay one should solve the time-dependent problem where, initially (at $t = 0$) one has an α-particle in the interior of the nuclear potential and finally one reaches a state where the α-particle moves away from the remaining nucleus.
According to perturbation theory, one should solve the equation

$$\left[-\frac{\hbar^2}{2m} \Delta + V(|\vec{r}|) \right] \psi(\vec{r}, t) = i\hbar \frac{\partial \psi(\vec{r}, t)}{\partial t}, \tag{4.7}$$

since we are considering a non-stationary problem. Following the methods outlined in chapter 3, we should finally solve the corresponding time-

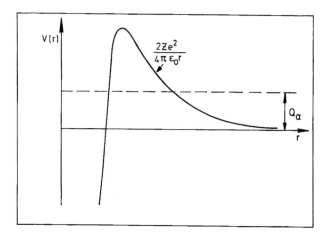

Figure 4.3. The Coulomb potential energy outside the nucleus for a given nucleus with charge Ze and an α-particle approaching this nucleus. In the internal part, the attractive nuclear part takes over. The Q_α value for a given decay is also indicated. In many cases Q_α is much smaller than the maximal Coulomb potential energy value.

independent problem

$$\left[-\frac{\hbar^2}{2m}\Delta + V(|\vec{r}|)\right]\varphi_n(\vec{r}) = E_n\varphi_n(\vec{r}), \tag{4.8}$$

and determine the internal solutions (α-particle inside the nucleus with energy $E_{n,i}$) and external solutions (α-particle moving outside the nucleus with energy $E_{m,f}$) as

$$\varphi_{n,i}(\vec{r}, t) = \varphi_{n,i}(\vec{r})e^{-iE_{n,i}t/\hbar}$$
$$\varphi_{m,f}(\vec{r}, t) = \varphi_{m,f}(\vec{r})e^{-iE_{m,f}t/\hbar}. \tag{4.9}$$

The full time-dependent solution can then be expanded on this basis as

$$\psi(\vec{r}, t) = \sum_n c_{n,i}(t)\varphi_{n,i}(\vec{r})e^{-iE_{n,i}t/\hbar} + \sum_m c_{m,f}(t)\varphi_{m,f}(\vec{r})e^{-iE_{m,f}t/\hbar}, \tag{4.10}$$

and $c_{n,i}(t)$ and $c_{m,f}(t)$ can be determined such that the time-dependent equation is fulfilled. This is possible by the choice

$$t = 0, \quad \text{all} \quad c_{m,f} = 0 \quad \text{and a single} \quad c_{n,i} = 1$$
$$t \gg 0, \quad \text{all} \quad c_{m,f} \neq 0, c_{n,i} \neq 0 \quad \text{and} \quad c_{n,i} \simeq 1,$$

where $\sum_n |c_{n,i}(t)|^2$ gives the probability that the α-particle is still within the nucleus and $\sum_m |c_{m,f}(t)|^2$, the probability that the α-particle is outside of the nucleus. This turns out to a time dependence in $\psi(\vec{r}, t)$ with a corresponding continuity equation

$$\frac{\partial}{\partial t} \psi^*(\vec{r}, t) \psi(\vec{r}, t) + \text{div } \vec{j}(\vec{r}, t) = 0. \tag{4.11}$$

Since a current of α-particles is leaving the nucleus, the density $\rho(\vec{r}, t)$ has to be a decreasing function with time. This is impossible for purely stationary solutions $\varphi_{n,i}(\vec{r}, t)$. It implies an energy term $e^{-iEt/\hbar}$ where E is now a complex variable if we wish to describe the state in the nucleus as an 'almost' stationary or 'quasi-stationary' state. Thereby, the level will obtain a linewidth (or corresponding lifetime) characteristic of the decay process and we can write, approximately

$$\psi(\vec{r}, t) \propto \psi(\vec{r}) e^{-i\left(E - i\frac{\lambda\hbar}{2}\right)t/\hbar}$$
$$\propto \psi(\vec{r}) e^{-iEt/\hbar} \cdot e^{-\lambda t/2}$$
$$|\psi(\vec{r}, t)|^2 \propto |\psi(\vec{r})|^2 e^{-\lambda t}. \tag{4.12}$$

Some examples are sketched in figure 4.4, where, for deep-bound states α-decay is very unlikely to occur but finally becomes a bound-state (with $\lambda \to 0$). For very weakly-bound states (near the top of the barrier), the state decays rapidly, characterized by a large λ and a corresponding large width ΔE since $\Delta E \Delta t \simeq \hbar$.

(ii) *Approximations.*

To simplify the above to a more more realistic case of α-decay, one can make a number of approximations that still keep the main quantum-mechanical tunnelling aspect, that characterize the slow process and move towards a stationary problem. This is because in a large number of α-decay cases, the process can be described as an 'almost' stationary problem (figure 4.5).

So, at level (a) we still consider the full Coulomb potential but study only real eigenvalues (no actual decay). At a simpler level (b), we approximate the potential by a nuclear square-well potential separated by a rectangular potential from the region where a free α-particle could exist. The most simple case to study (c) (Merzbacher 1970) is the simple transmission of a plane wave through a rectangular potential. Even the latter case (c) can be used to obtain estimates for penetrability in α-decay. This case will not be discussed here but we shall solve situation (b) in section 4.3.

4.3 Virtual levels: a stationary approach to α-decay

The potential with which we solve the α-decay problem is given in figure 4.6, where in region (I) stationary solutions describe the α-particle in the nucleus where the lifetime of the level in realistic cases is very long compared to the

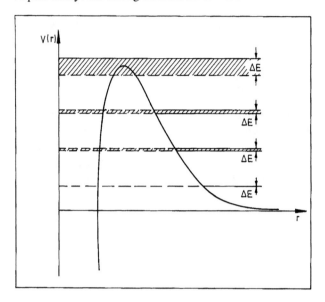

Figure 4.4. Schematic representation of the level width ΔE in α-decay according to the energy of the corresponding quasi-bound state in the potential constructed from the nuclear force (internal region) and the Coulomb energy (in the external region).

time that is typical of nucleon frequencies. In the external region, the α-particle can be considered as a free particle (region III).

At the point $x = 0$, the barrier goes to infinity, so, solutions in this interval, region (I), become

$$\psi_I = e^{ipx} - e^{-ipx},\tag{4.13}$$

where we have

$$p^2 = \frac{2mE}{\hbar^2}; \qquad q^2 = \frac{2m}{\hbar^2}(V_b - E),\tag{4.14}$$

and the differential equation in region (I), (III) reads

$$\frac{d^2\psi}{dx^2} + p^2\psi = 0.\tag{4.15}$$

Similarly, solutions in regions (II) and (III) become:

$$\psi_{II} = c_+e^{q(x-a)} + c_-e^{-q(x-a)}$$
$$\psi_{III} = c_1e^{ip(x-b)} + c_2e^{-ip(x-b)}.\tag{4.16}$$

Continuity conditions at the points a and b give for a

$$e^{ipa} - e^{-ipa} = c_+ + c_-$$
$$\frac{ip}{q}(e^{ipa} + e^{-ipa}) = c_+ - c_-,\tag{4.17}$$

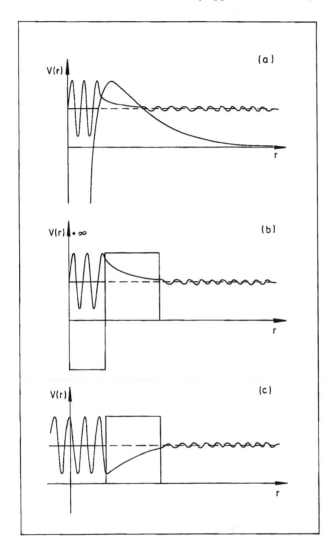

Figure 4.5. Various approximation methods used to describe decay probabilities for α-decay. From (a) to (c) the approximation to the potential describing the tunnelling simplifies the problem considerably from a realistic calculation (a) into the quantum mechanical tunnelling calculation through to a simple square potential (c). Wavefunctions (stationary solutions) are depicted in each case.

or

$$c_+ = \mathrm{i}\sin pa + \frac{\mathrm{i}p}{q}\cos pa;$$
(4.18)

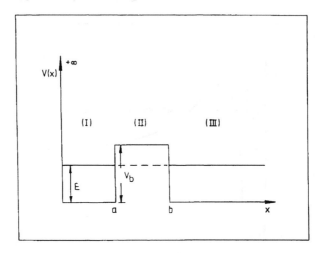

Figure 4.6. Potential model (schematic) used to study α-decay transition probabilities in a stationary description. The description in part (I) allows for quasi-bound states tunnelling through the rectangular potential well (II) into the external region (III). The stationary energy corresponds to the eigenvalue E.

and for b

$$c_1 + c_2 = c_+ e^{G/2} + c_- e^{-G/2}$$

$$c_1 - c_2 = \frac{q}{ip}(c_+ e^{G/2} - c_- e^{-G/2}), \qquad (4.19)$$

with $G = 2q(b-a)$. The last two equations give the values of c_1 and c_2, i.e.

$$c_1 = \frac{1}{2}\left\{\left(1 + \frac{q}{ip}\right)c_+ e^{G/2} + \left(1 - \frac{q}{ip}\right)c_- e^{-G/2}\right\}$$

$$c_2 = \frac{1}{2}\left\{\left(1 - \frac{q}{ip}\right)c_+ e^{G/2} + \left(1 + \frac{q}{ip}\right)c_- e^{-G/2}\right\}, \qquad (4.20)$$

and the wavefunction $\psi_{III}(x)$ reads

$$\psi_{III} = \frac{1}{2}c_+ e^{G/2}\left\{\left(1 + \frac{q}{ip}\right)e^{ip(x-b)} + \left(1 - \frac{q}{ip}\right)e^{-ip(x-b)}\right\}$$

$$+ \frac{1}{2}c_- e^{-G/2}\left\{\left(1 - \frac{q}{ip}\right)e^{ip(x-b)} + \left(1 + \frac{q}{ip}\right)e^{-ip(x-b)}\right\}, (4.21)$$

or, written in a slightly different way

$$\psi_{III} = c_+ e^{G/2}\left\{\cos p(x-b) + \frac{q}{p}\sin p(x-b)\right\}$$

$$+ c_- e^{-G/2}\left\{\cos p(x-b) - \frac{q}{p}\sin p(x-b)\right\}, \qquad (4.22)$$

and the ratio

$$\frac{\langle|\psi_{III}|^2\rangle}{\langle|\psi_I|^2\rangle} = \left(\frac{1}{4} + \frac{q^2}{4p^2}\right)\left(c_+^2 e^G + c_-^2 e^{-G}\right) + 2\left(\frac{1}{4} - \frac{q^2}{4p^2}\right)c_+c_-. \quad (4.23)$$

In almost all cases, $|\psi_{III}|$ is much larger than $|\psi_I|$. We are now interested in those situations where ψ_{III} is as small as possible, i.e. $c_+ = 0$ or,

$$c_+ = i\sin pa + \frac{ip}{q}\cos pa,$$

$$\tan(pa) = -\frac{p}{q}, \quad (4.24)$$

or

$$\tan\left(\sqrt{\frac{2mE}{\hbar^2}} \cdot a\right) = -\sqrt{\frac{E}{V_b - E}}. \quad (4.25)$$

Those values of $E(E_0)$ which fulfil the above equation correspond to solutions where a very small intensity occurs in the external region III. These energy eigenvalues E_0 correspond to 'virtual' levels in region (I) for which the α-particle are mainly localized *within* the nucleus with a very small penetrability through the barrier into the external region. The various possible solutions are given in figure 4.7 where, in the lower part (d), the virtual (mostly separated levels E_0) levels are drawn, these levels are embedded within the continuum of the free α-particle moving relative to the nucleus in the external region. The penetrability is then given by the expression

$$\left|\frac{\psi_{III}}{\psi_I}\right|^2 \propto e^{-G}$$

$$\propto \exp\left[-\frac{2}{\hbar}\sqrt{2m\,(V_b - E)}(b - a)\right]. \quad (4.26)$$

So, for $E = E_0$, α-decay can be described with the above penetrability; for $E \neq E_0(c_+ \neq 0)$, the wavefunction $\psi(x)$ describes the process where an α-particle is moving mainly on the right-hand side of the potential barrier.

4.4 Penetration through the Coulomb barrier

Using the more general barrier penetration problem as outlined in the WKB (Wentzel–Kramers–Brillouin) method, variations in the general potential can be handled (see Merzbacher 1990). The result, for a barrier $V(x)$ with classical turning points x_1 and x_2, turns out to be a simple generalization of the constant barrier result of e^{-G}, i.e. one gets

$$P \equiv \left|\frac{\psi_{III}}{\psi_I}\right|^2 \propto \exp\left[-2\int_{x_1}^{x_2} k(x)\,dx\right], \quad (4.27)$$

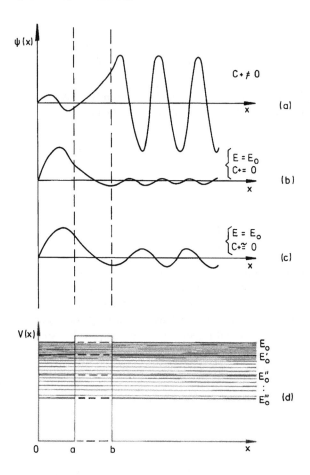

Figure 4.7. The various possible solutions to the eigenvalue equation with the potential as given in the lower part (*d*) of this figure. Solutions with $c_+ \neq 0$ correspond to the high density of states that describe a particle mainly localized in the external region $x \gg b$. For $c_+ \simeq 0$ and $E = E_0, E_0', \ldots$ one is approaching an almost bound state in the internal region $0 \leq x \leq a$, corresponding to a bound-state solution in this first interval (case (*b*)). In case (*a*), this separates into a wavefunction describing a quasi-bound state with only a small component in the external region, one arrives at the conditions for describing α-decay (virtual levels).

with

$$k(x) = \frac{1}{\hbar}\sqrt{2m(V(x) - E)}. \tag{4.28}$$

The case for an α-particle emitted from a daughter nucleus with charge Z_D, leads to the evaluation of the integral (figure 4.8)

$$\gamma = \frac{2}{\hbar} \int_R^b \left\{ 2m \left[\frac{z Z_D e^2}{4\pi \epsilon_0 \cdot r} - Q_\alpha \right] \right\}^{1/2} dr. \tag{4.29}$$

The value m is the (relative) reduced α-particle mass since the COM motion has been separated out. The energy released in the α-decay is the Q_α value. By using the notation

$$Q_\alpha = \frac{z Z_D e^2}{4\pi \epsilon_0 \cdot b}, \tag{4.30}$$

and thus

$$b = \frac{z Z_D e^2}{4\pi \epsilon_0 Q_\alpha}, \tag{4.31}$$

the integral reads

$$\gamma = \frac{2}{\hbar} \sqrt{2m Q_\alpha} \int_R^b \left(\frac{b}{r} - 1 \right)^{1/2} dr. \tag{4.32}$$

Using integral tables, or elementary integration techniques, the final result becomes

$$\gamma = \frac{2b}{\hbar} \sqrt{2m Q_\alpha} \left\{ \arccos \sqrt{\frac{R}{b}} - \sqrt{\frac{R}{b}} \cdot \sqrt{1 - \frac{R}{b}} \right\}. \tag{4.33}$$

If we substitute back for the value of b, using the fact that $Q_\alpha = \frac{1}{2}mv^2$ (with m the reduced α-mass and v the relative (α-daughter nucleus) velocity) and $2m Q_\alpha = m^2 v^2$, we can get the final result

$$\gamma = \frac{4z Z_D e^2}{4\pi \epsilon_0 \hbar v} \cdot \left\{ \arccos \sqrt{\frac{R}{b}} - \sqrt{\frac{R}{b}} \cdot \sqrt{1 - \frac{R}{b}} \right\}. \tag{4.34}$$

We also know (see figure 4.8) that $R/b = Q_\alpha/B$, where B is the height of the Coulomb barrier at the nuclear radius R

$$B = \frac{z Z_D e^2}{4\pi \epsilon_0 \cdot R}, $$

and

$$Q_\alpha = \frac{z Z_D e^2}{4\pi \epsilon_0 \cdot b}. \tag{4.35}$$

For thick barrier penetration $R/b \ll 1 (Q_\alpha/B \ll 1)$, we can approximate the $\arccos \sqrt{R/b}$ by $\pi/2 - \sqrt{R/b}$. Thereby we get

$$\gamma \simeq \frac{4z Z_D e^2}{4\pi \epsilon_0 \hbar v} \left(\frac{\pi}{2} - 2\sqrt{\frac{R}{b}} \right), \tag{4.36}$$

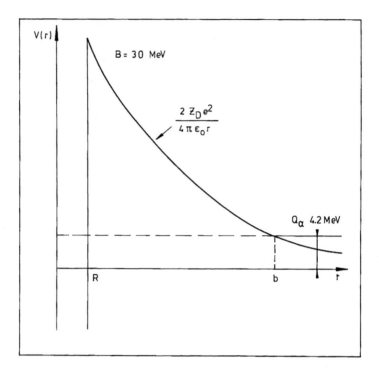

Figure 4.8. Actual description of α-decay with a decay energy of $Q_\alpha = 4.2$ MeV in ^{234}U where the Coulomb potential energy is depicted, coming up to a value of the barrier of 30 MeV at the nuclear radius $r = R$ ($R = r_0 A^{1/3}$ fm).

which can be rewritten in the form

$$\gamma \simeq \frac{z Z_D e^2}{2\epsilon_0 \hbar v} - \frac{1}{\hbar}\left(\frac{8 z Z_D e^2 m R}{\pi \epsilon_0}\right)^{1/2}. \tag{4.37}$$

The decay constant is now obtained by multiplying the probability of barrier penetration by a frequency which expresses, in a simple way, the frequency of impacts on the barrier. The latter value can be estimated as the 'internal' α-particle velocity divided by the nuclear radius, so we finally obtain for λ_α ($z = 2$)

$$\lambda_\alpha = \frac{v_{in}}{R}\exp\left[-\frac{Z_D e^2}{\epsilon_0 \hbar v} + \frac{4e}{\hbar}\left(\frac{Z_D m R}{\pi \epsilon_0}\right)^{1/2}\right]. \tag{4.38}$$

If we replace v by $\sqrt{2Q_\alpha/m}$ and take the log, a relation between $\log T$ and Q_α is obtained under the form

$$\log T = a + \frac{b}{\sqrt{Q_\alpha}}. \tag{4.39}$$

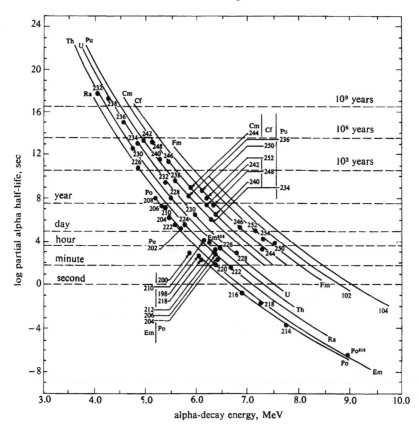

Figure 4.9. α-decay energy (in MeV). Experimental values of the half-life versus the α-decay energy for even–even nuclei. (Taken from Segré *Nuclei and Particles* 2nd edn © 1982 Addison-Wesley Publishing Company. Reprinted by permission.)

This law was empirically deduced as the Geiger–Nutall law of α-decay (Geiger and Nutall 1911, 1912) which is quite well verified by the quantum mechanical calculation carried out here. The data, expressing this variation are presented in figure 4.9, where the large span of $T(\alpha)$ values, ranges from beyond 10^9 y down to 10^{-6} s in ^{212}Po.

In this whole discussion we have always considered the α-particle to be present: of course the α-particle needs to be formed inside the nucleus and this formation probability can influence the decay rate (see Box 4b). Also, the α-particle can be emitted with non-zero orbital angular momentum which causes extra hindrance. These aspects will be discussed in the section 4.5.

We give a numerical example deriving the order of magnitude for the various quantities characterizing α-decay in Box 4a.

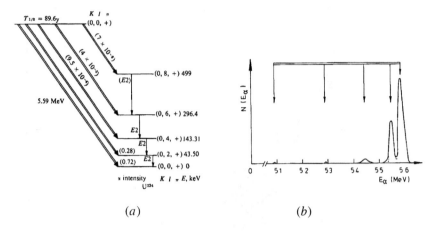

Figure 4.10. (*a*) Decay scheme of ^{238}Pu showing a number of α-transitions into the excited states (and ground state) of the final nucleus ^{234}U. The α-decay intensities are indicated (numbers between brackets). (Taken from Stephens 1960, Segré *Nuclei and Particles* 2nd edn © 1982 Addison-Wesley Publishing Company. Reprinted by permission.). (*b*) Corresponding spectrum of α-particles. The number $N(E_\alpha)$ is proportional to the various intensities. The maximum E_α value is 5.59 MeV in the present case.

4.5 Alpha-spectroscopy

4.5.1 Branching ratios

The alpha-decay probability (λ_α) derived in section 4.4 describes decay carrying away zero-angular momentum and involves decay of a nucleus in its ground state. Many α-particle emitters, though, show a line spectrum of α-groups corresponding to α-transitions to various nuclear excited states in the final nucleus. This is confirmed by the fact that gamma rays have been observed with energies corresponding to the energy differences between various α-groups.

A typical example, for the decay of ^{238}Pu to ^{234}U is shown in figure 4.10 where various groups are observed with a rapidly decreasing intensity as decay energy falls. Part of the decrease is understood since the corresponding Q_α is also decreasing and in the light of the dependence of $T(\alpha)$ on Q_α, a decrease in the partial decay probability results. A more serious effect is the fact that a large angular momentum is carried away when decay proceeds to the higher-lying excited states. In the example, we proceed from a $J_i^\pi = 0^+$ to $J_f^\pi = 0^+, 2^+, 4^+, 6^+$ and 8^+ states, implying that the angular momenta carried away by the α-particle are $L = 0, 2, 4, 6$ and 8, respectively. It is important to learn how increasing the L value influences the α-decay rate λ_α. The range of high-energy alpha particles is shown in figure 4.11.

Figure 4.11. A cloud chamber photograph of α-particles from the decay of ^{214}Bi $\rightarrow$ ^{214}Po $\rightarrow$ ^{210}Pb. An almost constant range for the 7.69 MeV α-particles is very well illustrated (taken from Philipp 1926).

4.5.2 Centrifugal barrier effects

The correct differential equation for solving the α-penetration problem (radial equation) for a general angular momentum eigenstate ℓ, characterizing the decay reads (with $R(r) = u(r)/r$)

$$-\frac{\hbar^2}{2m}\frac{d^2u(r)}{dr^2} + \left[\frac{2(Z-2)e^2}{4\pi\epsilon_0 \cdot r} + \frac{\ell(\ell+1)\hbar^2}{2mr^2}\right]u(r) = E, \qquad (4.40)$$

when the α-particle transmission is described by a wavefunction

$$\psi(\vec{r}) = \frac{u(r)}{r}Y_\ell^m(\theta, \varphi). \qquad (4.41)$$

This can be interpreted in a way that, for $\ell \neq 0$ values, the potential the α-particle has to tunnel through becomes bigger by the quantity $\ell(\ell+1)\hbar^2/2mr^2$ and gives rise to an effective potential

$$V^{\text{eff}}(r) = \frac{2(Z-2)e^2}{4\pi\epsilon_0 \cdot r} + \frac{\ell(\ell+1)\hbar^2}{2mr^2}. \qquad (4.42)$$

This is illustrated in figure 4.12 and the effect can be analyzed by its effect on the λ_α value. If we call σ the value of the increase in height of the centrifugal barrier at $r = R$ relative to the Coulomb barrier height at $r = R$, we obtain

$$\sigma = \frac{\ell(\ell+1)\hbar^2}{2mR^2} \cdot \frac{4\pi\epsilon_0 R}{2Z_D e^2} = \frac{\pi\epsilon_0\hbar\ell(\ell+1)}{mRZ_D e^2}, \qquad (4.43)$$

or

$$\sigma \simeq 0.002\ell(\ell+1).$$

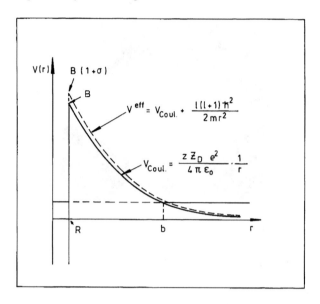

Figure 4.12. Illustration of the effect of the centrifugal to increase the height and width of the original Coulomb barrier, causing a decrease in the corresponding α-decay probability. The various parameters B, σ are explained in detail in the text.

Table 4.1. Ratio of λ_α probability for ℓ angular momentum compared to $\lambda_\alpha(\ell = 0)$ ($Z_D = 86$, $R = 9.87 \times 10^{-15}$ m and $Q_\alpha = 4.88$ MeV)

ℓ	$\dfrac{\lambda_\alpha(\ell)}{\lambda_\alpha(\ell=0)}$
0	1
1	0.7
2	0.37
3	0.137
4	0.037
5	7.1×10^{-3}
6	1.1×10^{-3}

The increase of the barrier height B to the value $B(1 + \sigma)$ leads to a small (but non-negligible) effect on the second term in the expression for λ_α as given in equation (4.38).

The analysis has to be changed since we now have

$$\frac{R}{b} = \frac{Q_\alpha}{B(1 + \sigma)},$$

and

$$\sqrt{\frac{R}{b}} = \sqrt{\frac{Q_\alpha}{B}}\frac{1}{\sqrt{1+\sigma}} \simeq \sqrt{\frac{Q_\alpha}{B}}\left(1 - \frac{\sigma}{2}\right). \tag{4.44}$$

This shows up in the new expression for λ_α,

$$\lambda_\alpha \cong \frac{zZ_\mathrm{D}e^2}{2\epsilon_0\hbar v} - \frac{1}{\hbar}\left(\frac{8zZ_\mathrm{D}e^2mR}{\pi\epsilon_0}\right)^{1/2}\left(1 - \frac{\sigma}{2}\right). \tag{4.45}$$

The numerical example discussed in Box 4a for $\ell = 2$, gives a correction in the exponent of

$$\exp\{83 \times \tfrac{1}{2} \times 0.002 \times 2(2+1)\} = \exp(0.498) \simeq 1.6.$$

This effect is still small, compared to changes in λ_α resulting from changes in Q_α and/or R. For large ℓ difference, though, a $\Delta\ell = 6$ value gives a reduction in the λ_α value of $\simeq 1000$ (compared with the $\ell = 0$ case).

4.5.3 Nuclear structure effects

In all the above arguments, even though the full Coulomb barrier and, eventually, centrifugal effects are taken into account, it has always been assumed that we study the α-particle penetration with the decay constant separated into a 'collision frequency' v times the penetration probability (for a single 'kick') $\lambda_\alpha = v \cdot P$. In fact we should take into account that the α-particle needs to be formed within the nuclear potential well and this formation probability can give an additional reduction in λ_α over the more simple approaches.

Here, one has to consider the nuclear wavefunctions in detail and this will be sensitive to the way the product of the final nuclear wavefunctions describing the daughter nucleus $\psi_f(A - 4)$, and the α-particle wavefunction φ_α resembles the initial A-particle nuclear wavefunction $\psi_i(A)$. This quantity, called the overlap integral or, the α-particle 'spectroscopic factor' A_α or α-reduced width, is defined as

$$A_\alpha^2 = |\langle\psi_i(A)|\psi_f(A - 4) \cdot \varphi_\alpha(4)\rangle|^2. \tag{4.46}$$

It will strongly depend on the particular shell-model orbitals of the final states of the protons and neutrons (see part C). A number of calculations have been performed by Mang (1960, 1964). Typical examples are those cases for strongly deformed nuclei where the internal structure of the final wavefunction is *not* changing but only the collective behaviour of the nucleus as a whole is changing (a collective rotational spectrum as is the case in the α-decay, presented in the $^{238}\mathrm{Pu} \rightarrow {}^{234}\mathrm{U}$ decay to the $0^+, 2^+, 4^+, 6^+, 8^+$ members of a rotational band (figure 4.10), see part D). In these cases, the value of A_α^2 remains constant but retardation results because of the centrifugal barrier acting on ℓ *and* of the decreasing Q_α value with increasing excitation energy in the final nucleus $^{234}\mathrm{U}$.

In Box 4b, we illustrate some of these reduced A_α^2 widths as well as the effect of α-decay connecting odd-mass nuclei.

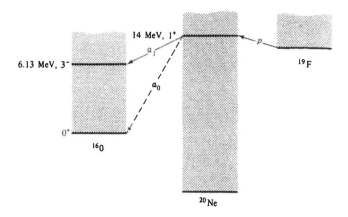

Figure 4.13. α-decay from a $J^\pi = 1^+$ level in ^{20}Ne. Only particular levels of interest that can be of use in testing parity non-conservation in α-decay are shown. (Taken from Frauenfelder and Henley (1991) *Subatomic Physics* © 1974. Reprinted by permission of Prentice-Hall, Englewood Cliffs, NJ.)

4.6   Conclusion

A full time-dependent study is not within our scope here; we could however find qualitative explanations for many of the observed α-decay properties. A nice discussion of time dependence is given by Fuda (1984) and by Bohm *et al* (1989). A study of α-decay spectra is presented by Desmarais and Duggan (1990).

We point out here a final, but very interesting, application of α-decay as a test of parity conservation in the strong decay. The angular momentum ℓ, carried away by the α-particle (since the α-particle itself has $J^\pi = 0^+$), also determines the total spin change if decay from (or to) a 0^+ state proceeds. The parity associated with the α-particle relative wavefunction, described by the $Y_\ell^m(\theta, \varphi)$ spherical harmonics, is $(-1)^\ell$. This brings in an extra selection rule in the sense that

$$\pi_i = \pi_f(-1)^\ell,$$

where ℓ is the α-decay angular momentum.

So, α-decay (even a very weak branch) in the decay of the 1^+ level in ^{20}Ne to the 0^+ ^{16}O ground state (proceeding via a $\ell = 1$ change) would be parity-forbidden. From this experiment, a limit on the parity mixed component contributing to the $1^+ \rightarrow 0^+$ α-decay, is derived as

$$\left| \frac{\text{amplitude (parity-odd)}}{\text{amplitude (parity-even)}} \right|^2 \lesssim 3 \times 10^{-13},$$

which is good evidence of parity conservation in the strong interaction amongst nucleons in the nucleus (the above is illustrated in figure 4.13).

Box 4a. α-emission in ${}^{238}_{92}U_{146}$

We consider a numerical example in detail for the emission of 4.2 MeV α-particles from ${}^{238}_{92}U_{146}$. We neglect, however, recoil effects since

$$m = \frac{m_\alpha M_D}{m_\alpha + M_D} = 3.932 \text{ amu,}$$

and

$$T_D = \frac{m_\alpha}{M_D} \cdot T_\alpha = 0.07 \text{ MeV.}$$

The relative α-particle velocity is obtained as

$$v \simeq \left(\frac{2Q_\alpha}{m}\right)^{1/2} = 1.42 \times 10^7 \text{ ms}^{-1}.$$

For the radius, we use $R = r_0 \cdot A^{1/3}$ (with $r_0 = 1.4$ fm), so for the daughter nucleus ${}^{234}_{90}Th_{144}$, we have

$$R = 8.6 \text{ fm.}$$

If we calculate the collision frequency of α-particles at the barrier we take $v_{in} \simeq v$ and derive the result

$$\frac{v_{in}}{R} \simeq 1.7 \times 10^{21} \text{ s}^{-1}.$$

The first term in the exponent for λ_α becomes

$$\frac{-Z_D e^2}{\epsilon_0 \hbar v} = \frac{90 \times (1.6 \times 10^{-19})^2}{8.8542 \times 10^{-12} \times 1.054 \times 10^{-34} \times 1.42 \times 10^7} = -173$$

(using $e = 1.6 \times 10^{-19}$ C; $\hbar = 1.054 \times 10^{-34}$ J s; $\epsilon_0 = 8.8542 \times 10^{-12}$ C^2/(N m^2)).
In a similar way, we obtain for the second term of λ_α

$$\frac{4e}{\hbar}\left(\frac{Z_D m R}{\pi \epsilon_0}\right)^{1/2} = \frac{4 \times 1.6 \times 10^{-19}}{1.054 \times 10^{-34}}$$

$$\times \left(\frac{90 \times 4 \times 1.66 \times 10^{-27} \times 8.6 \times 10^{-15}}{3.14 \times 8.8542 \times 10^{-12}}\right)^{1/2} = 83.$$

So, we obtain for the sum of the exponents

$$P = e^{-90} \simeq 10^{-39},$$

and, correspondingly for λ_α,

$$\lambda_\alpha = \left(\frac{v_{in}}{R}\right) \cdot P = 1.7 \times 10^{21} \times 10^{-39} \text{ s}^{-1} = 1.7 \times 10^{-18} \text{ s}^{-1},$$

or

$$T(\alpha) = 4.1 \times 10^{17} \text{ s} \cong 1.3 \times 10^{10} \text{ y}.$$

The experimental half-life in this particular case is 0.45×10^{10} y, a remarkable agreement, taking all simplifications into account.

From the expression for B, the barrier height at the nuclear radius R, we obtain the result

$$B = \frac{z Z_D e^2}{4\pi \epsilon_0 R} = 482 \times 10^{-14} \text{ J} \cong 30 \text{ MeV}$$

(using 1 MeV $= 1.60 \times 10^{-13}$ J). Similarly, we get a result for the exit radial value b since

$$b = \frac{R \cdot B}{Q_\alpha} = 61.$$

This indicates, indeed, a thick barrier ($R = 8.6$ fm) and the approximation $R/b \ll 1$ is fulfilled well here. The expression for λ_α is very sensitive in its dependence on the nuclear radius R. Only a 2% change in R results in changes of λ_α by a factor 2. The known α-decay results are all consistent with the use of a value $r_0 \simeq 1.4$–1.5 fm in calculating the nuclear radius R.

Box 4b. Alpha-particle formation in the nucleus: shell-model effects

The A_α^2 reduced α-widths ($\equiv \delta_\alpha^2$) in Po and At isotopes (calculated by Mang 1960, 1964) are illustrated (figure 4b.1) for (a) the even–even Po nuclei, (b) the odd-mass Po nuclei and (c) the odd-mass At nuclei. The closed circles are calculated values, the open circles represent data points. The calculated values have been normalized to ^{210}Po. Variations over one-order of magnitude are clearly recognizable and can be well explained using the simple spherical shell-model. The nucleus ^{212}Po differs from ^{208}Pb by two neutrons and two protons outside

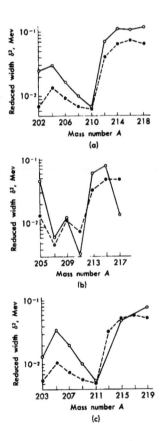

Figure 4b.1. Illustration of the α-decay width (expressed in equation (4.46) (in units MeV) as a function of the mass number A for the even–even Po nuclei (a), the even–odd Po nuclei (b) and the odd–even At nuclei (c)). In each case, the open circles are the experimental data points, the dots denote the calculated values from Mang (1960).

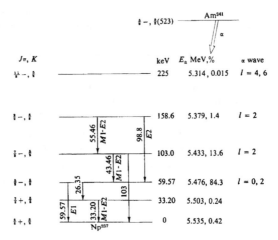

Figure 4b.2. Decay scheme of ^{241}Am, illustrating in particular the favoured α-decay to the band characterized by spin $\frac{5}{2}$. Both α-decay energies, intensities and probable orbital angular momentum value ℓ, taken away by the α-particle in the decay, are shown (taken from Stephens 1960).

of the $Z = 82$, $N = 126$ closed shells (see part C), whereas for ^{210}Po only two protons outside $Z = 82$ are present. In order to form the daughter nucleus ^{206}Pb and the α-particle, the neutron closed shell at $N = 126$ needs to be broken resulting in a large decrease of the α-reduced width A_α^2. This is the major effect used to explain the gross features in figure 4b.1. Finer details need more involved shell-model calculations.

In odd-A nuclei, the formation of the α-particle can put a stringent test on the shell-model orbits that the particular odd-particle may occupy. Whenever the unpaired nucleon remains in the same orbit in the initial and final nucleus, α-decay will proceed as it would in the even–even nucleus. If, though, the odd-particle is changing its state of motion, strong 'hindrance' can result (depending on the difference in the single-particle wavefunctions and on the particle's particular occupancy). So, it can happen that α-decay to an excited state proceeds faster than to the ground state, even overcoming Q_α differences that may be present (see figure 4b.2).

The odd-mass α-decay is presented schematically for $_{83}$Bi $\rightarrow$ $_{81}$Tl decays, in figure 4b.3. If the odd-particle (outside $Z = 82$) remains the odd-particle in the final nucleus, an excited state in Tl will be fed, though with large reduced α-width A_α^2. In proceeding to the odd-mass Tl ground state, the α-particle should have a more complicated structure and hindrance will result. This 'selective' method has been used, in particular by the group at IKS, Leuven to search for particle-hole excited 'intruder' states near closed shells. A very recent, but textbook, case is illustrated for the Po–Pb–Hg–(Bi–Tl) region in figures 4b.4 and 4b.5.

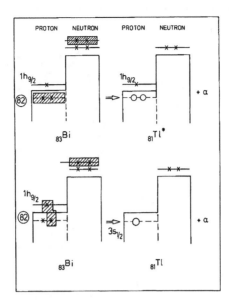

Figure 4b.3. Shell-model characterization of α-decay of $^{A}_{83}$Bi $\rightarrow$ $^{A-4}_{81}$Tl. The shaded boxes represent the α-particle being formed in the decay. This α-particle is preferentially formed between 2 protons and 2 neutrons in the same shell-model orbits (see chapter 9). Thus, the decay shown in the lower part will be strongly inhibited compared to α-decay leading to the $\frac{9}{2}^{-}$ excited state in the odd-mass Tl nuclei.

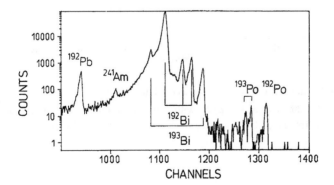

Figure 4b.4. Illustration of the α-spectrum in the reaction ^{20}Ne + ^{182}W at mass $A = 192$. Besides the α-lines from the isobars ^{192}Po, ^{192}Bi and ^{192}Pb, contamination lines from ^{241}Am are observed. The strongest line is the 6.06 MeV transition of ^{192}Bi. Two new lines at 6.245 MeV and 6.348 MeV can be assigned to the decay of ^{192}Bi, based on lifetime information (taken from Huyse 1991).

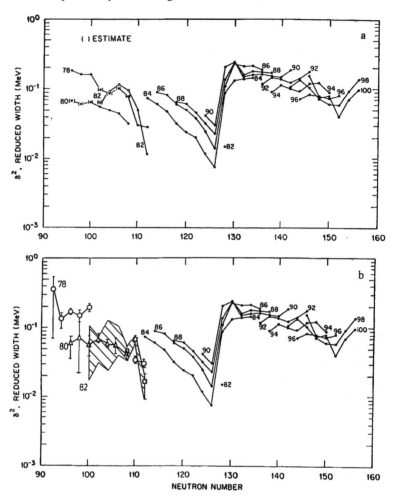

Figure 4b.5. An overview of the α-reduced width δ_α^2 in even–even nuclei near $^{208}_{82}\text{Pb}_{126}$. The reduced width is $\delta_\alpha^2 \equiv \lambda_\alpha \cdot h/P$ (with P the penetration factor). The figure (*a*) gives the data from Toth *et al* (1984) and (*b*) is taken from the thesis of Huyse (1991).

Chapter 5

Beta-decay: the weak interaction at work

5.1 The old beta-decay theory and the neutrino hypothesis

5.1.1 An historic introduction

Radioactivity was discovered in 1896 by Becquerel and it became clear within a few years that decaying nuclei could emit three types of radiation, called α, β and γ rays. An outstanding puzzle was related to the beta-decay process. The continuous energy distribution of beta decay electrons was a confusing experimental result in the 1920s. An example of such a beta spectrum is shown in figure 5.1. The energy distribution extends from zero to an upper limit (the endpoint energy) which is equal to the energy difference between the quantized initial and final nuclear states. A second, equally serious puzzle arose a few years later when it was realized that no electrons are present inside the nucleus. Where, then, do the electrons come from?

The first puzzle was solved by Pauli who suggested the existence of a new, very light uncharged and penetrating particle, the neutrino. In 1930 this was a revolutionary step. The neutrino carries the 'missing' energy. Conservation of electric charge requires the neutrino to be electrically neutral and angular momentum conservation and spin statistics considerations in the decay process require the neutrino to behave like a fermion of spin $\frac{1}{2}$. Experiment shows that in beta-decay processes, two types of neutrinos can be emitted. These are called the neutrino (ν) and antineutrino ($\bar{\nu}$).

To make the above arguments more quantitative, we discuss the beta-decay of the free neutron

$$n \rightarrow p + e^- + \bar{\nu}_e. \tag{5.1}$$

As discussed in chapter 1, section 1.8, we define the Q-value as the difference between the initial and final *nuclear* mass energies, i.e.

$$Q = (M_n - M_p - M_{e^-} - M_{\bar{\nu}_e})c^2, \tag{5.2}$$

131

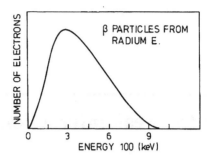

Figure 5.1. Simple example of a beta spectrum. This figure is taken from one of the classic papers, by Ellis and Wooster (1927) (reprinted with permission of the Royal Society).

(in what follows we denote the rest-mass of the electron $M_{e^-}c^2$ as m_0c^2). For the decay of the neutron at rest, this Q value equals

$$Q = T_p + T_{e^-} + T_{\bar{\nu}_e}. \tag{5.3}$$

For the moment, we ignore the proton recoil kinetic energy T_p which is only $\simeq 0.3$ keV. The electron and antineutrino will then share the decay energy (Q-value) which explains the observed electron spectrum shape. We also deduce that $Q \simeq (T_{e^-})_{max}$.

The measured maximum electron energy is 0.782 ± 0.013 MeV. From measured proton, neutron and electron masses, one derives

$$\begin{aligned} Q &= M_n c^2 - M_p c^2 - m_0 c^2 - M_{\bar{\nu}_e} c^2 \\ &= 939.573 \text{ MeV} - 938.280 \text{ MeV} - 0.511 \text{ MeV} - M_{\bar{\nu}_e} c^2 \\ &= 0.782 \text{ MeV} - M_{\bar{\nu}_e} c^2. \end{aligned} \tag{5.4}$$

So, within the precision of the measurement (13 keV), we can take the neutrino as massless (see section 5.4). Conservation of linear momentum can be used to identify beta-decay as a three-body process but this requires momenta measurements of the recoiling nucleus in coincidence with the emitted electron, registering the momentum. These experiments are difficult because of the very low recoiling energy of the final nucleus after beta-decay. In figure 5.2, we illustrate this using a picture obtained in a cloud chamber for the ^{6}He decay process

$$^6_2\text{He}_4 \rightarrow {}^6_3\text{Li}_3 + e^- + \bar{\nu}_e,$$

(Csikay and Szalay 1957). Whatever the mass of the neutrino might be, the existence of the additional particle is required by these experiments, from momentum addition.

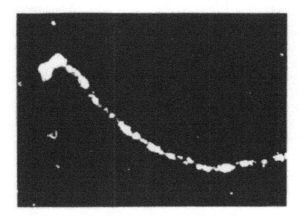

Figure 5.2. Recoil of a Li nucleus following the beta decay of ^{6}He. The lithium ion is at the left-hand side and the electron is the curved track. The photograph was taken in a low-pressure cloud chamber (taken from Csikay and Szalay 1957).

5.1.2 Energy relations and Q-values in beta-decay

In most decay processes, we shall use the notation T and E for kinetic and total energy, respectively. The masses denoted with the accent (M') give the nuclear mass whereas masses denoted by (M) give the total atomic mass. If we consider the neutrino massless, it moves with the speed of light like photons and thus

$$E_\nu = T_\nu = p_\nu c. \tag{5.5}$$

In describing the beta-decay processes, relativistic kinematics should be used in general. The nuclear recoil has such a low energy that non-relativistic methods apply.

We now consider the three basic beta-decay processes: β^--decay (n $\to$ p + e$^-$ + $\bar\nu_e$); β^+-decay (p $\to$ n + e$^+$ + ν_e) and electron capture EC (p + e$^-$ $\to$ n + ν_e) that may appear in a nucleus M_P (P: parent nucleus) with M_D (D: final or daughter nucleus) as the final nucleus.

(i) The Q_{β^-} value for the transmutation is

$$^A_Z X_N \to {}^A_{Z+1} Y_{N-1} + e^- + \bar\nu_e, \tag{5.6}$$

$$Q_{\beta^-} = T_{e^-} + T_{\bar\nu_e}$$
$$= M'_P c^2 - M'_D c^2 - m_0 c^2. \tag{5.7}$$

We can convert the nuclear masses into atomic masses, using the expression

$$M_P c^2 = M'_P c^2 + Z m_0 c^2 - \sum_{i=1}^{Z} B_i, \tag{5.8}$$

where the B_i represents the binding energy of the ith electron. So, one obtains for the $Q_{\beta-}$ value the result (neglecting the very small *differences* in electron binding energy between the daughter and parent atoms)

$$Q_{\beta-} = M_P c^2 - M_D c^2. \tag{5.9}$$

Thus, the β^- process is possible (exothermic with $Q_{\beta-} > 0$) with the emission of an electron and an antineutrino with positive kinetic energy whenever $M_P > M_D$.

(ii) The $Q_{\beta+}$ value for the transmutation

$$^A_Z X_N \rightarrow ^A_{Z-1} Y_{N+1} + e^+ + \nu_e, \tag{5.10}$$

is

$$\begin{aligned} Q_{\beta+} &= T_{e^+} + T_{\nu_e} \\ &= M'_P c^2 - M'_D c^2 - m_0 c^2. \end{aligned} \tag{5.11}$$

Using the same method as in equation 5.8 to transform nuclear masses in atomic masses, one obtains the result

$$Q_{\beta+} = M_P c^2 - (M_D c^2 + 2m_0 c^2). \tag{5.12}$$

So, the β^+ process has a threshold of $2m_0 c^2$ in order for the process to proceed spontaneously ($Q_{\beta+} > 0$).

(iii) A process, rather similar to β^+-decay by which the nuclear charge also decreases by one unit ($Z \rightarrow Z - 1$), is called electron capture, can also occur. An atomic electron is 'captured' by a proton, thereby transforming into a bound neutron and emitting a neutrino. This process leaves the final atom in an excited state, since a vacancy has been created in one of the inner electron shells. It is denoted by

$$^A_Z X_N + e^- \rightarrow ^A_{Z-1} Y_{N+1} + \nu_e, \tag{5.13}$$

and has a Q_{EC} value

$$Q_{EC} = M_P c^2 - (M_D c^2 + B_n), \tag{5.14}$$

with B_n, the electron binding energy of the nth electron ($n = $ K, L_I, L_{II}, L_{III}, M_I, ...) in the final atom. Here too, a constraint, albeit not very stringent, is present on the mass difference $M_P c^2 - M_D c^2$. Following the electron capture process, the vacancy created is very quickly filled as electrons from less bound orbitals make downward transitions thereby emitting characteristic x-rays. These processes will be discussed in more detail in section 5.2.

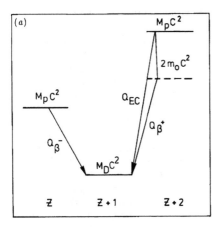
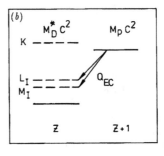

Figure 5.3. (*a*) The atomic masses for parent–daughter systems illustrating the energy relations in β^- and β^+ (EC) decay processes. The various Q-values are indicated with $2m_0c^2$ ($\simeq 1022$ keV) as a threshold for β^+ decay. (*b*) The mass relationships in electron capture between the parent and daughter atom. Since the final atom is left in an excited state $M_D^*c^2$, the various possibilities for capture in a $L_I, \ldots, M_I, \ldots$ configuration are schematically represented.

Table 5.1. (Taken from Krane, *Introductory Nuclear Physics* © 1987 John Wiley & Sons. Reprinted by permission.)

Decay	Type	Q (MeV)	T
$^{23}\text{Ne} \rightarrow {}^{23}\text{Na} + e^- + \bar{\nu}_e$	β^-	4.38	38 s
$^{99}\text{Tc} \rightarrow {}^{99}\text{Ru} + e^- + \bar{\nu}_e$	β^-	0.29	2.1×10^5 y
$^{25}\text{Al} \rightarrow {}^{25}\text{Mg} + e^+ + \nu_e$	β^+	3.26	7.2 s
$^{124}\text{I} \rightarrow {}^{124}\text{Te} + e^+ + \nu_e$	β^+	2.14	4.2 s
$^{15}\text{O} + e^- \rightarrow {}^{15}\text{N} + \nu_e$	EC	2.75	1.22 s
$^{41}\text{Ca} + e^- \rightarrow {}^{41}\text{K} + \nu_e$	EC	0.43	1.0×10^5 y

The various constraints on Q_{β^-}, Q_{β^+}, Q_{EC} are illustrated in figures 5.3(*a*) and (*b*). If beta-decay proceeds to an excited nuclear state as the final state, the appropriate Q-value has to be reduced from the ground state-to-ground state value.

In table 5.1 we present a few Q values for typical β-decay processes.

Table 5.2. Electron binding energies (in keV) in different shells (taken from Wapstra *et al* 1959).

	K	L_I	L_{II}	L_{III}	M_I	M_{II}	M_{III}	M_{IV}	M_V
37 Rb	15.201	2.067	1.866	1.806	0.324	0.250	0.240	0.114	0.112
38 Sr	16.107	2.217	2.008	1.941	0.359	0.281	0.270	0.137	0.134
39 Y	17.038	2.372	2.154	2.079	0.394	0.312	0.300	0.159	0.157
40 Zr	17.996	2.529	2.305	2.220	0.429	0.342	0.328	0.181	0.178
41 Nb	18.989	2.701	2.468	2.373	0.472	0.383	0.366	0.211	0.208
42 Mo	20.003	2.867	2.628	2.523	0.506	0.412	0.394	0.234	0.230
43 Tc	21.05	3.04	2.80	2.68	0.55	0.45	0.43	0.26	0.25
44 Ru	22.118	3.225	2.966	2.837	0.585	0.484	0.460	0.283	0.279
45 Rh	23.218	3.411	3.145	3.002	0.625	0.520	0.495	0.310	0.305
46 Pd	24.349	3.603	3.330	3.172	0.669	0.557	0.530	0.339	0.334
47 Ag	25.515	3.807	3.525	3.353	0.719	0.603	0.572	0.374	0.368
48 Cd	26.711	4.019	3.728	3.539	0.771	0.652	0.617	0.411	0.405
49 In	27.938	4.237	3.937	3.730	0.824	0.701	0.663	0.450	0.442
50 Sn	29.201	4.465	4.156	3.929	0.883	0.756	0.714	0.493	0.484
51 Sb	30.492	4.699	4.381	4.133	0.944	0.812	0.766	0.537	0.487
52 Te	31.815	4.939	4.613	4.342	1.006	0.869	0.818	0.583	0.572
53 I	33.171	5.190	4.854	4.559	1.074	0.932	0.876	0.633	0.621
54 Xe	34.588	5.453	5.101	4.782	1.15	1.00	0.941	0.67	0.67
55 Cs	35.984	5.721	5.359	5.011	1.216	1.071	1.003	0.738	0.724
56 Ba	37.441	5.994	5.624	5.247	1.292	1.141	1.066	0.795	0.780
57 La	38.931	6.270	5.896	5.490	1.368	1.207	1.126	0.853	0.838
58 Ce	40.448	6.554	6.170	5.729	1.440	1.276	1.188	0.906	0.889
59 Pr	41.996	6.839	6.446	5.969	1.514	1.340	1.246	0.955	0.935
60 Nd	43.571	7.128	6.723	6.209	1.576	1.405	1.298	1.001	0.978
61 Pm	45.19	7.42	7.02	6.46	1.65	1.47	1.36	1.05	1.03
62 Sm	46.843	7.739	7.313	6.717	1.723	1.542	1.420	1.108	1.080
63 Eu	48.520	8.055	7.620	6.980	1.803	1.615	1.483	1.164	1.134
64 Gd	50.221	8.379	7.930	7.243	1.881	1.691	1.546	1.217	1.186
65 Tb	51.990	8.713	8.252	7.515	1.968	1.772	1.617	1.277	1.242
66 Dy	53.777	9.044	8.580	7.790	2.045	1.857	1.674	1.332	1.294
67 Ho	55.601	9.395	8.912	8.067	2.124	1.924	1.743	1.387	1.347
68 Er	57.465	9.754	9.261	8.657	2.204	2.009	1.680	1.451	1.407
69 Tm	59.383	10.120	9.617	8.649	2.307	2.094	1.889	1.516	1.471
70 Yb	61.313	10.488	9.979	8.944	2.398	2.174	1.951	1.576	1.529
71 Lu	63.314	10.869	10.347	9.242	2.489	2.262	2.022	1.637	1.587
72 Hf	65.323	11.266	10.736	9.558	2.597	2.362	2.104	1.173	1.658
73 Ta	67.411	11.678	11.132	9.878	2.704	2.464	2.190	1.789	1.731
74 W	69.519	12.092	11.537	10.200	2.812	2.568	2.275	1.865	1.802
75 Re	71.673	12.524	11.957	10.533	2.929	2.677	2.364	1.947	1.880
76 Os	73.872	12.967	12.385	10.871	3.048	2.791	2.456	2.029	1.959
77 Ir	76.109	13.415	12.821	11.213	3.168	2.904	2.548	2.113	2.038
78 Pt	78.392	13.875	13.270	11.561	3.294	3.020	2.643	2.199	2.118
79 Au	80.726	14.355	13.735	11.291	3.428	3.151	2.745	2.294	2.208
80 Hg	83.119	14.843	14.214	12.287	3.564	3.282	2.848	2.389	2.298
81 Tl	85.531	15.349	14.699	12.659	3.707	3.419	2.959	2.487	2.391
82 Pb	88.015	15.873	15.210	13.046	3.862	3.568	3.079	2.596	2.495
83 Bi	90.536	16.396	15.719	13.426	4.007	3.705	3.187	2.695	2.588

Table 5.2. (Continued)

	K	L$_I$	L$_{II}$	L$_{III}$	M$_I$	M$_{II}$	M$_{III}$	M$_{IV}$	M$_V$
84 Po	93.11	16.94	16.24	13.82	4.16	3.85	3.30	2.80	2.69
85 At	95.74	17.49	16.79	14.22	4.32	4.01	3.42	2.91	2.79
86 Em	98.41	18.06	17.34	14.62	4.49	4.16	3.54	3.02	2.89
87 Fr	101.14	18.64	17.90	15.03	4.65	4.32	3.67	3.13	3.00
88 Ra	103.93	19.236	18.484	15.445	4.822	4.489	3.792	3.249	3.104
89 Ac	106.76	19.85	19.08	15.87	5.00	4.65	3.92	3.37	3.21
90 Th	109.648	20.463	19.691	16.299	5.182	4.822	4.039	3.489	3.331
91 Pa	112.60	21.105	20.314	16.374	5.367	5.002	4.175	3.612	3.441
92 U	115.610	21.756	20.946	17.168	5.549	5.182	4.303	3.726	3.551
93 Np	118.66	22.41	21.59	17.61	5.74	5.36	4.43	3.85	3.66
94 Pu	121.77	23.10	22.25	18.06	5.93	5.56	4.56	3.98	3.78
95 Am	124.94	23.80	22.94	18.52	6.14	5.75	4.70	4.11	3.90
96 Cm	128.16	24.50	23.63	18.99	6.35	5.95	4.84	4.24	4.02
97 Bk	131.45	25.23	24.34	19.47	6.56	6.16	4.99	4.38	4.15
98 Cf	134.79	25.98	25.07	19.95	6.78	6.37	5.13	4.51	4.28
99 Es	138.19	26.74	25.82	20.44	6.99	6.58	5.28	4.65	4.41
100 Fm	141.66	27.51	26.57	20.93	7.22	6.81	5.42	4.80	4.54

In table 5.2 we give the electron binding energies for nuclei with $37 \leq Z \leq 100$. These allow a precise determination of possible electron capture processes and the particular electron shells through which these can proceed (taken from Krane 1987).

5.2 Dynamics in beta-decay

5.2.1 The weak interaction: a closer look

One of the puzzles in understanding beta-decay (β^-, β^+) was the emission of particles (electron, positron, neutrino) that are not present in the atomic nucleus.

Fermi assumed that in the β^- process, the electron and antineutrino were created during the process of beta-decay (Fermi 1934). The act of creation is very similar to the process of photon emission in atomic and nuclear decay processes. By 1933 the quantum theory of radiation was well enough understood and Fermi constructed his theory of beta-decay after it. It was only with the fall of parity conservation in the weak interaction in 1957 that Fermi's theory had to be modified.

In figure 5.4(a), we show a diagram representing the β^--decay of the neutron. The analogy of this process with the electromagnetic interaction becomes clearer when we redraw it using the fact that antiparticles can be looked at as particles going backwards in time. The decay then appears as in figure 5.4(b).

In contrast to the electromagnetic interaction where a photon is exchanged between the charged particles (figure 5.5(a)), with zero rest mass ($m_\gamma = 0$) and an infinite interaction range, the β^--decay process appears to be short range. In quantum field theory, it can be shown that the range is inversely proportional

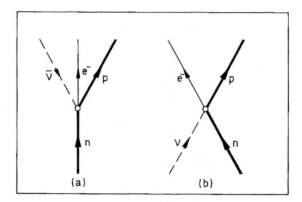

Figure 5.4. Example of neutron decay (*a*) and neutrino absorption (*b*). It is assumed that the absolute values of the matrix elements are equal. The diagrams give a pictorial description of the weak-interaction process.

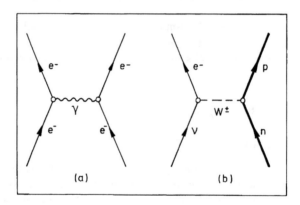

Figure 5.5. Comparison of the electromagnetic (*a*) and weak interaction vertices (*b*). (Taken from Frauenfelder and Henley (1991) *Subatomic Physics* © 1974. Reprinted by permission of Prentice-Hall, Englewood Cliffs, NJ.)

to the mass of the exchanged 'charge' carrier. Thus, a hypothetical particle, the intermediate boson ($W^{\pm}$) is exchanged. In this way, the symmetry aspects relating electromagnetic and weak interactions are introduced.

In Box 5a, we illustrate the recent work that led to the discovery of these intermediate bosons ($W^{\pm}$, Z^0) and to the construction of a unification in our understanding of the electromagnetic and weak interactions. This research at CERN is one of the most dramatic pieces of work where theory and state-of-the art particle detection techniques proved unification ideas (Salam 1980, Weinberg

1974, 1980, Glashow 1961, 1980) to be basically correct; ideas that have been developed on what is called the 'Standard Model'.

5.2.2 Time-dependent perturbation theory: the beta-decay spectrum shape and lifetime

In chapter 3 the general methods for evaluating the lifetimes of non-stationary states decaying through the interaction of certain residual interactions have been briefly outlined. We shall now apply the methods of time-dependent perturbation theory to evaluate various observables in the beta-decay process (T, beta spectrum shape, etc).

5.2.2.1 Phase space and two- and three-body processes

In the β^--decay process, where a neutron transforms into a proton in a nucleus $^A_Z X_N$ and an electron and an antineutrino are emitted, the transition rate can be determined from

$$\lambda = \frac{2\pi}{\hbar} |\langle \psi_f | H_{int} | \psi_i \rangle|^2 \frac{dn}{dE}(e^-, \bar{\nu}_e), \quad (5.15)$$

where the initial state ψ_i describes the wavefunction of the parent nucleus $\psi_P(\vec{r}_1, \vec{r}_2, \ldots, \vec{r}_A)$ and ψ_f the product wavefunction $\psi_D(\vec{r}_1, \vec{r}_2, \ldots, \vec{r}_A)$ $\psi_{e^-}(\vec{r}_{e^-}, Z)\psi_{\bar{\nu}_e}(\vec{r}_{\bar{\nu}_e})$. In both the parent and the daughter nucleus, A nucleons with coordinates $\vec{r}_1, \vec{r}_2, \ldots, \vec{r}_A$ are present but one of the neutrons has been changed into a proton through the interaction Hamiltonian, describing the decay process. At the same time, an electron, described by the wavefunction $\psi_{e^-}(\vec{r}_{e^-}, Z)$, is emitted from the nucleus. This wavefunction is the Coulomb function in the non-relativistic case. The antineutrino wavefunction $\psi_{\bar{\nu}_e}(\vec{r}_{\bar{\nu}_e})$ is given by a plane wave solution. By considering the interaction as a point interaction (see also in section 5.3) $g\delta(\vec{r}_p - \vec{r}_n)\delta(\vec{r}_n - \vec{r}_{e^-})\delta(\vec{r}_n - \vec{r}_{\bar{\nu}_e})$ with strength g, the integration in evaluating the transition matrix element is greatly simplified. We postpone this discussion to section 5.3 and concentrate on spectrum shape and lifetime; the value of λ can be rewritten as

$$\lambda = \frac{2\pi}{\hbar} |M_{fi}|^2 \cdot \frac{dn}{dE}(e^-, \bar{\nu}_e). \quad (5.16)$$

To evaluate both $|M_{fi}|^2$ and the density of final states

$$\frac{dn}{dE}(e^-, \bar{\nu}_e),$$

we use as a quantization volume a box normalized to a volume V for the electron and antineutrino wavefunctions.

In standard quantum mechanics texts, it is shown that, for a particle with one degree of freedom bound over the interval $(0, L)$, the number of levels n for a

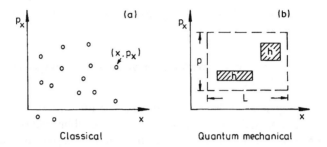

Figure 5.6. Comparison of the classical (*a*) and quantum mechanical (*b*) phase space extension for one-dimensional particle motion characterized by the coordinates (x, p_x). In the classical case, single points result whereas in the quantum mechanical situation the uncertainty principle involves 'volumes' of extension h.

particle with momentum in the interval $(0, p)$ is (figure 5.6(*b*))

$$n = \frac{L \cdot p}{2\pi\hbar}. \tag{5.17}$$

The three-dimensional case then becomes

$$n = \frac{1}{(2\pi\hbar)^3}\int d^3x \int d^3p,$$
$$n = \frac{V}{(2\pi\hbar)^3}\int d^3p, \tag{5.18}$$

with the density of states

$$\frac{dn}{dE} = \frac{V}{(2\pi\hbar)^3}4\pi p^2\frac{dp}{dE}. \tag{5.19}$$

For general relativistic motion where $E^2 = p^2c^2 + m^2c^4$ one gets $E\,dE = p\,dp \cdot c^2$, and

$$\frac{dn}{dE} = \frac{V}{(2\pi\hbar)^3}4\pi\frac{p^2}{c^2}\frac{E}{p},$$
$$\frac{dn}{dE} = \frac{4\pi V}{c^2(2\pi\hbar)^3}pE,$$

or

$$\frac{dn}{dE} = \frac{1}{2\pi^2}\frac{V}{c^3\hbar^3}E\sqrt{E^2 - m^2c^4}. \tag{5.20}$$

In the actual β^--decay process, an electron ($\vec{p}_{e^-}, E_{e^-}$), an antineutrino $\vec{p}_{\bar{\nu}_e}, E_{\bar{\nu}_e}$) and a recoiling nucleus $^A_{Z+1}Y_{N-1}(\vec{p}_Y, E_Y)$ contain the full kinematics (figure 5.7) such that in the three-body process, only two particles appear free and independent with the conditions

$$\vec{p}_{e^-} + \vec{p}_{\bar{\nu}_e} + \vec{p}_Y = 0$$
$$E = E_{e^-} + E_{\bar{\nu}_e} + E_Y. \tag{5.21}$$

The resulting density of states is then the product of the independent densities of final states of the electron and the neutrino and becomes

$$\frac{dn}{dE}(e^-, \bar{\nu}_e) = \frac{V^2}{(2\pi\hbar)^6} \frac{d}{dE} \int p_{e^-}^2 \, dp_e \, d\Omega_e p_{\bar{\nu}_e}^2 \, dp_{\bar{\nu}_e} \, d\Omega_{\bar{\nu}_e}. \tag{5.22}$$

The differentiation d/dE indicates how the integral changes under a variation of the total decay energy E. In order to evaluate the density

$$\frac{dn}{dE}(e^-, \bar{\nu}_e),$$

we need to specify the specific process we are interested in. If we evaluate the density of states for electrons emitted with an energy between E_{e^-} and $E_{e^-}+dE_{e^-}$ (independent of neutrino variable), E_{e^-} and consequently p_{e^-} are kept constant and the variation does *not* affect the electron observables. One obtains

$$\frac{dn}{dE}(e^-, \bar{\nu}_e) = \frac{V^2 \, d\Omega_{e^-} \, d\Omega_{\bar{\nu}_e}}{(2\pi\hbar)^6} p_{e^-}^2 \, dp_{e^-} p_{\bar{\nu}_e}^2 \frac{dp_{\bar{\nu}_e}}{dE}. \tag{5.23}$$

Since we use $E = E_{e^-} + E_{\bar{\nu}_e} + E_Y$ and neglect the very small nuclear recoil energy, for constant E_{e^-}, we obtain

$$\frac{d}{dE} = \frac{d}{dE_{\bar{\nu}_e}}.$$

Combining the above results, we obtain for the density of final states of an electron emitted with energy in the interval E_{e^-}, $E_{e^-}+dE_{e^-}$ and momentum in the interval p_{e^-}, $p_{e^-} + dp_{e^-}$ (integration is over the angles $d\Omega_{e^-}$ and $d\Omega_{\bar{\nu}_e}$)

$$\frac{dn}{dE}(e^-, \bar{\nu}_e) = \frac{V^2}{4\pi^4\hbar^6c^3} p_{e^-}^2 (E - E_{e^-})^2 \sqrt{1 - \frac{m_{\bar{\nu}_e}^2 c^4}{(E - E_{e^-})^2}} \, dp_{e^-}. \tag{5.24}$$

The above expression reduces to a much simpler form for $m_{\bar{\nu}_e} = 0$.

5.2.2.2 The beta spectrum shape

The partial decay probability, i.e. the probability of emitting an electron with energy in the interval E_{e^-}, $E_{e^-} +dE_{e^-}$ and momentum in the interval p_{e^-}, $p_{e^-} +$

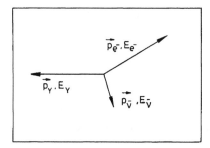

Figure 5.7. The energy and linear momenta $(E_{e^-}, \vec{p}_{e^-})$, $(E_{\bar{\nu}_e}, \vec{p}_{\bar{\nu}_e})$, $(E_Y, \vec{p}_Y)$ for the electron, antineutrino and recoiling nucleus in the β^--decay of a nucleus.

$\mathrm{d}p_{e^-}$, is

$$\Lambda(p_{e^-})\,\mathrm{d}p_{e^-} = \frac{2\pi}{\hbar}|M_{fi}|^2 \frac{V^2}{4\pi^4\hbar^6c^3}p_{e^-}^2(E - E_{e^-})^2$$

$$\times \sqrt{1 - \frac{m_{\bar{\nu}_e}^2 c^4}{(E - E_{e^-})^2}}\,\mathrm{d}p_{e^-}. \qquad (5.25)$$

The quantized volume of the normalized box will cancel in the electron and neutrino wavefunctions that appear in the matrix element $|M_{fi}|^2$ since, in general,

$$M_{fi} = \int \psi_D^*(\vec{r}_1, \vec{r}_2, \ldots, \vec{r}_A)\psi_{e^-}^*(\vec{r}_{e^-}, Z)\psi_{\bar{\nu}_e}^*(\vec{r}_{\bar{\nu}_e})$$

$$\times H_{int}\psi_P(\vec{r}_1, \vec{r}_2, \ldots, \vec{r}_A)\,\mathrm{d}\vec{r}_1\,\mathrm{d}\vec{r}_2 \ldots \mathrm{d}\vec{r}_A\,\mathrm{d}\vec{r}_{e^-}\,\mathrm{d}\vec{r}_{\bar{\nu}_e}. \qquad (5.26)$$

The neutrino wavefunction is written as

$$\frac{1}{\sqrt{V}}\exp(\mathrm{i}\vec{k}_{\bar{\nu}_e} \cdot \vec{r}_{\bar{\nu}_e}).$$

For the electron, a pure plane wave approximation is too crude and one has to take into account the distortion of the electron wavefunction caused by interaction with the electrostatic field of the nucleus. Quantitatively, the main effect is to alter the magnitude of the electron wavefunction at the origin such that

$$|\psi_{e^-}^*(0, Z)|^2 \simeq \frac{1}{V}\frac{2\pi\eta}{1 - e^{-2\pi\eta}} \cong \frac{1}{V}F(Z, p_{e^-}), \qquad (5.27)$$

with

$$\eta = \pm\frac{Ze^2}{4\pi\epsilon_0\hbar v_{e^-}},$$

and the positive (negative) sign is used for β^- (β^+)-decay, and v_{e^-} denotes the final velocity of the electron. The function $F(Z, p_{e^-})$ is called the Fermi function

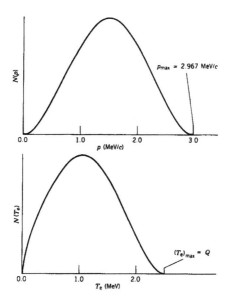

Figure 5.8. Electron energy and momentum distributions in beta-decay. These distributions are for a decay with $Q_B = 2.5$ MeV. (Taken from Krane, *Introductory Nuclear Physics* © 1987 John Wiley & Sons. Reprinted by permission.)

and slightly distorts the beta spectrum shape, as shown in figure 5.8. For small η values (small Z, large momentum p_{e^-}) the Fermi function approaches unity.

Combining all of the above contributions, the partial decay probability $\Lambda(p_{e^-})\,dp_{e^-}$ becomes (see figure 5.8 for a typical plot)

$$\Lambda(p_{e^-})\,dp_{e^-} = \frac{|M'_{fi}|^2}{2\pi^3\hbar^7 c^3} F(Z_{\mathrm{D}}, p_{e^-}) p_{e^-}^2 (E - E_{e^-})^2$$

$$\times \sqrt{1 - \frac{m_{\bar{\nu}_e}^2 c^4}{(E - E_{e^-})^2}}\, dp_{e^-}. \qquad (5.28)$$

In the expression (5.28), the matrix element M'_{fi} indicates that the precise structures of the electron and antineutrino wavefunction (i.e. normalization and Coulomb distortion) have been taken into account.

As an exercise, one can transform the expression $\Lambda(p_{e^-})\,dp_{e^-}$ to a differential in the electron energy $\Lambda(E_{e^-})\,dE_{e^-}$.

The expression (5.28) vanishes for $p_{e^-} = 0$ and $p_{e^-} = p_{\max}$, corresponding to a decay in which the electron carries all the energy, i.e. $E_{e^-} = E$. In figure 5.9, we show the momentum and kinetic energy ($E_{e^-} = T_{e^-} + m_0 c^2$) spectra for electrons and positrons emitted in the decay of ^{64}Cu (Evans 1955).

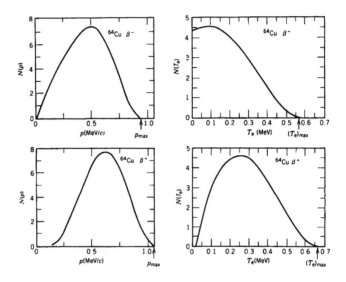

Figure 5.9. Momentum and kinetic energy distributions of electrons and positrons emitted in the decay of ^{64}Cu. A comparison with figure 5.8 illustrates the differences that arise from the Coulomb interactions within the daughter nucleus for electron and positron decay. (Taken from Evans 1955 *The Atomic Nucleus* © 1955 McGraw-Hill. Reprinted with permission.)

It is interesting to study the influence of the neutrino mass $m_{\bar{\nu}_e}$ on the beta spectrum shape. Therefore, we slightly change the representation of the electron decay into what is called a Fermi–Kurie plot. If the matrix element $|M'_{fi}|^2$ is totally independent of p_{e^-} and for vanishing neutrino mass $m_{\bar{\nu}_e}$, the quantity on the left-hand side of the expression

$$\sqrt{\frac{\Lambda(p_{e^-})}{p_{e^-}^2 \, F(Z_D, p_{e^-})}} \propto (E - E_{e^-})|M'_{fi}|, \tag{5.29}$$

should vary linearly with the total (or kinetic) electron energy.

The intercept with the energy (or momentum) axis is a convenient way to determine the decay endpoint energy (and so the Q-value). This procedure, in principle, applies to allowed transitions (see section 5.3). Forbidden transitions imply an extra $p_{e^-}(E_{e^-})$ dependence in the matrix element M'_{fi} that has to be taken into account. In figure 5.10, the Fermi–Kurie plot for a $0^+ \rightarrow 0^+$ allowed transition in the decay of ^{66}Ga is presented.

If the neutrino mass is assumed to be small but non-zero, interesting deviations from a straight line appear as exemplified for tritium β^--decay with $m_{\bar{\nu}_e} = 0$ and $m_{\bar{\nu}_e} = 30$ eV. A vertical asymptotical limit appears now at

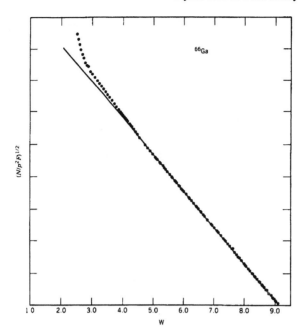

Figure 5.10. Fermi–Kurie plot (see equation (5.29)) for the allowed $0^+ \rightarrow 0^+$ beta-decay in ^{66}Ga. The horizontal axis gives the relativistic total energy $E_{e-} = T_{e-} + m_0c^2$ in units of m_0c^2 (called w). The deviations, for small energies, from a straight line arise from low-energy electron scattering within the radioactive source (from Camp and Langer 1963).

the maximum electron energy for $m_{\bar{\nu}_e} = 3$ eV as illustrated in figure 5.11. The analytical study of the Fermi–Kurie plot can be carried out by using equation (5.29), which contains small, but specific corrections implied by a neutrino mass.

5.2.2.3 Total half-life in beta-decay

The total decay probability and thus the half-life $T(\beta^-)$ can be derived by integrating over all possible momenta p_{e-} (or all possible E_{e-}) arising from β^--decay. We use $m_{\bar{\nu}_e} = 0$ and so

$$\lambda_{\beta^-} = \int_0^{p_{e-}\text{(max)}} \frac{|M'_{fi}|^2}{2\pi^3 \hbar^7 c^3} F(Z_D, p_{e-}) p_{e-}^2 (E - E_{e-})^2 \, dp_{e-}. \qquad (5.30)$$

We evaluate the integral by transforming the momentum (p_{e-}) and energy (E_{e-}) into reduced variables, i.e. $\eta \equiv p_{e-}/m_0c$; $w = E_{e-}/m_0c^2 = T_{e-}/m_0c^2 + 1$. The

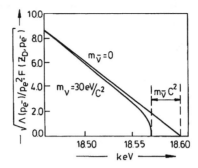

Figure 5.11. Illustration of the Fermi–Kurie plot for ^{3}H beta-decay to ^{3}He, in the hypothetical case of decay with a non-vanishing neutrino mass of $m_{\bar{\nu}_e}c^2 = 30$ eV and the zero-mass situation $m_{\bar{\nu}_e}c^2 = 0$ eV.

relativistic energy–mass relation then becomes

$$w^2 = \eta^2 + 1,$$

and

$$w \, dw = \eta \, d\eta.$$

Thus, the transition probability becomes

$$\lambda_{\beta^-} = \frac{m_0^5 c^4}{2\pi^3 \hbar^7} \int_1^{w_0} F\left(Z_D, \sqrt{w^2 - 1}\right) |M'_{fi}|^2 \sqrt{w^2 - 1}(w_0 - w)^2 w \, dw, \quad (5.31)$$

where w_0 is the reduced, maximum electron energy. If we take out the strength g of the weak interaction, a reduced matrix element $\bar{M}'_{fi}$ remains. The integral has to be evaluated, in general, numerically and is called the f-function, which still has a dependence on Z_D and w_0(or E)

$$f(Z_D, w_0) = \int_1^{w_0} F\left(Z_D, \sqrt{w^2 - 1}\right) \sqrt{w^2 - 1}(w_0 - w)^2 w \, dw. \quad (5.32)$$

Finally, this leads to the expression for the half-life

$$fT = 0.693 \frac{2\pi^3 \hbar^7}{g^2 m_0^5 c^4 |\bar{M}'_{fi}|^2}. \quad (5.33)$$

Before discussing this very interesting result we first point out that the f-function can be very well approximated for low Z_D and large beta-decay end-point energies w_0 ($w_0 \gg 1$), by the expression $f(Z_D, w_0) \sim w_0^5/5$. In figure 5.12 we show the quantity $f(Z_D, w_0)$ for both electrons and positrons.

From the expression for the fT value, it is possible to determine the strength g of the beta-decay process, if we know how to determine the reduced matrix

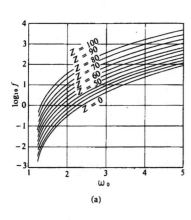 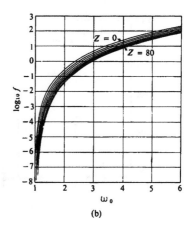

Figure 5.12. The quantity $\log_{10} f(Z_D, w_0)$ as a function of atomic number and total energy w_0 for (*a*) electrons and (*b*) positrons. (Taken from Feenberg and Trigg 1950.)

element $\bar{M}'_{fi}$ in a given beta-decay and the half-life. As we discuss in section 5.3; superallowed $0^+ \rightarrow 0^+$ transitions have $\bar{M}'_{fi} = \sqrt{2}$ and so the fT values should all be identical. Within experimental error, these values are indeed almost identical (table 5.3 and figure 5.13) and the β^--decay strength constant is

$$g = 0.88 \times 10^{-4} \text{ MeV fm}^3.$$

We can make this coupling strength more comparable to other fundamental constants by expressing it into a dimensionless form. Using the simplest form of a combination of the fundamental constants $m, \hbar, c$ we have

$$G = \frac{g}{m^i \hbar^j c^k}, \qquad (5.34)$$

which becomes dimensionless, when $i = -2$, $j = 3$, $k = -1$. Thus, the derived strength is

$$G = g \cdot \frac{m^2 c}{\hbar} = 1.026 \times 10^{-5}.$$

For comparison, the constants describing the pion–nucleon interaction in the strong interaction, the electromagnetic coupling strength $(e^2/\hbar c)$ and the gravitational force strength can be ranked, in decreasing order

Pion–nucleon (strong force) strength	1
Electromagnetic strength	$\frac{1}{137}$
Weak (beta-decay) coupling strength	10^{-5}
Gravitational coupling strength	10^{-39}

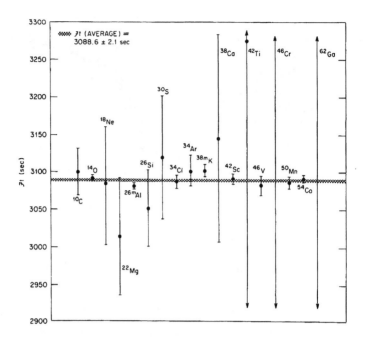

Figure 5.13. Superallowed fT values for the various decays. The fT average of 3088.6 ± 2.1 s is also indicated (taken from Raman *et al* 1975).

The adjective 'weak' here means weak relative to the strong and electromagnetic interactions (see the introductory chapter). The theory of beta-decay, in the above discussion is mainly due to Fermi, and has been considered in order to derive the beta spectrum shape and total half-life. In the next section, we shall discuss the implications of the matrix element $\bar{M}'_{fi}$ and the way that this leads to a set of selection rules to describe the beta-decay process.

5.3 Classification in beta-decay

5.3.1 The weak interaction: a spinless non-relativistic model

We have obtained a simple expression relating the fT value to the nuclear matrix element. Substituting the values of the constants $\hbar, g, m_0, c$, we obtain the result

$$fT \cong \frac{6000}{|\bar{M}'_{fi}|^2}. \tag{5.35}$$

From a study of the matrix element $\bar{M}'_{fi}$, we can obtain a better insight in the 'reduced' half-life for the beta-decay process, and at the same time a classification of various decay selection rules.

Table 5.3. fT values for superallowed $0^+ \rightarrow 0^+$ transitions. (Taken from Krane, *Introductory Nuclear Physics* © 1987 John Wiley & Sons. Reprinted by permission.)

Decay	fT (s)
$^{10}\text{C} \rightarrow {}^{10}\text{B}$	3100 ± 31
$^{14}\text{O} \rightarrow {}^{14}\text{N}$	3092 ± 4
$^{18}\text{Ne} \rightarrow {}^{18}\text{F}$	3084 ± 76
$^{22}\text{Mg} \rightarrow {}^{22}\text{Na}$	3014 ± 78
$^{26}\text{Al} \rightarrow {}^{26}\text{Mg}$	3081 ± 4
$^{26}\text{Si} \rightarrow {}^{26}\text{Al}$	3052 ± 51
$^{30}\text{S} \rightarrow {}^{30}\text{P}$	3120 ± 82
$^{34}\text{Cl} \rightarrow {}^{34}\text{S}$	3087 ± 9
$^{34}\text{Ar} \rightarrow {}^{34}\text{Cl}$	3101 ± 20
$^{38}\text{K} \rightarrow {}^{38}\text{Ar}$	3102 ± 8
$^{38}\text{Ca} \rightarrow {}^{38}\text{K}$	3145 ± 138
$^{42}\text{Sc} \rightarrow {}^{42}\text{Ca}$	3091 ± 7
$^{42}\text{Ti} \rightarrow {}^{42}\text{Si}$	3275 ± 1039
$^{46}\text{V} \rightarrow {}^{46}\text{Ti}$	3082 ± 13
$^{46}\text{Cr} \rightarrow {}^{46}\text{V}$	2834 ± 657
$^{50}\text{Mn} \rightarrow {}^{50}\text{Cr}$	3086 ± 8
$^{54}\text{Co} \rightarrow {}^{54}\text{Fe}$	3091 ± 5
$^{62}\text{Ga} \rightarrow {}^{62}\text{Zn}$	2549 ± 1280

Depicting the nucleus (see figure 5.14) with its A nucleons at the coordinates $(\vec{r}_1, \vec{r}_2, \ldots, \vec{r}_A)$, the β^--decay transforms a specific neutron into a proton accompanied by the emission of an electron and a antineutrino. We assume a system of non-relativistic spinless nucleons (zeroth-order approximation) and a point interaction represented by the Hamiltonian

$$H_{\text{int}} = g\delta(\vec{r}_n - \vec{r}_p)\delta(\vec{r}_n - \vec{r}_{e^-})\delta(\vec{r}_n - \vec{r}_{\bar{\nu}_e})(\hat{O}(\text{n} \rightarrow \text{p})), \quad (5.36)$$

where the operator $\hat{O}$ (n $\rightarrow$ p) changes a neutron into a proton. The initial, nuclear wavefunction for the nucleus $^A_Z X_N$ can be depicted as $\psi_P(\vec{r}_{p,1}, \vec{r}_{p,2}, \ldots, \vec{r}_{p,Z}; \vec{r}_{n,1}, \ldots, \vec{r}_{n,N})$ and that of the daughter by $\psi_D(\vec{r}_{p,1}, \vec{r}_{p,2}, \ldots, \vec{r}_{p,Z}, \vec{r}_{p,Z+1}; \vec{r}_{n,2}, \ldots, \vec{r}_{n,N})$.

Thus, we consider a process in which the neutron with coordinate $\vec{r}_{n,1}$ is transformed into a proton with coordinate $\vec{r}_{p,Z+1}$. The reduced matrix element (taking out the Coulomb factor from the electron wavefunction) becomes

$$M'_{fi} = \int \psi_D^*(\vec{r}_{p,1}, \ldots, \vec{r}_{p,Z+1}; \vec{r}_{n,2}, \ldots, \vec{r}_{n,N})\psi_{e^-}^*(\vec{r}_{e^-})\psi_{\bar{\nu}_e}^*(\vec{r}_{\bar{\nu}_e}) \cdot H_{\text{int}}$$
$$\times \psi_P(\vec{r}_{p,1}, \ldots, \vec{r}_{p,Z}; \vec{r}_{n,1}, \ldots, \vec{r}_{n,N})\,d\vec{r}_{e^-}\,d\vec{r}_{\bar{\nu}_e}\,d\vec{r}_i. \quad (5.37)$$

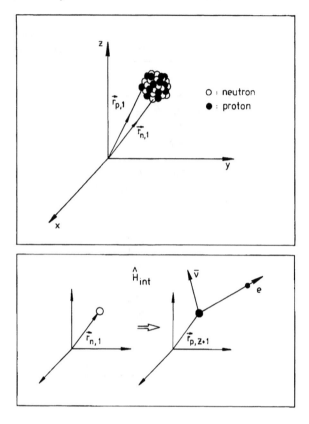

Figure 5.14. Schematic illustration of the neutron-to-proton transition, accompanied by the emission of an electron, antineutrino (e^-, $\bar{\nu}_e$) pair, in a non-relativistic spinless model description of the atomic nucleus. The coordinates of all nucleons are indicated by $\vec{r}_{p,i}$ (for protons) and $\vec{r}_{n,j}$ (for neutrons). In the lower part, the transformation of a neutron at coordinate $\vec{r}_{n,1}$ into a proton at coordinate $\vec{r}_{p,Z+1}$ is illustrated through the action of the weak interaction $\hat{H}_{int}$ as a zero-point interaction.

The integration is over all nucleonic coordinates and is denoted by $d\vec{r}_i$. The coordinates of the $A-1$ other nucleons remain unchanged, and using a simple single-particle product nuclear wavefunction, orthogonality simplifies the final expression very much. This means we use A-nucleon wavefunctions like

$$\psi_P(\vec{r}_i) = \prod_{i=1}^{Z} \varphi_P(\vec{r}_{p,i}) \cdot \prod_{j=1}^{N} \varphi_P(\vec{r}_{n,j})$$

$$\psi_D(\vec{r}_i) = \prod_{i=1}^{Z+1} \varphi_D(\vec{r}_{p,i}) \cdot \prod_{j=2}^{N} \varphi_D(\vec{r}_{n,j}), \qquad (5.38)$$

and the above matrix element, reduces to the form

$$M'_{fi} = g \int \varphi_D^*(\vec{r})_p \psi_{e^-}^*(\vec{r}) \psi_{\bar{\nu}_e}^*(\vec{r}) \hat{O}(n \to p) \varphi_P(\vec{r})_n \, d\vec{r}, \tag{5.39}$$

with the subscript $\varphi(\vec{r})_\rho$ ($\rho \equiv$ n, p) indicating a neutron or a proton.

Considering plane waves to describe the outgoing electron and antineutrino, then

$$M'_{fi} = g \int \varphi_D^*(\vec{r})_p e^{i(\vec{k}_{e^-} + \vec{k}_{\bar{\nu}_e}) \cdot \vec{r}} \hat{O}(n \to p) \varphi_P(\vec{r})_n \, d\vec{r}. \tag{5.40}$$

In most low-energy beta-decay processes, the electron and antineutrino wavelengths are small compared to the typical nuclear radius, i.e. R/λ_{e^-} and $R/\lambda_{\bar{\nu}_e} \ll 1$ and a series expansion can be performed, generating a series of possible contributions

$$M'_{fi} = g \left[\int \varphi_D^*(\vec{r})_p \hat{O}(n \to p) \varphi_P(\vec{r})_n \, d\vec{r} + i(\vec{k}_{e^-} + \vec{k}_{\bar{\nu}_e}) \right.$$
$$\left. \cdot \int \varphi_D^*(\vec{r})_p \vec{r} \hat{O}(n \to p) \varphi_P(\vec{r})_n \, d\vec{r} + \cdots \right]. \tag{5.41}$$

The first term means that a neutron is transformed into a proton and the integral just measures the overlap between both nuclear, single-particle wavefunctions. This term is only present if the parities of both the parent and daughter single-particle wavefunctions are identical. For the second term, because of the $\vec{r}$ factor, the wavefunctions must have opposite parity, etc . Thus, a parity selection rule appears.

Angular momentum selection can also be deduced from the structure of M'_{fi}, using the expansion of a plane wave into spherical harmonics, i.e.

$$e^{i(\vec{k}_{e^-} + \vec{k}_{\bar{\nu}_e}) \cdot \vec{r}} = \sum_{L,M} (4\pi) i^L j_L(kr) Y_L^M (\widehat{k_{e^-} + k_{\bar{\nu}_e}}) \cdot Y_L^{M^*}(\hat{r}), \tag{5.42}$$

where $k \equiv |\vec{k}_{e^-} + \vec{k}_{\bar{\nu}_e}|$ and $\hat{r} \equiv (\theta_r, \varphi_r)$, denote the angular variables. This gives rise to particular matrix elements of the type

$$\int \varphi_D^*(\vec{r})_p \hat{O}(n \to p) Y_L^{M^*}(\hat{r}) \varphi_P(\vec{r})_n j_L(kr) \, d\vec{r}. \tag{5.43}$$

The wavefunctions describing the single-particle motion in an average potential for both the parent (P) and daughter (D) nucleus separate into a radial and an angular part (neglecting the spin part, at present) as

$$\varphi_D(\vec{r}) = R_D(r) Y_{L_D}^{M_D}(\hat{r}) \qquad \varphi_P(\vec{r}) = R_P(r) Y_{L_P}^{M_P}(\hat{r}). \tag{5.44}$$

This reduces the particular L-matrix element into a product of a pure radial and an angular part, i.e.

$$\int Y_{L_D}^{M_D^*}(\hat{r}) Y_L^{M^*}(\hat{r}) Y_{L_P}^{M_P}(\hat{r}) \, d\Omega \cdot \int R_D(r) j_L(kr) R_P(r) r^2 \, dr. \tag{5.45}$$

The first part, using the Wigner–Eckart theorem (Heyde 1991) then immediately leads to angular momentum addition, such that $\vec{L}_P = \vec{L}_D + \vec{L}$, where $\vec{L}$ denotes the angular momentum carried by the $(e^-, \bar{\nu}_e)$ pair away from the nucleus.

Combining all the above results, one finally obtains

$$|\bar{M}_{fi}|^2 = \sum_L |\bar{M}'^L_{fi}|^2. \tag{5.46}$$

For a given initial state L_P, π_P and final state L_D, π_D, the selection rules, are to get a particular L-component contributing to the beta-decay process

$$\pi_P = \pi_D(-1)^L \qquad \vec{L}_P = \vec{L}_D + \vec{L}. \tag{5.47}$$

The relative values of the various L-terms, however, diminish very fast with increasing L because of the presence of the spherical Bessel function in the radial integral. A simple, more intuitive, argument indicates that each higher term reduces by a factor $(R/\lambda_{e^-}(\text{max}))^2$, the transition probability.

This indicates that for parity and angular momentum conservation, only the lowest L multipoles ($L = 0, 1$) will contribute in a significant way, except when a large difference occurs between the spins of initial and final states.

5.3.2 Introducing intrinsic spin

Nucleons, as well as the emitted electron and antineutrino, are fermions, so the spin degree of freedom has to be taken into account too. This means that in the weak-interaction operator, a term that can induce changes in the intrinsic spin orientation between the neutron and proton should be present. Moreover, the wavefunctions now become two-component (spin-up and spin-down) spinors.

In the matrix element $\bar{M}'_{fi}$, terms like

$$\bar{M}'_{fi} = \int \sum \psi^*_D \hat{\sigma} \psi_P \cdot \psi^*_{e^-} \hat{\sigma} \psi_{\nu_e} \, d\vec{r}, \tag{5.48}$$

occur. Here $\int \sum$ means integrating over the nuclear coordinates and summing over the possible intrinsic spin orientations. The corresponding selection rules becomes

$$\vec{J}_P = \vec{J}_D + \vec{J}_\beta \qquad (\vec{J}_\beta = \vec{L}_\beta + \vec{S}_\beta) \qquad \pi_P = \pi_D(-1)^{L_\beta}, \tag{5.49}$$

where $L_\beta(S_\beta)$ means, the angular momentum carried away by orbital (intrinsic) spin. Here $\vec{S}_\beta = \vec{0}, \vec{1}$ and the total matrix element can be obtained as a sum

$$|\bar{M}'_{fi}|^2 = \sum_{L_\beta, S_\beta} |\bar{M}'^{L_\beta, S_\beta}_{fi}|^2, \tag{5.50}$$

where, again, angular momentum and parity selection as well as the rapid decrease in importance of the various terms in the sum as L_β increases reduces the number of contributing terms.

In better, fully relativistic treatment, all wavefunctions become four-component spinors. Thereby the number of weak interaction terms appearing in the Hamiltonian greatly increases. In addition to the scalar (1) and spin ($\hat{\sigma}$) contributions, tensor and, also pseudoscalar and pseudovector terms can now contribute. Such terms give rise to parity violating contributions. A systematic development of relativistic beta-decay theory has to await a study of relativistic quantum mechanics.

5.3.3 Fermi and Gamow–Teller beta transitions

The strength of beta-decay differs for $S_\beta = 0$ or 1 processes. The first ones are called Fermi transitions, the latter Gamow–Teller and have strengths g_F and g_{GT}, respectively. A transition where both Fermi and Gamow–Teller processes can contribute is described by, in lowest order, the matrix element

$$|\bar{M}'_{fi}|^2 = |\bar{M}'_{fi}(F)|^2 + \frac{g_{GT}^2}{g_F^2}|\bar{M}'_{fi}(GT)|^2. \tag{5.51}$$

The best determination of the coupling strengths leads to a value of $g_{GT}/g_F = -1.259 \pm 0.004$.

The large dependence of the fT value on the matrix element $|\bar{M}'_{fi}|^2$, leads to the grouping of the various known beta-decay transition rates in terms of their log fT values.

For a pure Fermi transition ($S_\beta = 0$, $L_\beta = 0$), i.e. with the angular momentum restriction $\Delta J = 0$ and no change of parity, $|\bar{M}'_{fi}(F)|^2 = 1$. For a pure Gamow–Teller transition ($S_\beta = 1$, $L_\beta = 0$), i.e. with angular momentum change $\Delta J = 1$ and no change of parity, $|\bar{M}'_{fi}(GT)|^2 = 3$. The latter result is obtained by summing over all possible spin $\frac{1}{2}$ orientations, using the spin $\frac{1}{2}$ eigenvectors (spinors), i.e.

$$\sum_{\mu, m_f} |\langle \chi_{1/2}^{m_f} | \hat{\sigma}_\mu | \chi_{1/2}^{m_i} \rangle|^2 = 3. \tag{5.52}$$

The $0^+ \to 0^+$ transitions are uniquely allowed Fermi and are ideally suited to determine $g_F = g$, the weak-coupling strength because of the relation between measured T values and the pure Fermi matrix element. Examples are given in table 5.3.

A well-known, typical case is the $^{14}_8O_6 \to {}^{14}_7N_7 + e^+ + \nu_e$ decay, to the 0^+ excited state in ^{14}N at 2.311 MeV. The particular shell-model structure describing the 0^+ states in both nuclei is known, two protons outside an inert $^{12}_6C_6$ core can contribute to the decay process, we obtain the result

$$|\bar{M}'_{fi}(F; 0^+ \to 0^+)|^2 = 2.$$

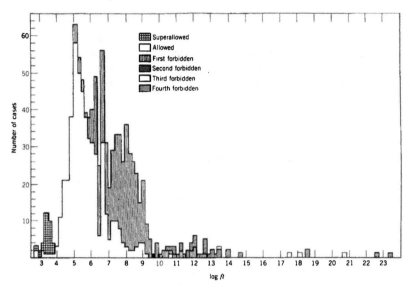

Figure 5.15. Systematics of experimental log ft values. (Taken from Meyerhof 1967 *Elements of Nuclear Physics* © 1967 McGraw-Hill. Reprinted with permission.)

5.3.4 Forbidden transitions

Allowed Fermi and Gamow–Teller transitions have a maximum spin change of one unit and *no* change in parity of the wavefunction.

Transitions in which parity changes occur and which have bigger spin changes have been observed. Such transitions are called forbidden because of the large reduction in transition probability compared to allowed cases. This comes from the $j_L(kr)$ dependence for an L-forbidden transition. For small values of the argument kr, which is usually the case in low-energy beta transitions

$$j_L(kr) \propto \frac{(kr)^L}{(2L+1)!!}. \tag{5.53}$$

So, for an $E = 1$ MeV transition in a nucleus with radius $R = 5$ fm, the reduction in transition probability is $\simeq 10^4$. One then speaks of first ($L_\beta = 1$), second ($L_\beta = 2$), ..., L_β forbidden beta transitions with the log fT value increasing by about 4 units in each step. Typical log fT values are given in table 5.4 and a more extensive distribution of log fT values is shown in figure 5.15.

Table 5.5 contains a summary of how to classify the various beta-decay processes.

Table 5.4.

Transition type	$\log fT$
Superallowed	2.9–3.7
Allowed	4.4–6.0
First forbidden	6–10
Second forbidden	10–13
Third forbidden	>15

Table 5.5.

Classification in β-decay

$\vec{J}_P = \vec{J}_D + \vec{L}_\beta + \vec{S}_\beta$

$\pi_P = \pi_D(-1)^{L_\beta}$

Allowed transitions	$\vec{L}_\beta = \vec{0}$
1st forbidden transitions	$\vec{L}_\beta = \vec{1}$
2nd forbidden transitions	$\vec{L}_\beta = \vec{2}$
Fermi transitions	$\vec{S}_\beta = \vec{0}$
Gamow–Teller transitions	$\vec{S}_\beta = \vec{1}$

I. Allowed transitions ($\vec{L}_\beta = 0, \pi_P = \pi_D$)

Fermi-type ($\vec{S}_\beta = \vec{0}$)

$\vec{J}_P = \vec{J}_D$

$|\Delta J| = 0$

$0^+ \to 0^+$: superallowed

Gamow–Teller type ($\vec{S}_\beta = \vec{1}$)

$\vec{J}_P = \vec{J}_D + \vec{1}$

$|\Delta J| = 0, 1$: no $0^+ \to 0^+$

$0^+ \to 1^+$: unique Gamow–Teller

II. 1st forbidden transitions ($\vec{L}_\beta = \vec{1}, \pi_P = -\pi_D$)

Fermi-type ($\vec{S}_\beta = \vec{0}$)

$\vec{J}_P = \vec{J}_D + \vec{1}$

$|\Delta J| = 0, 1$

no $0^- \to 0^+$

Gamow–Teller type ($\vec{S}_\beta = \vec{1}$)

$\vec{J}_P = \vec{J}_D + \underbrace{\vec{1} + \vec{1}}_{\vec{0}, \vec{1}, \vec{2}}$

3 types

(i) $|\Delta J| = 0$

(ii) $|\Delta J| = 0, 1$;   no $0^+ \to 0^+$

(iii) $|\Delta J| = 0, 1, 2$;   no $0^- \to 0^+$

no $1^+ \to 0^-$

no $\frac{1}{2}^+ \to \frac{1}{2}^-$

5.3.5 Electron-capture processes

If the energy difference between parent and daughter nuclei is less than $2m_0c^2$, then rather than β^+-decay, electrons from the atomic bound states (K, L, M, ...)

can be captured by a proton in the nucleus and a neutron is formed and neutrino emitted. The Q-value, neglecting small electron binding energy differences, is

$$Q_{EC} = M_P c^2 - M_D^* c^2$$
$$= M_P c^2 - (M_D c^2 + B_n). \qquad (5.54)$$

Electron capture leads to a vacancy being created in one of the strongest bound atomic states, and secondary processes will be observed such as the emission of x-rays and Auger electrons. Auger electrons are electrons emitted from one of the outer electron shells, and take away some of the remaining energy. The process could be compared with internal conversion of x-rays (see chapter 6). Whenever L-Auger electrons are emitted through K x-ray interval conversion, the L-electron kinetic energy will be given by

$$T_{e^-} = h\nu_K - B_L$$
$$= B_K - 2B_L. \qquad (5.55)$$

The electron capture decay constant λ_{EC} can be evaluated in a straightforward way by determining both the phase space factor, i.e. the level density for the final states dn/dE, and the transition matrix element M_{fi}. Since in electron capture a single particle (the neutrino) is emitted, the density of final states becomes

$$\frac{dn_{\nu_e}}{dE} = \frac{V \cdot p_{\nu_e}^2}{2\pi^2 \hbar^3} \frac{dp_{\nu_e}}{dE}, \qquad (5.56)$$

with p_{ν_e} the neutrino momentum. Neglecting the small recoil energy and taking the neutrino with zero mass $m_{\nu_e} = 0$; one derives easily that $dp_{\nu_e}/dE = 1/c$ and that $p_{\nu_e}^2 = E_{\nu_e}^2/c^2$. So, one obtains

$$\frac{dn_{\nu_e}}{dE} = \frac{V \cdot E_{\nu_e}^2}{2\pi^2 c^3 \hbar^3}. \qquad (5.57)$$

The transition matrix element M_{fi}, in a single-particle evaluation as was outlined in section 5.3.1, is

$$M_{fi} = g \cdot \int \psi_\nu^*(0) \psi_{e^-}(0) \psi_n^*(\vec{r}) \psi_p(\vec{r}) \, d^3\vec{r}. \qquad (5.58)$$

Capture is most likely for a 1s-state electron since the K-electron wavefunction at the origin is maximal and is given by

$$\psi_{e^-}(0) = \frac{1}{\sqrt{\pi}} \left(\frac{Zm_0 e^2}{4\pi \epsilon_0 \hbar^2} \right)^{3/2}. \qquad (5.59)$$

Combining the results obtained above, the K-electron capture transition probability is

$$\lambda_{EC} = \frac{E_{\nu_e}^2}{\pi^2 c^3 \hbar^4} \cdot g^2 |\bar{M}'_{fi}|^2 \left(\frac{Zm_0 c^2}{4\pi \epsilon_0 \hbar^2} \right)^3, \qquad (5.60)$$

and has a quadratic dependence on the total neutrino energy E_{ν_e}.

5.4 The neutrino in beta-decay

The neutrino is one of the most pervasive forms of matter in the universe, yet is also one of the most elusive. We have seen that it has no electric charge, little or no mass and behaves like a fermion with intrinsic spin $\hbar/2$. The neutrino was suggested by Pauli to solve a number of problems related to standard beta-decay processes. In this section we concentrate on a number of properties related to the neutrino (namely interaction with matter, neutrino mass, double beta-decay and different types of neutrinos).

In recent years (in particular since the second edition of this book was published) 2000–2004, very important breakthrough results concerning the small neutrino mass (and the subsequent implication for neutrino oscillation properties among the different electron, muon and tau-neutrino flavours) have been reported. Experiments have now solved, in an unambiguous way, the long-standing neutrino problem and experiments at Kamiokande (KamLAND) have shown, over a long baseline, the disappearance of the electron-type antineutrinos ($\bar{\nu}_e$) that are produced in large quantities in the β^- decay from neutron-rich fission products produced at nuclear reactor plants in Japan. Recently, a number of review papers on neutrino physics have appeared and I cite those that have appeared since the second edition. The reader can, thus, find a wealth of new information on the field of neutrino physics in (Zuber 1998, Wolfenstein 1999, Fisher *et al* 1999, Kirsten 1999, Haxton and Holstein 2000, Kajita and Totsuka 2001, Bemporad C *et al* 2002, Gonzales-Garcia and Nir 2003).

5.4.1 Inverse beta processes

Shortly after the publication of the Fermi theory of beta decay, Bethe and Peierls pointed out the possibility of inverse beta-decay. Here, the nucleus captures a neutrino or antineutrino and ejects an electron or positron. These processes can be depicted as

$$\begin{aligned}
{}^{A}_{Z}X_N + \nu_e &\rightarrow {}^{A}_{Z+1}Y_{N-1} + e^- \\
{}^{A}_{Z}X_N + \bar{\nu}_e &\rightarrow {}^{A}_{Z-1}Y_{N+1} + e^+.
\end{aligned} \tag{5.61}$$

The cross-sections for such inverse processes are expected to be extremely small because of the characteristic weakness of the beta interaction.

The first process we consider is the inverse of the neutron decay, i.e.

$$\bar{\nu}_e + p \rightarrow n + e^+. \tag{5.62}$$

As written, such a process satisfies the conservation of leptons (see later) in contrast to a process we shall discuss later, i.e.

$$\nu_e + p \rightarrow n + e^+.$$

The transition probability for the inverse reaction is given by

$$\lambda = \frac{2\pi}{\hbar}|M_{fi}|^2 \frac{dn}{dE}, \tag{5.63}$$

and the cross-section is defined as the transition probability divided by the flux of incoming neutrinos. So, we obtain

$$\sigma_c = \frac{\lambda}{c}V = \frac{2\pi V}{\hbar c}|M_{fi}|^2 \frac{dn}{dE}. \tag{5.64}$$

The essential difference resides within the density of final states of the emitted positrons. Neglecting recoil but treating the positron as a relativistic particle, we obtain

$$\begin{aligned}
\frac{dn_{e^+}}{dE} &= \frac{p_{e^+}^2}{2\pi^2\hbar^3}V\frac{dp_{e^+}}{dE} \\
&= \frac{p_{e^+}}{2\pi^2\hbar^3 c^2}V E_{e^+} \\
&= \frac{m_0^2 c V}{2\pi^2\hbar^3}w\sqrt{w^2-1},
\end{aligned} \tag{5.65}$$

where we have used the reduced energy $w = E/m_0 c^2$. The nuclear matrix element corresponds to that in the neutron beta-decay, a transition in which both the Fermi and Gamow–Teller contribution arise

$$|\bar{M}'_{fi}|^2 = |\bar{M}'_{fi}(F)|^2 + \frac{g_{GT}^2}{g_F^2}|\bar{M}'_{fi}(GT)|^2. \tag{5.66}$$

We obtain the cross-section

$$\begin{aligned}
\sigma_c &= \frac{g_F^2 m_0^2}{\pi\hbar^4}\left(|\bar{M}'_{fi}(F)|^2 + \left(\frac{g_{GT}}{g_F}\right)^2 |\bar{M}'_{fi}(GT)|^2\right)w\sqrt{w^2-1} \\
&= \left(\frac{\hbar}{m_0 c}\right)^2 G^2 \left\{|\bar{M}'_{fi}(F)|^2 + \left(\frac{g_{GT}}{g_F}\right)^2 |\bar{M}'_{fi}(GT)|^2\right\}w\sqrt{w^2-1},
\end{aligned} \tag{5.67}$$

where the Fermi coupling constant g_F^2 is expressed in terms of the dimensionless quantity G^2. The above total cross-section is obtained by integrating over all angles between the positron and antineutrino. Using a value of the matrix element $\simeq 5$, and using the numerical value of G, the calculated cross-sections (in units 10^{-44} cm^2) are given in Table 5.6.

For the capture reaction, $\bar{\nu}_e + p \rightarrow n + e^+$, the threshold energy is $E_{\bar{\nu}_e} = 18$ MeV. For $E_{\bar{\nu}_e} = 2.8$ MeV antineutrinos, one has positrons with an energy $E_{e^+} = 1$ MeV. A relation between cross-section and mean-free path is $\ell = 1/n \cdot \sigma_c$ where n is the number of nuclei per cm^3. For protons in water $n \simeq 3 \times 10^{22}$, giving a value of $\ell \simeq 3 \times 10^{20}$ cm or $\simeq 300$ light years.

Table 5.6.

$E_{\bar{\nu}_e}/m_0c^2$	E_{e+}/m_0c^2	$\sigma/10^{-44}$ cm^2
4.5	2.0	8
5.5	3.3	20
10.8	8.3	180

The above reaction was studied by Cowan and Reines in 1958 and 1959, using a 1000 MW reactor as a source of antineutrinos. With a flux of $\sim 10^{13}$/cm^2 s passing through a target of water containing $\simeq 10^{28}$ protons in which some CdCl$_2$ was dissolved, reactions were observed. A positron produced in the capture reaction $\bar{\nu}_e + p \rightarrow n + e^+$ quickly annihilates with an electron within $\simeq 10^{-9}$ s producing two 511 keV gamma rays travelling in opposite directions. The recoiling neutron slows down through collisions with protons and is then captured in the Cd nuclei by the reaction Cd(n, γ)Cd*, producing gamma rays of total energy of $\simeq 8$ MeV (see figure 5.16 for the set-up). This neutron capture process takes about 10^{-5} s, so a characteristic signature of two simultaneous 511 keV gamma-rays, followed by the neutron capture gamma rays, a few microseconds later indicates a capture reaction. The expected event rate is $F \cdot N \cdot \sigma_c \cdot \epsilon$ where F is the $\bar{\nu}_e$ flux, N the number of protons, σ_c the capture cross-section and ϵ the efficiency of the detector which was about $\simeq 3 \times 10^{-2}$. Substitution of these values gives an event rate of $\simeq 1$ per hour which was actually observed.

The equality of the neutrino and antineutrino could be tested by studying the inverse reaction of electron capture

$$^{37}\text{Ar} + e^- \rightarrow {}^{37}\text{Cl} + \nu_e,$$

that is

$$^{37}\text{Cl} + \nu_e \rightarrow {}^{37}\text{Ar} + e^-,$$

but, using antineutrinos. It should be noted that the neutrinos are emitted in the fission process where neutron rich nuclei lower their neutron number through the β^--decay process n $\rightarrow$ p + e$^-$ + $\bar{\nu}_e$. The experiment of Davis (1955) gave a negative result pointing towards an experimental way of distinguishing between the neutrino and antineutrino. In section 5.4.2 we shall discuss yet another process allowing one to discriminate between neutrino and the antineutrino double beta-decay.

Even though the antineutrino capture of ^{37}Cl was a forbidden process, the reaction

$$\nu_e + {}^{37}\text{Cl} \rightarrow {}^{37}\text{Ar} + e^- \qquad (Q = -0.814 \text{ MeV})$$

could be used to detect a neutrino flux. The ^{37}Ar formed is unstable and decays back to ^{37}Cl with a half-life of $T = 35$ days. This electron capture produces

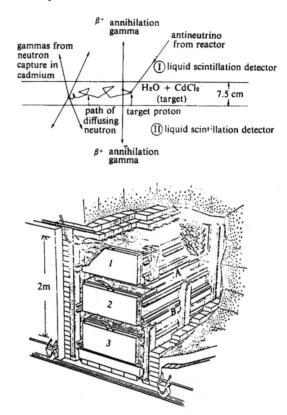

Figure 5.16. In the upper part are the schematics of neutrino detection. An antineutrino from a reactor produces a neutron and a positron. The positron is detected via its annihilation radiation. The neutron is moderated and detected via the capture gamma-ray (after Cowan and Reines 1953). In the lower part, the detector set-up is given. The detector tanks 1, 2 and 3 contained 1400 litres of liquid scintillator solution. They were viewed by 110 5-inch photomultiplier tubes. The tanks A and B each contained 200 litres of water–CdCl target. (Taken from Segré, *Nuclei and Particles*, 2nd edn © 1982 Addison-Wesley Publishing Company. Reprinted by permission.)

an excited ^{37}Cl atom with a K-shell vacancy and subsequent emission of Auger electrons and x-rays.

This neutrino capture reaction has been used over the years in order to detect neutrinos emitted in nuclear, stellar reactions (our sun, stars, etc) such as

$$^{7}_{4}\text{Be} + p \rightarrow {}^{8}_{5}B + \gamma$$
$$^{8}_{5}B \rightarrow {}^{8}_{4}\text{Be}^* + e^+ + \nu_e, \qquad (14.04 \text{ MeV})$$

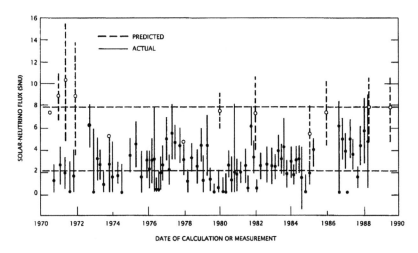

Figure 5.17. Measurements of the solar-neutrino flux (given in solar-neutrino units (SNU), defined as one neutrino interaction per 10^{36} atoms per second) (full lines) gives an average value of 2.1 ± 0.3 SNU (lower broken line). Theoretical calculations since 1970 (vertical dashed lines) have a consistent predicted average flux of 7.9 ± 2.6 SNU (upper, horizontal dashed line). The important difference dramatically illustrates the solar neutrino problem (adapted from Bahcall 1990, *Scientific American* (May)).

which produce only a small number of neutrinos in the sun burning process but with enough energy to initiate the neutrino capture in ^{37}Cl. Experiments were performed by Davis with the surprising result (see figure 5.17) that the observed flux of solar neutrinos is about one third of the predicted value. The calculated value clearly depends on models for the solar interior that are considered to be sufficiently well understood. This problem has been known as the 'solar neutrino puzzle'.

An experimental program started by Davies (Davies 1968) produced the surprising result (see figure 5.17) that the observed flux of solar neutrinos in this capture reaction is about one-third of the predicted value. These calculated values depend on models of the solar interior but these are considered to be sufficiently well understood (Bahcall 1968). This then gave rise to the solar neutrino problem (see section 5.4.5)

5.4.2 Double beta-decay

Situations can occur near the bottom part of the mass valley (see chapter 7) where a given nucleus $^{A}_{Z}X_N$ has an adjacent nucleus $^{A}_{Z+1}Y_{N-1}$, with higher mass while the nucleus $^{A}_{Z+2}Y'_{N-2}$ has a lower mass corresponding to an energy difference ΔE (see figure 5.18).

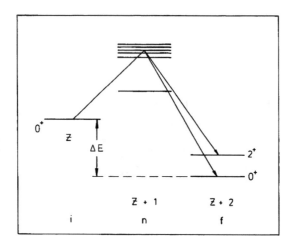

Figure 5.18. Simplified level energies is a double beta-decay process from a nucleus $^A_Z X_N$ to the nucleus $^A_{Z+2} Y_{N-2}$ which proceeds via the intermediate states in the nucleus $Z + 1$ (see text).

This means that the direct transition $_Z X \rightarrow _{Z+2} Y'$ is forbidden since the beta-decay transition operator only changes a neutron into a proton via a one-body operator. Decay possibilities do exist as a second-order decay process, indicating that the intermediate states $|n\rangle$ are virtual states and also that energy conservation is not satisfied in these contributions, whose amplitudes sum to give a total matrix element.

One then obtains

$$\langle f|H_{\text{int}}|i\rangle = \sum_n \frac{\langle f|H_{\text{int}}|n\rangle \langle n|H_{\text{int}}|i\rangle}{E_i - E_n}, \tag{5.68}$$

and a simple estimate for the double-beta-decay process is given by the square of the standard beta-decay matrix element. In the detailed calculations, the individual matrix elements need to be evaluated and weighted with the energy denominator but this study is also outside the scope of the present discussion.

Double-beta-decay thus is a second-order process (see figure 5.19(*a*)).

$$^A_Z X_N \rightarrow {}^A_{Z+2} Y'_{N-2} + e^- + e^- + \bar{\nu}_e + \bar{\nu}_e,$$

with the emission of two electrons and two antineutrinos. A typical example is the pair ^{48}Ca and ^{48}Ti. Other cases are the parent nuclei ^{96}Zr, ^{100}Mo, ^{116}Cd, ^{124}Sn, ^{130}Te, ^{150}Nd and ^{238}U.

The lifetime estimates are typically of the order $T \simeq 10^{20}$ y for an energy difference of $\delta E \cong 5 m_0 c^2$, and double beta-decay has been observed both by

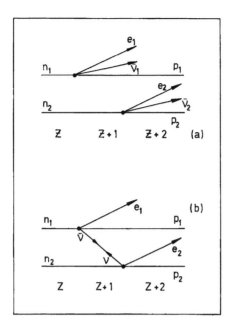

Figure 5.19. (*a*) Diagrammatic illustration of a normal two-neutron double beta-decay process, accompanied by the emission of two (e⁻, $\bar{\nu}_e$) pairs. (*b*) Diagrammatic illustration of a neutrinoless two-neutron double beta-decay process. The antineutrino emitted in one of the interactions is absorbed again at the other interaction point.

geochemical techniques (extraction of the daughter nuclei $(Z+2)$ from the parent (Z) in an old ore) and with counters. In table 5.7, we present a summary of some selected double beta-decay results.

Double beta-decay, however, carries a number of very intriguing aspects related to the neutrino properties. It was suggested that in relativistic quantum mechanics a neutrino particle (fermion) might exist for which the charge conjugate state of the original neutrino is identical to (Majorana) or different from (Dirac) the neutrino itself, i.e.

$$C|\nu_e\rangle \equiv |\bar{\nu}_e\rangle = |\nu_e\rangle \quad \text{Majorana}, \qquad C|\nu_e\rangle \equiv |\bar{\nu}_e\rangle \neq |\nu_e\rangle \quad \text{Dirac.} \quad (5.69)$$

Such Majorana particles appear in a natural way in theories (GUT-theories) that unify the strong and electroweak interactions with the possibility that the lepton number is no longer conserved, since now (figure 5.19(*b*)), a double-beta-decay process could be imagined in which the emitted antineutrino is absorbed as a neutrino, thereby giving rise to a process

$$^A_Z X_N \rightarrow \, ^A_{Z+2} Y'_{N-2} + e^- + e^-,$$

Table 5.7. Summary of selected double beta-decay results

	Experiment		Calculation	
	Geochemistry	Laboratory	Doi *et al* (1983)	Haxton *et al* (1981, 1982)
^{76}Ge				
$T_{1/2}(2\nu)$ (y)			2.3×10^{21}	3.7×10^{20}
$T_{1/2}(0\nu)$ (y)		$> 3.7 \times 10^{22}$	9.4×10^{22}	
m_ν (eV)			< 16	< 7
^{82}Se				
$T_{1/2}(2\nu)$ (y)	1.5×20^{20}	$(1.0 \pm 0.4) \times 10^{19}$	1.5×10^{20}	1.7×10^{19}
$T_{1/2}(0\nu)$ (y)		$> 3.1 \times 10^{21}$	3.2×10^{22}	
m_ν (eV)			< 33	< 12
^{130}Te				
$T_{1/2}(2\nu)$ (y)	2.6×10^{21}		2.6×10^{21}	1.7×10^{19}
$T_{1/2}(0\nu)$ (y)			2.5×10^{23}	
m_ν (eV)			< 130	
$^{130/128}$Te				
$T_{1/2}^{130/128}$	$(1.0 \pm 1.1) \times 10^{-4}$			
m_ν (eV)			< 5	< 5

with the creation of just *two* electrons. The energy sum of these two electrons $E_{e_1^-} + E_{e_2^-}$ is constant in this case, in contrast to the more usual (classic) process. The transition probability in this situation turns out to be $\simeq 10^5$ shorter than in the $(2e^-, 2\bar{\nu}_e)$ process. The two-electron energy spectrum $(E_{e_1^-}, E_{e_2^-})$ is indicated in figure 5.20.

The distinction between Majorana and Dirac particles as noted earlier was tested in the Davis experiment as discussed in subsection 5.4.1. The reactions

$$n \rightarrow p + e^- + \bar{\nu}_e, \qquad (\bar{\nu}_e = \nu_e) + n \rightarrow p + e^-,$$

were not observed. Subsequently, Davis (1955) concluded that the neutrino was different from its antiparticle so not a Majorana particle. This discussion had to be changed in the light of an experiment carried out by Wu (1957) which indicated that the weak interaction violated parity conservation and implied specific helicities for the neutrino (left-handed) (see section 5.5) and for the antineutrino (right-handed). In the light of this, the Davis experiment and the forbidden character of neutrinoless double beta-decay can more easily be explained. The neutrino and antineutrino never match because of the difference

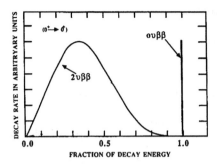

Figure 5.20. Two-electron energy spectrum for a $0^+ \rightarrow 0^+$ transition for the two double-beta-decay possibilities $2\nu\beta\beta$ and $0\nu\beta\beta$. (Taken from Avignone and Brodzinski, © 1988 reprinted with kind permission from Pergamon Press Ltd, Headington Hill Hall, Oxford OX3 0BW, UK.)

in helicity needed in the inverse reaction, i.e. one has

$$n \rightarrow p + e^- + \bar{\nu}(R), \qquad \nu(L) + n \rightarrow p + e^-.$$

Likewise, the neutrinoless double beta-decay, $nn \rightarrow pp + e^- e^-$ is forbidden, even if the neutrino has a Majorana character. Note that these statements are valid only if the neutrino is massless since for massive particle helicity is not a fixed quantum number (see relativistic quantum mechanics texts).

It should now be clear that the study of double beta-decay under laboratory conditions is of the utmost importance. The first evidence was presented by Elliott *et al* (1987) and is highlighted in Box 5b.

Experiments searching for neutrino-less double β-decay, in the reaction

$$^{76}\text{Se} \rightarrow {}^{76}\text{Ge} + 2e^- + (2\nu),$$

have been carried out at the Gran Sasso laboratory (Klapdor-Kleingrothaus *et al* 2001) with the highest sensitivity. A review of many of the studies carried out in the field of double β-decay is presented by Zdesenko (Zdesenko 2002).

5.4.3 The neutrino mass

It should be clear by now that a knowledge of the neutrino mass, if different from zero, is a fundamental problem in nuclear and particle physics. We have also recognized that a precise knowledge, with an upper *and* lower limit is a very difficult experimental problem.

One of the most straightforward methods relies on scrutinizing the beta spectrum shape (or the Fermi–Kurie plot) at and near the maximum electron energy. The β-spectrum shape is

$$\Lambda(p_{e^-})\mathrm{d}p_{e^-} = \frac{|M'_{fi}|^2}{2\pi^3\hbar^7 c^3} F(Z_\mathrm{D}, p_{e^-}) p_{e^-}^2 (E - E_{e^-})^2 \sqrt{1 - \frac{m_{\bar{\nu}_e}^2 c^4}{(E - E_{e^-})^2}}, \quad (5.70)$$

or, transforming to the electron energy E_{e^-}, is

$$\Lambda(E_{e^-})\,\mathrm{d}E_{e^-} = \frac{|M'_{fi}|^2}{2\pi^3\hbar^7 c^3} F(Z_\mathrm{D}, p_{e^-}) p_{e^-} E_{e^-} (E - E_{e^-})^2$$

$$\times \sqrt{1 - \frac{m_{\bar{\nu}_e}^2 c^4}{(E - E_{e^-})^2}} \,\mathrm{d}E_{e^-}. \quad (5.71)$$

The effect of a finite, but small neutrino mass will be best observed for beta-decay processes with a very small Q value. The β^--decay of tritium (^{3}H) is an attractive candidate for such a search since

$$^3\mathrm{H} \rightarrow {}^3\mathrm{He} + e^- + \bar{\nu}_e, \qquad (Q_\beta = 18.6\,\mathrm{keV})$$

because of the very low Q_{β^-} value and the possibility of detailed calculations of the electron wavefunction in the resulting ^{3}He ion.

In figure 5.21, we illustrate the results (left-hand side) for the Fermi–Kurie plot obtained by Bergkvist (1972). The more recent data of Lubimov (1980) seem to indicate a non-zero rest mass of about 30 eV, with error limits between 14 eV and 46 eV. These measurements have been criticized since, the small effect one is searching for, requires the utmost care in extracting a finite neutrino mass and many corrections are necessary.

The more recent data from Kündig *et al* (1986) have better statistics and the analysis yields, including systematic errors (figure 5.22),

$$m_{\bar{\nu}_e} c^2 \lesssim 18\,\mathrm{eV}.$$

One can question the need for all this effort. One of the main reasons relates to testing models of the universe. The original Big Bang theory predicts that the universe should contain neutrinos with a concentration of $\simeq 10^8/\mathrm{m}^3$. If these particles have a finite but small rest mass, this matter content might be large enough to make our universe collapse under the force of gravity. The limit on neutrino mass for this condition may be as low as 5 eV. Precise answers are therefore very important and have an immediate bearing on our basic understanding of profound questions in cosmology (figure 5.23).

Recently, on the occasion of the 1987A supernova that was first seen on 23rd February of that year in the Large Magellanic Cloud, optical signals as well as neutrino signals were detected at the same time. They had been travelling through space for 170 000 years and the neutrinos were recorded on earth about three hours before arrival of the optical signal. The neutrino emission spectrum

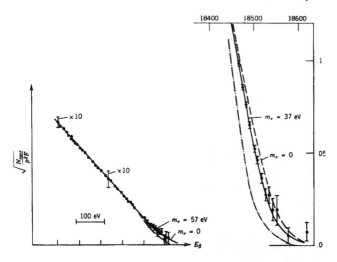

Figure 5.21. Fermi–Kurie plots of tritium beta-decay. The data (left) are from Bergkvist (1972) and are consistent with a zero mass, indicating an upper limit of 57 eV. The more recent data (right) by Lubimov (1980) seem to indicate a non-zero neutrino mass of about 30 eV.

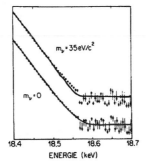

Figure 5.22. Upper end of the electron spectrum for the Kündig data. The solid lines indicate fits with neutrino masses of $m_{\nu_e} = 0$ and 35 eV/c^2. The non-zero mass indicates a fit with net deviation from the data (taken from Kündig *et al* 1986).

is expected to be a thermal spectrum corresponding to a temperature of about 5 MeV and the neutrinos corresponding to a different energy E_{ν_e} spread out over a certain time span in their motion from the star towards the Earth. This effect was observed by the experimental groups: a burst of neutrino events with energies of $\simeq$10 MeV over a time span of about $\sim$10 s (figure 5.24).

The mass-dependence of the neutrino is contained in the following expressions: for a relativistic particle, moving with almost the speed of light one

Figure 5.23. Cartoon illustrating the importance of a knowledge of the neutrino mass in order to answer a number of questions in cosmology. (Taken from de Rujula 1981. Reprinted with permission of CERN and A de Rujula.)

obtains the velocity

$$v_{\nu_e} = \frac{c}{\sqrt{1 + (m_{\nu_e}^2 c^2 / p_{\nu_e}^2)}}$$

$$v_{\nu_e} \simeq c - \frac{m_{\nu_e}^2 c^3}{2 p_{\nu_e}^2}$$

$$v_{\nu_e} \simeq c - \frac{m_{\nu_e}^2 c^5}{2 E_{\nu_e}^2}. \tag{5.72}$$

The arrival time of high-energy neutrinos will be earlier than for the low-energy neutrinos, i.e.

$$\frac{\delta t}{t} \sim \frac{\delta v}{v} = \frac{m_{\nu_e}^2 c^4}{E_{\nu_e}^2} \frac{\delta E_{\nu_e}}{E_{\nu_e}}. \tag{5.73}$$

The recorded events, seen by the Kamiokande detector and IBM group, have an energy spread $\delta E_{\nu_e} \sim 10 \, \text{MeV}$ and the neutrinos arrive over a time span $\delta t \sim 10 \, \text{s}$.

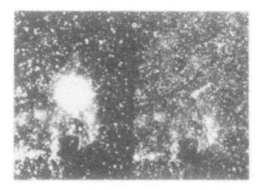

Figure 5.24. Supernovae 1987 shown (before and after) the explosion which appeared on the night of 27 February 1987 in the Large Magellanic Cloud. This SN 1987a event is about 170 000 light years distant and was used to obtain estimates on the emitted neutrino energy and on the possible neutrino mass. (Taken from Holstein 1989, *Weak Interactions in Nuclei*, © 1989 Princeton University Press. Reprinted with permission of Princeton University Press.)

So, we simply deduce

$$m_{\nu_e} c^2 \lesssim E_{\nu_e} \left(\frac{\delta t}{t} \cdot \frac{E}{\delta E_{\nu_e}} \right)^{1/2}$$

$$\lesssim 10\,\text{MeV} \left(\frac{10\,\text{s}}{10^{13}\,\text{s}} \right)^{1/2}$$

$$\lesssim 10\,\text{eV}. \tag{5.74}$$

Using a more careful model for time and energy distribution, one obtains as a realistic limit

$$m_{\nu_e} c^2 \lesssim 20\,\text{eV}.$$

A recent review on various direct neutrino mass measurements has been given by Robertson and Knapp (1988).

5.4.4 Different types of neutrinos: the two neutrino experiment

The neutrino that we have discussed up until now is produced in beta-decay transformations. It was known, in the higher energy regime, that neutrinos could also be produced in the decay of pions according to the reaction

$$\pi^+ \to \mu^+ + \nu_\mu, \qquad \pi^- \to \mu^- + \bar{\nu}_\mu.$$

The idea occurred that the above neutrinos might have completely different properties from the neutrinos accompanying the neutron and proton beta-decay.

Table 5.8.

	Mass (MeV/c^2)	Spin	Charge	Lepton number	Mean life τ (s)
ν_e	0?	$\frac{1}{2}$	0	$\ell_e = +1$	Stable
ν_μ	0?	$\frac{1}{2}$	0	$\ell_\mu = +1$	Stable
ν_τ	0?	$\frac{1}{2}$	0	$\ell_\tau = +1$	Stable
e^-	0.511004	$\frac{1}{2}$	-1	$\ell_e = +1$	Stable
μ^-	105.658	$\frac{1}{2}$	-1	$\ell_\mu = +1$	2.198×10^{-6}
τ^-	$1784.1 \begin{smallmatrix}+2.7\\-3.6\end{smallmatrix}$	$\frac{1}{2}$	-1	$\ell_\tau = +1$	$(3.04 \pm 0.09)10^{-13}$

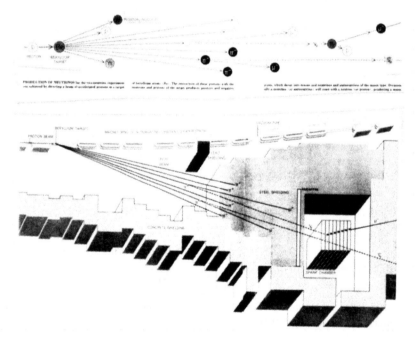

Figure 5.25. Production of neutrinos for the two-neutrino experiment by colliding high-energy protons with a Be target. Here, $\pi^\pm$-particles are produced which decay in $\mu^\pm$ accompanied by neutrinos and antineutrinos of the muon type. Occasionally, a neutrino (antineutrino) interacts with a neutron (proton) producing a muon. Lower part: the experimental set-up at the BNL-AGS illustrating the processes occurring in the upper part, but now in a more realistic situation (adapted from Lederman 1963 from *Scientific American* (March)).

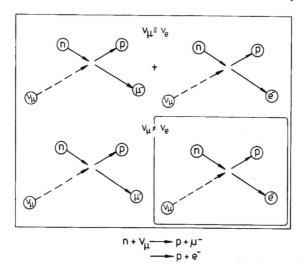

Figure 5.26. The logic of the two-neutrino experiment depends on the identity (top) or non-identity (bottom) of electron-type and muon-type neutrino. If they are identical, the reaction of muon-type neutrinos and neutrons should produce muons and electrons in equal numbers. If the two types of neutrino are different, the same reaction should produce muons but not electrons (adapted from Lederman 1963 from *Scientific American* (March)).

In 1959 suggestions were made to create intense beams of neutrinos at the AGS at Brookhaven National Laboratory, USA (Pontecorvo, Dubna (USSR) and Schwarz, Columbia University (USA). The 15 GeV protons were aimed at a Be target producing various particles: protons, neutrons and also creating unstable pions, which decay to muons and neutrinos. In the experiment, one had to remove all the accompanying particles. This required huge amounts of screening and stopping material in the form of a massive wall of iron (figure 5.25). The beam of neutrinos then enters a spark chamber and signals the very rare interactions of neutrinos with normal matter, i.e. the neutrons always produce muons and never electrons. As shown in figure 5.26, if $\nu_\mu \equiv \nu_e$ an equal number of muons and electrons should be produced. The observed asymmetry in particle production dramatically proves the existence of two-types of neutrinos: the electron type of neutrino ν_e and the muon type of neutrino ν_μ. This leads to a classification (table 5.8) of some of the more important leptons (weak interacting particles). In addition, the question about the precise number of neutrino families is clearly posed (see Box 5c).

Table 5.9. Neutrino-producing reactions in the sun. (Reprinted table from Davies R Jr 2003 *Rev. Mod. Phys.* **75** 985 © 2003 by the American Physical Society.)

	Reaction	Frequency	Energy (MeV)	Name
PPI	$p + p \rightarrow {}^2H + e^+ + \nu_e$	99.75 %	0.0 - 0.42	pp
	$p + e^- + p \rightarrow {}^2H + \nu_e$	0.25 %	1.44	pep
	${}^2H + p \rightarrow {}^3He + \gamma$	100 %		
	${}^3He + {}^3He \rightarrow {}^4He + 2p$	85 %		
PPII	${}^3He + {}^4He \rightarrow {}^7Be + \gamma$	15 %		
	$e^- + {}^7Be \rightarrow {}^7Li + \nu_e$	99.99 %	0.86, 0.38	${}^7Be$
	$p + {}^7Li \rightarrow {}^4He + {}^4He$	100 %		
PPIII	$p + {}^7Be \rightarrow {}^8Be + \gamma$	0.01 %		
	${}^8B \rightarrow {}^4He + {}^4He + e^+ + \nu_e$	100 %	0 - 14.1	${}^8B$

5.4.5   Neutrino oscillations: the final verdict

5.4.5.1   The solar neutrino problem and its solution

As discussed at the end of section 5.4.1, Davies' experiments on the detection of the electron-type neutrino ν_e in the capture process ${}^{37}Cl + \nu_e \rightarrow {}^{37}Ar + e^-$, using a 100 000 gallon chlorine–argon neutrino detector system in the Homestake Gold Mine in Lead, South Dakota constructed in the period 1965–66, started data taking in 1967. These data started the solar neutrino problem. A very personal account of the early experiments and set-up is presented by Davies himself in his 2002 Noble prize talk (Davies 2003).

The neutrinos needed to induce this capture reaction must have an energy of at least 0.814 MeV. Therefore, this reaction is sensitive to neutrinos produced in reactions producing the higher-energy PPIII neutrinos (see table 5.9).

It took a long time before other studies of the neutrino problem such as the study of neutrino interactions inside the ultra-pure water detector with the Kamiokande experiment (Kamio Nuclear Decay Experiment—originally set up to study the possible decay of protons) were taken up in Japan. Here, the elastic scattering of neutrinos produces recoiling electrons that can be detected through their Cerenkov radiation. Directional information is present through the momentum of the incoming neutrino and clear evidence for a solar origin was pointed out (Hirata *et al* 1990, Fukuda *et al* 2001).

Two other radiochemical reaction studies were performed in the 1990s, studying the capture reaction of electron-neutrinos on ${}^{71}Ga$ producing the element ${}^{71}Ge$. The SAGE and GALLEX experiments were carried out in the Baksa

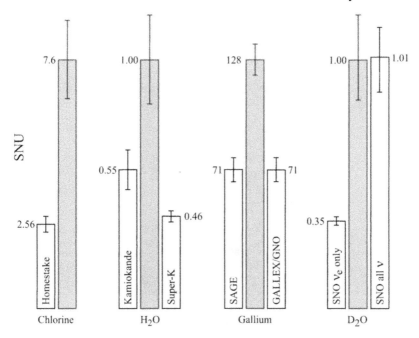

Figure 5.27. A comparison of measured solar neutrino fluxes from various types of experiments, such as indicated in the open bars (Homestake, Kamiokande, SuperKamiokande, SAGE, GALLEX/GNO, SNO), compared with the results of solar model predictions. (Reprinted table from Davies R Jr 2003 *Rev. Mod. Phys.* **75** 985 © 2003 by the American Physical Society.)

(Russia) and in the Gran Sasso Mine (Italy) and are sensitive to the neutrinos produced in the pp fusion reaction, producing essentially low-energy neutrinos (see table 5.9) but here, too, only about 50% of the expected solar neutrino flux was detected. An overview of the solar problem is most clearly presented in figure 5.27.

It was clear that to solve this problem, there was a need to test the idea that neutrinos might oscillate in flavour if the mass of neutrinos were not exactly zero. This idea of neutrino oscillations had already been put forward quite some time before (Gribov and Pontecorvo 1969, Wolfenstein 1978, Mikheyev and Smirnov 1985). So, the point at stake was to find whether the solar neutrino flux contained other flavours besides the electron neutrino flavour. The neutrino problem was hinting at an affirmative direction but one needed a dedicated detector where a sensitivity to the various flavours of neutrinos could be independently tested. A special detector (SNO for Sudbury Neutrino Observatory) was built in a deep nickel mine in Canada and looks very much like the Kamiokande Cerenkov detector, except that the team was now using heavy water with a 2.2 MeV neutrino

CHARGED CURRENT

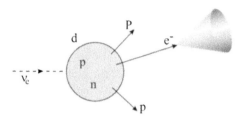

ELASTIC SCATTERING

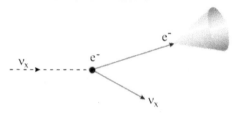

NEUTRAL CURRENT

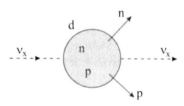

Figure 5.28. Neutrino interactions with deuterium in the SNO detector: (*a*) the charge–current interaction, sensitive to ν_e only, (*b*) elastic scattering and (*c*) the neutral–current interaction. The latter two reactions are sensitive to all flavours. The reactions (*a*) and (*b*) are characterized by the e^- Cerenkov radiation

detection threshold (Falk 2001, Heeger 2001). Here, the following reactions can be seen (see figure 5.28):

$$\nu_e + d \rightarrow e^- + p + p, \tag{5.75}$$

$$\nu_x + d \rightarrow \nu_x + n + p, \tag{5.76}$$

$$\nu_x + e^- \rightarrow \nu_x + e^-. \tag{5.77}$$

The first reaction is the charged current (CC) reaction, the next one the neutral current (NC) reaction and the third denotes elastic scattering of the neutrino (ES). The first reaction (CC) is only sensitive to electron-type neutrinos, the elastic scattering (ES) is essentially sensitive to ν_e with a reduced sensitivity to ν_μ and ν_τ. Comparing these two fluxes can give indications of flavour transformations without even resorting to a particular solar model. The NC reaction is then sensitive to all three flavours. In a series of ground-breaking experiments (Ahmad 2001, Ahmad 2002), it was clear that

(i) through CC and ES reactions, there was unambiguous shortage of electron neutrinos compared to the solar model predictions (see also figure 5.27) since $\phi^{CC}(\nu_e) < \phi^{ES}(\nu_x)$: and

(ii) combining these results with recent results (Ahmad 2002) on the NC reaction (see figure 5.27), it becomes clear that the neutrino flux at the detector must contain other flavours and the total flux is fully consistent with the solar model predictions. The combination of the results of reactions (5.75) and (5.76) is unambiguous and indicates the solution of a long-standing puzzle that was noted back in 1967 by Davies.

These results also point towards the fact that the flavours change between the production site in the sun and the detector on earth, consistent with an oscillatory behaviour that itself points towards non-zero mass differences for the neutrino flavours. For a simple two-flavour mixing system, the probability that a neutrino of flavour ν_α has changed into a neutrino of flavour type ν_β, can be described as

$$P(\nu_\alpha \rightarrow \nu_\beta) = \sin^2 2\theta \sin^2(\pi x/L) \tag{5.78}$$

with appropriate units and L defined as

$$L = \frac{4\pi E\hbar}{\Delta m^2 c^3} = 2.48 \left(\frac{E}{MeV}\right)\left(\frac{eV^2}{\Delta m^2}\right) m, \tag{5.79}$$

so that L is expressed in units of metre and where we have used $\Delta m_{ij}^2 = |m_i^2 - m_j^2|$. Further technical details on the issue of neutrino oscillations are discussed in Zuber (1998), Wolfenstein (1999), Fisher *et al* (1999), Kirsten (1999), Haxton and Holstein (2000), Kajita and Totsuka (2001), Bemporad C *et al* (2002) and Gonzales-Garcia and Nir (2003).

5.4.5.2 *Atmospheric neutrinos*

Indications for flavour changes in the neutrino family have been obtained from the study of cosmic rays. In the phase where high-energy cosmic rays enter the atmosphere, they tend to interact with the nitrogen and oxygen nuclei thereby producing π and K mesons. These mesons quickly decay into muons and the corresponding neutrino ν_μ. The secondary μ then decays into an electron e$^-$, an additional neutrino ν_e and antineutrino $\bar\nu_\mu$. This process results in two muon-type neutrinos ν_μ, $\bar\nu_\mu$ and one electron-type ν_e (see figure 5.29). Thus, the ratio $N(\nu_\mu)/N(\nu_e)$ becomes 2. For high-energy processes, the muons with the longer lifetime cannot decay before the detector is reached and this ratio will increase above the value of 2 (see (Koshiba 2003) for more details).

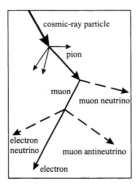 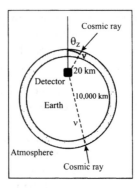 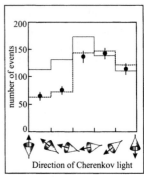

Figure 5.29. Process for the detection of atmospheric neutrinos. Left-hand part: cosmic rays give rise to pion production in the upper atmosphere, the pion decaying into a muon and finally an electron also producing muon neutrinos, muon anti-neutrinos and electron neutrinos along the path. Middle part: the production mechanism and the geometry for the detection of atmospheric neutrinos. Right-hand part: prediction of oscillation hypothesis results (dashed line) as compared with the SuperKamiokande data (data points) as well as with the results of a no-oscillation hypothesis (full line). The little cones below indicate the direction of the Cherenkov light (left-hand side—upward-going neutrinos, right-hand side—downward-going neutrinos). A net deficiency with respect to the zenith orientation is observed. (Taken from 'Neutrino Oscillations' (Kaneyuki and Scholberg), American Scientist, May–June 1999, © Sigma Xi, The Scientific Research Society.)

These atmospheric neutrinos then pass through the earth into a detector and, in this process, some neutrinos will travel only about 10–30 km whereas other ones, coming through the earth, may travel a distance of some 13 000 km (see figure 5.29). If the ideas for possible neutrino flavour oscillations as put forward by Gribov and Pontecorvo (Gribov and Pontecorvo 1969) and Wolfenstein (Wolfenstein 1978) turned out to be correct, then oscillations might change some of the flavours within the longer path. So, putting up a detector for the neutrinos could resolve this oscillation problem. Because the neutrino interaction with regular matter is particularly weak and because of the high energies of the incoming neutrinos (typically of the order of 1 GeV), reactions of the type

$$\nu_e + n \rightarrow e^- + p, \qquad \nu_\mu + n \rightarrow \mu^- + p$$

as well as the reactions of anti-neutrinos in the entrance channels create rapidly moving charged particles e^-, μ^- and their antiparticles. These charged particles can then be detected through Cerenkov radiation in the appropriate medium. The massive SuperKamiokande detector, containing 50 000 ton of pure water, replacing the former Kamiokande detector that contained only 3000 ton of water, led to about one event every 90 min. How do we now discriminate between electrons and muons? It appears that the Cerenkov ring produced by an energetic electron is more diffuse compared to the relatively clean rings of a muon producing Cerenkov radiation. Since the charged leptons essentially move forward in the direction of the momentum vector of the incoming neutrino, *both the direction and the flavour of the neutrinos* that react with water can be obtained.

The very precise data taken at SuperKamiokande (Hatakeyama *et al* 1998, Fukuda *et al* 1999, Boezio *et al* 1999, Futagami *et al* 1999) came up with a result $N(\nu_{\mu})/N(\nu_{e})$ of approximately 1—which was totally unexpected. This gives a clear indication of missing muon neutrinos (ν_{μ}) with the number of electron neutrinos (ν_{e}) very much as expected. If this is correct, there should be a very clear directional correlation with the largest suppression for the muon neutrinos reaching the earth. This is indeed borne out by the SuperKamiokande results (see previous references and also the review paper (Kajita and Totsuka 2001) (see figure 5.29). The most plausible explanation then looks as follows: atmospheric ν_{μ} neutrinos oscillate into ν_{τ} neutrinos which cannot be observed since the energy is too low to produce τ leptons. The strong suppression in the ν_{μ} neutrino flux, moreover, is consistent with a maximal mixing angle $\theta \sim \pi/4$ and also gives evidence for an oscillation length of the order of the diameter of the earth. In 2002, Koshiba received the Noble prize in physics (Koshiba 2003) for his work on neutrino astrophysics through building this particular detector in Japan. More recently, at SNO (Ahmad *et al* 2002a), studying the day–night neutrino energy spectra and rates in the subset of CC processes, a night-minus-day rate difference of 14%±6.3% and an electron neutrino ν_{e} asymmetry of 7.0%±4.9% is obtained. These results are consistent with large-angle mixing in the two active flavours. These results are consistent with analyses carried out at SuperKamiokande of the recoil electron energy spectrum and the zenith dependence of the solar neutrino flux in neutrino–electron scattering data (Fukuda *et al* 2001).

5.4.5.3 *Earth-based experiments*

The ideal and definitive test for neutrino oscillations is, of course, to create, under controlled conditions, a given flux and flavour of neutrino-beams and study the oscillations through (i) experimental verification of a reduced flux of the initial flavour (disappearance experiments) and/or (ii) the appearance of a given flux of neutrinos of a different flavour (appearance experiments).

Accelerator neutrino experiments

In experiments of this type, a typical set-up consists of a proton accelerator in which an intensive beam of neutrinos is generated after stopping the protons in a beam-stopper. Thereby, depending on the energy of the initial protons, a large number of ν_{μ}, $\bar{\nu}_{\mu}$ and ν_{e} neutrinos (only very few $\bar{\nu}_{e}$ anti-neutrinos are produced) are produced through the reactions

$$\pi^{+} \rightarrow \mu^{+} + \nu_{\mu}; \qquad \mu^{+} \rightarrow e^{+} + \nu_{e} + \bar{\nu}_{\mu}, \tag{5.80}$$

and with the antiparticle π^-,

$$\pi^- \to \mu^- + \bar{\nu}_\mu; \qquad \mu^- + (A, Z) \to (A, Z - 1) + \nu_\mu. \tag{5.81}$$

The latter reaction is very important and to a large extent prevents the free decay of the negative muon through the reaction $\mu^- \to e^- + \bar{\nu}_e + \nu_\mu$. So, very few $\bar{\nu}_e$ appear and sensitive searches for oscillations of the type $\nu_\mu \to \nu_e$ as well as for the antineutrinos can be carried out by looking for the appearance of electron-type neutrinos downstream of the beam stop. With a detector placed at a given distance L from the beam stop (neutrino creation), data for a given mixing angle and neutrino mass difference should follow a modulation given by the expression

$$P(L) \approx \sin^2 2\theta \sin^2 \left(\frac{\Delta m^2 L}{4 E_\nu} \right).$$

No net results (or upper limits) can as yet put constraints on the mixing parameters or mass difference (Zuber 1998, Wolfenstein 1999, Fisher *et al* 1999, Kirsten 1999, Haxton and Holstein 2000). Of course, very small values of Δm^2 implies a large L and, thus, very intense neutrino fluxes are needed.

By now, most experiments of this type have been carried out with rather small L values. Experimental searches at the SPS accelerator at CERN in the period 1993–98 (CHORUS—Cern Hybrid Oscillation Research apparatUS and NOMAD—Neutrino Oscillation Magnetic Detector), aiming at oscillation events of the type $\nu_\mu \to \nu_\tau$, have not given any conclusive results. Experiments at Los Alamos (LSND experiment—Liquid Scintillation Neutrino Detector), using middle-energy protons, have reported on a positive signal downstream of the proton beam stop ($L \simeq 30$ m), as a result of an oscillation of the type $\bar{\nu}_\mu \to \bar{\nu}_e$ and the subsequent reaction $\bar{\nu}_e + p \to n + e^+$. The collection of Cerenkov radiation for the outgoing positron and the 2.2 MeV γ from the capture reaction $n + p \to d + \gamma$ are indicative of the appearance of electron anti-neutrinos. This is a very difficult experiment, hampered by the large background resulting from cosmic-ray events. After a careful analysis, 22 $\bar{\nu}_e$ events were observed compared to an anticipated background of 4.6 ± 0.6 events (Athanassopoulos *et al* 1995). This excess is consistent with the mixing parameters $\Delta m^2 \approx 0.2$–2 eV^2 and $\sin^2 2\theta \approx 0.03$–0.003. Later experiments at the Rutherford Laboratory (KARMEN for KArlsruhe Rutherford interMediate Energy Neutrino experiment), using an 800 MeV proton beam, have found no evidence. This issue is very important since it signals the appearance of a new flavour that was not originally present in the beam (appearance-type experiments) and new tests are to be set up.

Very recent experiments in Japan, producing an almost pure muon neutrino beam (98% purity) produced at the 12 GeV KEK proton accelerator (Japanese National High Energy Accelerator Laboratory), directed the neutrino beam to the Super-Kamiokande underground detector (that also indicated evidence for oscillations in atmospheric muon neutrinos) located 250 km away from KEK (Ahn *et al* 2003). Data have been taken during the period 1999–2001. This is the first real long-baseline experiment in which a neutrino beam was directed to an off-site detector. Normally, one expected 80.1 (+6.2; −5.4) induced neutrino interactions at Super-Kamiokande in the absence of any oscillations. The detected number was 56 events. Since the 'near'-detector neutrino flux and the initial neutrino energy spectrum were measured, the probability that the observed data are consistent with a 'no-oscillation' hypothesis is less then 1%. Moreover, these K2K (KEK to Super-Kamiokande) experimental results are fully consistent with the atmospheric neutrino data taken at SK.

More experiments using long-baseline distances are being planned and/or set-up such as the CERN to Gran Sasso project and the Minos experiment where detectors are placed in the Canadian Soudan Mine with a neutrino beam generated at Fermilab. In this latter experiment, with a baseline of 730 km, one aims to study disappearing effects on the ν_μ beam, oscillating into the ν_τ flavour.

Note added in proof. The latter experiment just started at the present stage of proofreading.

Reactor experiments

A totally different approach makes use of the fact that in the nuclear fission process, neutron-rich nuclei are produced which then decay towards the line of β-stability through the reaction $(A, Z) \rightarrow (A, Z+1) + e^- + \bar{\nu}_e$, thereby acting as an intense source of electron anti-neutrinos. Experiments searching for possible oscillations of the type $\bar{\nu}_e \rightarrow \bar{\nu}_x$ have been performed at the reactors at Bugey and Chooz. Results on possible solutions in the mixing angle versus mass difference phase space have been discussed at some length (Zuber 1998, Wolfenstein 1999, Fisher *et al* 1999, Kirsten 1999, Haxton and Holstein 2000). Very recent experiments, carried out at the Kamioka mine (called KamLAND for Kamioka Liquid Scintillation Anti-Neutrino Detector), where the largest low-energy anti-neutrino detector has been built, have recently given fascinating results with respect to the disappearance of anti-neutrinos from the original electron flavour (see Box 5d). These results (Eguchi *et al* 2003) support the large mixing-angle solution as obtained from the SNO experiments.

It is clear that the field of neutrino physics has proven to be a most active research field where experiments should be able to give improved results on both the mixing model (mixing angle, mass differences for various flavours) as well on the consequences of the presence of non-vanishing neutrino masses in physics in general.

5.5   Symmetry breaking in beta-decay

5.5.1   Symmetries and conservation laws

One of the main features relating to symmetries in dynamical systems is the fact that whenever a law is invariant under a certain symmetry operation, a corresponding conserved quantity exists.

The quantum mechanical formulation of the above statement is best given by calculating the time dependence of the expectation value of a given observable $\hat{O}$. Using the time dependence of the wavefunction, corresponding to a given Hamiltonian $\hat{H}$, one obtains the equation

$$\frac{\mathrm{d}}{\mathrm{d}t}\langle\hat{F}\rangle = \frac{\mathrm{i}}{\hbar}\langle[\hat{H}, \hat{F}]\rangle, \qquad (5.82)$$

where the observable $\hat{F}$ does not explicitly depend on t. The above equation tells us that one has a conserved quantity $\mathrm{d}/\mathrm{d}t\langle\hat{F}\rangle = 0$, whenever the operators $\hat{H}$ and $\hat{F}$ commute. The latter commutation relation can also be expressed via the invariance of the quantum mechanical wavefunction under the action of certain transformations generated by the $\hat{F}$ operator. A typical example is that invariance under translation leads to the conservation of the corresponding momentum. A number of symmetries and related conservation laws are depicted in table 5.10. In this table one recognizes some of the well-known continuous symmetry operations (translations in a spatial direction or in the time coordinate, rotations, Lorentz transformations). Besides these, a number of important discrete symmetry operations are occurring such as time reversal, parity (reflection) operation, charge or particle–antiparticle transformation. Some symmetry operations belong in particular to the world of quantum physics such as the symmetries in the wavefunction related to interchanging coordinates of identical particles in a many-particle wavefunction and gauge transformation.

In general, one can use the equation for $\mathrm{d}/\mathrm{d}t\langle\hat{F}\rangle$ to conclude that whenever a quantity is *not* conserved, the Hamiltonian should contain particular terms that do not commute with the operator $\hat{F}$. We shall make use of this result as a key-argument when discussing the parity non-conservation in beta-decay processes.

A very readable text on symmetries in physical laws within a more general context has been written by Feynmann in his Physics Lectures (Feynmann *et al* 1965, chapter 52). In the next section we shall concentrate on the parity symmetry operation and its consequences on various properties, both at the level of classical physics as well as concerning its application to quantum mechanical systems.

5.5.2   The parity operation: relevance of pseudoscalar quantities

The parity operator acting on a certain object (active transformation) changes all coordinates into the corresponding reflected values with respect to the origin O of the (x, y, z) coordinate system.

Table 5.10.

Symmetry	Conservation Law
Translation in space	Linear momentum
Translation in time	Energy
Rotation	Angular momentum
Uniform velocity	
Transformation (Lorentz tr.)	
Time reversal	
Reflection	Parity
Particle–antiparticle	
Exchange of identical particles	
Quantum mechanical phase	

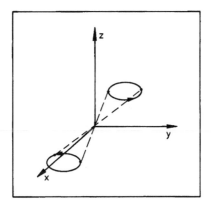

Figure 5.30. Effect of the parity operator $\hat{P}$ acting on the circular motion of a particle. It also shows that the rotational direction remains unchanged under the point parity operation.

Acting on vector quantities, the position vector $\vec{r}$ is changed into $-\vec{r}$, the linear momentum is changed into $-\vec{p}, \ldots$. These vectors are called *polar* vectors since these vectors are related to certain directions. Vector quantities, like angular momentum or circular current motion are not changed under the parity operation, i.e. $\vec{\ell} = \vec{r} \times \vec{p}$ transforms into

$$\hat{P}(\vec{\ell}) = \hat{P}(\vec{r} \times \vec{p}) = -\vec{r} \times (-\vec{p}) = \vec{\ell}. \tag{5.83}$$

This applies to any angular momentum (see figure 5.30) and the corresponding vectors: intrinsic spin $\vec{s}$, total angular momentum $\vec{J}, \ldots$ are called *axial* vectors.
    In addition, scalar properties (numbers $a, x, \ldots$ or scalar products of polar or axial vectors $\vec{r} \cdot \vec{r}; \vec{\ell} \cdot \vec{\ell}, \ldots$) are invariant under the parity operation. One can,

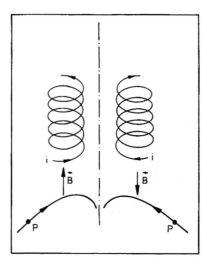

Figure 5.31. Parity (reflection in mirror indicated by the dash-dotted line) operation for the Lorentz force $\vec{F} = q(\vec{E} + \vec{v} \times \vec{B})$. The mirror image and the original set-up of a charged particle moving in a magnetic field caused by the solenoid are shown.

however, construct quantities, called pseudoscalar, which are formed by the scalar product of a polar and an axial vector $\vec{p} \cdot \vec{J}$, $\vec{p} \cdot \vec{s}$. Such quantities change sign under parity transformations.

It was a very natural suggestion that *all* laws in physics are invariant under the parity operation: the force law (figure 5.31)

$$\vec{F} = q(\vec{E} + \vec{v} \times \vec{B}), \tag{5.84}$$

holds also after space reflection since $\vec{F} \rightarrow -\vec{F}$, $\vec{E} \rightarrow -\vec{E}$; $\vec{v} \rightarrow -\vec{v}$, $\vec{B} \rightarrow \vec{B}$. The Newton law of motion

$$\vec{F} = m \frac{d\vec{p}}{dt}, \tag{5.85}$$

holds also after space reflection.

In the same way the observables such as energy density are real scalar quantities and are expressed via scalar products like $\vec{E} \cdot \vec{E}$, $\vec{B} \cdot \vec{B}$ and terms like $\vec{E} \cdot \vec{B}$ cannot appear. This invariance under parity transformation for any possible experiment that may be realized expresses a fundamental symmetry of the basic laws of motion on a macroscopic level.

The implications for quantum mechanics, where the wavefunction $\psi(\vec{r})$ carries all dynamical information are as follows. Starting from a Hamiltonian which is invariant under the parity operator $[\hat{H}, \hat{P}] = 0$, one can show that wavefunctions are characterized by a good parity quantum number, i.e.

$$\hat{P}\psi(\vec{r}) = \pi \psi(\vec{r}), \tag{5.86}$$

with $\pi = \pm 1$. As an example, the spherical harmonics $Y_\ell^m(\theta, \varphi)$ have the parity $(-1)^\ell$.

In a reaction

$$a + b \rightarrow c + d,$$

total parity has to be conserved if $[\hat{H}, \hat{P}] = 0$. If we call $\pi_a, \pi_b, \pi_c, \pi_d$ the parity of the objects a, b, c, d and if the relative motion of a and b (c versus d) is described by a relative angular momentum $\ell_{ab}(\ell_{cd})$, one has the conservation law

$$\pi_a \pi_b (-1)^{\ell_{ab}} = \pi_c \pi_d (-1)^{\ell_{cd}}. \tag{5.87}$$

In a system where parity is not conserved, the total wavefunction should contain a parity-even and a parity-odd component, i.e.

$$|\alpha\rangle = c_1 |\text{even}\rangle + c_2 |\text{odd}\rangle, \tag{5.88}$$

with

$$|c_1|^2 + |c_2|^2 = 1.$$

One has

$$\hat{P}|\alpha\rangle = c_1 |\text{even}\rangle - c_2 |\text{odd}\rangle \neq |\alpha\rangle, \tag{5.89}$$

and c_2/c_1 is a measure of parity-breaking. We now state an important result. Whenever the commutator $[\hat{H}, \hat{P}] \neq 0$, i.e. indicating that the Hamiltonian *should* contain pseudoscalar quantities, parity is no longer a conserved quantity. In that case, the expectation value of *any* pseudoscalar operator $\hat{O}_{PS}$ will not disappear and so the setting up of experiments where a pseudoscalar quantity is measured is fundamentally important.

For a system, described by a fixed parity $\hat{P}\psi = \pi\psi$, the expectation value of any pseudoscalar quantity becomes zero since in the integral the integrand is an odd function under space inversion and integrating over *all* space, $\langle \hat{O}_{PS} \rangle = 0$.

$$\langle \hat{O}_{PS} \rangle = \int \psi^*(\vec{r}) \hat{O}_{PS} \psi(\vec{r}) \, d\vec{r}. \tag{5.90}$$

Some typical examples are $\vec{s} \cdot \vec{p}$ where $\vec{p}$ describes the linear momentum of a particle and $\vec{s}$ its intrinsic spin. The operator describes the helicity $\hat{h}$ or handedness of the particle. We shall discuss this in some detail later, in section 5.5.4.

5.5.3 The Wu–Ambler experiment and the fall of parity conservation

In the light of the above discussion, and in the light of some theoretical suggestions by Lee and Yang (1956) several experimental groups started investigations on the conservation of parity in weak interaction processes. It was Wu *et al* (1957) who set out to measure a possible asymmetry in the electron emission of ^{60}Co (a β^- emitter, decaying into ^{60}Ni) relative to the orientation of the original initial spin orientation of the ^{60}Co nuclei which decays by a $5^+ \rightarrow 4^+$ Gamow–Teller transition. The measurement of the pseudoscalar quantity

$$\langle \vec{p}_{e^-} \cdot \vec{J}_i \rangle, \tag{5.91}$$

could give a unique answer to the occurrence of pseudoscalar quantities in the interaction describing the beta-decay processes.

The $5^+ \rightarrow 4^+$ Gamow–Teller decay is described in a somewhat schematic way in figure 5.32; one unit of angular momentum is taken away in the decay process; the sum of the electron ($\frac{1}{2}$) and antineutrino ($\frac{1}{2}$) spins is shown by the thick arrows. Since the net momentum carried away is $\vec{L}_\beta = \vec{0}$, the linear momenta $\vec{p}_{e^-}$ and $\vec{p}_{\bar{\nu}_e}$ can be drawn parallel to each other. If we consider the process after reflection in the mirror (any mirror image represents a possible experiment if parity is a conserved quantity), we can conclude that the electron emission should have no preferential direction of emission, relative to the initial ^{60}Co spin (5^+) orientation.

The actual experiment had to implement conditions for aligning all ^{60}Co initial spins in the same direction, i.e. polarized, by the action of an external magnetic field at low temperature ($T \sim 0.01°$ K). The Boltzmann factor describing the relative population of the various magnetic substates M (with a magnetic field B and considering a positive g-factor), gives the expression for the population distribution within the hyperfine levels

$$N(M_J) \propto \exp\left(-\frac{E_0 - gM_J B\mu_N}{kT}\right), \tag{5.92}$$

and so, at low enough temperature and using a high value for the magnetic field, approximately all nuclei are in the $M=+5$ substate at the lowest energy, thereby inducing an ensemble of polarized nuclei (see figure 5.33). The actual set-up is shown in figure 5.34 and we refer to the original article (Wu *et al* (1957)) for a succinct description of the experiment (see Box 5e). A reversal of the B-field should have no influence on the count rate of the electrons emitted from polarized nuclei. The actual experiment showed a dramatic change in the number of electrons counted indicating an asymmetry in the emission of electrons relative to the initial ^{60}Co spin: electrons were emitted preferentially opposite to the 5^+ spin orientation (see also the schematic representation in figure 5.32).

The experiment monitored the orientation of the ^{60}Co nuclei as a function of time by measuring the anisotropy of gamma radiation, emitted after the beta-decay, in an equatorial and a polar detector. Details can again be found in the original article.

This experiment gave an unambiguous indication of the violation of parity through the observation of a non-zero value of the pseudoscalar measurement $\langle \vec{p}_{e^-} \cdot \vec{J}_i \rangle$. In the light of our earlier discussion in section 5.2.2 the conclusion is non-conservation of parity and thus, the appearance of pseudoscalar terms in the

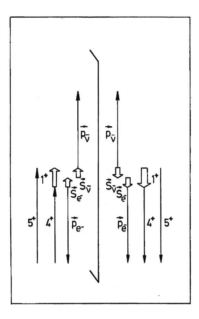

Figure 5.32. The various angular momentum vectors associated with the $5^+ \rightarrow 4^+$ Gamow–Teller decay of ^{60}Co to ^{60}Ni. Both the original and mirror image are given. The nuclear spin vectors $(\vec{5}^+, \vec{4}^+)$, the linear momentum $(\vec{p})$ and spin $(\vec{s})$ vectors are drawn. The total angular momentum removed is one unit of spin (units $\hbar$) and is constructed from the individual intrinsic spin $\frac{1}{2}$ of the electron (e^-) and antineutrino $(\bar{\nu}_e)$. The linear momentum vectors are drawn such that no net orbital angular momentum is carried away from the nucleus.

interaction describing the weak interaction since now

$$[\hat{H}, \hat{P}] \neq 0. \tag{5.93}$$

The consequences of the above conclusion can only be fully appreciated within the framework of relativistic quantum mechanics for describing the beta-decay process.

The Wu experiment has been redone by Chirovsky *et al* (1980) with the aim of measuring the full angular distribution of electrons being emitted in the β^--decay of ^{60}Co. These results are presented in figure 5.35, indicating an intensity variation of the form

$$I(\theta) = A + B \cdot \cos\theta, \tag{5.94}$$

and this expresses all details of parity violation.

The experiment can be analysed in a schematic way using the diagrams in figure 5.36. Here, the original experimental set-up and the reflected (mirror)

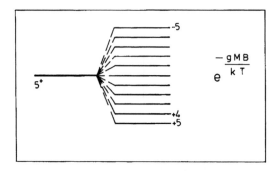

Figure 5.33. Effect on the various magnetic substates for the initial ^{60}Co nuclei with spin $J = 5$ ($M = +5, \ldots, -5$) when placed into a strong external magnetic $\vec{B}$ field at very low temperature T. The Boltzmann factor regulates the precise distribution over the various M-substates.

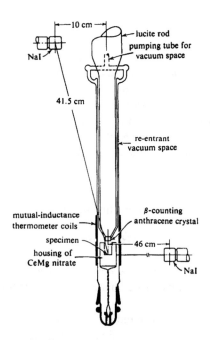

Figure 5.34. A schematic drawing of the lower part of the cryostat used in the Wu *et al* set-up (taken from Wu *et al* 1957).

image are put together. Now, the mirror image represents a non-observable situation indicating the breakdown of parity!

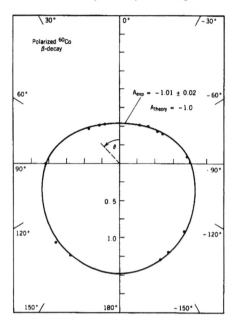

Figure 5.35. The data points on the polar diagram illustrate the observed beta intensity in the ^{60}Co decay at an angle θ relative to the polarization direction. The solid curve is the result from the Fermi theory of beta-decay and conforms to the expression $1 + AP\cos\theta$. The asymmetry for $0°$ and $180°$ is well illustrated (taken from Chirovsky *et al* 1980).

In the next section we concentrate on the consequence for the neutrino (or antineutrino) emitted in the above decay process. Before explaining this subject, we remark that if we carry out the parity operation (the mirror reflection) $\hat{P}$ and if at the same time we change all particles into antiparticles, i.e. ^{60}Co becomes anti-^{60}Co; $e^- \rightarrow e^+$, $\bar{\nu}_e \rightarrow \nu_e$ through the charge conjugation operation $\hat{C}$; the $\hat{C}\hat{P}$ operation transforms the original experimental situation in another possible experimental situation. It is understood up till now that, although beta decay violates the parity $\hat{P}$ symmetry, it conserves the $\hat{C}\hat{P}$ symmetry. Certain elementary particle processes related to K-meson decay (Christensen *et al* 1964) violate even this $\hat{C}\hat{P}$ symmetry. No good understanding of these processes has been obtained as yet! (See figure 5.37.)

5.5.4 The neutrino intrinsic properties: helicity

The above experiments on beta-decay unambiguously showed that the antineutrino (or neutrino in β^+ processes) has a very definite handedness. In the processes that occur in nature, only right-handed antineutrinos and left-handed neutrinos appear showing a basic asymmetry in the reflection ($\hat{P}$) of a given

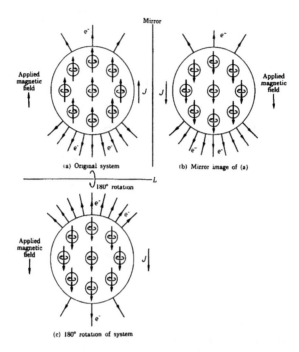

Figure 5.36. The ^{60}Co beta-decay experiment, drawn in a schematic way, to illustrate the essentials of the test of parity conservation in weak processes. Both the original system (a), the mirror image (b) and the rotated (rotation by 180°) system (c) are shown. Since the mirror image (b) is not a realizable experiment and the rotated (c) is; a clear proof of the break-down of reflection symmetry in beta-decay is imposed.

beta-decay. The helicity operator, given by $\hat{h} \equiv \hat{\vec{\sigma}} \cdot \hat{\vec{p}}/|\vec{p}|$, when acting on a fermion described by a plane wave, moving in the positive z-axis direction and characterized by a single spin (spin-up) component, has an eigenvalue $+1$. This can easily be seen by evaluating

$$\frac{\hat{\sigma}_z \hat{p}_z}{|p_z|} \left\{ e^{ikz} \begin{pmatrix} 1 \\ 0 \end{pmatrix} \right\}. \tag{5.95}$$

With $p_z = -i\hbar\partial/\partial z$ and

$$\hat{\sigma}_z = \frac{\hbar}{2} \begin{pmatrix} 1 & 0 \\ 0 & -1 \end{pmatrix},$$

the eigenvalue of $\hat{\sigma}_z \hat{p}_z/|p_z|$ becomes $+1$.

A direct measurement of the neutrino helicity was carried out in an elegant experiment by Goldhaber *et al* (1958). Since the analysis uses concepts relating

Figure 5.37. Symmetry prospects on CP violation. (Taken from Fabergé, CERN Courier 1966. Reprinted with permission of CERN.)

to gamma-decay, we do not give the full discussion here. We refer to Jelley (1990) for a simple but still careful analysis. The outcome verifies a helicity eigenvalue of $h = -1$ for the neutrino, consistent with complete polarization.

It is this helicity property that causes the antineutrino capture process

$$\bar{\nu}_e + {}^{37}Cl \rightarrow {}^{37}Ar + e^-,$$

to be forbidden and also forbids neutrinoless double beta decay to occur, even though the neutrino would be a Majorana particle.

We would like to end this section and chapter by pointing out that the questions related to the neutrino properties in beta-decay processes, are not yet solved in a convincing way. The Dirac versus Majorana interpretation is still open; the double beta-decay processes have to be studied more carefully and the search for a finite mass for the neutrino are both issues that are very much open to question.

Box 5a. Discovering the W and Z bosons: detective work at CERN and the construction of a theory

---

As discussed in the introductory section on the weak decay processes ($\beta^{\mp}$ decay, EC), it became clear that the 'range' of interaction has to be very short. The idea was put forward of bosons carrying the weak force and coupling on one side to the hadronic current (changing a proton into a neutron, or the other way round) and on the other side to the weak current (creating, for example, the electron and electron antineutrino in the $n \rightarrow p + e^- + \bar{\nu}_{e^-}$ reaction). Rest masses of the order of 80–100 GeV were suggested and in order to eventually prove the existence of such particles (called intermediate bosons), one had to obtain at least twice the rest mass energy in the COM system when colliding particles at the big accelerators.

This project was performed with the CERN SPS, using collisions between protons and accumulated antiprotons ($\bar{p}$) accelerated in the same machine to give 540 GeV COM energy. The layout of the interconnected system of accelerators at CERN producing the necessary beam of particles is presented in figure 5a.1. It is the work of Rubbia in particular, combined with the expertise at CERN in producing and accelerating intense beams of $\bar{p}$ particles (in particular through the work of Van Der Meer) that made the building of the p–$\bar{p}$ collider a success.

What was discovered? Normally, when p–$\bar{p}$ collisions occur, large showers of particles are formed. In nine events out of thousands of millions of others, the collision produced a track with an electron (or proton) with high energy

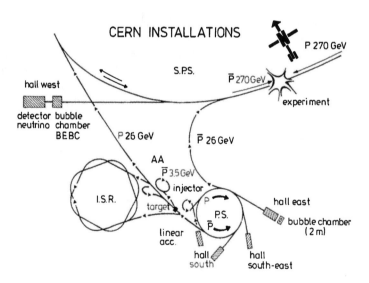

Figure 5a.1. The complex network connecting the various CERN accelerators that made the search for the intermediate boson possible.

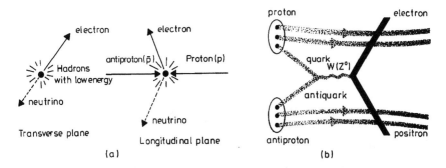

Figure 5a.2. (*a*) The p–p̄ interaction process resulting in the emission of an electron and a neutrino from the decay of the carrier of the weak force. (*b*) Description of the same process but now presented at the level of the quarks constituting the proton and antiproton in creating the weak force carrier and its subsequent decay processes.

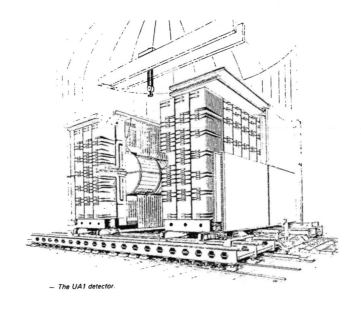

— *The UA1 detector.*

Figure 5a.3. Schematic drawing of the UA-1 detector, the first to show the decay process $p + \bar{p} \to W^{\pm} + X$ with $X^{\pm} \to e^{\pm} + \nu$ (Brianti and Gabathuler 1983).

(20–40 GeV) moving at almost right angles to most other tracks of particles and seemingly unbalanced by momentum conservation. The interpretation (see figures 5a.2(*a*), (*b*)) suggested that one of the three valence quarks in the proton had collided 'head-on' with an antiquark of the antiproton. Through the interaction via the weak force, a *W* particle of $\simeq 80$ GeV was created, which

Volume 122B, number 1   PHYSICS LETTERS   24 February 1983

EXPERIMENTAL OBSERVATION OF ISOLATED LARGE TRANSVERSE ENERGY ELECTRONS
WITH ASSOCIATED MISSING ENERGY AT $\sqrt{s}$ = 540 GeV

UA1 Collaboration, CERN, Geneva, Switzerland

G. ARNISON [j], A. ASTBURY [j], B. AUBERT [b], C. BACCI [i], G. BAUER [i], A. BÉZAGUET [d], R. BÖCK [d],
T.J.V. BOWCOCK [i], M. CALVETTI [d], T. CARROLL [d], P. CATZ [b], P. CENNINI [d], S. CENTRO [d],
F. CERADINI [d], S. CITTOLIN [d], D. CLINE [i], C. COCHET [k], J. COLAS [b], M. CORDEN [c], D. DALLMAN [d],
M. DeBEER [k], M. DELLA NEGRA [b], M. DEMOULIN [d], D. DENEGRI [k], A. Di CIACCIO [i],
D. DiBITONTO [d], L. DOBRZYNSKI [g], J.D. DOWELL [c], M. EDWARDS [c], K. EGGERT [a],
E. EISENHANDLER [f], N. ELLIS [d], P. ERHARD [a], H. FAISSNER [a], G. FONTAINE [g], R. FREY [h],
R. FRÜHWIRTH [l], J. GARVEY [c], S. GEER [g], C. GHESQUIÈRE [g], P. GHEZ [b], K.L. GIBONI [a],
W.R. GIBSON [f], Y. GIRAUD-HÉRAUD [g], A. GIVERNAUD [k], A. GONIDEC [b], G. GRAYER [j],
P. GUTIERREZ [h], T. HANSL-KOZANECKA [a], W.J. HAYNES [j], L.O. HERTZBERGER [2], C. HODGES [h],
D. HOFFMANN [a], H. HOFFMANN [d], D.J. HOLTHUIZEN [2], R.J. HOMER [c], A. HONMA [f], W. JANK [d],
G. JORAT [d], P.I.P. KALMUS [f], V. KARIMÄKI [e], R. KEELER [f], I. KENYON [c], A. KERNAN [h],
R. KINNUNEN [e], H. KOWALSKI [d], W. KOZANECKI [h], D. KRYN [d], F. LACAVA [d], J.-P. LAUGIER [k],
J.-P. LEES [b], H. LEHMANN [a], K. LEUCHS [a], A. LÉVÊQUE [k], D. LINGLIN [b], E. LOCCI [k], M. LORET [k],
J.-J. MALOSSE [k], T. MARKIEWICZ [d], G. MAURIN [d], T. McMAHON [c], J.-P. MENDIBURU [g],
M.-N. MINARD [b], M. MORICCA [i], H. MUIRHEAD [d], F. MULLER [d], A.K. NANDI [j], L. NAUMANN [d],
A. NORTON [d], A. ORKIN-LECOURTOIS [g], L. PAOLUZI [i], G. PETRUCCI [d], G. PIANO MORTARI [i],
M. PIMIÄ [e], A. PLACCI [d], E. RADERMACHER [a], J. RANSDELL [h], H. REITHLER [a], J.-P. REVOL [d],
J. RICH [k], M. RIJSSENBEEK [d], C. ROBERTS [j], J. ROHLF [d], P. ROSSI [d], C. RUBBIA [d], B. SADOULET [d],
G. SAJOT [g], G. SALVI [f], G. SALVINI [i], J. SASS [k], J. SAUDRAIX [k], A. SAVOY-NAVARRO [k],
D. SCHINZEL [i], W. SCOTT [j], T.P. SHAH [j], M. SPIRO [k], J. STRAUSS [i], K. SUMOROK [d], F. SZONCSO [i],
S. Van der MEER [d], J.-P. VIALLE [d], J. VRANA [g], V. VUILLEMIN [d], H.D. WAHL [i], P. WATKINS [c],
J. WILSON [c], Y.G. XIE [d], M. YVERT [b] and E. ZURFLUH [d]

*Aachen [a] – Annecy (LAPP) [b] – Birmingham [c] – CERN [d] – Helsinki [e] – Queen Mary College, London [f] – Paris (Coll. de France) [g]
– Riverside [h] – Rome [i] – Rutherford Appleton Lab. [j] – Saclay (CEN) [k] – Vienna [l] Collaboration*

Received 23 January 1983

We report the results of two searches made on data recorded at the CERN SPS Proton–Antiproton Collider: one for isolated large-E_T electrons, the other for large-E_T neutrinos using the technique of missing transverse energy. Both searches converge to the same events, which have the signature of a two-body decay of a particle of mass ~80 GeV/c^2. The topology as well as the number of events fits well the hypothesis that they are produced by the process $\bar{p} + p \rightarrow W^\pm + X$, with $W^\pm \rightarrow e^\pm + \nu$; where $W^\pm$ is the Intermediate Vector Boson postulated by the unified theory of weak and electromagnetic interactions.

[1] University of Wisconsin, Madison, WI, USA.
[2] NIKHEF, Amsterdam, The Netherlands.

Figure 5a.4. First page of the original article presenting the first observation of a decay of the $W^\pm$ boson. The title page presents this observation in a very cautious way as '... isolated large transverse energy electrons with ... missing energy at $s = 540$ GeV'. (Taken from *Phys. Lett.* **122B** 1983. Reprinted with permission of Elsevier Science Publishers.)

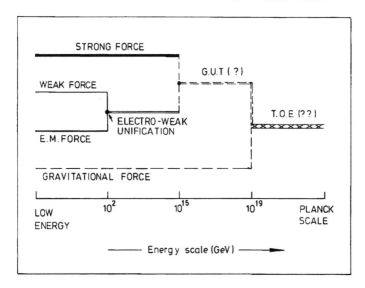

Figure 5a.5. A diagram illustrating a possible unification scheme for all forces. The various linking lines where forces join together and the corresponding energy scale where these processes might occur are illustrated. Here G.U.T. represents Grand Unification Theory and T.O.E. Theory of Everything.

subsequently decayed into an electron and an electron–antineutrino (or positron and electron–neutrino). The two particles share the energy and momentum liberated and mainly go on in opposite directions since the W particles are formed almost at rest. The neutrino went on undetected leaving an apparent imbalance in momentum! This was the signature of a very important event verifying the electroweak theory.

The detectors needed to identify the events (UA-1 and UA-2 located at the SPS collision points) are huge in terms of dimensions, and advanced electronic, calorimetric and other detecting methods were used to observe just those 'golden' events. The UA-1 detector, as well as the first page of the article announcing the detection of $(e^-, \bar{\nu}_e)$ and (e^+, ν_e) events are illustrated in figures 5a.3 and 5a.4. A very nice account of this accomplishment is discussed in the book by Close, Martin and Sutton, *The Particle Explosion* (Close *et al* 1987).

A more precise value for the mass from these few events is calculated as 83 ± 5 GeV in accord with the predicted value of 82 ± 2 GeV from the Salam–Weinberg model. The production rate is also correctly predicted. The neutral partner, the Z^0 particle, which is produced at a rate ten times smaller, has been observed more recently both at Stanford and CERN.

The above data illustrate magnificently the correctness of ideas that were put forward by Weinberg, Salam and Glashow who pointed out that the electromagnetic force and the weak force can actually be unified into a single electroweak theory. The development of the theory, along a steady line of unification of the two forces has thereby obtained an important test and might hint towards a higher hierarchy of forces unified in even more encompassing theories. A diagram illustrating this possible unification on an energy scale over which these unifications should become observable is presented in figure 5a.5.

The idea of unifying and 'assembling' a theory is presented in the *CERN Courier*, Vol 23, November 1983.

Box 5b. First laboratory observation of double beta-decay

The first direct evidence of the double beta-decay of a nucleus has been reported by Elliott, Hahn and Moe in the issue of *Physical Review Letters* **59** 2020 (1987). Double beta-decay is an extremely rare process in which two electrons are emitted. Elliott *et al* observed the decay of ^{82}Se into ^{82}Kr plus two electrons and two electron antineutrinos with a measure of $T = 1.1^{+0.8}_{-0.3} \times 10^{20}$ y, making it the rarest natural decay process ever observed under laboratory conditions. More details are described in a review article on double beta-decay (Moe and Rosen, Scientific American, November 1989) and in figure 5b.1 we present the salient features that led to the above decay processes. The captions discuss the most important steps of the experimental set-up, as well as the observed electron energy spectra.

The ultimate goal is to find out if double beta-decay *without* neutrinos can be observed. Such decay modes would violate lepton-number conservation, one of the few conservation laws thought to be rigidly fulfilled. The Lepton number is defined as $+1$ for an electron and neutrino, and -1 for their antiparticles! Therefore, in the standard model of particle physics, the emission of an electron must be accompanied by an antineutrino: the neutrino in this model is called a Dirac neutrino.

There exists a Majorana theory of neutrinos (Furry 1939) in which the neutrino and antineutrino are the *same* particle. The only distinction is that neutrinos are left-handed and antineutrinos right-handed. If neutrinos are exactly massless, there is no way to distinguish the Dirac neutrino from the Majorana neutrino. Thus, the observation of neutrinoless beta-decay would not only demonstrate lepton-number non-conservation but would prove that the neutrino's mass is *not* exactly zero!

The right- and left-handedness of the various beta-decay processes is illustrated in figure 5b.2, and is discussed in the section on parity violation in beta-decay.

The present stage of theory and experiment suggests that if electron neutrinos are Majorana particles, the effective mass is $\lesssim 1$ eV. Now, a direct two-neutrino measurement exists, and as long as the possibility of a measurable neutrino mass exists, the quest for neutrinoless double beta-decay will go on.

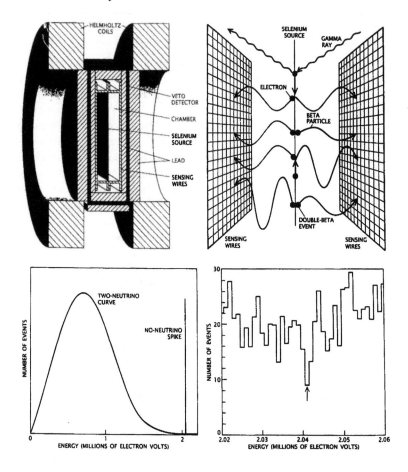

Figure 5b.1. The chamber (upper left) in which the first direct evidence for double beta-decay was obtained. A sample of ^{82}Se is supported in the central plane of the detector. Surrounding is a chamber filled with He gas. The chamber is protected from outside radioactivity by Pb shielding and a 'veto' detector which discriminates against incoming cosmic radiation. The Helmoltz coils generate a magnetic field causing the electrons to follow helical paths and ionize the He gas in its passage (upper right). An applied electric field causes the resulting free electrons to drift to sensing wires, which register arrival time and position. The size and pitch of a helix determines the beta-ray energy. The double-beta decay event can be imitated by a few rare background events as shown (lower left). The energy spectrum associated with ^{76}Ge decay for the two-neutrino and no-neutrino possibilities is shown lower left. The latter spike should appear at 2.041 MeV (arrow in lower right). The spectrum in this region is largely the result of statistical fluctuations in the background. If the neutrinoless double-beta-decay contribution is assumed to be less than the size of the statistical fluctuations, the half-life of ^{76}Ge should be larger than 2.3×10^{24} y (taken from Moe and Rosen 1989 from *Scientific American* (November)).

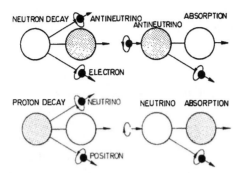

Figure 5b.2. When a neutron (top left) decays by a beta transition it emits an $(e^-, \bar{\nu}_e)$ pair with a right handed anti neutrino. For the corresponding (bottom left) proton decay, a (e^+, ν_e) pair with a left-handed neutrino is created. On the right-hand side, the corresponding absorption processes are shown.

Box 5c. The width of the Z^0 particle: measuring the number of neutrino families

---

The LEP accelerator at CERN has been operating since 14 July 1990. One of the major goals for experiments in the early stages was the production of tens of millions of Z^0 particles and the study of their subsequent decay modes.

The quantity of interest is the lifetime or the level width $\Gamma(Z^0)$. This quantity is measured and at the same time the mass of the Z^0 particle is obtained: the scattering cross-section is determined for a number of energies near the exact mass (for the $e^+e^- \rightarrow$ hadrons. A Lorentzian shape is obtained with its maximum at M_{Z^0} and characterized by a width $\Gamma(Z^0)$ (figure 5c.1). In explaining the shape *and* width correctly, various processes need to be taken into account. A table with decay possibilities, partial widths and branching ratios, is given (table 5c.1).

It is standard practice to assume three families of leptons, as follows

$$e^-, \quad e^+, \quad \nu_e, \quad \bar{\nu}_e$$
$$\mu^-, \quad \mu^+, \quad \nu_\mu, \quad \bar{\nu}_\mu$$
$$\tau^-, \quad \tau^+, \quad \nu_\tau, \quad \bar{\nu}_\tau.$$

They are classified according to the electron, mu-meson (muon) and tau-meson particles. If these were the only three neutrino families, a good fit to the $\Gamma(Z^0)$ width would be obtained.

Calculations using $\sin^2 \theta_W = 0.230$ and $M_{Z^0} = 91.9$ GeV/c^2, result in a total width $\Gamma(Z^0) = 2.56$ GeV (assuming $t\bar{t}$ is forbidden and using *three* lepton families) and a mean-life $\tau(Z^0) \simeq 2.6 \times 10^{-25}$ s.

One can reverse the argument and derive a constraint on the number of neutrino (lepton) families since each massless extra pair contributes another

Table 5c.1.

Decay mode			Partial width (MeV)	Branching ratio
$Z^0 \rightarrow$	e^+e^-		88	0.034
	$\mu^+\mu^-$		88	0.034
	$\tau^+\tau^-$		88	0.034
	$\nu_{e^-}\bar{\nu}_{e^-}$		175	0.068
	$\nu_\mu\bar{\nu}_\mu$		175	0.068
	$\nu_\tau\bar{\nu}_\tau$		175	0.068
	$u\bar{u}$		302	0.118
	$d\bar{d}$		388	0.152
	$s\bar{s}$		388	0.152
	$c\bar{c}$	hadrons	302	0.118
	$b\bar{b}$		388	0.152
	$t\bar{t}$		?	

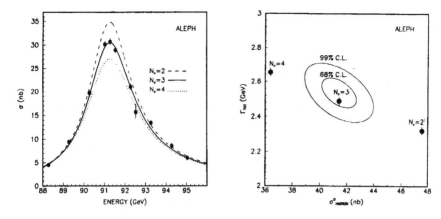

Figure 5c.1. Illustration of the first results on a precise determination of the number of light neutrinos and the Z^0 boson partial widths. Besides the basic figures from the ALEPH collaboration (taken from Decamp *et al* 1990), the first page of the SLAC publication that appeared in the vol 63, no 20 issue of *Physical Review Letters* of 13 November 1989 as well as the CERN (LEP) results announced in the 16 November 1989 *Physics Letters* issue (**231B**) are also shown.

175 MeV to the width. The data, at present, are clearly consistent with $N_\nu = 3$. The original data were published by groups at Stanford (13 November 1989) and CERN (16 November 1989) within a week and both confirm the number of families as 3. Better statistics obtained at CERN (1991) reinforced this constraint to $N_\nu = 3$. We give, as an illustration of the size of the groups, and a measure of the work involved in these experiments, the title pages of the two original experiments.

VOLUME 63, NUMBER 20   PHYSICAL REVIEW LETTERS   13 NOVEMBER 1989

Measurements of Z-Boson Resonance Parameters in e^+e^- Annihilation

G. S. Abrams,[1] C. E. Adolphsen,[2] D. Averill,[4] J. Ballam,[3] B. C. Barish,[5] T. Barklow,[3] B. A.
Barnett,[6] J. Bartelt,[3] S. Bethke,[1] D. Blockus,[4] G. Bonvicini,[7] A. Boyarski,[3] B. Brabson,[4]
A. Breakstone,[8] J. M. Brom,[4] F. Bulos,[3] P. R. Burchat,[2] D. L. Burke,[3] R. J. Cence,[8]
J. Chapman,[7] M. Chmeissani,[7] D. Cords,[3] D. P. Coupal,[3] P. Dauncey,[6] H. C. DeStaebler,[3] D. E.
Dorfan,[2] J. M. Dorfan,[3] D. C. Drewer,[6] R. Elia,[3] G. J. Feldman,[3] D. Fernandes,[3] R. C. Field,[3]
W. T. Ford,[9] C. Fordham,[3] R. Frey,[7] D. Fujino,[3] K. K. Gan,[3] E. Gero,[7] G. Gidal,[1]
T. Glanzman,[3] G. Goldhaber,[1] J. J. Gomez Cadenas,[2] G. Gratta,[2] G. Grindhammer,[3]
P. Grosse-Wiesmann,[3] G. Hanson,[3] R. Harr,[1] B. Harral,[6] F. A. Harris,[8] C. M. Hawkes,[5]
K. Hayes,[3] C. Hearty,[1] C. A. Heusch,[2] M. D. Hildreth,[3] T. Himel,[3] D. A. Hinshaw,[9] S. J.
Hong,[7] D. Hutchinson,[3] J. Hylen,[6] W. R. Innes,[3] R. G. Jacobsen,[3] J. A. Jaros,[3] C. K. Jung,[3]
J. A. Kadyk,[1] J. Kent,[2] M. King,[2] S. R. Klein,[3] D. S. Koetke,[3] S. Komamiya,[3] W. Koska,[7]
L. A. Kowalski,[3] W. Kozanecki,[3] J. F. Kral,[1] M. Kuhlen,[5] L. Labarga,[2] A. J. Lankford,[3] R. R.
Larsen,[3] F. Le Diberder,[3] M. E. Levi,[1] A. M. Litke,[2] X. C. Lou,[4] V. Lüth,[3] J. A. McKenna,[5]
J. A. J. Matthews,[6] T. Mattison,[3] B. D. Milliken,[5] K. C. Moffeit,[3] C. T. Munger,[3] W. N.
Murray,[4] J. Nash,[3] H. Ogren,[4] K. F. O'Shaughnessy,[3] S. I. Parker,[8] C. Peck,[5] M. L. Perl,[3]
F. Perrier,[3] M. Petradza,[3] R. Pitthan,[3] F. C. Porter,[5] P. Rankin,[9] K. Riles,[3] F. R. Rouse,[3] D. R.
Rust,[4] H. F. W. Sadrozinski,[2] M. W. Schaad,[1] B. A. Schumm,[1] A. Seiden,[2] J. G. Smith,[9]
A. Snyder,[4] E. Soderstrom,[5] D. P. Stoker,[6] R. Stroynowski,[5] M. Swartz,[3] R. Thun,[7] G. H.
Trilling,[1] R. Van Kooten,[3] P. Voruganti,[3] S. R. Wagner,[3] S. Watson,[3] P. Weber,[3] A. Weigend,[3]
A. J. Weinstein,[2] A. J. Weir,[5] E. Wicklund,[5] M. Woods,[3] D. Y. Wu,[5] M. Yurko,[4]
C. Zaccardelli,[2] and C. von Zanthier[2]

[1]*Lawrence Berkeley Laboratory and Department of Physics,
University of California, Berkeley, California 94720*
[2]*University of California at Santa Cruz, Santa Cruz, California 95064*
[3]*Stanford Linear Accelerator Center, Stanford University, Stanford, California 94309*
[4]*Indiana University, Bloomington, Indiana 47405*
[5]*California Institute of Technology, Pasadena, California 91125*
[6]*Johns Hopkins University, Baltimore, Maryland 21218*
[7]*University of Michigan, Ann Arbor, Michigan 48109*
[8]*University of Hawaii, Honolulu, Hawaii 96822*
[9]*University of Colorado, Boulder, Colorado 80309*
(Received 12 October 1989)

We have measured the mass of the Z boson to be 91.14 ± 0.12 GeV/c^2, and its width to be $2.42^{+0.45}_{-0.35}$ GeV. If we constrain the visible width to its standard-model value, we find the partial width to invisible decay modes to be 0.46 ± 0.10 GeV, corresponding to 2.8 ± 0.6 neutrino species, with a 95%-confidence-level upper limit of 3.9.

PACS numbers: 14.80.Er, 13.38.+c, 13.65.+i

We present an improved measurement of the Z-boson resonance parameters. The measurement is based on a total of 19 nb^{-1} of data recorded at ten different center-of-mass energies between 89.2 and 93.0 GeV by the Mark II detector at the SLAC Linear Collider. This data sample represents approximately 3 times the data integrated luminosity presented in an earlier Letter.[1] The statistical significance of the luminosity measurement is further improved by including a detector component — the mini-small-angle monitor (MiniSAM) — not used in the previous analysis. The larger data sample and improved luminosity measurement result in a significant reduction in the resonance-parameter uncertainties. In particular, our observations exclude the presence of a fourth standard-model massless neutrino species at a confidence level of 95%.

The Mark II drift chamber and calorimeters provide the principal information used to identify Z decays.[2] Charged particles are detected and momentum analyzed in a 72-layer cylindrical drift chamber in a 4.75-kG axial magnetic field. The drift chamber tracks charged particles with $|\cos\theta| < 0.92$, where θ is the angle to the incident beams. Photons are detected in electromagnetic calorimeters that cover the region $|\cos\theta| < 0.96$. The calorimeters in the central region (barrel calorimeters) are lead–liquid-argon ionization chambers, while the end-cap calorimeters are lead–proportional-tube counters.

There are two detectors for the small-angle e^+e^- (Bhabha) events used to measure the integrated luminosity. The small-angle monitors (SAM's) cover the angular region of $50 < \theta < 160$ mrad. Each SAM consists of

PHYSICS LETTERS B

Volume 231, number 4   16 November 1989

CONTENTS

A determination of the properties of the neutral
intermediate vector boson Z^0
L3 Collaboration. B. Adeva. O. Adriani.
M. Aguilar-Benitez, H. Akbari. J. Alcaraz. A. Aloisio.
M.G. Alviggi, Q. An. H. Anderhub, A.L. Andersson.
L. Antonov, D. Antreasyan. A. Arefiev. T. Azemoon.
T. Aziz. P.V.K.S. Baba. P. Bagnaia, J.A. Bakken.
L. Baksay. R.C. Ball. S. Banerjee. J. Bao. L. Barone.
A. Bay. U. Becker. S. Beingessner. Gy.L. Bencze.
G. Bencze. J. Berdugo. P. Berges. B. Bertucci.
B.L. Betev. A. Biland. R. Bizzarri. J.J. Blaising.
P. Blömeke. G.J. Bobbink. M. Bocciolini. W. Böhlen.
A. Böhm. T. Böhringer. B. Borgia. D. Bourilkov.
M. Bourquin. D. Bouigny. J.G. Branson. I.C. Brock.
F. Bruyant, J.D. Burger. J.P. Burq. X.D. Cai.
D. Campana. C. Camps. M. Capell. F. Carbonara.
A.M. Cartacci. M. Cerrada. F. Cesaroni. Y.H. Chang.
U.K. Chaturvedi. M. Chemarin. A. Chen. C. Chen.
G.M. Chen. H.F. Chen. H.S. Chen. M. Chen.
M.L. Chen, G. Chiefari. C.Y. Chien. C. Civinini.
I. Clare. R. Clare. G. Coignet. N. Colino.
V. Commichau. G. Conforto. A. Contin. F. Crijns.
X.Y. Cui. T.S. Dai. R. D'Alessandro, X. De Bouard.
A. Degre. K. Deiters. E. Dénes. P. Denes.
F. DeNotaristefani. M. Dhina. M. Diemoz.
H.R. Dimitrov. C. Dionisi. F. Dittus, R. Dolin.
E. Drago. T. Driever. P. Duinker. I. Duran.
A. Engler. F.J. Eppling. F.C. Erne. P. Extermann.
R. Fabbretti. G. Faber. S. Falciano. S.J. Fan.
M. Fabre. J. Fay. J. Fehlmann. H. Fenker.
T. Ferguson. G. Fernandez. F. Ferroni. H. Fesefeldt.
J. Field. G. Forconi. T. Foreman. K. Freudenreich.
W. Friebel. M. Fukushima. M. Gailloud.

Yu. Galaktionov. E. Gallo, S.N. Ganguli, S.S. Gau.
G. Gavrilov, S. Gentile, M. Gettner, M. Glaubman.
S. Goldfarb. Z.F. Gong, E. Gonzalez, A. Gordeev,
P. Göttlicher, C. Goy, G. Gratta, A. Grimes.
C. Grinnell, M. Gruenewald. M. Guanziroli.
A. Gurtu, D. Güsewell, H. Haan, K. Hangarter.
S. Hancke, M. Harris, D. Harting, F.G. Hartjes,
C.F. He, A. Heavey, T. Hebbeker, M. Hebert.
G. Herten, U. Herten. A. Hervé, K. Hilgers.
H. Hofer, L.S. Hsu, G. Hu, G.Q. Hu. B. Ille.
M.M. Ilyas, V. Innocente, E. Isiksal, E. Jagel,
B.N. Jin, L.W. Jones, M. Jongmanns, H. Jung.
P. Kaaret, R.A. Khan, Yu. Kamyshkov, D. Kaplan,
W. Karpinski, Y. Karyotakis, V. Khoze, D. Kirkby,
W. Kittel. A. Klimentov, P.F. Klok, M. Kollek.
M. Koller. A.C. König, O. Kornadt, V. Koutsenko.
R.W. Kraemer, V.R. Krastev, W. Krenz, A. Kuhn,
V. Kumar. A. Kunin, S. Kwan, G. Landi, K. Lanius.
D. Lanske, S. Lanzano. M. Lebeau, P. Lebrun,
P. Lecomte, P. Lecoq. P. Le Coultre, I. Leedom,
J.M. Le Goff. L. Leistam, R. Leiste, J. Lettry.
X. Leytens. C. Li, H.T. Li, J.F. Li. L. Li, P.J. Li,
X.G. Li, J.Y. Liao. R. Liu, Y. Liu, Z.Y. Lin,
F.L. Linde. B. Lindemann, D. Linnhofer,
W. Lohmann, S. Lökós, E. Longo, Y.S. Lu,
J.M. Lubbers, K. Lübelsmeyer, C. Luci, D. Luckey,
L. Ludovici. X. Lue, L. Luminari, W.G. Ma,
M. MacDermott, R. Magahiz, M. Maire,
P.K. Malhotra, A. Malinin, C. Maña, D.N. Mao,
Y.F. Mao. M. Maolinbay. P. Marchesini,
A. Marchionni, J.P. Martin, L. Martinez,
H.U. Martyn. F. Marzano, G.G.G. Massaro,
T. Matsuda, K. Mazumdar, P. McBride,
Th. Meinholz, M. Merk, R. Mermod, L. Merola,
M. Meschini, W.J. Metzger, Y. Mi, M. Micke.
U. Micke, G.B. Mills, Y. Mir, G. Mirabelli, J. Mnich.
M. Moeller. L. Montanet, B. Monteleoni,
G. Morand. R. Morand. S. Morganti, V. Morgunov.
R. Mount. E. Nagy. M. Napolitano, H. Newman,
L. Niessen. W.D. Nowak, J. Onvlee, J. Ossmann.
D. Pandoulas. G. Paternoster, S. Patricelli, Y.J. Pei,
Y. Peng, D. Perret-Gallix, J. Perrier, E. Perrin,
A. Pevsner, M. Pieri. V. Pieri, P.A. Piroué,
V. Plyaskin, M. Pohl, V. Pojidaev, C.L.A. Pols,
N. Produit. P. Prokofiev. J.M. Qian, K.N.Qureshi.
R. Raghavan. G. Rahal-Callot. P. Razis, K. Read.
D. Ren. Z. Ren. S. Reucroft. A. Ricker. T. Riemann.
C. Rippich. S. Rodriguez. B.P. Roe. M. Röhner.
S. Röhner. L. Romero. J. Rose. S. Rosier-Lees.
Ph. Rosselet. J.A. Rubio. W. Ruckstuhl.
H. Rykaczewski. M. Sachwitz. J. Salicio.
M. Sassowsky. G. Sauvage. A. Savin. V. Schegelsky.
A. Schetkovsky. P. Schmitt. D. Schmitz. P. Schmitz.

(Continued on preceding page)

PHYSICS LETTERS B

Volume 231, number 4   16 November 1989

CONTENTS

(Continued on preceding page)

Box 5d. The neutrino vanishing act: disappearance of electron anti-neutrinos from Japanese reactors

The neutrino vanishing act: electron-antineutrinos from Japanese reactors disappear on their way to the Kamioka mine The neutrino-oscillation problem has been an issue that was most dramatically illustrated by Davies' experiment at Homestake (Davies *et al* 1968), signaling a missing neutrino problem for solar neutrinos, produced in the fusion reactions in the sun. This solar neutrino problem has essentially been solved at the SNO detector in Canada very recently (Ahmad *et al* 2001, Ahmad *et al* 2002). Starting from fission in nuclear reactors producing neutron-rich fragments, a large flux of electron anti-neutrinos is produced through the β-decay process as they decay towards stable nuclei. Using the inverse reaction to detect the electron anti-neutrinos i.e. $\bar{\nu}_e + p \rightarrow e^+ + n$, one is able to get a unique signal by detecting both the positron and the delayed 2.2 MeV γ-ray from neutron capture on a proton. Such a detector, KamLAND (standing for Kamiokande Liquid Scintillator Anti-Neutrino Detector), occupies the earlier Kamiokande site and consists of a kilotonne of ultra-pure scintillation liquid held in a weather balloon. This balloon is surrounded by an array of 1879 photomultiplier tubes which pick up the light resulting from the neutrino reaction with protons in the scintillator. Data, discussed by Eguchi *et al* (Eguchi *et al*

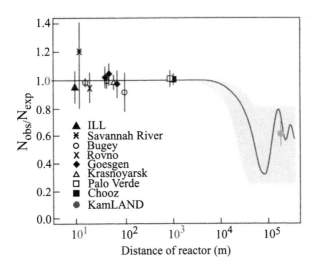

Figure 5d.1. First results from the KamLAND neutrino oscillation experiment, plotting the observed number of $\bar{\nu}_e$ ($N_{observed}$) (flux at the detector site) over the expected number of $\bar{\nu}_e$ ($N_{expected}$) (flux if no oscillations are present). The results of other reactor experiments (see figure) are also incorporated. (Reprinted figure from Eguchi *et al* 2003 *Phys. Rev. Lett.* **90** 021802 © 2003 by American Physical Society.)

2003), have been collected over the period March through October 2002. The detailed analyses in order to extract the electron-antineutrino reactions expected during this time span are described in (Eguchi *et al* 2003). The expected number of electron anti-neutrinos in the absence of any oscillation amounts to 86.8 ± 5.6 events. These anti-neutrinos are provided by the many nuclear reactors but the flux is essentially dominated by a few powerful reactors at an average distance of $\approx$180 km. More than 79% of the flux comes from 26 reactors between 138–214 km away. One close reactor at 88 km contributes up to 6.7% of the total. The relative narrow band of distances allows KamLAND to be sensitive to distortions in the neutrino energy spectrum for a certain set of oscillation parameters. The final outcome of the analyses results in 54 events, and in figure 5d.1, we show the ratio of the expected flux for KamLAND as well as for earlier reactor experiments as a function of the average distance from the source. This result with the disappearance of 32 electron anti-neutrinos is very spectacular and indicates a region in the oscillation parameter space corresponding to a large mixing angle (LMA) solution $\sin^2 2\theta = 1.0$ and $\Delta m^2 = 6.9 \times 10^{-5}$ eV^2 consistent with results derived from SNO experiments. The upshot here is that the source of neutrinos, in the present experiment, is totally free of any ambiguity since it is produced from nuclear β-decay. Results from new runs over extended time periods will bring in extra information on the finer details of neutrino oscillation behaviour and neutrino masses.

Box 5e. Experimental test of parity conservation in beta-decay: the original paper

Experimental Test of Parity Conservation in Beta Decay*

C. S. Wu, *Columbia University, New York, New York*

AND

E. Ambler, R. W. Hayward, D. D. Hoppes, AND R. P. Hudson,
National Bureau of Standards, Washington, D. C.

(Received January 15, 1957)

IN a recent paper[1] on the question of parity in weak interactions, Lee and Yang critically surveyed the experimental information concerning this question and reached the conclusion that there is no existing evidence either to support or to refute parity conservation in weak interactions. They proposed a number of experiments on beta decays and hyperon and meson decays which would provide the necessary evidence for parity conservation or nonconservation. In beta decay, one could measure the angular distribution of the electrons coming from beta decays of polarized nuclei. If an asymmetry in the distribution between θ and $180° - \theta$ (where θ is the angle between the orientation of the parent nuclei and the momentum of the electrons) is observed, it provides unequivocal proof that parity is not conserved in beta decay. This asymmetry effect has been observed in the case of oriented Co^{60}.

It has been known for some time that Co^{60} nuclei can be polarized by the Rose-Gorter method in cerium magnesium (cobalt) nitrate, and the degree of polarization detected by measuring the anisotropy of the succeeding gamma rays.[2] To apply this technique to the present problem, two major difficulties had to be overcome. The beta-particle counter should be placed *inside* the demagnetization cryostat, and the radioactive nuclei must be located in a *thin surface* layer and polarized. The schematic diagram of the cryostat is shown in Fig. 1.

To detect beta particles, a thin anthracene crystal $\frac{3}{8}$ in. in diameter$\times\frac{1}{16}$ in. thick is located inside the vacuum chamber about 2 cm above the Co^{60} source. The scintillations are transmitted through a glass window and a Lucite light pipe 4 feet long to a photo-multiplier (6292) which is located at the top of the cryostat. The Lucite head is machined to a logarithmic spiral shape for maximum light collection. Under this condition, the Cs^{137} conversion line (624 kev) still retains a resolution of 17%. The stability of the beta counter was carefully checked for any magnetic or temperature effects and none were found. To measure the amount of polarization of Co^{60}, two additional NaI gamma scintillation counters were installed, one in the equatorial plane and one near the polar position. The observed gamma-ray anisotropy was used as a measure of polarization, and, effectively, temperature. The bulk susceptibility was also monitored but this is of secondary significance due to surface heating effects, and the gamma-ray anisotropy alone provides a reliable measure of nuclear polarization. Specimens were made by taking good single crystals of cerium magnesium nitrate and growing

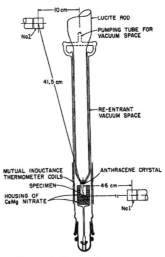

Fig. 1. Schematic drawing of the lower part of the cryostat.

on the upper surface only an additional crystalline layer containing Co^{60}. One might point out here that since the allowed beta decay of Co^{60} involves a change of spin of one unit and no change of parity, it can be given only by the Gamow-Teller interaction. This is almost imperative for this experiment. The thickness of the radioactive layer used was about 0.002 inch and contained a few microcuries of activity. Upon demagnetization, the magnet is opened and a vertical solenoid is raised around the lower part of the cryostat. The whole process takes about 20 sec. The beta and gamma counting is then started. The beta pulses are analyzed on a 10-channel pulse-height analyzer with a counting interval of 1 minute, and a recording interval of about 40 seconds. The two gamma counters are biased to accept only the pulses from the photopeaks in order to discriminate against pulses from Compton scattering.

A large beta asymmetry was observed. In Fig. 2 we have plotted the gamma anisotropy and beta asymmetry *vs* time for polarizing field pointing up and pointing down. The time for disappearance of the beta asymmetry coincides well with that of gamma anisotropy. The warm-up time is generally about 6 minutes, and the warm counting rates are independent of the field direction. The observed beta asymmetry does not change sign with reversal of the direction of the demagnetization field, indicating that it is not caused by remanent magnetization in the sample.

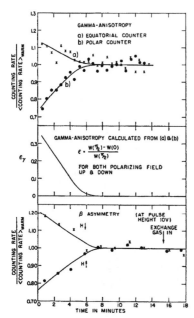

FIG. 2. Gamma anisotropy and beta asymmetry for polarizing field pointing up and pointing down.

The double nitrate cooling salt has a highly anisotropic g value. If the symmetry axis of a crystal is not set parallel to the polarizing field, a small magnetic field will be produced perpendicular to the latter. To check whether the beta asymmetry could be caused by such a magnetic field distortion, we allowed a drop of $CoCl_2$ solution to dry on a thin plastic disk and cemented the disk to the bottom of the same housing. In this way the cobalt nuclei should not be cooled sufficiently to produce an appreciable nuclear polarization, whereas the housing will behave as before. The large beta asymmetry was not observed. Furthermore, to investigate possible internal magnetic effects on the paths of the electrons as they find their way to the surface of the crystal, we prepared another source by rubbing $CoCl_2$ solution on the surface of the cooling salt until a reasonable amount of the crystal was dissolved. We then allowed the solution to dry. No beta asymmetry was observed with this specimen.

More rigorous experimental checks are being initiated, but in view of the important implications of these observations, we report them now in the hope that they may stimulate and encourage further experimental investigations on the parity question in either beta or hyperon and meson decays.

The inspiring discussions held with Professor T. D. Lee and Professor C. N. Yang by one of us (C. S. Wu) are gratefully acknowledged.

* Work partially supported by the U. S. Atomic Energy Commission.
[1] T. D. Lee and C. N. Yang, Phys. Rev. 104, 254 (1956).
[2] Ambler, Grace, Halban, Kurti, Durand, and Johnson, Phil. Mag. 44, 216 (1953).
[3] Lee, Oehme, and Yang, Phys. Rev. (to be published).

The sign of the asymmetry coefficient, α, is negative, that is, the emission of beta particles is more favored in the direction opposite to that of the nuclear spin. This naturally implies that the sign for C_T and C_T' (parity conserved and parity not conserved) must be opposite. The exact evaluation of α is difficult because of the many effects involved. The lower limit of α can be estimated roughly, however, from the observed value of asymmetry corrected for backscattering. At velocity $v/c \approx 0.6$, the value of α is about 0.4. The value of $\langle I_z \rangle / I$ can be calculated from the observed anisotropy of the gamma radiation to be about 0.6. These two quantities give the lower limit of the asymmetry parameter $\beta (\alpha = \beta \langle I_z \rangle / I)$ approximately equal to 0.7. In order to evaluate α accurately, many supplementary experiments must be carried out to determine the various correction factors. It is estimated here only to show the large asymmetry effect. According to Lee and Yang[3] the present experiment indicates not only that conservation of parity is violated but also that invariance under charge conjugation is violated.[4] Furthermore, the invariance under time reversal can also be decided from the momentum dependence of the asymmetry parameter β. This effect will be studied later.

(Reprinted with permission of The American Physical Society.)

Chapter 6

Gamma decay: the electromagnetic interaction at work

6.1 The classical theory of radiation: a summary

Before describing the quantized radiation process, in which the nucleus de-excites from an initial configuration to a lower excited state or to the ground state, we briefly recall the basic results obtained by solving Maxwell's equations for the free radiation field. For this we shall start from the electric field configuration generated by an electric dipole (with dipole moment Π).

The electric field configuration surrounding a dipole (see figure 6.1) as a function of (r, θ) is derived from the potential

$$V = \frac{\Pi_0 \cos \theta}{4\pi \epsilon_0 r^2},$$

(6.1)

giving rise to the radial (E_r) and angular (E_θ) electric field components

$$E_r = -\frac{\partial V}{\partial r} = \frac{2\Pi_0 \cos \theta}{4\pi \epsilon_0 r^3},$$
$$E_\theta = -\frac{1}{r}\frac{\partial V}{\partial \theta} = \frac{\Pi_0 \sin \theta}{4\pi \epsilon_0 r^3}.$$

(6.2)

If now the dipole has a periodic variation; the changing dipole electric field will generate (through Maxwell's equations) a varying magnetic field (figure 6.2).

In the near-source zone, one can (not including any retardation effects) depict E_r and E_θ as

$$E_r = \frac{2\Pi_0 \cos \theta}{4\pi \epsilon_0 r^3} \sin \omega t$$
$$E_\theta = \frac{\Pi_0 \sin \theta}{4\pi \epsilon_0 r^3} \sin \omega t.$$

(6.3)

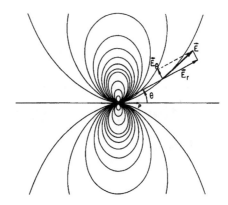

Figure 6.1. Force lines for an electric dipole. At any point, the electric field vector $\vec{E}$ is at a tangent to the given force line. Both the radial ($\vec{E}_r$) and polar ($\vec{E}_\theta$) components are illustrated. (Adapted from Alonso and Finn 1971, *Fundamental University Physics, vol II: Fields and Waves* © 1967 Addison-Wesley Publishing Company. Reprinted by permission.)

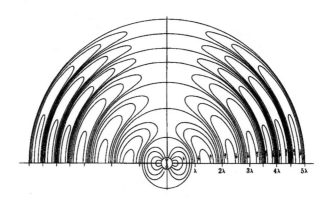

Figure 6.2. Electric-field configuration as produced by an oscillating electric dipole placed at the origin. (Adapted from Alonso and Finn 1971, *Fundamental University Physics, vol II: Fields and Waves* © 1967 Addison-Wesley Publishing Company. Reprinted by permission.)

Far away from the dipole region, the outgoing wave will eventually behave like a spherical wave of the form e^{ikr}/r and the radial electric field component will tend to zero, $E_r \to 0$, whereas the angular E_θ component becomes equal to the total electric field

$$E_\theta = |\vec{E}| = \frac{\Pi_0 \sin\theta}{4\pi\epsilon_0 r}\left(\frac{\omega}{c}\right)^2 \sin(kr - \omega t). \tag{6.4}$$

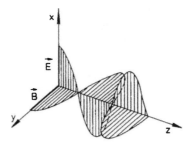

Figure 6.3. Electric and magnetic oscillating vector fields, illustrated in a schematic way, in the radiation zone where a description using plane waves can be used.

In the radiation zone the wave will approach the plane wave structure (see figure 6.3) with the following relation determining the magnetic field:

$$|\vec{B}| = \frac{1}{c}|\vec{E}|.$$

In this region one has

$$\vec{E} = \vec{E}_0 \sin(kz - \omega t)$$
$$\vec{B} = \vec{B}_0 \sin(kz - \omega t), \tag{6.5}$$

with $\omega = 2\pi\nu$, $\lambda = 2\pi/k$ and $|\vec{B}_0| = \frac{1}{c}|\vec{E}_0|$.

The energy densities in the electric and magnetic field become

$$E_E = \frac{1}{2}\epsilon_0|\vec{E}|^2$$
$$E_B = \frac{1}{2\mu_0}|\vec{B}|^2 = \frac{1}{2\mu_0 c^2}|\vec{E}|^2 = \frac{1}{2}\epsilon_0|\vec{E}|^2, \tag{6.6}$$

(where $c = 1/\sqrt{\epsilon_0\mu_0}$), and the total energy density is

$$E = E_E + E_B = \epsilon_0|\vec{E}|^2. \tag{6.7}$$

The intensity then reads

$$I = c\epsilon_0|\vec{E}|^2 \tag{6.8}$$

or, time averaged,

$$\langle I \rangle = \frac{1}{2}c\epsilon_0|\vec{E}|^2. \tag{6.9}$$

The Poynting vector $\vec{S} = c^2\epsilon_0\vec{E} \times \vec{B}$ has a magnitude with the value $\frac{1}{2}c\epsilon_0|\vec{E}|^2$ and thus equals the intensity of the electromagnetic radiation field (energy/cm²/s). The total flux through a surface S is then given by the integral

$$\int_S c^2\epsilon_0|\vec{E} \times \vec{B}| \cdot \vec{1}_n \, dS = \frac{dE}{dt}, \tag{6.10}$$

and equals the energy passing through the surface S per unit of time, and which we express as dE/dt. Using the expression for $\vec{E}$ (and the fact that $|\vec{B}| = \frac{1}{c}|\vec{E}|$), the total energy density in the 'radiation regime' becomes

$$E = \epsilon_0 |\vec{E}|^2 = \frac{\Pi_0^2}{16\pi^2} \frac{\sin^2\theta}{\epsilon_0 r^2} \frac{\omega^4}{c^4} \sin^2(kr - \omega t), \tag{6.11}$$

and averaging over time,

$$\langle E \rangle = \frac{\Pi_0^2 \omega^4}{32\pi^2 c^4 \epsilon_0 r^2} \sin^2\theta. \tag{6.12}$$

The intensity then reads

$$\langle I(\theta) \rangle = c\langle E \rangle = \frac{\Pi_0^2 \omega^4}{32\pi^2 c^3 \epsilon_0 r^2} \sin^2\theta,$$

and the total flux through a surface S (equal to the energy passing through this surface per unit of time) is

$$\left\langle \frac{dE}{dt} \right\rangle = \int_S I(\theta)\, dS$$

$$\left\langle \frac{dE}{dt} \right\rangle = \frac{\Pi_0^2 \omega^4}{12\pi c^3 \epsilon_0}. \tag{6.13}$$

Since the dipole moment is expressed by $\Pi_0 = ez_0$ with e the magnitude of the oscillating charge and z_0 the amplitude of the oscillation placed along the z-axis, we obtain (for e.g. an oscillating proton)

$$\left\langle \frac{dE}{dt} \right\rangle = \frac{e^2 z_0^2 \omega^4}{12\pi c^3 \epsilon_0}. \tag{6.14}$$

For the oscillating dipole $\Pi = ez_0 \sin\omega t$, the time-averaged acceleration becomes $\langle a^2 \rangle = z_0^2 \omega^4/2$ and the above equation can be written more generally

$$\left\langle \frac{dE}{dt} \right\rangle = \frac{e^2 \langle a^2 \rangle}{6\pi \epsilon_0 c^3}. \tag{6.15}$$

This is the famous Larmor equation, relating the radiated energy $\langle dE/dt \rangle$ to the acceleration $\langle a^2 \rangle$ of particle with charge e.

To conclude this summary, we present in figures 6.4(a) and (b) the $\vec{E}$, $\vec{B}$, $\vec{S}$ configurations as resulting from an electric oscillating dipole as well as for an oscillating magnetic dipole (resulting from a small circular current at the origin).

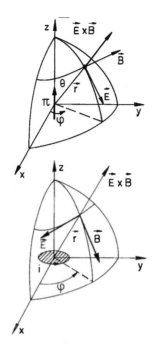

Figure 6.4. The relationships between the various vectors characterizing electric and magnetic dipole configurations. The figure is drawn in the near-source region where spherical symmetry is still present. The Poynting vector, is in the direction which coincides with the vector product $\vec{E} \times \vec{B}$ for an outward flow of radiation energy, is also indicated.

6.2 Kinematics of photon emission

Emission of energetic photons—gamma radiation—is typical for a nucleus de-exciting from some high-lying excited state to the ground-state configuration. These transmutations take place within the same nucleus $^A_Z X_N$ in contrast to the beta decay and alpha decay processes. They merely represent a re-ordering of the nucleons within the nucleus with a lowering of mass from the excited $(M_0^* c^2)$ to the lowest $(M_0 c^2)$ value.

The total energy balance then reads (figure 6.5)

$$M_0^* c^2 = M_0 c^2 + E_\gamma + T_0, \tag{6.16}$$

with E_γ the energy of the emitted photon and T_0 the kinetic energy of the recoiling nucleus. Linear momentum conservation leads to an expression

$$\vec{p}_\gamma + \vec{p}_0 = 0. \tag{6.17}$$

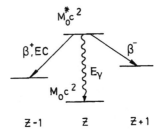

Figure 6.5. The mass–energy relationships for a nucleus in an excited state $M_0^* c^2 (^A_Z X_N)$ which can undergo β^-, β^+, EC and gamma decay, with the emission of gamma radiation with energy E_γ to the ground state $M_0 c^2$.

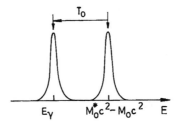

Figure 6.6. Illustration of the fact that in the gamma emission process the initial nucleus is at rest and receives a certain recoil energy T_0. Thereby the gamma energy E_γ is slightly different from the nuclear energy (mass) difference $M_0^* c^2 - M_0 c^2$.

The recoil energy is very small so non-relativistic expressions can be used, i.e.

$$T_0 = \frac{p_0^2}{2M_0} = \frac{p_0^2 c^2}{2M_0 c^2} = \frac{p_\gamma^2 c^2}{2M_0 c^2} = \frac{E_\gamma^2}{2M_0 c^2}. \tag{6.18}$$

For a 1 MeV photon energy and a nucleus with $A \simeq 100$, the recoil energy is $\simeq 5$ eV. Even though this energy is very small (figure 6.6), the recoil shifts the gamma radiation out of resonance condition since the natural linewidth of the radiation is even smaller.

The emission of photons without recoil is possible if one implants the nucleus in a lattice such that the recoil is taken by the whole lattice and not by a single nucleon. If the energy needed to excite the whole system into its first excited state ($\hbar\omega_{\text{lattice}}$) is much larger than the single-nucleon recoil energy

$$\hbar\omega_{\text{lattice}} \gg T_0, \tag{6.19}$$

then, due to quantum mechanical effects the energy of the emitted gamma radiation takes away the total energy difference $M_0^* c^2 - M_0 c^2$. This effect was found by Mössbauer and gave rise to a nuclear technique which proved to be very important in various domains of physics. The Mössbauer effect is described in great detail in Segré, *Nuclei and Particles*, Benjamin (1977).

6.3 The electromagnetic interaction Hamiltonian: minimum coupling

6.3.1 Constructing the electromagnetic interaction Hamiltonian

In order to derive the interaction Hamiltonian inducing electromagnetic transitions in the atomic nucleus within a quantum-mechanical framework, we start from a consideration of the classical Lagrangian describing the motion of a relativistic particle in an electromagnetic field.

The equation of motion describing this situation and to which Lagrange's equation applies is

$$\frac{\mathrm{d}}{\mathrm{d}t} \frac{m\vec{v}}{\sqrt{1 - v^2/c^2}} = q(\vec{E} + \vec{v} \times \vec{B}), \tag{6.20}$$

where we describe a particle with mass m and charge q and moving with velocity $\vec{v}$. The fields can be derived from a vector potential $\vec{A}$, using

$$\vec{B} = \vec{\nabla} \times \vec{A}$$

$$\vec{E} = -\vec{\nabla}\Phi - \frac{\partial \vec{A}}{\partial t}. \tag{6.21}$$

The equation of motion then becomes

$$\begin{aligned}
\frac{\mathrm{d}}{\mathrm{d}t} \frac{m\vec{v}}{\sqrt{1 - v^2/c^2}} &= q\left(-\vec{\nabla}\Phi - \frac{\partial \vec{A}}{\partial t} + \vec{v} \times (\vec{\nabla} \times \vec{A})\right) \\
&= \vec{\nabla}(-q\Phi + q\vec{v} \cdot \vec{A}) - q\left(\frac{\partial}{\partial t} + \vec{v} \cdot \vec{\nabla}\right)\vec{A} \\
&= \vec{\nabla}(-q\Phi + q\vec{v} \cdot \vec{A}) - q\frac{\mathrm{d}\vec{A}}{\mathrm{d}t}.
\end{aligned} \tag{6.22}$$

We can then rewrite the equation of motion, using the kinetic momentum

$$\vec{p}_K = \frac{m\vec{v}}{\sqrt{1 - v^2/c^2}}, \tag{6.23}$$

as

$$\frac{\mathrm{d}}{\mathrm{d}t}\left(\frac{m\vec{v}}{\sqrt{1 - v^2/c^2}} + q\vec{A}\right) + \vec{\nabla}(q\Phi - q\vec{v} \cdot \vec{A}) = 0. \tag{6.24}$$

Making use of Lagrange's equation, which should lead to the above equation of motion, one can deduce a suitable Lagrangian as

$$L(\vec{r}, \vec{v}) = -mc^2\sqrt{1 - v^2/c^2} + q\vec{v} \cdot \vec{A}(\vec{r}, t) - q\Phi(\vec{r}, t). \quad (6.25)$$

The canonical momentum $\vec{p}_c \equiv \vec{\nabla}_v L(\vec{r}, \vec{v})$ then becomes

$$\vec{p}_c = \frac{m\vec{v}}{\sqrt{1 - v^2/c^2}} + q\vec{A}(\vec{r}, t) = \vec{p}_K + q\vec{A}(\vec{r}, t). \quad (6.26)$$

The canonical momentum is that variable which has to be replaced by $-i\hbar\vec{\nabla}$ when adopting quantum mechanics in order to obtain the appropriate Schrödinger equation. The corresponding Hamiltonian

$$H = \vec{p}_c \cdot \vec{v} - L,$$

then becomes

$$H = c\{[\vec{p}_c - q\vec{A}]^2 + (mc)^2\}^{1/2} + q\Phi, \quad (6.27)$$

and expresses the total energy of the system. For the non-relativistic limit when the kinetic energy is small compared to the rest mass energy, one can expand the square root expression and obtain

$$H = mc^2 + \frac{1}{2m}(\vec{p}_c - q\vec{A})^2 + q\Phi. \quad (6.28)$$

The first term simply expresses the rest mass and taking out the quadratic term on $\vec{A}$ i.e. $(q^2/2mc^2)\vec{A}^2$; the 'interaction' Hamiltonian becomes

$$H_{\text{int}}^{\text{em}} = q\Phi - q\vec{v} \cdot \vec{A} = q\Phi - \frac{q}{m}\vec{p} \cdot \vec{A}. \quad (6.29)$$

(From here on, in order to simplify the notation, we replace $\vec{p}_c$ by $\vec{p}$.) For systems involving a continuous charge density $\rho(\vec{r}, t)$ and a current density $\vec{j}(\vec{r}, t)$, the extension of the above interaction Hamiltonian becomes

$$H_{\text{int}}^{\text{em}} = \int [\rho\Phi - \vec{j} \cdot \vec{A}] \, d\vec{r}. \quad (6.30)$$

The Hamiltonian of equation (6.28) (neglecting the rest energy mc^2) is obtained from the free particle Hamiltonian by the substitution (and adding the potential energy $q\Phi$)

$$\vec{p} \rightarrow \vec{p} - \frac{q}{m}\vec{A}, \quad (6.31)$$

which is called 'minimum electromagnetic coupling'.

In the next subsection we shall derive the transition probability for one-photon emission (or absorption) using time-dependent perturbation theory and the above interaction Hamiltonian $H_{\text{int}}^{\text{em}}$.

6.3.2   One-photon emission and absorption: the dipole approximation

The previous discussion was a classical one and in proceeding towards the quantum mechanical description the above interaction Hamiltonian has to be rewritten as an operator.   A major difficulty lies in the correct, quantum mechanical description of the radiation field, described by the classical vector potential $\vec{A}(\vec{r}, t)$. A proper discussion uses techniques of quantum field theory: quantum electrodynamics is not discussed here.

It will be possible, though, to describe the basic one-photon emission process as depicted in figure 6.7, where an initial quantum mechanical system (e.g. atom, nucleus etc) $|i\rangle$ ends up, through the electromagnetic interaction operator, a final state $|f\rangle$. One can use time-dependent perturbation theory to derive

$$\lambda_{if} = \frac{2\pi}{\hbar} |\langle f|\hat{H}_{\text{int}}^{\text{em}}|i\rangle|^2 \frac{dn}{dE}(E_\gamma), \tag{6.32}$$

where dn/dE is the density of final states to which a photon is emitted with momentum $\vec{p}_\gamma$ and energy E_γ. This latter part was discussed in chapter 3, and is

$$\frac{dn}{dE}(E_\gamma) = \frac{V E_\gamma^2}{(2\pi\hbar c)^3} d\Omega, \tag{6.33}$$

for emission of the photon in the solid angle element $d\Omega$. The electromagnetic interaction Hamiltonian, for an electron in the interaction process ($q = -e$ (with $e > 0$)) becomes

$$\hat{H}_{\text{int}}^{\text{em}} = \frac{e}{m}\vec{p} \cdot \vec{A}, \tag{6.34}$$

with $\vec{p} \rightarrow -i\hbar\vec{\nabla}$ to transform to quantum mechanics.   The expression for $\vec{A}$ in describing emission cannot be obtained using non-relativistic methods since photons always move with the speed of light.   The quantum mechanics of describing the transition from a state containing no photon towards one (final state) where a photon has been created in a state with momentum $\vec{p}_\gamma$ and polarization vector $\vec{\epsilon}$ needs a second-quantized version of the electromagnetic field ($\vec{A}$ vector potential).   One can, loosely, postulate that $\vec{A}$ is describing the wavefunction of the created photon (see Greiner 1980) in the form

$$\vec{A} = a_0\vec{\epsilon}\cos(\vec{k} \cdot \vec{r} - \omega t), \tag{6.35}$$

where $\vec{\epsilon}$ is the polarization vector and a_0 the amplitude of one photon in the plane wave, normalized to volume V. One can determine a_0 by equating the energy in this volume V to $E_\gamma = \hbar\omega$. Then, one gets

$$E = \epsilon_0\overline{|\vec{E}|^2} \cdot V = \frac{\epsilon_0 a_0^2 \omega^2}{2} \cdot V, \tag{6.36}$$

and, equating to $\hbar\omega$, the value of a_0 is

$$a_0 = \sqrt{\frac{2\hbar^2}{\epsilon_0 E_\gamma V}}. \tag{6.37}$$

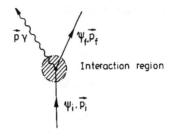

Figure 6.7. Emission process of a photon, characterized by momentum $\vec{p}_\gamma$, from a nucleus characterized by wavefunction ψ_i, momentum $\vec{p}_i$ decaying to the final state described by ψ_f and $\vec{p}_f$. The interaction region is schematically illustrated by the shaded region. (Taken from Frauenfelder and Henley (1991) *Subatomic Physics* © 1974. Reprinted by permission of Prentice-Hall, Englewood Cliffs, NJ.)

It is convenient, though, to rewrite the expression of $\vec{A}$ as

$$\vec{A}(\text{one photon}) = \sqrt{\frac{\hbar^2}{2\epsilon_0 E_\gamma V}}\,\vec{\epsilon}\,(\exp[\mathrm{i}(\vec{p}_\gamma \cdot \vec{r} - E_\gamma t)/\hbar] + \exp[-\mathrm{i}(\vec{p}_\gamma \cdot \vec{r} - E_\gamma t)/\hbar]).$$

(6.38)

This expression is quite appropriate, expressing the two independent polarization components of a transversal, radiation field in the plane-wave regime.

We now have to evaluate the matrix element, according to figure 6.7,

$$\langle f | \hat{H}^{\text{em}}_{\text{int}} | i \rangle \equiv \int \Psi_f^* \hat{H}^{\text{int}}_{\text{em}} \Psi_i \, \mathrm{d}\vec{r}$$

$$= -\frac{\mathrm{i}e\hbar}{m} \int \Psi_f^* \vec{\nabla} \Psi_i \cdot \vec{A} \, \mathrm{d}\vec{r}.$$

(6.39)

In the calculation below, we make the 'dipole approximation' so that in the plane-wave expression we only consider the lowest-order contributions, i.e.

$$\exp\left(\pm \frac{\mathrm{i}\vec{p}_\gamma \cdot \vec{r}}{\hbar}\right) = 1 \pm \mathrm{i}\frac{\vec{p}_\gamma \cdot \vec{r}}{\hbar} + \cdots.$$

(6.40)

The validity of the approximation implies that

$$E_\gamma \ll \frac{\hbar c}{R} \simeq \frac{197\,\text{MeV fm}}{R(\text{in fm})},$$

(6.41)

and is fulfilled when the dipole radiation is allowed. In the atomic nucleus with $R \simeq 5\text{–}8$ fm, this condition is fulfilled for most low-energy transitions of a few MeV. In the further evaluation of the nuclear matrix element, we consider wavefunctions describing spinless systems with intial (E_i) and final

(E_f) stationary energy values. So we obtain

$$\langle f|\hat{H}_{\text{int}}^{\text{em}}|i\rangle = -\mathrm{i}\frac{\hbar^2 e}{m}\sqrt{\frac{1}{2\epsilon_0 E_\gamma V}}(\exp[\mathrm{i}(E_f - E_\gamma - E_i)t/\hbar]$$

$$+ \exp[\mathrm{i}(E_f + E_\gamma - E_i)t/\hbar])\vec{\epsilon}\cdot\int \psi_f^*\vec{\nabla}\psi_i\,\mathrm{d}\vec{r}. \qquad (6.42)$$

The second exponential becomes unity because of energy conservation in the decay process

$$E_i = E_f + E_\gamma, \qquad (6.43)$$

whereas the first term reduces to $\exp(-2\mathrm{i}E_\gamma t/\hbar)$. With the time $t \gg 2\pi\hbar/E_\gamma$ (validity condition for perturbation theory to apply), this exponential is a rapid oscillatory function which averages out the contribution to λ_{if} from this term. One finally obtains the result

$$\langle f|\hat{H}_{\text{int}}^{\text{em}}|i\rangle = -\mathrm{i}\frac{\hbar^2 e}{m}\sqrt{\frac{1}{2\epsilon_0 E_\gamma V}}\vec{\epsilon}\cdot\int \psi_f^*\vec{\nabla}\psi_i\,\mathrm{d}\vec{r}, \qquad (6.44)$$

and, subsequently,

$$\lambda_{if} = \frac{e^2 E_\gamma}{8\pi^2 m^2 c^3}\left|\vec{\epsilon}\cdot\int \psi_f^*\vec{\nabla}\psi_i\,\mathrm{d}\vec{r}\right|^2 \mathrm{d}\Omega. \qquad (6.45)$$

The expression (6.45) can be evaluated taking into account the fact that ψ_i and ψ_f are eigenfunctions of the Hamiltonian describing the particle system (without $\hat{H}_{\text{int}}^{\text{em}}$) and

$$\hat{H}_0 = \frac{\hat{p}^2}{2m} + U(\vec{r}). \qquad (6.46)$$

Since one has

$$\hat{H}_0\psi_i = E_i\psi_i, \qquad \hat{H}_0\psi_f = E_f\psi_f, \qquad (6.47)$$

it is possible to express the commutator $[\vec{r}, \hat{H}_0]$ as $[\vec{r}, \hat{H}_0] = (\hbar^2/m)\vec{\nabla}$. The integral can thereby be transformed into the expression

$$\int \psi_f^*\vec{\nabla}\psi_i\,\mathrm{d}\vec{r} = \frac{m}{\hbar^2}E_\gamma\int \psi_f^*\vec{r}\psi_i\,\mathrm{d}\vec{r}, \qquad (6.48)$$

and we use a shorthand notation for the $\vec{r}$ integral as $\langle f|\vec{r}|i\rangle$. This gives rise to the transition probability for one-photon emission within the solid angle $\mathrm{d}\Omega$ as

$$\lambda_{if}(\Omega) = \frac{e^2}{8\pi^2\hbar^4 c^3}\frac{E_\gamma^3}{\epsilon_0}|\vec{\epsilon}\cdot\langle f|\vec{r}|i\rangle|^2\,\mathrm{d}\Omega. \qquad (6.49)$$

Using a specific choice of the (x, y, z)-axis, the vectors $\langle f|\vec{r}|i\rangle$, $\vec{p}_\gamma$ and $\vec{\epsilon}$ are drawn in figure 6.8, where we have chosen $\vec{p}_\gamma$ pointing in the z-direction and the

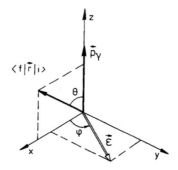

Figure 6.8. Relation amongst various vector quantities characterizing the gamma emission process. The photon polarization vector is given by $\vec{\epsilon}$ with the photon emitted in the direction of the positive z-axis ($\vec{p}_\gamma$). The vector characterizing the dynamics of the decaying nucleus $\langle f|\vec{r}|z\rangle$ is taken to be in the (x, z) plane. (Taken from Frauenfelder and Henley (1991) *Subatomic Physics* © 1974. Reprinted by permission of Prentice-Hall, Englewood Cliffs, NJ.)

vector $\langle f|\vec{r}|i\rangle$ to determine the (x, z) plane. All directions are now fixed and one gets

$$\lambda_{if}(\Omega) = \frac{e^2}{8\pi^2\hbar^4c^3} \frac{E_\gamma^3}{\epsilon_0} |\langle f|\vec{r}|i\rangle|^2 \sin^2\theta \cos^2\varphi \, d\Omega. \tag{6.50}$$

If one sums over *all* directions (integrating over $d\Omega$) and sums over the two independent polarization states for $\vec{\epsilon}$, the transition rate for unpolarized photons becomes

$$\lambda_{if} = \frac{4}{3}\alpha \left(\frac{E_\gamma}{\hbar c}\right)^3 c|\langle f|\vec{r}|i\rangle|^2, \tag{6.51}$$

where E_γ is the photon energy, and

$$\alpha = \frac{e^2}{\hbar c} \frac{1}{4\pi\epsilon_0} = \frac{1}{137.04},$$

the fine-structure constant and $\langle f|\vec{r}|i\rangle$ the dipole matrix element. One immediately deduces that the parities of the initial and final nuclear stationary states ψ_i and ψ_f have to have opposite values in order to obtain a non-vanishing matrix element: the parity selection rule easily results. One obtains a rough estimate for this electric dipole transition probability using some typical values: $R = 1.2A^{1/3}$ in approximating the nuclear matrix element, $E_\gamma = 1$ MeV as

$$\lambda_{if}(\text{E1}) \simeq 5.5 \times 10^{14} A^{2/3} \text{ s}^{-1}, \tag{6.52}$$

which, for $A \simeq 100$, gives a mean-life of about 8×10^{-17} s. The factor E_γ^3 indicates the density of final states and means that for various transitions, chararacterized by the same nuclear matrix element, the lifetime changes with E_γ^{-3}.

The factor $\alpha = (e^2/\hbar c)(1/4\pi\epsilon_0)$ (using SI units) and $e^2/\hbar c$ (using Gaussian units) expresses the interaction strength in the electromagnetic coupling of the radiation field to the matter field and plays an identical role as the strength G in beta decay. It also is a dimensionless constant.

6.3.3 Multipole radiation

In the study of one-photon emission we have been using the long-wavelength or dipole approximation, i.e. the factor $e^{i\vec{p}_\gamma \cdot \vec{r}}$ was approximated by unity. This electromagnetic transition is characterized by:

(i) a change of parity between the initial and final nuclear wavefunctions $\psi_i(\vec{r})$ and $\psi_j(\vec{r})$;

(ii) using the same analysis as was carried out in beta-decay, angular momentum selection implies the condition

$$\vec{J}_i = \vec{J}_f + \vec{1},$$

since the vector potential (describing the photon wavefunction) carries an angular momentum of one unit ($\hbar$). In a number of situations, the angular momentum difference between the initial and final state can exceed one unit of angular momentum (in units $\hbar$), implying that one should take higher-order terms in the expansion of the plane wave into account. This will give rise to the matrix element

$$\langle f | \hat{H}_{\text{int}}^{\text{em}}(\text{order } L) | i \rangle = -i\frac{e\hbar}{m} \int \psi_f^* \vec{\nabla} \psi_i \cdot \left(\frac{\vec{p}_\gamma \cdot \vec{r}}{\hbar} \right)^L (\pm i)^L \, \mathrm{d}\vec{r}. \qquad (6.53)$$

This matrix element is reduced relative to the dipole matrix element by a factor $(k_\gamma R)^L$ where R characterizes the nuclear radius. For the example used before, taking $E_\gamma = 1$ MeV and a nuclear radius of $R = 5$ fm; one has

$$(k_\gamma R)^L = \left(\frac{E_\gamma R}{\hbar c} \right)^L \simeq \left(\frac{5 \text{ MeV fm}}{197 \text{ MeV fm}} \right)^L \simeq \left(\frac{1}{40} \right)^L. \qquad (6.54)$$

The lifetime e.g. for an $L = 2$ transition is greater by a factor $\simeq 1600$ relative to the dipole lifetime. The electric transition, corresponding to $L = 2$ is called a quadrupole transition and, inspecting the integrandum implies the parity selection rule $\pi_i = \pi_f$. At the same time, the angular momentum selection rule constrains the values to

$$\vec{J}_i = \vec{J}_f + \vec{2}.$$

Generally, for electric L-pole radiation the selection rules become (EL)

$$\pi_i = \pi_f(-1)^L, \qquad \vec{J}_i = \vec{J}_f + \vec{L}. \qquad (6.55)$$

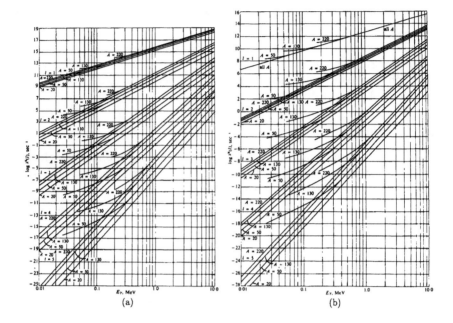

Figure 6.9. (*a*) The transition probability for gamma transitions, as a function of E_γ (in MeV units), based on the single-particle model (see chapter 9) is shown for electric multipole transitions ($L = 1, 2, \ldots, 5$) for various mass regions. In (*b*) similar data is shown for the magnetic multipole transitions. (Taken from Condon and Odishaw, *Handbook on Physics*, 2nd edn © 1967 McGraw-Hill. Reprinted with permission.)

It is also possible, though it is more difficult, to consider the analogous radiation patterns corresponding to periodic variations in the current distribution within the atomic nucleus. Without further proof, we give the corresponding selection rules for magnetic L-pole radiation as (ML)

$$\pi_i = \pi_f(-1)^{L+1}, \qquad \vec{J}_1 = \vec{J}_f + \vec{L}. \tag{6.56}$$

Using slightly more realistic estimates for the initial and final nuclear wavefunctions $\psi_i(\vec{r})$, $\psi_f(\vec{r})$; still rather simple estimates (Weisskopf estimates) can be derived for the total electric and magnetic L-pole radiation. These results are presented in figures 6.9(*a*) and (*b*) for electric $L = 1, \ldots, 5$ and magnetic $L = 1, \ldots, 5$ transitions respectively. For magnetic transitions, μ_p denotes the magnetic moment of the proton as $\mu_p = 2.79$ (in nuclear magnetons).

An example of the decay of levels in the nucleus ^{190}Hg shows a realistic situation for various competing transitions according to the angular momentum values. In the cascade $8^+ \to 6^+ \to 4^+ \to 2^+ \to 0^+$, pure E2 transitions result. For $2^+ \to 2^+$, $4^+ \to 4^+$ transitions, on the other hand, both M1 and

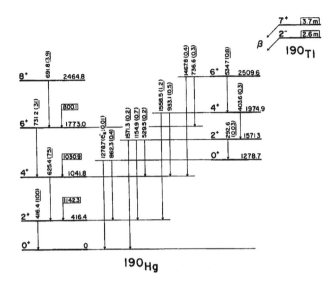

Figure 6.10. Illustration of a realistic gamma decay scheme of even–even nucleus ^{190}Hg. The level energies, gamma transition energies and corresponding intensities are illustrated in all cases. Within the specific bands, transitions are of E2 type; while inter-band transitions can be of both E2 and M1 type depending on the spin difference between initial and final state. (Taken from Kortelahti *et al* 1991.)

E2 transitions can contribute (M3 and E4 will be negligible relative to the more important M1 and E2 transitions) (see figure 6.10).

In a number of cases (see figure 6.11), the spin difference between the first excited state and the ground state can become quite large such that a high multipolarity is the first allowed component in the electromagnetic transition matrix element, due to the restrictions

$$\vec{J}_i = \vec{J}_f + \vec{L}. \tag{6.57}$$

In the case of ^{117}In and ^{117}Sn, M4 transitions occur for the transitions $\frac{1}{2}^- \to \frac{9}{2}^+$; $\frac{11}{2}^- \to \frac{3}{2}^+$ with lifetimes that can become of the order of hours and sometimes days. These transitions are called isomeric transitions and the corresponding, metastable configurations, are called 'isomeric states'. A systematic compilation for all observed M4 transitions has been carried out by Wood (private communication): the reduced half-life T is presented in figure 6.12 and is compared with the M3, M4 and E5 single particle estimates of Weisskopf and Moszkowski. The data points agree very well with the M4 estimates.

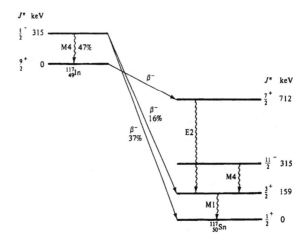

Figure 6.11. Diagram showing some high-multipole transitions in ^{117}In and ^{117}Sn. The spin and parity and excitation energy of the excited nuclear states are given. The beta and gamma decay branches from the isomeric state in ^{117}In ($\frac{1}{2}^{-}$) are also shown. (Taken from Jelley 1990.)

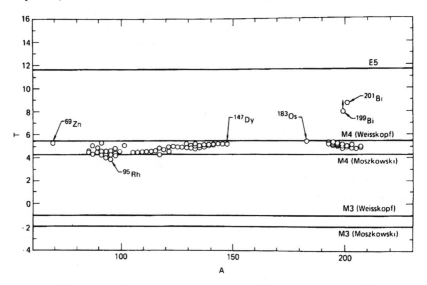

Figure 6.12. The reduced half-life T in s and gamma energy in MeV for all known M4 transitions between simple single-particle states in odd-mass nuclei (76 cases) plotted against mass number A. The Weisskopf and Moszkowski single-particle estimates of M3, M4 and E5 transitions are drawn as horizontal lines. (Taken from Heyde *et al* 1983.)

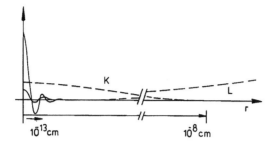

Figure 6.13. Pictorial presentation of the different length scales associated with the nuclear (full lines at 10^{-13} cm scale) and electronic (dashed lines on a 10^{-8} cm scale) wavefunctions. Only the K (or s-wave) electrons have wavefunction with non-vanishing amplitudes at the origin and will cause electron conversion to occur mainly via K-electron emission.

6.3.4 Internal electron conversion coefficients

In the preceding section we have discussed the de-excitation of the atomic nucleus via the emission of photons. The basic interaction process is as depicted in figure 6.7 and described in section 6.3.2. There now exists a different process by which the nucleus 'transfers' its excitation energy to a bound electron, causing electrons to move into an unbound state with an energy balance of

$$T_{e^-} = (E_i - E_f) - B_n, \qquad (6.58)$$

where $E_i - E_f$ is the nuclear excitation energy, B_n the corresponding electron binding energy and T_{e^-} the electron kinetic energy. This process does not occur as an internal photon effect since no actual photons take part in the transition. The energy is transmitted mainly through the Coulomb interaction and a larger probability for K electrons will result because K electrons have a non-vanishing probability of coming into the nuclear interior (see figure 6.13).

The basic process, with exchange of virtual photons, mediating the Coulomb force is illustrated in figure 6.14(*a*) and conversion electrons can be observed superimposed on the continuous beta spectrum if the decay follows a beta-decay transition, as presented schematically in figure 6.14(*b*). The transition matrix element, when a 1s electron is converted into a plane wave state can then be written as

$$M_{if} \propto \sum_i \int \exp(-\mathrm{i}\vec{k}_e \cdot \vec{r}_e) \psi_f^*(\vec{r}_i) \frac{e^2}{4\pi\epsilon_0 |\vec{r}_e - \vec{r}_i|} b_0^{-3/2} \exp\left(-\frac{r_e}{b_0}\right) \psi_i(\vec{r}_i) \,\mathrm{d}\vec{r}_i \,\mathrm{d}\vec{r}_e,$$
$$(6.59)$$

where the sum goes over *all* protons in the nucleus; $\vec{k}_{e^-}$ describes the wavevector of the outgoing electron and $\psi_f(\vec{r}_i)$ and $\psi_i(\vec{r}_i)$ denote the nuclear final and

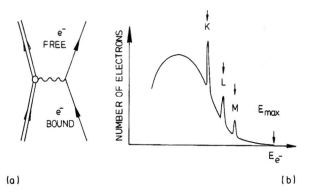

Figure 6.14. (*a*) The electromagnetic interaction process indicating the transition from a bound into a continuum state for the electron. The double line represents the atomic nucleus and the wavy line the Coulomb interaction via exchange of virtual photons. (*b*) Discrete electron energies associated with discrete transitions of electrons from the $K, L_I, M_I, \ldots$ electronic shells. The specific transitions are superimposed on the continuous background of the beta-decay process.

initial wavefunctions with b_0 being the Bohr radius for the 1s electron, i.e. $b_0 = \hbar^2 4\pi\epsilon_0/me^2$ when $Z = 1$. The Coulomb interaction indicates a drop in transition probability with increasing distance $|\vec{r}_e - \vec{r}_i|$ i.e. K-electron conversion will be the most important case. For $r_e > r_i$ one can use the expansion

$$\sum_i \frac{1}{|\vec{r}_e - \vec{r}_i|} = \frac{1}{r_e} \sum_i \sum_l \left(\frac{r_i}{r_e}\right)^l P_l(\cos\theta), \tag{6.60}$$

with $\vec{r}_i \cdot \vec{r}_e = r_i r_e \cos\theta$.

This matrix element can be separated into an electron part times a nuclear part, making use of the expansion

$$P_l(\cos\theta) = \sum_m \frac{4\pi}{2l+1} Y_l^{m*}(\hat{r}_e) Y_l^m(\hat{r}_i), \tag{6.61}$$

and using the shorthand notation for the nuclear matrix element

$$M_{if}^N(l, m) \equiv \sum_i \int \psi_f^*(\vec{r}_i) r_i^l Y_l^m(\hat{r}_i) \psi_i(\vec{r}_i) \, d\vec{r}_i, \tag{6.62}$$

since

$$M_{if} \propto \sum_{l,m} \frac{e^2 k_e^{l-2}}{b_0^{3/2}} \frac{4\pi}{2l+1} \int \exp(-i\vec{k}_e \cdot \vec{r}_e) \left(\frac{1}{k_e r_e}\right)^{l+1} Y_e^{m*}(\hat{r}_e)$$

$$\times \exp\left(-\frac{r_e}{b_0}\right) d(k\vec{r}_e) \cdot M_{if}^N(l, m). \tag{6.63}$$

Here, it becomes immediately clear that, in contrast to the situation where photons are emitted. Electron conversion processes can occur in situations when $l = 0$ or when no change occurs in spin between initial and final states including $0^+ \rightarrow 0^+$ transitions.

A detailed derivation of the electron conversion transition probability is outside the scope of this text. If we consider, however, the lowest order for which both photons and electrons can be emitted i.e. dipole processes, from a 1s bound electron state to a p-wave outgoing electron, we can evaluate the conversion transition probability and the corresponding conversion coefficient, defined as

$$\alpha_K = \frac{\lambda_{if}(e)}{\lambda_{if}(\gamma)}. \tag{6.64}$$

A detailed discussion on the calculation of electron conversion coefficients as well as an appreciation of present-day methods used to measure conversion coefficients and their importance in nuclear structure are presented in Box 6b.

6.3.5   E0—monopole transitions

Single-photon transitions are strictly forbidden between excited 0^+ states and the 0^+ ground state in even–even nuclei. The most common de-excitation process is electron conversion as discussed in detail in section 6.3.4. The $0^+ \rightarrow 0^+$ transitions contain, however, a very specific piece of information relating to the nuclear radius and deformation changes in the nucleus. The transition process is again one where the Coulomb field is acting and only vertical photons are exchanged.

For s-electrons, a finite probability exists for the electron to be within the nucleus for which $r_e < r_i$ so that

$$\frac{1}{|\vec{r}_e - \vec{r}_i|} = \frac{1}{r_i} \sum_l \left(\frac{r_e}{r_i} \right)^l P_l(\cos \theta), \tag{6.65}$$

which, for the $l = 0$ part (monopole part) becomes $1/r_i$. The transition matrix element can be evaluated as

$$M_{if} \propto \sum_i \int_0^R \psi_f^* \psi_i \left(\int_0^{r_i} \varphi_f^* \frac{1}{r_i} \varphi_i \, d\vec{r}_e \right) d\vec{r}_i, \tag{6.66}$$

where R denotes the nuclear radius, ψ_f and ψ_i are the nuclear final and initial state wavefunctions. The electron wavefunctions φ_i and φ_f describe an initial bound s-electron wavefunction and the outgoing electron wave, which also has to be an $l = 0$ or s-wave to conserve angular momentum. To a very good approximation $\varphi_f^*(r_e) = \varphi_f^*(0)$ and $\varphi_i(r_e) = \varphi_i(0)$, so that the integral over the electron coordinates reduces to

$$I_e = \varphi_f^*(0)\varphi_i(0) \int_0^{r_i} \frac{1}{r_i} 4\pi r_e^2 \, dr_e, \tag{6.67}$$

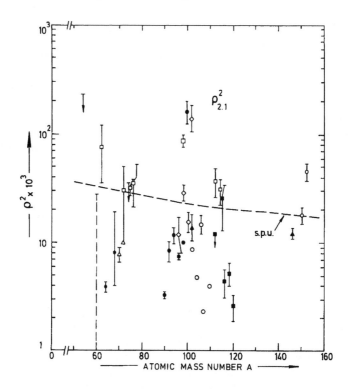

Figure 6.15. Compilation of $0^+ \rightarrow 0^+$ E0 matrix elements (expressed as ρ^2) for the mass region $60 < A < 160$. The dashed line corresponds to the single-particle ρ^2 value. The symbols correspond to various isotope chains. (Taken from Heyde and Meyer 1988.)

and the total matrix element M_{if} becomes

$$M_{if} \propto \sum_i \int \psi_f^* r_i^2 \psi_i \, \mathrm{d}\vec{r}_i. \tag{6.68}$$

This matrix element is a measure of the nuclear radius whenever $\psi_f \simeq \psi_i$.

Many E0 ($0^+ \rightarrow 0^+$) transitions have been measured and give interesting information on nuclear structure. In figure 6.15, an up-to-date collection of E0 transition probabilities (denoted by ρ^2 which is proportional to $|M_{if}|^2$) is shown for $0_2^+ \rightarrow 0_1^+$ transitions where the dashed line represents the single-particle E0 estimate.

Whenever the transition energy in the nucleus is larger than $2m_0c^2$, there exists the possibility that the Coulomb field (virtual photon) can convert into a pair of an electron and a positron. The best known example is the 6 MeV $0_2^+ \rightarrow 0_1^+$ transition in ^{16}O. This pair creation process is a competitive de-excitation mechanism for high-lying 0^+ excited states (see figure 6.16).

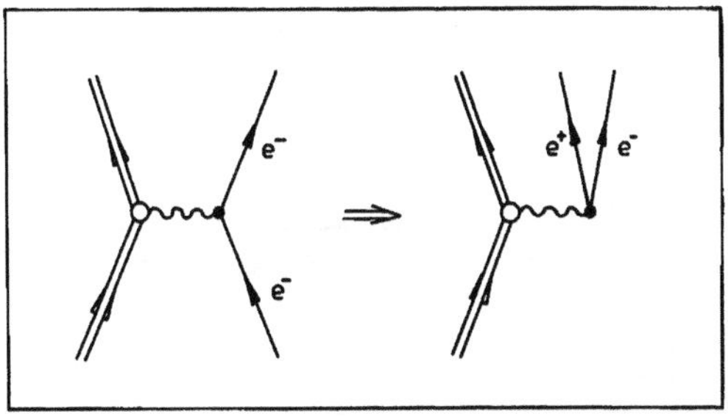

Figure 6.16. The pair creation (e^+e^- creation) process in the field of an atomic nucleus (right-hand vertex) compared to the vertex where an electron is scattered off an atomic nucleus. The double line indicates the nucleus; the wavy line the exchange of virtual photons.

6.3.6 Conclusion

In studying the electromagnetic interaction and its effect on the nucleus, using time-dependent perturbation theory, a number of interesting results have been obtained. A more rigorous treatment however needs a consistent treatment of quantized radiation and matter fields and their interactions.

Box 6a. Alternative derivation of the electric dipole radiation fields

Alternative derivation of the electric dipole radiation fields

J. A. Souza

Instituto de Fisica. Universidade Federal Fluminense. 24.000 Niteroi. RJ. Brazil

(Received 23 November 1981; accepted for publication 26 February 1982)

We propose an alterative derivation of the fields of an oscillating electric dipole, which makes explicit reference to the dipole from the beginning, is mathematically simple, and involves no approximations.

The derivations of the oscillating electric dipole fields which the student usually encounters in his undergraduate course on electromagnetic theory are not, in general, very enlightening. They usually involve approximations on the behavior of oscillating current elements (for example, that the wavelength of the radiation be much greater than the racteristic dimensions of the radiating system),[1] and the students often feel an uncomfortable sensation in the lack of logic, since only at the end of the calculations is the electric dipole explicitly invoked.

On the other hand, formalisms which make an explicit reference to the dipole, like the multipole solution of the Maxwell equations or the Hertz vector formalism are mathematically too lengthy (or too cumbersome) to be presented in a first contact with the electromagnetic radiation theory, unless one omits some intermediate steps (which brings the uncomfortable sensations back). We propose, in this paper, a derivation of the potentials of an oscillating electric dipole which starts from the dipole concept and which is mathematically simple. The reasoning is as follows: we assume a "point dipole" (a concept already known by the student from electrostatics) at the origin of the coordinates, with the dipole moment pointing in the z direction. The dipole is assumed to be oscillating harmonically with frequency ω. This will constitute a small current element, in the z direction, at the origin. Now, it is already known by the student, too, the vector potential A has the ie direction as the current, so the only non-vanishing Cartesian component of A in this case will be A_z. Thus we write

$$A_z(r,t) = f(r)e^{i\omega t} \quad \tiny{r/c}, \qquad (1)$$

where we have assumed only a radial dependence for A_z, since from magnetostatics we know that A for a small current element has no angular dependence, and the harmonic oscillation of the charges cannot, of course, alter this situation. The factor $t - r/c$ in the exponential represents the fact that electromagnetic signals propagate, in vacuum, with finite velocity c, a concept that should be very familiar to the student at this stage of the course.

We have two problems now: how to find the unknown function $f(r)$, and how to obtain the scalar potential ϕ, since we need to know it in order to compute the fields.

Now, we know that in the Lorentz (radiation) gauge, A and ϕ are linked by (we are using Gaussian units)

$$\nabla\cdot A + \frac{1}{c}\frac{\partial\phi}{\partial t} = 0. \qquad (2)$$

and that, very near the dipole, one must observe an electric field oscillating harmonically in time, but with a spatial dependence equal to that of the static case, since the effects due to the finite velocity of propagation then become negligible. So, the scalar potential must behave for small r, as

$$\phi(r) = p\cos\theta/r^2, \qquad (3)$$

where p is the amplitude of the dipole moment.

Now using (2), with

$$\nabla\cdot A = \frac{\partial A_z}{\partial z} = \frac{\partial A_z}{\partial r}\frac{\partial r}{\partial z} = \frac{z}{r}\frac{\partial A_z}{\partial z} = \cos\theta\frac{\partial A_z}{\partial r}.$$

and with A_z given by (1), we obtain (a prime means derivative with respect to r)

$$-ik\phi = \cos\theta\, e^{i\omega t} \quad {}^{kr}[f' - ikf]. \qquad (4)$$

where $k = \omega/c$, and we have taken into account the time dependence of ϕ in the form $e^{i\omega t}$. Hence

$$\phi(r,\theta,t) = \cos\theta\, e^{i\omega t} \quad {}^{kr}[(i/k)f' + f]. \qquad (5)$$

We now have two cases, depending on whether f' or f dominates as $r\to 0$:

(a) The dominant term is given by f; we have $\cos\theta f = p\cos\theta/r^2$, which gives

$$f(r) = p/r^2. \qquad (6)$$

But with this $f(r)$, we get $f'(r) = -2p/r^3$, and this leads to a contradiction.

(b) The dominant term when $r\to 0$ is given by f'; in this case if $\cos\theta = p\cos\theta/r^2$, which gives

$$f'(r) = -ikp/r^2. \qquad (7)$$

and we will have for $f(r)$:

$$f(r) = ikp/r. \qquad (8)$$

In this case we have no inconsistency. So the potentials will be

$$\phi(r,\theta,t) = p\cos\theta\, e^{i\omega t} \quad {}^{kr}[(1/r^2) + (ik/r)], \qquad (9)$$

$$A_z(r,t) = (ikp/r)e^{i\omega t} \quad {}^{kr}. \qquad (10)$$

We can now pass to spherical coordinates r, θ, ϕ, so that we have

$$A_r = A_z\cos\theta, \quad A_\theta = A_z\sin\theta, \quad A_\phi = 0. \qquad (11)$$

and by using

$$E = -\nabla\phi - \frac{1}{c}\frac{\partial A}{\partial t}, \quad B = \nabla\times A, \qquad (12)$$

obtain the radiation fields following the usual pattern

[1]See any standard undergraduate textbook on electromagnetic theory for example, J. Reitz and F. Milford, *Foundations of Electromagnetic Theory* (Addison-Wesley, Reading, MA, 1960). Sec. 16-5

Box 6b. How to calculate conversion coefficients and their use in determining nuclear strucure information

---

K conversion is the ejection of a bound 1s electron into an outgoing p-wave state, and if we consider that the conversion occurs through the dipole component of the Coulomb interaction, then a more detailed evaluation of α_K, the K-shell is possible.

The p-wave component present in the expansion of an outgoing plane wave, used to describe the electron emitted, is

$$\psi_f(\vec{r}_e) = N \frac{\cos\theta}{(k_e r_e)^{1/2}} J_{3/2}(k_e r_e), \tag{6b.1}$$

or, asymptotically ,for large $k_e r_e$,

$$\psi_f(\vec{r}_e) = -N\cos\theta \left(\frac{2}{\pi k_e^2 r_e^2}\right)^{1/2} \cos(k_e r_e). \tag{6b.2}$$

The normalization can be fixed by enclosing the outgoing wave in a very large sphere with radius R and using the above asymptotic wave, which leads to

$$N = k_e \left(\frac{3}{4R}\right)^{1/2}. \tag{6b.3}$$

The initial 1s-electron wavefunction reads

$$\psi_i(\vec{r}_e) = \frac{1}{\pi^{1/2}} \left(\frac{Z}{b_0}\right)^{3/2} e^{-(Zr_e/b_0)}, \tag{6b.4}$$

with

$$b_0 = \frac{\hbar^2 \cdot 4\pi\epsilon_0}{me^2}.$$

If we start from the classical argument that the interaction process is caused by the varying dipole potential (see equation (6.1)), one can write this perturbation as

$$eV(\vec{r}_e) = \frac{1}{2} \frac{e\Pi_0 \cos\theta}{4\pi\epsilon_0} \frac{1}{r_e^2}, \tag{6b.5}$$

since

$$\mathcal{V}(\vec{r}_e, t) = eV(\vec{r}_e)(e^{i\omega t} + e^{-i\omega t}).$$

The transiton matrix element then becomes

$$M_{fi} = \int \psi_f^*(\vec{r}_e)eV(\vec{r}_e, t)\psi_i(\vec{r}_e)\, d\vec{r}_e$$

$$= \frac{\Pi_0}{4\pi\epsilon_0} ek_e \left(\frac{3}{4R}\right)^{1/2} \frac{1}{\pi^{1/2}} \left(\frac{Z}{b_0}\right)^{3/2}$$

$$\times \int e^{-(Zr_e/b_0)} \frac{\cos\theta}{r_e^2} \frac{J_{3/2}(k_e r_e)}{(k_e r_e)^{1/2}} \cos\theta \, d\vec{r}_e, \tag{6b.6}$$

$$M_{fi} = \frac{\Pi_0}{2} \frac{1}{4\pi\epsilon_0} \left(\frac{4\pi}{3R}\right)^{1/2} e k_e \left(\frac{Z}{b_0}\right)^{3/2} I(Z, b_0, k_e),$$

with

$$I(Z, b_0, k_e) = \int e^{-(Zr_e/b_0)} \frac{J_{3/2}(k_e r_e)}{(k_e r_e)^{1/2}} \, dr_e. \tag{6b.7}$$

The density of final states for p-wave electron emission follows from the asymptotic expression and the condition that $\psi_f(R) = 0$ or

$$k_e R = (n + \tfrac{1}{2})\pi, \tag{6b.8}$$

from which one easily derives

$$\frac{dn}{dE} = \frac{R}{\hbar\pi v_e}. \tag{6b.9}$$

Combining all terms one obtains the transition probability

$$\lambda_K = \frac{2\pi}{\hbar} \frac{\Pi_0^2}{(4\pi\epsilon_0)^2} \frac{e^2 k_e^2}{3} \left(\frac{Z}{b_0}\right)^3 \frac{I^2(Z, b_0, k_e)}{\hbar v_e}. \tag{6b.10}$$

The electric dipole transition photon emission transition probability has been derived as (see equation (6.51))

$$\lambda_{if}^\gamma = \frac{4}{3} \frac{e^2}{\hbar c \cdot 4\pi\epsilon_0} \left(\frac{E_\gamma}{\hbar c}\right)^3 c|\langle f|\vec{r}|i\rangle|^2. \tag{6b.11}$$

Using the correspondence between the quantum mechanical electric dipole matrix element and the dipole moment Π_0

$$\Pi_0^2 \Longleftrightarrow 4|\langle f|\vec{r}|i\rangle|^2 e^2, \tag{6b.12}$$

the transition E1 gamma transition probability becomes

$$\lambda_{if}^\gamma = \frac{\Pi_0^2}{3} \frac{E_\gamma^3}{\hbar^4 c^3} \frac{1}{4\pi\epsilon_0}, \tag{6b.13}$$

giving a ratio

$$\alpha_K = 2\frac{\lambda_K}{\lambda_{if}^\gamma} = \frac{4\pi}{\hbar} \frac{k_e^2 e^2}{v_e} \left(\frac{Z}{b_0}\right)^3 \frac{(\hbar c)^3}{E_\gamma^3} I^2(Z, b_0, k_e) \frac{1}{4\pi\epsilon_0}. \tag{6b.14}$$

A closed form can be obtained under the conditions $Z/b_0 \ll k_e$, i.e. transition energy large compared to K-electron binding energy and assuming non-relativistic motion for the ejected electron, leading to $v_e = \sqrt{2\hbar\omega/m}$. The integral then is evaluated as

$$\int J_{3/2}(k_e r_e) \frac{dr_e}{(k r_e)^{1/2}} = \left(\frac{2}{\pi k_e^2}\right)^{1/2}. \tag{6b.15}$$

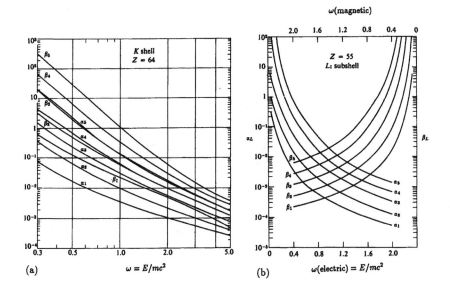

Figure 6b.1. Electric (α_L) and magnetic (β_L) conversion coefficients for the K-shell at $Z = 64$. The energy scale is given in relative units $E/m_0c^2 = \omega$. (Taken from Rose 1965 in α, β and γ *Spectroscopy*, ed K Sieghbahn.)

And the final expression is

$$\alpha_K = \tfrac{1}{2} Z^3 \left(\frac{e^2}{\hbar c \cdot 4\pi\epsilon_0} \right)^4 \left(\frac{2m_0c^2}{E_\gamma} \right)^{7/2}. \qquad (6b.16)$$

It can be shown that the above expression can be extended to general EL conversion with the result

$$\alpha_K^L = Z^3 \left(\frac{L}{L+1} \right) \alpha^4 \left(\frac{2m_0c^2}{E_\gamma} \right)^{L+(5/2)}. \qquad (6b.17)$$

Using relativistic correct wavefunctions and other refinements, the calculations soon become complex. We show the behaviour of electric (EL) and magnetic (ML) conversion through the K-shell at $Z = 64$ in figure 6b.1.

In contrast to the early measurement of conversion electrons using Geiger-Müller counters and beta magnetic spectrographs, present-day detection of electrons uses solid-state detectors with very good energy resolution. A typical electron spectrum for mass $A = 196$, corresponding to the conversion process of a number of gamma transitions in ^{196}Pb, is shown in figure 6b.2, where K, L, M lines are presented. The energy scale is given in keV units.

Determining the α_K coefficients, which have high sensitivity to the gamma multipolarity, gives a method for characterizing gamma transitions with a unique

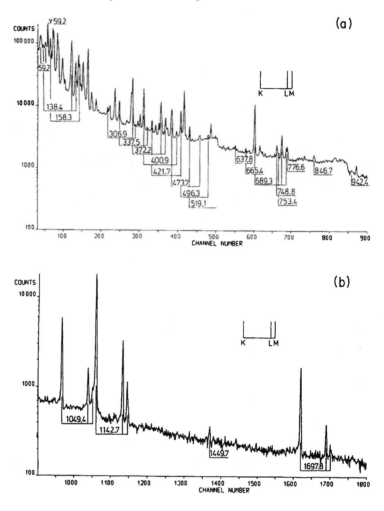

Figure 6b.2. A typical electron spectrum obtained for mass $A = 196$. The electron transitions, corresponding to the decay of ^{196}Bi are indicated with the corresponding transition energy (in keV units). (Taken from Huyse 1991.)

multipolarity. The α_K values, measured for ^{190}Hg (see figure 6b.3 for these α_K values) then determine a number of gamma multipolarities, giving direct information on spin changes and on possible level spins in the nucleus ^{190}Hg.

In a number of cases, for $0^+ \to 0^+$ transitions where photon emission is forbidden, conversion electrons can still be observed. By combining the γ-ray spectrum and the corresponding electron spectrum, as is carried out for ^{198}Pb, two transitions with a $0^+ \to 0^+$ signature are clearly observable for the 1392.1 keV and 1734.7 keV transitions (see figure 6b.4). This technique is a very useful and

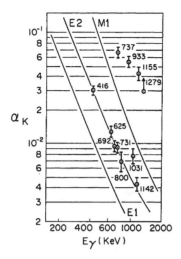

Figure 6b.3. A plot of the α_K values observed for the transitions as shown in figure 6.10. The solid lines are the theoretical values for E1, E2 and M1 multipolarities. The 416 keV E2 $2^+ \rightarrow 0^+$ transition was used to normalize the experimental conversion-electron and gamma-ray intensities. (Taken from Kortelahti *et al* 1991.)

unique method to identify $0^+ \rightarrow 0^+$ transitions and, subsequently, to obtain nuclear structure information on excited 0^+ levels in the atomic nucleus. This method was used successfully in the observation of the first excited 0^+ states in neutron-deficient even–even Pb nuclei by the Lisol group in Leuven.

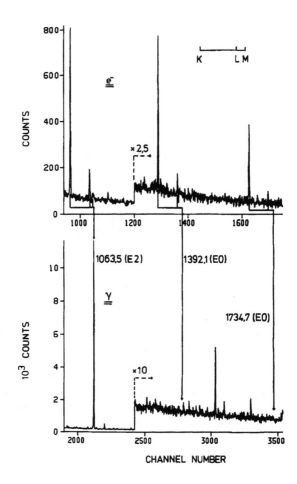

Figure 6b.4. Comparison of gamma-ray and electron spectra of transitions in ^{198}Pb. In this way, a unique signature for $0^+ \rightarrow 0^+$ transitions is obtained for those cases which have no corresponding γ-ray component. The $0^+ \rightarrow 0^+$ 1392.1 keV and 1734.7 keV transitions are clearly identified in the upper spectrum. (Taken from Huyse 1991.)

Problem set—Part B

1. In the study of beta-decay, the transition probability can be reduced into the particular combination $\log(f T_{1/2})$. This function is inversely proportional to the beta transition matrix element $|M|^2$. Show that by studying this matrix element and by using a particular one-point interaction describing the beta-decay process given as $g\delta(\vec{r}_p - \vec{r}_n) \cdot \delta(\vec{r}_p - \vec{r}_{e^-}) \cdot \delta(\vec{r}_p - \vec{r}_{\bar{\nu}})$, one can set up a classification within the beta-decay process.

Discuss the order of magnitude of the successive terms as well as the parity selection rule for the beta-decay process.

2. Determine how the original beta-spectrum shape for an allowed transition $\Lambda(p_{e^-})$ becomes modified considering a one-time forbidden beta emission in which an electron is emitted with momentum $\vec{p}_{e^-}$ and an antineutrino with momentum $\vec{p}_{\bar{\nu}}$ within the solid angles $d\Omega_{e^-}$ and $d\Omega_{\bar{\nu}}$, respectively.

Hint: You can consider the situation in which angular integration over the electron and antineutrino emission gives an average of the scalar product $\vec{p}_e \cdot \vec{p}_{\bar{\nu}}$ equal to zero.

3. Show that in the beta-decay emission, using non-relativistic kinematics, the mean electron kinetic energy amounts to $\frac{1}{3}$ of the maximal electron kinetic energy (replace the Fermi function by the value 1 throughout). How does this result change when relativistic kinematics are considered.

4. Show that by making the appropriate combination of the half-life in beta-decay, i.e. forming the product of the f function and the half-life $f(Z, \omega_0) \cdot T_{1/2}$, one can determine the nuclear beta-decay matrix element.

5. Classify and discuss in detail, the twice-forbidden Fermi and Gamow–Teller beta transitions. Present in each case (a) the restrictions on angular momentum and (b) the restrictions on parity for the initial and final nuclear states.

Present a number of examples from actual beta-decay processes.

6. Derive the Q-value for a number of beta-decay processes, making use of atomic binding energy values:

(a) for β^+-decay,
(b) for K-electron capture,

(c) for neutrino capture,

(d) for neutrino-less double β^--decay.

(e) Derive the limiting value for electron capture in $^{148}_{66}$Dy in which K- and L-capture are just forbidden. This concerns a $0^+ \rightarrow 2^-$ transition. What type of transition is this?

(f) Derive an approximate value for the half-life in a double β^--decay process, i.e. in the transition $Z \rightarrow Z + 2$, starting from second-order perturbation theory and assuming the half-lives for the separate $Z \rightarrow Z + 1$ and $Z + 1 \rightarrow Z + 2$ beta-decay transitions are known.

7. Show that the total beta-decay constant λ_{β^-} is proportional to $Q^5_{\beta^-}$ (with Q_{β^-} the Q-value characterizing the β^--decay process), in the situation that the electron kinetic energy $T_{e^-} \gg m_0 c^2$ (with $m_0 c^2$ the electron rest energy).

8. Explain how in the study of beta-decay experiments, the measurement of a pseudo-scalar observable can give a proof of parity non-conservation of the wavefunctions considered.

Show that in the β^--decay of ^{60}Co (with initial angular momentum and parity 5^+) into ^{60}Ni (with final angular momentum and parity 4^+), when taking into account the presence of only right-handed neutrinos (or left-handed antineutrinos), all electrons are emitted at $180°$ with respect to the orientation of the originally polarized nuclear spin 5 in the ^{60}Co nuclei. Classify this particular beta-transition.

9. We give the atomic binding energy in the following elements: ^{163}Ho $(Z = 67) = 1329\,604$ keV and ^{163}Dy $(Z = 66) = 1330\,389$ keV. Which decay processes can occur in this particular case?

Data given: $(M(^1\text{n}) - M(^1\text{H})) \cdot c^2 = 782$ keV.

10. Express the Q-value of beta-decay using the binding energy of the parent and daughter nuclei. We give the following values: $(m_n - m_{1\text{H}}) \cdot c^2 = 0.782$ MeV, $2m_0 c^2 = 1.022$ MeV.

We consider a given mass chain $(A = 141)$ and give binding energy (in units MeV), spin and parity of the ground state for the various elements

$Z = 54$	1164.82	$7/2^+$
$Z = 55$	1170.04	$7/2^+$
$Z = 56$	1174.24	$5/2^-$
$Z = 57$	1176.48	$5/2^+$
$Z = 58$	1178.13	$7/2^-$
$Z = 59$	1177.93	$5/2^+$
$Z = 60$	1175.33	$3/2^+$
$Z = 61$	1170.82	$3/2^+$

Describe, moreover, which type of beta-decay results in each case $(\beta^-, \beta^+,$ electron capture, etc) as well as the nature (allowed, first-forbidden, etc) of the decay process.

11. Discuss the recoil energy for α-, β- and γ-decay.

Calculate and compare the order of magnitude in each case for a nucleus with mass number $A = 40$, $Z = 20$ and an equal Q-value ($Q_\alpha = Q_{\beta^-} = Q_\gamma = 5$ MeV). In which decay process can one neglect this recoil energy and still have a satisfactory energy balance.

12.   Derive an approximate half-life for an electric dipole transition corresponding to a gamma emission of 1 MeV for a nucleus with mass number $A = 140$.

13.   In the discussion in section 6.3 when deriving the electromagnetic transition matrix element, the dipole approximation is used (equation (6.40)). Show that if we go beyond this approximation and also consider the next term in the expansion of the plane wave, an operator results that is able to describe both M1 (magnetic dipole) and E2 (electric quadrupole) transitions. Explain why both an M1 and E2 transition may occur. Using a full expansion of the plane wave in spherical harmonics, it is possible to derive a general expression for the electromagnetic transition operators.

14. Modify the electric dipole decay constant, as derived in equation (6.51) (where a matrix element over the nuclear coordinate r results), in order to show the equivalence with the Larmor 'radiation' formula (equation (6.15)) explicitly.

PART C

NUCLEAR STRUCTURE:
·AN INTRODUCTION

In this part, C, we address the nucleus as a many-body system from a number of viewpoints namely collective and independent-particle model approaches. In chapter 7, we consider the atomic nucleus as a charged liquid drop in order to describe a number of smoothly varying properties, in particular the nuclear binding energy. In chapter 8, taking an opposite approach, the most simple independent particle motion in which nucleons are described as a non-interacting gas of fermions (protons and neutrons), i.e. the Fermi gas model, is presented. In chapter 9 then, we go much further in describing the atomic nucleus as depicted through the independent-particle shell model. We present the salient features of the model using a harmonic oscillator average potential as a first approximation to the actual mean field and also discuss a number of experimental facts that corroborate the concept of independent-particle motion to lowest order. Using a number of specific boxes, we highlight the material presented in part C with recent features of nuclear structure used to describe the atomic nucleus. At the same time, we indicate a number of contact points with other domains in physics such as astrophysics and advanced numerical methods to model many-body systems.

We stress that this part contains the essential elements in order to cope with the nucleus and the underlying model structures. This part, therefore, forms an essential element, together with parts A and B, in carrying out a first but extensive study of the field of nuclear physics.

Chapter 7

The liquid drop model approach: a semi-empirical method

7.1 Introduction

The liquid drop model was historically the first model to describe nuclear properties. The idea came primarily from the observation that nuclear forces exhibit saturation properties. The binding energy per nucleon, $BE(A, Z)/A$ is a clear indication of this observation (figure 7.1): a non-saturated force would lead to a binding energy given by the $A(A - 1)/2$ nucleon 2-body interaction energy, in total contradiction with the observation of figure 7.1. Also, the nucleus presents a low compressibility and so a well defined nuclear surface. It is soon clear though, that the liquid drop model taken as a fully classical model, cannot be extrapolated too far in the atomic nucleus

(i) In the inset in figure 7.1, spikes on $BE(A, Z)/A$ appear at values $A = 4n$ ($n = 1$ ^{4}He, $n = 2$ ^{8}Be, $n = 3$ ^{12}C, $n = 4$ ^{16}O$_1$, ...) and present a favoured n α-particle-like structure reflecting aspects of the nucleon–nucleon force.

(ii) In a liquid, the average distance between two fluid 'particles' is about equal to that value where the potential interaction energy is a minimum, which, for the nuclear interaction is $\simeq 0.7$ fm (figure 7.2). The nucleons are, on average, much farther apart. An important reason is the fact that fermions are Fermi–Dirac particles so we are considering a Fermi liquid. The Pauli principle cuts out a large part of nucleon 2-body interactions since the nearby orbitals in the nuclear potential are occupied (up to the Fermi level). Scattering within the Fermi liquid is a rare process compared to collision phenomena in real, macroscopic fluid systems. Thus, the mean free path of a nucleon, travelling within the nucleus easily becomes as large as the nuclear radius (see figures 7.3(a) and (b)) and we are working mainly with a weakly interacting Fermi gas (see chapter 8).

(iii) Saturation of binding energy results in a value of $BE(A, Z)/A \simeq$ 8 MeV independent of A and Z and represents charge independence of the nuclear interaction in the nucleus. Each nucleon interacts with a limited number of

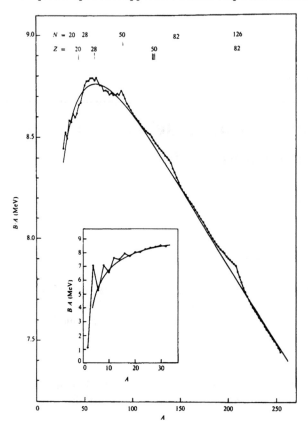

Figure 7.1. Binding energy per nucleon as a function of the atomic mass number A. The smooth curve represents a pure liquid drop model calculation. Deviations from the curve occur at various specific proton (Z) and neutron (N) numbers. In the inset, the very low mass region is presented in an enlarged figure. (Taken from Valentin 1981.)

nucleons, a conclusion that can be derived by combining the Pauli principle with Heisenberg's uncertainty principle, and the short-range character of the nucleon–nucleon force. The simple argument runs as follows:

$$\Delta E \cdot \Delta t \simeq \hbar$$
$$d \simeq \Delta t \cdot c \simeq \frac{\hbar}{\Delta E} \cdot c \simeq \frac{\hbar \cdot mc^2}{mc \cdot \Delta E}. \tag{7.1}$$

With ΔE for a 2-body interaction in the central zone of the nucleus, $\Delta E \simeq$ 200 MeV and $d \simeq 1$ fm. The total binding energy is now the subtle difference between the total kinetic and potential energy (figure 7.2): the kinetic energy rapidly increases with decreasing internucleon distance whereas the potential

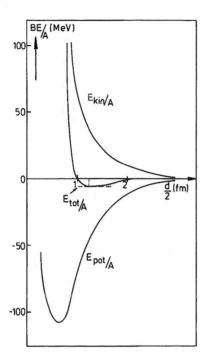

Figure 7.2. Balancing effect between the repulsive kinetic and attractive potential energy contribution to the total nuclear binding energy, all expressed as MeV per nucleon. It is shown that the total binding energy is a rather small, negative value ($\simeq 8$ MeV/nucleon) and results as a fine balance between the two above terms. The various energy contributions are given as a function of the internuclear distance. (Taken from Ring and Schuck 1980.)

energy becomes negative. For small distances ($d \lesssim 1$ fm), the kinetic energy dominates giving a total positive energy. For large distances ($d \gtrsim 3$ fm) there is almost no nuclear interaction remaining. A weak minimum develops at $r_0 \simeq 1.2$ fm; the equilibrium value. So, this explains the limited number of nucleons interacting through the nucleon–nucleon interaction in the nucleus.

7.2 The semi-empirical mass formula: coupling the shell model and the collective model

As is clear from the introductory section, the binding energy shows a smooth variation with A and Z and a liquid drop model (or collective model) approach can be assumed to describe this smooth variation. A number of indications, however, show the need for specific shell model corrections to this liquid drop model. As is clear in figure 7.3(a), the motion of nucleons in the average potential $U(r)$ will

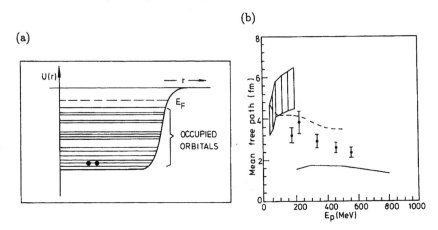

Figure 7.3. (*a*) Illustration of two interacting nucleons moving in deeply-bound orbitals in the nuclear potential. Very few final states, due to the Pauli principle, are available to the scattered nucleons and a rather large, free mean path results. (*b*) The nucleon mean free path in nuclear matter. The shaded band indicates the data for ^{40}Ca, ^{90}Zr and ^{208}Pb. The various data points given are explained by Rego (1991). The theoretical curves, dashed curve for Dirac–Brueckner and full line for an optical potential, are discussed by Rego (1991) also.

be given by the Schrödinger equation

$$-\frac{\hbar^2}{2m}\Delta\varphi(\vec{r}) + U(r)\varphi(\vec{r}) = E\varphi(\vec{r}). \tag{7.2}$$

In parametrizing the main effects of the nuclear single-particle motion we will use the fact that nucleons move in a simpler potential and postpone a more detailed shell model study until chapter 9.

The collective, or liquid drop, model aspects come from (in a simplest approach) a spherical, liquid drop with a surface described by $R(\theta, \varphi) = R_0$. If we proceed further, and allowing deformation effects to influence the nuclear, collective dynamics, a surface can be depicted as (figure 7.4)

$$R(\theta, \varphi) = R_0\left(1 + \sum_{\lambda\mu}\alpha_{\lambda\mu}Y^*_{\lambda\mu}(\theta, \varphi)\right), \tag{7.3}$$

where $Y_{\lambda\mu}(\theta, \varphi)$ describe the spherical harmonics. The small amplitude oscillations when $\alpha_{\lambda\mu} = \alpha_{\lambda\mu}(t)$ then give rise to a harmonic oscillator approximation resulting in the collective dynamics with a Hamiltonian (figure 7.5)

$$H = \sum_{\lambda\mu}\left(\frac{B_\lambda}{2}|\dot{\alpha}_{\lambda\mu}|^2 + \frac{C_\lambda}{2}|\alpha_{\lambda\mu}|^2\right). \tag{7.4}$$

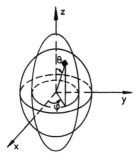

Figure 7.4. Nuclear, collective model description. A description of the nuclear surface $R(\theta, \varphi)$ is given in terms of the spherical coordinates θ, φ for each point. Excursions from a spherical into an ellipsoidal shape are indicated.

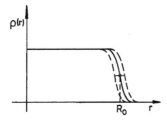

Figure 7.5. Nuclear density variations are mainly situated in the nuclear surface region for a nuclear radius R_0. The density variation leads to oscillatory quantum mechanical motion described by the various multipole components λ which are described by the $Y_{\lambda\mu}(\theta, \varphi)$ spherical harmonics.

Quantization then gives the nuclear, collective model of motion. This discussion is taken up again in Part D. For the liquid drop behaviour we examine the static potential energy to find a correct parametrization over a large A interval.

7.2.1 Volume, surface and Coulomb contributions

The three terms discussed represent the energy contributions for a charged drop and their dependence on the number of nucleons A and protons Z.

(i) The volume term expresses the fact that the nuclear force is saturated and thus, a certain part of the nuclear interior represents a given binding energy contribution. Expressing the large, overall constancy of $BE(A, Z)/A$ this term gives

$$BE(A, Z) = a_{\mathrm{v}} A, \tag{7.5}$$

as the major dependence.

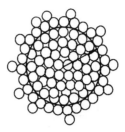

Figure 7.6. Pictorial representation of nuclear surface effects on binding energy. Nucleons at or near to the nuclear surface are less strongly bound than a nucleon within the nuclear interior.

The various liquid drop corrections will reduce the binding energy to the more realistic value of $\simeq 8$ MeV per nucleon.

(ii) The surface effect takes into account, in a simple way, the fact that nucleons at, or close to, the nuclear surface will have a reduced binding energy since only partial surrounding with nucleons is possible, for a nucleus with a finite radius. This is pictorially represented in figure 7.6. The correction will be proportional to the nuclear surface area $(4\pi R^2)$, and we obtain

$$BE(A, Z) = a_v A - a_s A^{2/3}, \tag{7.6}$$

as a revised value.

(iii) Coulomb effects result since a charge of Ze is present within the nuclear volume. For a homogeneously charged liquid drop with sharp radius R and density

$$\rho_c = \frac{Ze}{\frac{4}{3}\pi R^3}, \tag{7.7}$$

one can evaluate the Coulomb contribution to the nuclear binding energy by a classical argument. The Coulomb energy needed to add a spherical shell, to the outside of the sphere with radius r, to give an increment dr becomes

$$U_c' = \frac{1}{4\pi\varepsilon_0} \int_0^R \frac{\frac{4}{3}\pi r^3 \rho_c 4\pi r^2 \rho_c}{r} \, dr. \tag{7.8}$$

Using the above charge density, the integral becomes (figure 7.7)

$$U_c' = \frac{3}{5} \frac{Z^2 e^2}{R} \frac{1}{4\pi\varepsilon_0}. \tag{7.9}$$

In the argument used above we have smoothed out the charge of Z nucleons over the whole nucleus. In this evaluation we have counted a self-energy Coulomb

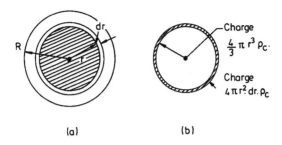

Figure 7.7. Evaluation of the Coulomb energy of a liquid, spherical charged drop. The Coulomb energy is evaluated by calculating the energy needed to constitute the full nuclear charge in terms of spherical shells dr filled in one after the other (a). The electrostatic potential energy calculation reduces to the Coulomb energy between a central charge $\frac{4}{3}\pi r^3 \rho_c$ and the charge in an infinitesimal shell $4\pi r^2 \, dr \rho_c$ shown as (b).

interaction which is spurious and we should correct for the effect of Z protons. Using the same method as before but now with the proton smeared charge density

$$\rho_p = \frac{e}{\frac{4}{3}\pi R^3}, \tag{7.10}$$

and a self-Coulomb energy, for the Z protons, as

$$U_c'' = \frac{3}{5}\frac{Ze^2}{R}\frac{1}{4\pi\varepsilon_0}, \tag{7.11}$$

the total Coulomb energy correction becomes

$$U_c = U_c' - U_c'' = \frac{3}{5}Z\frac{(Z-1)}{R}\frac{1}{4\pi\varepsilon_0}, \tag{7.12}$$

or, parametrized using the variables A, Z,

$$U_c = a_c Z(Z-1)A^{-1/3}. \tag{7.13}$$

The combined effect then gives the binding energy

$$BE(A, Z) = a_v A - a_s A^{2/3} - a_c Z(Z-1)A^{-1/3}, \tag{7.14}$$

or, per nucleon,

$$BE(A, Z)/A = a_v - a_s A^{-1/3} - a_c Z(Z-1)A^{-4/3}, \tag{7.15}$$

the behaviour of which is shown in figure 7.8. The surface energy gives the largest correction for smallest A ($A^{-1/3}$ effect) whereas the Coulomb energy correction,

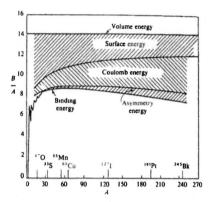

Figure 7.8.   The various (volume, surface, Coulomb, asymmetry) energy terms contributing to the nuclear binding energy (per nucleon). (Taken from Valentin 1981.)

taking a simple estimate of $Z = A/2$, is largest for heavy nuclei with many protons ($A^{2/3}$ effect). The two terms taken together produce a maximum in the $BE(A, Z)/A$ curve, already very close to the region of most strongly bound (per nucleon) nuclei. It is also clear from this figure that fusion of light nuclei towards $A \cong 56$ and fission of very heavy (actinide nuclei) will be processes liberating a large amount of energy.

For nuclei, which deviate from a spherical shape, both the surface and Coulomb energy corrections will change in a specific way. If we denote the nucleus with its lowest deformation multipoles via the (θ, φ) expansion

$$R = R_0(1 + \alpha_2 P_2(\cos \theta) + \alpha_4 P_4(\cos \theta)), \qquad (7.16)$$

correction functions for surface energy $g(\alpha_2, \alpha_4)$ and for Coulomb energy $f(\alpha_2, \alpha_4)$ appear in the expression for $BE(A, Z)$, i.e. we obtain a result

$$BE(A, Z) = a_v A - a_s g(\alpha_2, \alpha_4)A^{2/3} - a_c f(\alpha_2, \alpha_4)Z(Z - 1)A^{-1/3}. \quad (7.17)$$

A simple illustration of the above modification is obtained for an ellipsoidal deformation and, retaining a constant volume for the deformation, we get the major axis

$$a = R(1 + \epsilon), \qquad \left(\epsilon = \sqrt{1 - (b^2/a^2)}\right), \qquad (7.18)$$

and the minor axis

$$b = R(1 + \epsilon)^{-1/2}, \qquad (7.19)$$

with volume $V = \frac{4}{3}\pi ab^2 \simeq \frac{4}{3}\pi R^3$. Using the parameter of deformation ϵ, the surface and Coulomb energy terms become

$$E_s = a_s A^{2/3}(1 + \tfrac{2}{5}\epsilon^2)$$
$$E_c = a_c Z(Z - 1)A^{-1/3}(1 - \tfrac{1}{5}\epsilon^2). \qquad (7.20)$$

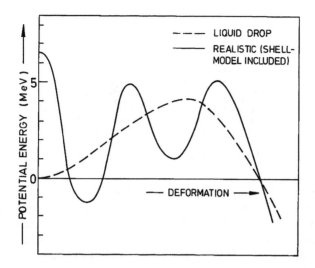

Figure 7.9. Potential energy function for deformations leading to fission. The liquid drop model variation along a path leading to nuclear fission (dashed line) as well as a more realistic determination of the total energy (using the Strutinsky method as explained in chapter 13) are shown as a function of a general deformation parameter.

The total energy charge, due to deformation, reads

$$\Delta E = \Delta E_s + \Delta E_c = \epsilon^2[\tfrac{2}{5}A^{2/3}a_s - \tfrac{1}{5}a_c Z(Z-1)A^{-1/3}]. \tag{7.21}$$

We use the simplification $Z(Z-1) \to Z^2$ and the best fit values for a_s and a_c; 17.2 MeV and 0.70 MeV, respectively. Then if $\Delta E > 0$, the spherical shape is stable, and results in the limit $Z^2/A < 49$.

The curve, describing the potential energy versus nuclear distance between two nuclei has the qualitative shape, given in figure 7.9 by the dashed line. The point at $\epsilon = 0$ corresponds to the spherical nucleus. For large separation, calling $r = R_1 + R_2$ (with R_1 and R_2 the radii of both fragments), the energy varies according to the Coulomb energy

$$U_c = \frac{Z_1 Z_2 e^2}{4\pi\varepsilon_0 r}. \tag{7.22}$$

Deformation complicates the precise evaluation of the full total potential energy and requires complicated fission calculations.

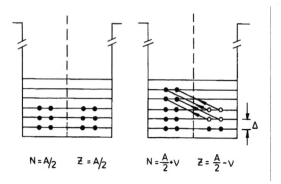

Figure 7.10. Schematic single-particle model description (Fermi–gas model, see chapter 8) for evaluating the nuclear asymmetry energy contribution for the total nuclear system. Two different distributions of A nucleons over the proton and neutron orbitals with twofold (spin) degeneracy ($N = Z = A/2$ and $N = A/2 + v$, $Z = A/2 - v$) are shown. The average nuclear level energy separation is denoted by the quantity Δ (see text).

7.2.2 Shell model corrections: symmetry energy, pairing and shell corrections

As explained before, even though the nuclear binding energy systematics mimics the energy of a charged, liquid drop to a large extent, the specific nucleon (Dirac–Fermi statistics) properties of the nuclear interior modify a number of results. They represent manifestations of the Pauli principle governing the occupation of the single-particle orbitals in the nuclear, average field *and* the nucleonic residual interactions that try to pair-off identical nucleons to 0^+ coupled pairs.

(i) *Symmetry energy* In considering the partition of A nucleons (Z protons, N neutrons) over the single-particle orbitals in a simple potential which describes the nuclear average field, a distribution of various nuclei for a given A, but varying (Z, N), will result. The binding energy of these nuclei will be maximum when nucleons occupy the lowest possible orbitals. The Pauli principle, however, prevents the occupation of a certain orbital by more than two identical nucleons with opposite intrinsic spin orientations. The symmetric distribution $Z = N = A/2$ proves to be the energetically most favoured (if only *this* term is considered!). Any other repartition, $N = (A/2) + v$, $Z = (A/2) - v$, will involve lifting particles from occupied into empty orbitals. If the average energy separation between adjacent orbitals amounts to Δ, replacing v nucleons will cost an energy loss of (figure 7.10)

$$\Delta E_{\text{binding}} = v\left(\Delta\frac{v}{2}\right), \tag{7.23}$$

and, with $v = (N - Z)/2$, this becomes

$$\Delta E_{\text{binding}} = \tfrac{1}{8}(N - Z)^2\Delta. \tag{7.24}$$

The potential depth U_0, describing the nuclear well does not vary much with changing mean number: for the two extremes ^{16}O and ^{208}Pb, the depth does not change by more than 10% (see Bohr and Mottelson 1969, vol 1) and thus, the average energy spacing between the single particles, Δ, should vary inversely proportionally to A, or,

$$\Delta \propto A^{-1}. \tag{7.25}$$

The final result, expressing the loss of symmetry energy due to the Pauli effect which blocks the occupation of those levels that already contain two identical nucleons, becomes

$$BE(A, Z) = a_v A - a_s A^{2/3} - a_c Z(Z - 1)A^{-1/3} - a_A(A - 2Z)^2 A^{-1}. \tag{7.26}$$

The relative importance of this symmetry (or asymmetry) term is illustrated in figure 7.8, where the volume term, surface term and Coulomb term are also drawn. A better study of the symmetry energy can be evaluated using a Fermi gas model description as the most simple independent particle model for nucleons moving within the nuclear potential. This will be discussed in chapter 8 giving rise to a numerical derivation of the coefficient a_A.

(ii) *Pairing energy contribution* Nucleons preferentially form pairs (proton pairs, neutron pairs) in the nucleus under the influence of the short-range nucleon–nucleon attractive force. This effect is best illustrated by studying nucleon separation energies.

In chapter 1, we have expressed the nuclear binding energy as the energy difference between the rest mass, corresponding to A free nucleons and A nucleons, bound in the nucleus. Similarly, we can define each of the various separation energies as the energy needed to take a particle out of the nucleus and so this separation energy becomes equal to the energy with which a particular particle (or cluster) is bound in the nucleus.

Generally, we have expressed the binding energy as

$$BE(A, Z) = ZM_pc^2 + NM_nc^2 - M'(^A_Z X_N)c^2. \tag{7.27}$$

The 'binding energy' with which the cluster $^{A'}_{Z'}Y_{N'}$ is bound in the nucleus $^A_Z X_N$ (with in general $A' \ll A$, $Z' \ll Z$, $N' \ll N$) becomes ($A'' = A - A'$, $Z'' = Z - Z'$, $N'' = N - N'$)

$$S_Y = M'(^{A'}_{Z'}Y_{N'})c^2 + M'(^{A''}_{Z''}U_{N''})c^2 - M'(^A_Z X_N)c^2, \tag{7.28}$$

or, rewritten, using the binding energy for the nuclei X, Y and U,

$$S_Y = BE(A, Z) - [BE(A', Z') + BE(A'', Z'')]. \tag{7.29}$$

For a proton and a neutron, this becomes,

$$S_p = BE(A, Z) - BE(A - 1, Z - 1)$$
$$S_n = BE(A, Z) - BE(A - 1, Z), \tag{7.30}$$

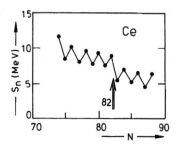

Figure 7.11. Single-neutron separation energy S_n for the even–even and even–odd Ce ($Z = 58$) nuclei. Besides a sudden lowering in the average S_n at $N = 82$, a specific odd–even staggering, proving a nucleon pairing effect, is well illustrated.

respectively. The separation energy for an α-particle reads

$$S_\alpha = BE(A, Z) - (BE(A - 4, Z - 2) + BE(4, 2)). \qquad (7.31)$$

In mapping S_p and S_n values, a specific saw-tooth figure results (see figure 7.11 for S_n values in the Ce isotopes for $70 \le N \le 90$). This figure very clearly expresses the fact that it costs more energy (on average 1.2–1.5 MeV) to separate a neutron (for Ce one has $Z = 58$) from a nucleus with even neutron number, than for the adjacent odd-neutron number nuclei. There is an overall trend which is not discussed at present, but the above point proves an odd–even effect showing that even–even nuclei are more bound than odd–even nuclei by an amount which we call δ. Proceeding to an odd–odd nucleus we have to break a pair, relative to the odd–even case and lose an amount δ of binding energy. Taking the odd–even nucleus as a reference point we can then express the extra pairing energy correction as

$$\Delta E_{\text{pair}} = \begin{array}{ll} +\delta & \text{(e–e)} \\ 0 & \text{(o–e)} \\ -\delta & \text{(o–o)}. \end{array} \qquad (7.32)$$

Combining all of the above results, derived in 7.2.1 and 7.2.2, we obtain a semi-empirical mass equation

$$BE(A, Z) = a_v A - a_s A^{2/3} - a_c Z(Z-1)A^{-1/3} - a_A(A-2Z)^2 A^{-1} \quad \begin{array}{l} +\delta \\ +0 \\ -\delta. \end{array} \qquad (7.33)$$

This final form is commonly known as the Bethe–Weizsäcker mass equation. Using fits of known masses to this equation one can determine the coefficients a_v, a_s, a_c, a_A and a_p (when expressing $\delta \simeq a_p A^{-1/2}$ which is obtained in an empirical way). A fit by Wapstra gives the values (Wapstra 1971)

$$a_v = 15.85 \text{ MeV}$$

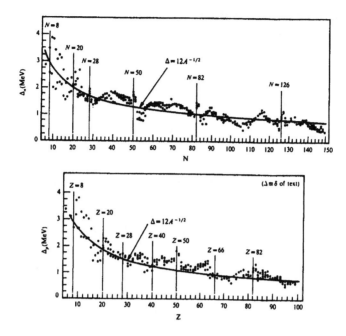

Figure 7.12. The odd–even mass differences for neutrons (upper part) and protons (lower part) ($\Delta \equiv \delta$ of text), which are based on the analysis of Zeldes *et al* (1967).

$$a_s = 18.34 \text{ MeV}$$
$$a_c = 0.71 \text{ MeV}$$
$$a_A = 23.21 \text{ MeV}$$
$$a_p = 12 \text{ MeV}.$$

The volume energy amounts to almost 16 MeV per nucleon: the various corrections then steadily bring this value towards the empirical value as observed throughout the nuclear mass region. The pairing coefficient is perhaps less well determined: we illustrate its behaviour in figure 7.12.

In this figure the smooth curve represents a fit to the data with a very good correspondence. Slight, systematic variations, however, occur. Plotting the difference $M_{\text{exp}} - M_{\text{liq.drop}}$ (in MeV units) a set of specific results becomes clear: nuclei are more bound than the liquid drop model predicts near proton (Z) and/or neutron number (N): 20, 28, 50, 82, 126. This points towards a shell structure within the nuclear single particle motion evidenced by figure 7.13 in a most dramatic way. This has implications to be discussed later in chapter 9.

7.3   Nuclear stability: the mass surface and the line of stability

From the Bethe–Weizsäcker binding energy relation, we can obtain a mass equation by equating the nuclear binding energy in equations (7.33) and (7.27). This results in the nuclear mass equation

$$M'(_Z^A X_N)c^2 = ZM_pc^2 + (A - Z)M_nc^2 - a_vA + a_sA^{2/3}$$

$$+ a_cZ(Z - 1)A^{-1/3} + a_A(A - 2Z)^2A^{-1} \begin{matrix} +\delta \\ +0 \\ -\delta \end{matrix} . \quad (7.34)$$

For each A value this represents a quadratic equation in the proton number Z i.e.

$$M'(_Z^A X_N)c^2 = xA + yZ + zZ^2 \begin{matrix} +\delta \\ +0 \\ -\delta \end{matrix} . \quad (7.35)$$

In this simplified form, one has

$$\begin{aligned} x &= M_nc^2 - a_v + a_A + a_sA^{-1/3} \\ y &= (M_p - M_n)c^2 - 4a_A - a_cA^{-1/3} \\ z &= a_cA^{-1/3} + 4a_AA^{-1}. \end{aligned} \quad (7.36)$$

We can now determine, for each A value, the nucleus with the lowest mass (largest binding energy), by solving the equation

$$\frac{\partial}{\partial Z}(M'(_Z^A X_N)) = 0, \quad (7.37)$$

or

$$Z_0 = \frac{-y}{2z}. \quad (7.38)$$

Feeding in the specific y and z values, this results in the most stable Z_0 value

$$Z_0 = \frac{4a_A + (M_n - M_p)c^2 + a_cA^{-1/3}}{8a_AA^{-1} + 2a_cA^{-1/3}}. \quad (7.39)$$

This can be brought into the more transparent form

$$Z_0 = \frac{A/2 + (M_n - M_p)c^2A/8a_A + a_cA^{2/3}/8a_A}{1 + \frac{1}{4}(a_c/a_A)A^{2/3}}. \quad (7.40)$$

The second and third terms in the numerator both become negligible relative to the $A/2$ factor, which gives the main effect. The approximate, but very useful expression, reads

$$Z_0 = \frac{A/2}{1 + \frac{1}{4}(a_c/a_A)A^{2/3}} = \frac{A/2}{1 + 0.0077A^{2/3}}. \quad (7.41)$$

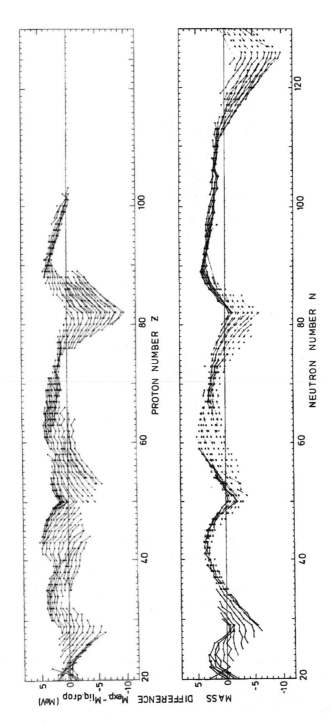

Figure 7.13. Nuclear (liquid drop) masses and the deviations with respect to the nuclear data and this, as a function of proton and neutron number. The shell closure effects are shown most dramatically (adapted from Myers 1966).

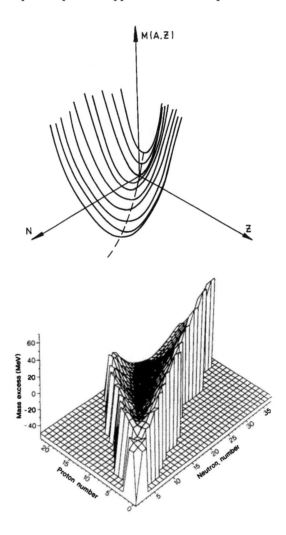

Figure 7.14. Nuclear mass surface $M(A, Z)$ as a function of proton and neutron numbers. The dashed line along the value connects the nuclei with smallest mass and thus the most strongly bound nuclei for each A value. In the lower part, a more realistic view of the nuclear mass surface, here represented by the mass excess $M'(A, Z) - A$, for light nuclei is presented. The latter figure is taken from Hall (1989).

This curve $Z_0 = Z_0(A, a_c, a_A)$ gives the projection of the minimum points of the nuclear mass surface (figure 7.14) on the (N, Z) plane and proceeds, initially, as $Z_0 \simeq A/2$ (for small A values) or, along the diagonal. For large A, deviations towards N values, larger than Z, become important. A realistic plot, representing

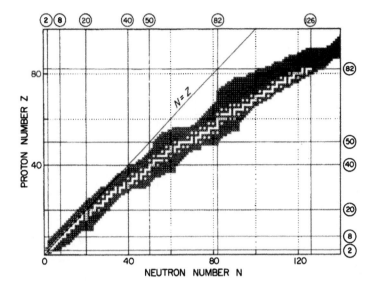

Figure 7.15. Distribution of the observed nuclei as a function of N and Z. Stable nuclei are shown as empty squares and occur in between the nuclei that are unstable against beta-decay, shown as the region of partially filled squares. Unstable nuclei can also decay by various other decay modes (nucleon emission, alpha-decay, fission, etc). (Taken from Wong, *Introductory Nuclear Physics* © 1990. Reprinted by permission of Prentice-Hall, Englewood Cliffs, NJ.)

the nuclei in the (N, Z) plane (figure 7.15) shows this drastic deviation in a very clear way.

A number of interesting results can be derived from this mass surface and from the semi-empirical mass equation.

(i) If we make cuts through the surface, for a given A value (see figure 7.16); a parabolic behaviour of nuclear masses shows up. For odd A values, in the $A = 101$ chain as an example, only one stable nucleus results. The various elements then have $\beta^{\pm}$ (or EC) decay towards the only stable element at (or near) the bottom of the parabola. For even A nuclei, on the other hand, both even–even and odd–odd nuclei can occur and (because of the $\pm\delta$ value) two parabolae are implied by the mass equation. For $A = 106$, we present the various decay possibilities. The two parabolae are shifted over the interval 2δ. In many cases for $A = $ even, more than one stable element results. These are cases where double-beta decay processes could occur. There are even cases with three 'stable' elements which depend on the specific curvature of the parabola and the precise location of the integer Z values.

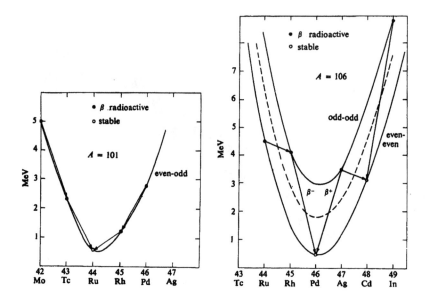

Figure 7.16. Left: the energy (in MeV) of various nuclei with mass number of $A = 101$. The zero point on the energy scale is chosen in an arbitrary way. Right: a similar figure but for $A = $ even and $A = 106$. Various decay possibilities are presented in each case by full lines with an arrow. (Taken from Segré, *Nuclei and Particles*, 2nd edn © 1982 Addison-Wesley Publishing Company. Reprinted by permission.)

(ii) The semi-empirical mass equation, shows various constraints on the range of possible A, Z values and stability against a number of radioactive decay processes can be determined. Spontaneous α-decay ($S_\alpha = 0$) follows from the equation

$$BE(^A_Z X_N) - [BE(^{A-4}_{Z-2}Y_{N-2}) + BE(^4_2 He_2)] = 0. \qquad (7.42)$$

The limit to the region for spontaneous α-emitters can be obtained from equation (7.33). In general, lifetimes for α emission become very short in the actinide region so that the last stable elements occur around $A \simeq 210$. The condition $S_n = 0$ ($S_p = 0$), similarly indicates the borderline where a neutron (proton) is no longer bound in the nucleus: this line is called the neutron (proton) drip line and is schematically drawn in Box 7b.

(iii) The energy released in nuclear fission, in the simple case of symmetric fission of the element (A, Z) into two nuclei $(A/2, Z/2)$ is

$$E_{\text{fission}} = M'(^A_Z X_N)c^2 - 2M'(^{A/2}_{Z/2}Y_{N/2})c^2. \qquad (7.43)$$

Using a simplified mass equation (replacing $Z(Z-1) \rightarrow Z^2$ and neglecting the pairing correction δ) one obtains

$$E_{\text{fission}} = [a_s A^{2/3}(1 - 2^{1/3}) + a_c Z^2 A^{-1/3}(1 - 2^{-2/3})]c^2$$

$$= (-5.12A^{2/3} + 0.28Z^2 A^{-1/3}]c^2. \tag{7.44}$$

This value becomes positive near $A \simeq 90$ and reaches a value of about 185 MeV for ^{236}U. For these fission products, neutron-rich nuclei are obtained which will decay by the emission of n $\rightarrow$ p + e$^-$ + $\bar{\nu}_e$. So, a fission process is a good generator of $\bar{\nu}_e$ antineutrinos of the electron type; the fusion process on the other hand will mainly give rise to ν_e electron neutrino beta-decay transitions.

7.4 Two-neutron separation energies

Nuclear masses and a derived quantity, the two-neutron separation energy, S_{2n}, form important indicators that may reveal the presence of extra correlations on top of a smooth liquid-drop behaviour. S_{2n} is defined as

$$S_{2n}(A, Z) = BE(A, Z) - BE(A - 2, Z), \tag{7.45}$$

where $BE(A, Z)$ is the binding energy defined as positive, i.e. it is the positive of the energy of the ground state of the atomic nucleus for a nucleus with A nucleons and Z protons.

The simplest liquid-drop model (LDM) (Wapstra 1958, Wapstra and Gove 1971) will be used throughout because it allows the overall description of the binding energy along the whole table of masses or for long series of isotopes.

To see how the values of S_{2n} evolve globally, through the complete mass chart, we can start from the semi-empirical mass formula (Weizsäcker 1935, Bethe and Bacher 1936),

$$BE(A, Z) = a_V A - a_S A^{2/3} - a_C Z(Z - 1)A^{-1/3} - a_A(A - 2Z)^2 A^{-1}. \tag{7.46}$$

Even though there exist much refined macroscopic models, the present simple LDM (7.46) is a suitable starting point for our purpose in analysing the physics behind the S_{2n} systematics. The S_{2n} value can be written as

$$S_{2n} \approx 2(a_V - a_A) - \frac{4}{3}a_S A^{-1/3} + \frac{2}{3}a_C Z(Z - 1)A^{-4/3} + 8a_A \frac{Z^2}{A(A - 2)}, \tag{7.47}$$

where the surface and Coulomb terms are only approximated expressions. If one inserts the particular value of Z, Z_0, that maximizes the binding energy for each given A (this is the definition for the valley of stability),

$$Z_0 = \frac{A/2}{1 + 0.0077A^{2/3}}, \tag{7.48}$$

we obtain, for large values of A, the result

$$S_{2n} = 2(a_V - a_A) - \frac{4}{3}a_S A^{-1/3} + \left(8a_A + \frac{2}{3}a_C A^{2/3}\right)\frac{1}{4 + 0.06A^{2/3}}. \tag{7.49}$$

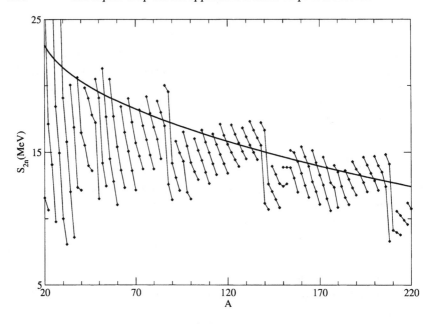

Figure 7.17. Comparison between the experimental S_{2n} (diamonds) values and the LDM prediction (bold curve) along the valley of stability. The experimental data correspond to even–even nuclei around the line of maximum stability. Points connected with lines correspond to nuclei with equal Z. (Reprinted from Fossion R *et al* 2002 *Nucl. Phys.* A **697** 703 © 2002 with permission from Elsevier.)

In the present form, we use the following values for the LDM parameters: $a_V = 15.85$ MeV, $a_C = 0.71$ MeV, $a_S = 18.34$ MeV and $a_A = 23.21$ MeV (Wapstra 1958, Wapstra and Gove 1971).

In figure 7.17, we illustrate the behaviour of S_{2n} along the valley of stability (7.49) for even–even nuclei, together with the experimental data. The experimental data correspond to a range of Z between Z_0+1 and Z_0-2. It appears that the overall decrease and the specific mass dependence are well contained within the LDM.

We can also see how well the experimental two-neutron separation energy, through a chain of isotopes, is reproduced using the LDM. From figure 7.17, it is clear that, besides the sudden variations near mass number $A = 90$ (presence of shell closure at $N = 50$) and near mass number $A = 140$ (presence of the shell closure at $N = 82$), the specific mass dependence for a series of isotopes comes closer to specific sets of straight lines.

Next, we observe that the mass formula in equation (7.46) is able to describe the observed almost-linear behaviour of S_{2n} for series of isotopes. The more appropriate way of carrying out this analysis is to expand the different terms in equation (7.46) around a particular value of A (or N, because Z is fixed),

$A_0 = Z + N_0$, and to keep the main orders. Therefore, we define $X = A - A_0$ and $\varepsilon = X/A_0$. Let us start with the volume term

$$BE_V(A) - BE_V(A_0) = a_V X. \qquad (7.50)$$

The surface term gives rise to

$$BE_S(A) - BE_S(A_0) \approx -a_S A_0^{2/3} \left(\frac{2}{3}\varepsilon - \frac{1}{9}\varepsilon^2 \right) = -a_S \frac{2}{3} \frac{X}{A_0^{1/3}} + a_S \frac{1}{9} \frac{X^2}{A_0^{4/3}}, \qquad (7.51)$$

and the contribution of the Coulomb term is

$$BE_C(A) - BE_C(A_0) \approx -a_C Z(Z-1) A_0^{1/3} \left(-\frac{1}{3}\varepsilon + \frac{2}{9}\varepsilon^2 \right)$$

$$= \frac{a_C}{3} Z(Z-1) \frac{X}{A_0^{4/3}} - \frac{2a_C}{9} Z(Z-1) \frac{X^2}{A_0^{7/3}}. \qquad (7.52)$$

Finally, the asymmetry contribution is,

$$BE_A(A) - BE_A(A_0) \approx -a_A \left(A_0 - \frac{4Z^2}{A_0} \right) \varepsilon - a_A \frac{4Z^2}{A_0} \varepsilon$$

$$= -a_A \left(1 - \frac{4Z^2}{A_0^2} \right) X - a_A \frac{4Z^2}{A_0^3} X^2. \qquad (7.53)$$

First, it is clear that the coefficients of the linear part are (for $A_0 \approx 100$ and $Z \approx 50$ and taking for a_V, a_C, a_S and a_A the values given in the previous section) about two orders of magnitude larger than the coefficients of the quadratic contribution. With respect to the second-order terms, it is interesting to see the value of each of them: the surface term gives 0.0044 MeV, the Coulomb term −0.0083 MeV and the asymmetry term −0.23 MeV. As a consequence, in this case, the leading term is the asymmetry one and it is essentially the main source of nonlinearities in the BE and, therefore, the source of the slope of the S_{2n}. In order to illustrate these results, we present, in figure 7.18 the different contributions of the LDM (volume, volume plus surface, volume plus surface plus Coulomb and volume plus surface plus Coulomb plus asymmetry term) to the BE and S_{2n} for different families of isotopes. It thus appears that only the asymmetry term induces the quadratic behaviour in BE and the linear one in S_{2n}.

In equation (7.49), we make use of the classical values as discussed in the early papers of Wapstra (Wapstra 1958, Wapstra and Gove 1971). This fit, of course, was constrained to a relatively small set of experimental data points. More recent fits, including (i) higher-order terms in the LDM, (e.g. the surface symmetry correction, a finite-range surface term, droplet correction, etc) and (ii) the much extended range of experimental data points, seem to give rise to an increased asymmetry energy coefficient a_A with values of the order of ≈ 30 MeV

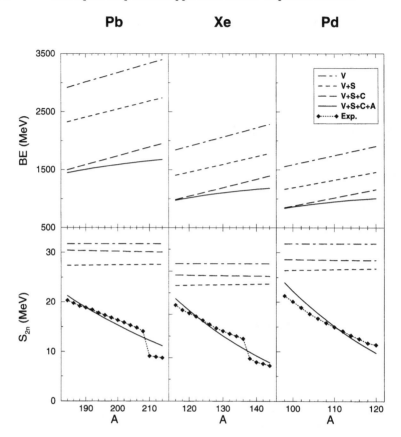

Figure 7.18. Contributions of the different terms of the mass formula to the BE (top row) and S_{2n} (bottom row) for Pb, Xe and Pd. In the S_{2n} panels the experimental data are also shown. (Reprinted from Fossion R *et al* 2002 *Nucl. Phys.* A **697** 703 © 2002 with permission from Elsevier.)

(Kruppa *et al* 2000). We show, in figure 7.18, the results obtained using an increased a_A coefficient, which gives rise to an increase in the absolute value of the S_{2n} slope.

In the present analysis, we have left out the effect of the pairing energy, with a dependence $BE_{\text{pairing}} = -11.46A^{-1/2}$ (Wapstra 1958, Wapstra and Gove 1971), because the net result is an overall shift in the BE, but the relative variation in the S_{2n} values, over a mass span of $\Delta A = 20$ units, is of the order of $\approx 100\,\text{keV}$ and, as such, is not essential for the plot of figure 7.18.

In figure 7.19, we like to illustrate the particular values of S_{2n} that were derived making use of the atomic mass evaluation data ((Audi and Wapstra 1995, Audi *et al* 1997)—Ame95 (upper part of the figure)), compared to a re-evaluation

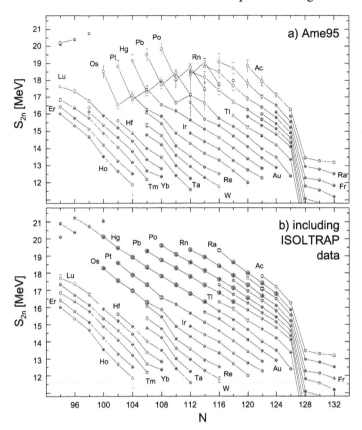

Figure 7.19. The results for the experimental S_{2n} values as taken from the atomic and mass evaluation of Audi and Wapstra (Audi and Wapstra 1995, Audi *et al* 1997) (upper part) and after an analyses including the recent data taken in the Pb region by the ISOLTRAP measurements (Bollen *et al* 1996, Beck *et al* 2000) (lower part). (Reprinted from Fossion R *et al* 2002 *Nucl. Phys.* A **697** 703 © 2002 with permission from Elsevier.)

including the data that were taken using the ISOLDE trap (ISOLTRAP) where data have been collected with much greater precision in recent years (Bollen *et al* 1996, Beck *et al* 2000).

Finally, it should be stressed that, far-from-stability and for very neutron-rich nuclei, the asymmetry term is at the origin of the decreasing trend in the BE when A increases further. This actually corresponds to the well-known drip-line phenomenon but here within a pure LDM approach.

Box 7a. Neutron star stability: a bold extrapolation

By adding a gravitational binding energy term to the regular binding energy expression of equation (7.33), negligible corrections are added for any normal nucleus. The binding energy for a nucleus $^A_Z X_N$ now reads

$$BE(A, Z) = a_v A - a_S A^{2/3} - a_c Z(Z-1)A^{-1/3} - a_A(A-2Z)^2 A^{-1} \begin{matrix} +\delta \\ +0 \\ -\delta \end{matrix}$$

$$+ \frac{3}{5}\frac{G}{r_0} M^2 A^{-1/3}. \tag{7a.1}$$

We can, however, use this simple equation to carry out a bold extrapolation using the parameters a_v, a_s, a_c, a_A and δ as determined from known masses.

Applying the equation to a hypothetical 'neutron' nucleus, where Coulomb energy becomes zero and neglecting the pairing energy δ and even the surface term $A^{2/3}$ is negligible with respect to the volume term, we look to see if such an object can exist in a bound state, i.e. have positive binding energy $BE(A, Z)$. We look to the limiting condition, i.e. $BE(A, Z) = 0$ which gives rise to

$$a_v A - a_A A + \frac{3}{5} G \left(\frac{M^2}{r_0} A^{-1/3} \right) = 0. \tag{7a.2}$$

Since $M = A M_n$, $R = r_0 A^{1/3}$ with $r_0 = 1.2$ fm, the resulting condition is

$$\tfrac{3}{5} G (M_n^2 / r_0) A^{2/3} = 7.5 \text{ MeV}. \tag{7a.3}$$

Using known constants i.e. $G = 6.7 \times 10^{-11}$ Jm kg^{-2}, and that 1 MeV = 1.602×10^{-13} J and further using $M_n = 1.67 \times 10^{-27}$ kg and $r_0 = 1.2 \times 10^{-15}$ m, the following are the limiting conditions for the 'neutron' nucleus

$$A \cong 5 \times 10^{55}$$
$$R \simeq 4.3 \text{ km}$$
$$M \cong 0.045 M_\odot.$$

More precise calculations give as a result for the mass, a value of about $0.1 M_\odot$.

The object we have obtained corresponds to a neutron star. Even though we have made a very naïve extrapolation, it is very remarkable that extrapolating the semi-empirical mass equation produces the correct order of magnitude for known neutron stars. The extrapolation goes from $1 \le A \le 250$ to $A \simeq 5 \times 10^{55}$, or extrapolating over 53 orders of magnitude. This clearly explains, though, that neutron stars are bound according to the rules given by the strong interaction; in particular the volume and asymmetry terms.

Box 7b. Beyond the neutron drip line

APPROXIMATELY 288 stable or near-stable nuclear species, characterized by the number of neutrons, N, and the number of protons, Z, occur in nature. The total number of nuclear systems (N, Z), radioactive but stable against prompt particle emission, is predicted to be about 6,000, but only about one-third of these have been observed. As shown in the figure, which does not include heavy and superheavy elements, the neutron-rich nuclei, many of which have yet to be identified, present a major challenge. For most elements, we simply do not know of reactions that will allow us to probe the region of the neutron drip line (see figure legend). Within the past few years there has, however, been remarkable progress in the techniques for studying the lightest elements, as exemplified by the recent work of Seth and co-workers. In a recent paper[1] they describe a spectroscopic study of the nucleus ^{9}He, which has five more neutrons than the usual ^{4}He nucleus, taking it beyond the neutron drip line.

Working at the Los Alamos Meson Physics Facility, Seth and co-workers used a special pion spectrometer (EPICS) to observe double-charge-exchange reactions of the type ^{9}Be (π^-, π^+) ^{9}He, in which a negatively charged pion impinging on the stable beryllium nucleus is converted into a positive pion, and two protons in the nucleus become neutrons, to give the helium nucleus. The spectroscopy involves determining the intensity of the scattered pions as a function of their loss of energy (having allowed for the recoil of the target nucleus). A resonance in the spectrum corresponds to a bound or quasi-stationary state of the product nucleus.

Early results reported at the 1981 Helsingør conference[2] showed that the spectrum is dominated by a broad distribution, representing all the possible states of the non-resonant break-up products, and also a peak attributed to the ^{9}He ground state. The new experiment, with improved resolution and statistics, confirms the analysis and also reveals additional peaks representing excited states in ^{9}He. The ground state is unbound by 1.13 ± 0.10 MeV with respect to neutron emission. Because the resonances have no measurable intrinsic width, it can be deduced from Heisenberg's uncertainty principle that the ^{9}He states live for at least 10^{-20} s, so that this nucleus survives for at least several periods of its intrinsic motion before breaking up.

It is interesting that the measured ground-state masses of the neutron-rich helium nuclei turn out to be lower, which is to say that the systems are more bound than semi-empirical estimates ('Garvey–Kelson relations') based on measured masses from near-stability suggest. This could mean that these nuclei adjust to the neutron excess by assuming a structure that is different from the one encountered near stability. If so, theoretical extrapolations to the neutron drip line are much more uncertain than normally assumed.

The double-charge-exchange reaction is of limited value for studying the drip line for heavier systems. Seth *et al.*[1] point out that a search for the ^{10}He ground state would require a target of long-lived, radioactive ^{12}Be, not easily available in usable quantities. The heaviest unbound system that can be reached from a natural target is ^{13}Be, so the prospects for progress in this direction are limited.

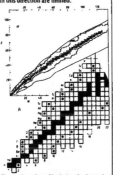

The nuclear chart. Vertical scale, increasing nuclear charge, Z, on which the chemical identity of the nucleus depends; horizontal scale, neutron number, N. Black squares, stable isotopes. In a, the zig-zag lines are the experimental limits of known isotopes; the two smooth lines indicate the estimated limits towards prompt instability, the proton (upper) and neutron (lower) 'drip lines'. b shows the status for the light nuclei. The heavy line surrounds the (shaded) area of isotopes detected experimentally with half lives longer than 10^{-7}; external squares as thin lines, predicted but as-yet unobserved isotopes. Stars, heaviest and lightest isotopes for which the half life is known experimentally. All isotopes of all elements up to nitrogen now seem to have been detected. Note also that the proton drip line has been reached for many medium-weight and heavy elements and, apparently, for all light elements up to scandium, where $Z = 21$.

Fortunately, fragments from heavy-ion reactions at intermediate energy can be identified in flight, making possible studies of all bound isotopes of the lighter elements[11]. In the past two years, this method has reached a new level of perfection in a series of experiments[3-6] carried out at GANIL, the French heavy-ion laboratory in Caen, where energies of 30–100 MeV per nucleon are available. With the observation[10] of the nuclides ^{22}C and ^{23}N there, all particle-stable isotopes of the seven elements up to nitrogen have been produced in the laboratory.

The essential tool in the GANIL experiments is the magnetic spectrometer LISE. Two dipole magnets select the mass-to-charge ratio A/Z, and a degrader, placed between the magnets, causes an energy loss proportional to Z^2. The net effect is a complete separation by mass and element, allowing the study[7] of radiations from the stopped radioactive fragments. The flight time through the apparatus, about 10^{-3} s, limits the technique to nuclei with half lives longer than this value.

An interesting hint of an open problem in nuclear physics emerges from the pattern of particle-stable nuclei shown in the lower part of the figure, which combines information from many sources. The preponderance of even-N isotopes near the neutron drip line suggests that neutron pairing is essential for nuclear stability in this region. Pairing in nuclear parlance denotes a structure in which the neutrons (or protons) are coupled in spin-zero pairs, where the two nucleons have the same orbital quantum numbers, and where a unique combination of pairs gives rise to a favoured, low-lying spin-zero ground state in even–even nuclei. For example, the helium isotopes are stable for masses $A = 4,6,8$, and unstable for $A = 5,7,9$. The seven elements up to nitrogen show the same tendency: in each case the heaviest bound isotope has even $N (= 2n)$ and in all but two cases the isotope with $N = 2n - 1$ is unbound. In two cases (helium and boron), the isotope with $N = 2n - 3$ is also unbound.

Superficially, the importance of pairing agrees well with the conventional wisdom that even-N nuclei are more strongly bound by $12A^{-1/2}$ MeV than odd-N nuclei (for even Z). The pairing strength, however, decreases strongly with increasing neutron excess, as expressed by three approximately equivalent, semi-empirical relations[8-11] including an isospin- (or charge-) symmetry term. Extrapolations to the neutron drip line of these relations predict that pairing vanishes there. This is manifestly not the case. Neither the old treatment of pairing, which works well near the bottom of the stability valley, nor the new, improved approximation seems to hold near the drip line. The upshot is that experiments, in particular very accurate determinations of mass, will again

have the last word, and our ideas about stability and structure of very neutron-rich nuclei will change in the next few years. □

1 Seth K K *et al Phys Rev Lett* 58 1930–1933 (1987)
2 Seth K K *Proc 4th Int Conf Nuclei Far from Stability* CERN 81-09, 655–663 (1981)
3 Artukh A G *et al Nucl Phys* A176 284–298 (1971)
4 Symons, T J M *et al Phys Rev Lett* 42 40–43 (1979)
5 Langevin M *et al Phys Lett* 150B 71–74 (1985)
6 Pougheon F *et al Europhys Lett* 2 505–509 (1986)

7 Dufour J P *et al Z Phys* A334 417–xxx xxxx (1986)
8 Dufour J P *et al Nucl Instr Meth* A248 xxx xxx (1986)
9 Vogel P Jonson B & Hansen P G *Phys Lett* 139B 227–230 (1984)
10 Jensen A S Hansen P G & Jonson B xxx xxx A431 393–418 (1984)
11 Madland D G & Nix J R *Nucl Phys* xxx xxx xxx xxx

P G Hansen is at the Institute of Physics, University of Aarhus, DK 8000 Aarhus C, Denmark

Chapter 8

The simplest independent particle model: the Fermi-gas model

Different model approaches try to accentuate various aspects of nuclear structure in a simple and schematic way. No single model, as yet, is detailed enough to encompass *all* aspects of the nucleus. At present we concentrate on a very simple, independent particle model: the Fermi-gas model where we consider all nucleons to move as elements of a fermion gas within the nuclear volume $V (\simeq \frac{4}{3}\pi r_0^3 A)$. Because of the strong spatial confinement, the energy levels in this approach will be widely spaced. Only the lowest levels will be occupied, except for very high energies. For typical excitation energies of up to $\simeq 10$ MeV, we can use a degenerate Fermi-gas model description.

8.1 The degenerate fermion gas

In figure 8.1, we present the two potential wells, one for neutrons and one for protons. The least bound nucleons have equal energy and the zero coincides. The two potentials, though, have slightly different shapes, mainly because of the Coulomb part: the well for protons is less deep because of the Coulomb potential by the amount E_c and externally, the $1/r$ dependence of the Coulomb potential extends the range.

Each level (here drawn as equidistant which slightly deviates from the actual level spacing in such a square well) can contain two identical fermions with different spin orientations. For low excitation energy, called the temperature $T = 0$ limit, the levels are filled pairwise up to a Fermi level beyond which levels are fully unoccupied: this situation we call the degenerate Fermi gas. For quite high energies (temperature T_1), a smooth redistribution of the occupancy results and only for very high excitation energies (T_2) are all levels only partially occupied (figure 8.2). From the density of states obtained from a free Fermi gas

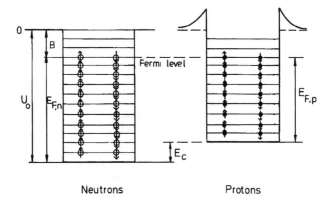

Figure 8.1. Nuclear square well potentials containing the available neutrons (left part) and protons (right part). The double-degeneracy, associated with nucleon intrinsic spin is presented. The various quantities refer to the neutron and proton Fermi energy, $E_{F,n}$ and $E_{F,p}$ respectively. The nuclear potential depth is given by U_0, the nucleon binding energy of the last nucleon by B and the Coulomb energy E_C. The Fermi level of both systems is drawn at the same energy.

confined to stay within a volume V; the nuclear volume, we have

$$n = \frac{2V}{(2\pi\hbar)^3} \int_0^{p_F} \mathrm{d}^3 p, \tag{8.1}$$

or

$$n = \frac{V p_F^3}{3\pi^2 \hbar^3}, \tag{8.2}$$

and an expression for the Fermi momentum p_F is observed as

$$p_F = \hbar \left(3\pi^2 \frac{n}{V}\right)^{1/3} = h \left(\frac{3}{8\pi} \frac{n}{V}\right)^{1/3}. \tag{8.3}$$

where n/V denotes the density and contributes to a quantum-mechanical pressure.

Applying the above arguments, first to the atomic nucleus, in order to obtain an idea of the well depth U_0, we know that we have a rather small number of protons (Z) and neutrons (N) to be distributed.

We can derive

$$p_{F,n} = \frac{\hbar}{r_0} \left(\frac{9\pi N}{4A}\right)^{1/3}, \qquad p_{F,p} = \frac{\hbar}{r_0} \left(\frac{9\pi Z}{4A}\right)^{1/3}, \tag{8.4}$$

for the neutron and proton Fermi momentum, respectively. In deriving this result we used for the nuclear volume, $V = \frac{4}{3}\pi r_0^3 A$. A simple estimate of

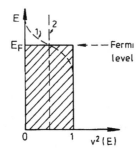

Figure 8.2. Occupation $v^2(E)$ as a function of increasing level energy. The square distribution corresponds to a degenerate $(T = 0)$ fermion gas distribution. Distribution (1) corresponds to a well excited $(T \neq 0)$ situation whereas distribution (2), where the occupation is mainly independent of the level energy, corresponds to a very high excitation of the fermion assembly.

these Fermi momenta can be reached by considering self-conjugate nuclei with $N = Z = A/2$, resulting in

$$p_{F,n} \cong p_{F,p} \cong \frac{\hbar}{r_0} \left(\frac{9\pi}{8} \right)^{1/3}. \tag{8.5}$$

Using the combination, $\hbar c = 197$ MeV fm, this becomes

$$p_{F,n} = p_{F,p} \simeq \frac{297}{r_0} \text{ MeV}/c. \tag{8.6}$$

The corresponding Fermi kinetic energy, using a value of $r_0 = 1.2$ fm, becomes

$$E_F = \frac{p_{F,p}^2}{2m} = \frac{p_{F,n}^2}{2m} \simeq 33 \text{ MeV}. \tag{8.7}$$

This energy corresponds to the kinetic energy of the highest occupied orbit (smallest binding energy). Given the average binding energy $B = 8$ MeV, we can make a good estimate of the nuclear well depth of $U_0 \simeq 41$ MeV. In realistic calculations (see part D), values of this order are obtained.

One can also derive an average kinetic energy per nucleon and this gives (for non-relativistic motion)

$$\langle E \rangle_p = \frac{\int_0^{p_{F,p}} E_{kin} d^3 p}{\int_0^{p_F} d^3 p} = \frac{3}{5} \frac{p_{F,p}^2}{2m} = \frac{3}{5} E_{F,p} \simeq 20 \text{ MeV}. \tag{8.8}$$

From the above discussion in the degenerate Fermi-gas model, even at zero excitation energy, a large amount of 'zero-point' energy is present and, owing to the Pauli principle, a quantum-mechanical 'pressure' results. We shall discuss this when presenting an extensive application to astrophysics in section 8.3.

8.2   The nuclear symmetry potential in the Fermi gas

Starting from the above results, and given a nucleus with Z protons and N neutrons, the total, average kinetic energy becomes

$$\langle E(A, Z) \rangle = Z\langle E \rangle_{\rm p} + N\langle E \rangle_{\rm n}$$

$$= \frac{3}{10m}(Z p_{\rm F,p}^2 + N p_{\rm F,n}^2), \tag{8.9}$$

or, filling in the particular values for $p_{\rm F,p}$ and $p_{\rm F,n}$, derived in section 8.1, the average total kinetic energy reads

$$\langle E(A, Z) \rangle = \frac{3}{10m}\frac{\hbar^2}{r_0^2}\left(\frac{9\pi}{4}\right)^{2/3}\frac{(N^{5/3} + Z^{5/3})}{A^{2/3}}. \tag{8.10}$$

This result (derived under the constraints that protons and neutrons move independently from each other and that the nucleon–nucleon interactions are considered to be present through the average potential well) can now be expanded around the symmetric case with $N = Z = A/2$. This expansion will lead to an expression for the symmetry energy and we shall be able to derive a value for a_A. Calling

$$Z - N = \epsilon, \qquad Z + N = A, \tag{8.11}$$

we can substitute in equation (8.10), $Z = A/2(1 + \epsilon/A)$; $N = A/2(1 - \epsilon/A)$ and with $\epsilon/A \ll 1$ obtain

$$\langle E(A, Z) \rangle = \frac{3}{10m}\frac{\hbar^2}{r_0^2}\left(\frac{9\pi}{4}\right)^{2/3}A\left(\frac{1}{2}\right)^{5/3}[(1 + \epsilon/A)^{5/3} + (1 - \epsilon/A)^{5/3}]. \tag{8.12}$$

Inserting the binomial expansion

$$(1 + x)^n = 1 + nx + \frac{n(n - 1)}{2}x^2 + \cdots, \tag{8.13}$$

and again replacing ϵ by $Z - N$ and keeping only the lowest non-vanishing terms in $(Z - N)$, we obtain finally

$$\langle E(A, Z) \rangle = \frac{3}{10m}\frac{\hbar^2}{r_0^2}\left(\frac{9\pi}{8}\right)^{2/3}\left\{A + \frac{5}{9}\frac{(N - Z)^2}{A} + \cdots\right\}. \tag{8.14}$$

The first term is proportional to A and contributes to the volume energy. The next term has exactly the form of the symmetry energy A, Z dependent term in the Bethe–Weizsäcker mass equation. Inserting the values of the various constants $m, \hbar, r_0, \pi$, we obtain the result

$$\langle E(A, Z) \rangle = \langle E(A, Z = A/2) \rangle + \Delta E_{\rm symm}$$

$$\Delta E_{\rm symm} \cong 11 \text{ MeV}(N - Z)^2 A^{-1}. \tag{8.15}$$

The numerical value is, however, about half the value of a_A as determined in chapter 7. Without discussing this point here in detail, the difference arises from the fact that the nuclear well depth U_0 is itself *also* dependent on the neutron excess $N - Z$ and makes up for the missing contribution to the symmetry energy.

Higher-order terms with a dependence $(N - Z)^4/A^3, \ldots$ naturally derive from the more general $\langle E(A, Z) \rangle$ value.

8.3 Temperature $T = 0$ pressure: degenerate Fermi-gas stability

The above study of quantum-mechanical effects, resulting from the occupation of a set of single-particle levels according to the Fermi–Dirac statistics, can be applied to various other domains of physics. In this section we study an extensive application to the stability of Fermi-gas assemblies: in particular a degenerate electron gas and a degenerate neutron gas.

We have derived the expression for the dependence of the Fermi momentum on the particle density as

$$p_F = h \left(\frac{3}{8\pi} \frac{n}{V} \right)^{1/3}, \tag{8.16}$$

and of the Fermi energy (in a non-relativistic case)

$$E_F = \frac{h^2}{8m} \left(\frac{3}{\pi} \frac{n}{V} \right)^{2/3}. \tag{8.17}$$

This Fermi-gas accounts for a pressure which has a quantum-mechanical origin. In statistical physics, one derives

$$pV = n \left(\frac{m\bar{v}^2}{3} \right) = \frac{2}{3} n \langle E \rangle, \tag{8.18}$$

or

$$p = \frac{2}{5} n \frac{E_F}{V} = \frac{h^2}{20m} \left(\frac{3}{\pi} \right)^{2/3} \left(\frac{n}{V} \right)^{5/3}. \tag{8.19}$$

In the extreme relativistic case, where $E_F = p_F c$, one can make an analogous calculation.

(i) *Degenerate electron gas–white dwarf stars.* We shall derive the total energy $E(r)$, as a function of the radius r of the object. The equilibrium condition arises from balancing the gravitational energy and the quantum-mechanical Fermi-gas energy for an electron system. The gravitational energy for a uniform sphere is

$$E_G = -\frac{3}{5} G \frac{M^2}{r}. \tag{8.20}$$

Consider a star to made be up of atoms with mass number A. If n is the total number of nucleons, we have n/A nuclei with Z protons. If we call x the ratio $x = Z/A$, then the total number of electrons (and protons) will be $n_e = xn$. The total mass of the star becomes $M \simeq nM_p$ (M_p: proton mass). We shall see that all electrons (and of course the nuclei) move non-relativistically in a star with small mass that is contracting after having exhausted its thermonuclear fuel. In the total kinetic energy, the nuclear part is negligible and thus, the total energy of the star becomes (if the thermal energy has become vanishingly small)

$$E(r) = n_e\langle E\rangle_e + E_G$$
$$= \frac{3}{5}n_e E_{F,e} - \frac{3}{5}\frac{GM^2}{r}, \tag{8.21}$$

assuming uniform density. Using the equation (8.17) but now substituting m_e for m, xn for n_e and $V = \frac{4}{3}\pi r^3$, we obtain

$$E(r) = \frac{3}{5}\frac{xn}{r^2}\frac{h^2}{8m_e}\left(\frac{9}{4\pi^2}xn\right)^{2/3} - \frac{3}{5}G\frac{n^2 M_p^2}{r}. \tag{8.22}$$

This relation $E(r)$ is plotted in figure 8.3, where we also show the particle energy (quantum-mechanical part) and the gravitational contribution. The lowest value of $E(r)$ can be obtained from the plot, where $dE(r)/dr = 0$ and gives

$$r_{eq} = \frac{xh^2}{4m_e}\left(\frac{9}{4\pi^2}xn\right)^{2/3}\frac{1}{GnM_p^2}. \tag{8.23}$$

At the equilibrium value the energy $|E_G| = 2|E_{QM}|$ holds. For a typical white dwarf star $n \simeq 10^{57}$ and $x = 1/2$. The radius r_{eq} then becomes $\simeq 800$ km, corresponding to a density $\rho \simeq 3 \times 10^6$ g cm^{-3}. Jupiter comes close to the line giving r_{eq} (km) as a function of the mass of the object. The mass of the object, however, must be large enough so that the electrons are in a common Fermi system (figure 8.4).

A star cannot shrink below the radius given in equation (8.23). Then $r_{eq} \propto n^{-1/3}$ and the electron Fermi energy increases rapidly with n so, the expression for $E(r)$ becomes incorrect since $E_{F,e} \simeq m_0 c^2$ and we should use relativistic models. Using the extreme limit for *all* electrons, we obtain

$$E_{F,e} = p_{F,e}c = \frac{hc}{2}\left(\frac{3}{\pi}\frac{n}{V}\right)^{1/3}, \tag{8.24}$$

and

$$\langle E_F\rangle_e = \frac{3}{4}E_{F,e}. \tag{8.25}$$

By the same procedure as before we derive the result

$$E(r) = \frac{3}{4}\frac{xn}{r}\frac{hc}{2}\left(\frac{9}{4\pi^2}xn\right)^{1/3} - \frac{3}{5}G\frac{n^2 M_p^2}{r}. \tag{8.26}$$

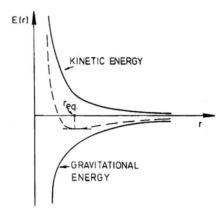

Figure 8.3. Contribution to the energy of a white dwarf. The kinetic, gravitational potential and total energy $E(r)$ are shown. A stable minimum occurs at a given radius. (Adapted from Orear 1979. Reprinted with the permission of MacMillan College Publishing Company from *Physics* by J Orear © 1979 by MacMillan College Publishing Company Inc.)

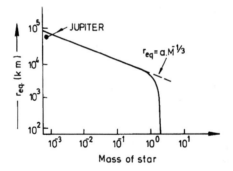

Figure 8.4. This value of r, minimizing the sum of the electron Fermi energy and the gravitational energy as a function of the mass of the star. A uniform density is assumed and the exact relativistic expression for the Fermi energy is used. (Adapted from Orear 1979. Reprinted with the permission of MacMillan College Publishing Company from *Physics* by J Orear © 1979 by MacMillan College Publishing Company Inc.)

Here, both terms have now the same $1/r$ dependence and the second dominates if the mass nM_p becomes sufficiently large. So, for $n > n_{crit}$, the energy $E(r)$ continues to decrease and no stable configuration at any radius can be determined.

Even though in this regime we should use a slightly more realistic model, we can however try determining an estimate of n_{crit} by equating both terms in equation (8.26). This gives

$$\frac{3}{4}x\frac{n_{\text{crit}}}{r}\frac{hc}{2}\left(\frac{9}{4\pi^2}xn_{\text{crit}}\right)^{1/3} = \frac{3}{5}G\frac{n_{\text{crit}}^2 M_p^2}{r},$$

or

$$n_{\text{crit}} \simeq \frac{(125\pi)^{1/2}}{4}x^2\left(\frac{\hbar c}{GM_p^2}\right)^{3/2}. \tag{8.27}$$

Here, the dimensionless quantity $n_0 \equiv (\hbar c/GM_p^2)^{3/2} = 2.4 \times 10^{57}$ occurs. The corresponding critical star mass i.e. $n_{\text{crit}}M_p$ is called the 'Chandrasekhar' limit and is the largest equilibrium mass that can exist as a quantum-mechanical, relativistic electron gas. This mass is only about 40% larger than the mass of the sun $M_\odot$ ($M_\odot = 0.49n_0M_p$).

(ii) *Neutron stars.* In all of the above derivations, both the total energy $E(r)$ and the equilibrium radius r_{eq} could be reduced considerably if the value of x ($\sim 1/2$) could be reduced to a very small number.

If the electron energy becomes very high, the possibility exists for inverse beta decay

$$e^- + p \to n + \nu_e.$$

The detailed conversion of normal matter into neutron-rich matter is complicated but calculations show that at densities $\rho > 10^{11}$ g cm^{-3}, neutrons become much more abundant than protons. We are dealing with neutron stars.

We now try to determine the equilibrium values of r and x for a given number of nucleons with $n_p = n_e = xn$. In this regime, the electrons will move fully relativistically for any neutron star. The neutrons, however, with their much larger mass M_p remain non-relativistic as long as the density of the star is less than that density ρ_n for which $p_{F,n} = M_pc$, or

$$\rho_n = \frac{8\pi M_p}{3(1-x)}\left(\frac{M_pc}{h}\right)^3 = \frac{6\times 10^{15}}{1-x}\text{ g cm}^{-3}. \tag{8.28}$$

The total neutron star energy then becomes

$E(r,x) = $ neutron energy (N.R.) + electron energy (R) + gravitational energy

$$E(r,x) = \frac{3}{5}\frac{(1-x)n}{r^2}\frac{h^2}{8M_p}\left[\frac{9}{4\pi^2}(1-x)n\right]^{2/3}$$

$$+ \frac{3}{4}\frac{xn}{r}\frac{hc}{2}\left[\frac{9}{4\pi^2}xn\right]^{1/3} - \frac{3}{5}G\frac{n^2M_p^2}{r}. \tag{8.29}$$

The values of r and x that minimize $E(r,x)$ can be obtained by solving the equations

$$\frac{\partial E(r,x)}{\partial r} = 0, \qquad \frac{\partial E(r,x)}{\partial x} = 0. \tag{8.30}$$

Because x is very small, we can obtain an approximate solution as follows: first use $x = 0$ in the equation $\partial E(r, x)/\partial r = 0$ to obtain r_{eq} and then use that value of r_{eq} in the equation for $\partial E(r, x)/\partial x = 0$ to obtain x. In the first step we obtain

$$r_{eq} = \frac{h^2}{4M_p}\left(\frac{9}{4\pi^2}n\right)^{2/3}\frac{1}{GnM_p^2}. \tag{8.31}$$

This expression is very similar to the value r_{eq} for the degenerate electron gas in a white dwarf star, but since now M_p appears instead of m_e, a neutron star has a smaller radius by $\simeq 1000$ times. For a star with a mass $M = M_\odot$, $n = 1.2 \times 10^{57}$ and $r_{eq} = 12.6$ km with a density of $\rho = 2.4 \times 10^{14}$ g cm^{-3}. This density is about the density of the nucleus as discussed in chapters 1 and 7. In order now to obtain a value for x, we put $r_{eq} = 12.6$ km into the equation for $\partial E(r, x)/\partial x = 0$ and solve for x with a result $x \simeq 0.005$. This indicates that 99.5% of all nucleons in this 'star nucleus' are neutrons.

The study of neutron stars and very compact objects in the universe requires a full treatment of nuclear interaction combined with relativity. A full account is presented in the book *Black Holes, White Dwarfs and Neutron Stars* (Shapiro and Teukolsky 1983).

Chapter 9

The nuclear shell model

In the discussion in chapters 7 and 8, it became clear that single-particle motion cannot be completely replaced by a collective approach where the dynamics is contained in collective, small amplitude vibrations and/or rotations of the nucleus as a whole. On the other hand, a simple independent-particle approach required by the Fermi-gas model does not contain enough detailed features of the nucleon–nucleon forces, active in the nucleus. The specific shell-model structure is evident from a number of experimental facts which we discuss first, before developing the shell model in more detail. Finally, we relate the nuclear average field to the underlying nuclear interaction making use of a Hartree–Fock approach. At the same time, we point out the increasing complexity of the systems under study that require extensive computations. This aspect too is indicated.

9.1 Evidence for nuclear shell structure

One of the most obvious indications for the existence of a shell structure is obtained by comparison with the analogous picture of electrons moving in the atom. In studying the ionization energy as a function of the number of electrons (as a function of Z), a clear indication for 'magic' numbers at 2, 10, 18, 36, 54, 86 was obtained (figure 9.1). The behaviour can be explained since, going from atom Z to $Z + 1$, the charge of the nucleus increases. The ionization energy will thus increase on the average and in a rather smooth way when filling the electron shells. When starting to fill a new electron shell, the last electron occurs in a *less* strongly bound orbit but a screening effect of the central, nuclear charge by a number of filled electron orbit presents an effective smaller Coulomb potential. So, the electron ionization energy reflects rapid variations whenever an electron shell is closed.

The analogous quantity in the atomic nucleus is the neutron (proton) separation energy $S_n(S_p)$. In chapter 7, when indicating the need for a strong pairing effect between nucleons, an odd–even effect became clear. There, we studied the Ce $(Z = 58)$ nuclei for different neutron numbers N in the interval

275

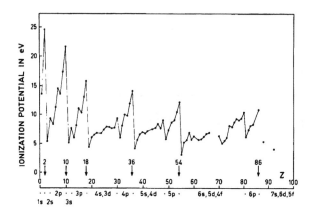

Figure 9.1. The values of the atomic ionization potentials are taken from the compilation by Moore (1949). The various configurations and closed-shell configurations are shown. (Taken from Bohr and Mottelson, *Nuclear Structure*, vol. 2, © 1982 Addison-Wesley Publishing Company. Reprinted by permission.)

$70 \leq N \leq 88$. At $N = 82$, a clear discontinuity shows up in S_n. Neutrons beyond $N = 82$ are clearly less strongly bound in the nucleus compared to the inner neutrons. This difference is of the order of $\simeq$2–2.5 MeV and can be best observed by concentrating on the even–even *or* odd–even nuclei. The detailed behaviour over a much larger mass region for $S_n(A, Z)$ and $S_p(A, Z)$ are presented in figures 9.2 and 9.3, respectively. For most nuclei, $S_p \simeq S_n \simeq 8$ MeV: the experimental value which also follows from the semi-empirical mass equation. This separation energy is maximum at the 'magic' nuclear numbers 2, 8, 20, 28, 50, 82, 126 and are clearly different with the electron numbers. For S_p, no direct increase shows up; we see mainly strong discontinuities in the mass dependence.

The energy difference between the liquid-drop nuclear mass and the experimental nuclear mass, discussed in figure 7.13, in chapter 7 is another indicator of the presence of certain configurations at which the nucleus is more strongly bound than is indicated by a uniform filling of orbitals in a Fermi-gas model.

A number of other indicators are obtained by studying the Z, N variations of various nuclear observables, e.g. the excitation energy $E_x(1)$ of the first excited state in all even–even nuclei. The extra stability at neutron number 8, 20, 28, 50, 82, 126 is very clearly present as illustrated in figure 9.4.

So, even though a number of distinct differences between the atomic nucleus and the electron motion in an atom appear, the presence of an average nuclear potential acting on nucleons in an averaged way is very clear. In the final section of this chapter we shall concentrate more on the precise arguments and methods

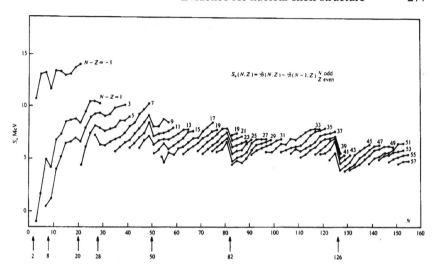

Figure 9.2. The neutron separation energies S_n (MeV), taken from the compilation by Mattauch, Thiele and Wapstra (1965).

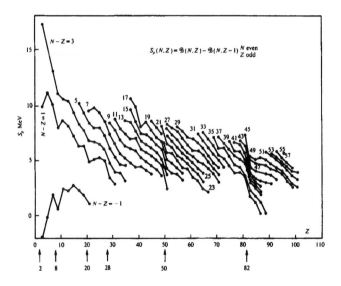

Figure 9.3. As for figure 9.2 but now for proton separation energies S_p (MeV).

used to define and determine such an average single-particle potential starting from the two-body nuclear interaction.

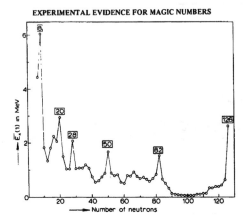

Figure 9.4. The occurrence of magic numbers as demonstrated by the average excitation energy of the first excited state in doubly-even nuclei, as a function of neutron number N (taken from Brussaard and Glaudemans 1977).

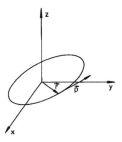

Figure 9.5. Schematic illustration of a particle, moving with linear momentum $\vec{p}$ (at a radial distance $|\vec{r}|$) in a central field characterized by a potential energy $U(|\vec{r}|)$.

9.2   The three-dimensional central Schrödinger equation

Assuming nucleons move in a central potential (that depends only on the distance of the nucleon to the centre of mass of the nucleus where the (x, y, z)-axis system is mounted) with potential energy $U(|\vec{r}|) = U(r)$, we may write the one-body Schrödinger equation that describes the situation, shown in figure 9.5, as

$$\left[-\frac{\hbar^2}{2m}\Delta + U(r) \right] \varphi(\vec{r}) = E\varphi(\vec{r}). \tag{9.1}$$

(i) If the potential energy is defined as a function of the radial coordinate r, an expression of this problem in spherical coordinates (r, θ, φ) is preferable.

Using the relations

$$\Delta = \frac{\partial^2}{\partial r^2} + \frac{2}{r}\frac{\partial}{\partial r} - \frac{\hat{l}^2}{r^2}\frac{1}{\hbar^2},$$

$$\hat{l}^2 = -\left[\frac{1}{\sin\theta}\frac{\partial}{\partial\theta}\left(\sin\theta\frac{\partial}{\partial\theta}\right) + \frac{1}{\sin^2\theta}\frac{\partial^2}{\partial\varphi^2}\right]\hbar^2, \qquad (9.2)$$

the Schrödinger equation becomes

$$\left[\frac{\partial^2}{\partial r^2} + \frac{2}{r}\frac{\partial}{\partial r} - \frac{1}{\hbar^2}\frac{\hat{l}^2}{r^2} - \frac{2m}{\hbar^2}U(r)\right]\varphi(\vec{r}) = -\frac{2m}{\hbar^2}E\varphi(\vec{r}). \qquad (9.3)$$

Solutions that separate the radial variable r, from the angular variables (θ, φ) gives solutions of the type

$$\varphi(\vec{r}) = R(r)F(\theta, \varphi). \qquad (9.4)$$

This leads to the equation

$$F(\theta, \varphi)\left\{\frac{d^2}{dr^2} + \frac{2}{r}\frac{d}{dr} + \frac{2m}{\hbar^2}(E - U(r))\right\}R(r) = \frac{R(r)}{r^2}\frac{\hat{l}^2}{\hbar^2}F(\theta, \varphi), \qquad (9.5)$$

or, separating the r and (θ, φ) variables by dividing both sides by the product $R(r)F(\theta, \varphi)$, we obtain

$$\frac{\left\{\dfrac{d^2}{dr^2} + \dfrac{2}{r}\dfrac{d}{dr} + \dfrac{2m}{\hbar^2}(E - U(r))\right\}R(r)}{R(r)/r^2} = \frac{\dfrac{\hat{l}^2}{\hbar^2}F(\theta, \varphi)}{F(\theta, \varphi)} = K^2, \qquad (9.6)$$

where K^2 is a constant.

The above equalities give rise to two equations, separating the radial and angular variables

$$\hat{l}^2 F(\theta, \varphi) = \hbar^2 K^2 F(\theta, \varphi) \qquad (9.7a)$$

$$\left\{\frac{d^2}{dr^2} + \frac{2}{r}\frac{d}{dr} + \frac{2m}{\hbar^2}(E - U(r))\right\}R(r) = \frac{K^2}{r^2}R(r). \qquad (9.7b)$$

Starting from the structure of the $\hat{l}^2$ operator, as depicted above, the equation (9.7a) is also separable in θ and φ. So, we can write

$$F(\theta, \varphi) = \Phi(\varphi)\Theta(\theta), \qquad (9.8)$$

and

$$\frac{\dfrac{1}{\sin\theta}\dfrac{\partial}{\partial\theta}\left(\sin\theta\dfrac{\partial}{\partial\theta}\right)}{\Theta(\theta)/\sin^2\theta} - K^2\sin^2\theta = -\frac{1}{\Phi(\varphi)}\frac{\partial^2\Phi(\varphi)}{\partial\varphi^2} = m^2. \qquad (9.9)$$

The equation in the azimuthal angle φ

$$\frac{d^2\Phi(\varphi)}{d\varphi^2} + m^2\Phi(\varphi) = 0, \qquad (9.10)$$

has a solution

$$\Phi(\varphi) = Ce^{im\varphi} + Be^{-im\varphi}. \qquad (9.11)$$

From the condition that $\Phi(\varphi + 2\pi) = \Phi(\varphi)$, integer values of m only need be considered. The equation for $\Theta(\theta)$ then becomes

$$\left[\frac{1}{\sin\theta}\frac{d}{d\theta}\left(\sin\theta\frac{d}{d\theta}\right) - \frac{m^2}{\sin^2\theta} \right]\Theta(\theta) = K^2\Theta(\theta). \qquad (9.12)$$

One can show that this equation contains regular solutions in the domain $0 \le \theta \le \pi$ when $K^2 = l(l+1)$ with $l \ge |m|$, and is an integer and positive. The solutions are polynomial functions in $\cos\theta$, depending on l and m so we can denote the solution as

$$F(\theta, \varphi) = Ce^{im\varphi}\Theta_l^m(\theta), \qquad (9.13)$$

Normalizing the $F(\theta, \varphi)$ on the unit sphere, the spherical harmonics result in

$$Y_l^m(\theta, \varphi) = (-1)^m \frac{e^{im\varphi}}{\sqrt{2\pi}}\left[\frac{2l+1}{2}\frac{(l-m)!}{(l+m)!}\right]^{1/2}P_l^m(\cos\theta), \qquad (9.14)$$

with

$$P_l^m(\cos\theta) = \frac{1}{2^l l!}(\sin\theta)^m\frac{d^{l+m}}{(d\cos\theta)^{l+m}}(\cos^2\theta - 1)^l. \qquad (9.15)$$

The remaining, radial differential equation which contains the dynamics (the forces and their influence on the particle motion) becomes

$$\left[\frac{d^2}{dr^2} + \frac{2}{r}\frac{d}{dr} - \frac{l(l+1)}{r^2} + \frac{2m}{\hbar^2}(E - U(r))\right]R(r) = 0. \qquad (9.16)$$

This radial equation, for given l value, using the boundary conditions $R(r) \to 0$ (for $r \to \infty$) contains a number of regular solutions, characterized by a radial quantum number n (the number of roots of this radial wavefunction in the interval $(0, \infty)$). The conventions used to label this radial behaviour will be discussed in the following sections.

Finally, we obtain

$$\varphi_{nlm}(\vec{r}) = R_{nl}(r)Y_l^m(\theta, \varphi), \qquad (9.17)$$

with $-l \le m \le l$, as the solutions of the central force Schrödinger equation, giving rise to eigenvalues E_{nl}.

It is these solutions that we shall concentrate on in order to classify the motion of nucleons (protons, neutrons) in the nuclear, average potential and study the energy spectra so that we can understand the 'magic' numbers which come from the empirical evidence for shell structure.

(ii) Sometimes the potential function can affect the particle motion differently according to the specific direction (x, y, z) in which it is moving. So, more generally than under (i), the potential energy can be separable as

$$U(x, y, z) = U(x) + U'(y) + U''(z). \tag{9.18}$$

The corresponding one-particle Schrödinger equation then becomes

$$\left[-\frac{\hbar^2}{2m} \Delta + U(x) + U'(y) + U''(z) \right] \varphi(x, y, z) = E\varphi(x, y, z). \tag{9.19}$$

since

$$\Delta = \frac{\partial^2}{\partial x^2} + \frac{\partial^2}{\partial y^2} + \frac{\partial^2}{\partial z^2},$$

separable solutions $\varphi(x, y, z) = \varphi_1(x)\varphi_2(y)\varphi_3(z)$ can be obtained and the Schrödinger equation reduces to three equations

$$\left[-\frac{\hbar^2}{2m} \frac{d^2}{dx^2} + U(x) \right] \varphi_1(x) = E_1\varphi_1(x)$$

$$\left[-\frac{\hbar^2}{2m} \frac{d^2}{dy^2} + U'(y) \right] \varphi_2(y) = E_2\varphi_2(y) \tag{9.20}$$

$$\left[-\frac{\hbar^2}{2m} \frac{d^2}{dz^2} + U''(z) \right] \varphi_3(z) = E_3\varphi_3(z),$$

with

$$E = E_1 + E_2 + E_3. \tag{9.21}$$

Then, a large degeneracy results in the energy eigenvalues. The above method can be used e.g. to study the motion in an anisotropic harmonic oscillator potential with

$$U(x, y, z) = \tfrac{1}{2}m\omega_x^2 x^2 + \tfrac{1}{2}m\omega_y^2 y^2 + \tfrac{1}{2}m\omega_z^2 z^2. \tag{9.22}$$

We shall come back to this potential in section 9.4.

9.3 The square-well potential: the energy eigenvalue problem for bound states

According to the short-range characteristics of the nuclear strong force acting between nucleons in the nucleus, we can try to study a number of simplified models that allow an exact treatment. The aim is to find out the stable configurations when putting nucleons into the independent-particle solutions and to study under which conditions these agree with the data, as discussed in section 9.1.

For the square-well potential model (figure 9.6) with

$$U(r) = -U_0 \qquad r \le R$$
$$U(r) = \infty \qquad r > R, \tag{9.23}$$

the corresponding Schrödinger one-particle equation becomes

$$\left[-\frac{\hbar^2}{2m} \Delta + U(r) \right] \varphi(\vec{r}) = E\varphi(\vec{r}), \tag{9.24}$$

with

$$\Delta = \frac{d^2}{dr^2} + \frac{2}{r} \frac{d}{dr} - \frac{\hat{l}^2}{\hbar^2} \frac{1}{r^2}.$$

Here, as discussed in section 9.2, $\hat{l}^2$ denotes the angular momentum operator, with eigenfunctions $Y_l^m(\theta, \varphi)$ and corresponding eigenvalues $l(l+1)\hbar^2$. Separating the wavefunction as

$$\varphi(\vec{r}) = R_{nl}(r) Y_l^m(\theta, \varphi), \tag{9.25}$$

the corresponding radial equation becomes

$$\frac{d^2 R(r)}{dr^2} + \frac{2}{r} \frac{dR(r)}{dr} + \left[\frac{2m}{\hbar^2}(E - U(r)) - \frac{l(l+1)}{r^2} \right] R(r) = 0. \tag{9.26}$$

We define the quantity (in the interval $0 \le r \le R$)

$$k^2 = \frac{2m}{\hbar^2}(E + U_0),$$

and the solutions become the solutions of the Bessel differential equation (for the regular ones which we concentrate on)

$$R_{nl}(r) = \frac{A}{\sqrt{kr}} J_{l+1/2}(kr). \tag{9.27}$$

These solutions are regular at the origin (the spherical Bessel functions of the first kind) with

$$j_l(kr) = \sqrt{\frac{\pi}{2kr}} J_{l+1/2}(kr). \tag{9.28}$$

The regular and irregular solutions are presented in figure 9.7. The energy, as measured from the bottom of the potential well is expressed as

$$E' = \frac{\hbar^2 k^2}{2m}, \tag{9.29}$$

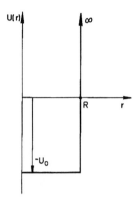

Figure 9.6. Characterization of the square-well potential with constant depth $-U_0$ (in the interval $0 < r \le R$) within the nuclear interior and infinite for $r > R$.

and describes the kinetic energy of the nucleon moving in the potential well.

The energy eigenvalues then follow from the condition that the wavefunction disappears at the value of $r = R$, i.e.

$$R_{nl}(R) = 0 \qquad \text{or} \qquad J_{l+1/2}(k_{nl}R) = 0. \tag{9.30}$$

For these values, for which $k_{nl}R$ coincides with the roots of the spherical Bessel function, one derives

$$E'_{nl} = \frac{\hbar^2 k_{nl}^2}{2m} = \frac{X_{nl}^2 \hbar^2}{2mR^2}, \tag{9.31}$$

with $X_{nl} = k_{nl}R$. By counting the different roots (figure 9.7) for $j_0(X)$, $j_1(X)$, $j_2(X)$, ... one obtains an energy spectrum shown in Box 9b (figure 9b.1) where we give both the energy eigenvalues, the partial occupancy which, for an orbit characterized by l is $2(2l+1)$ (counting the various magnetic degenerate solutions with $-l \le m \le l$ and the two possible spin orientations $m_s = \pm 1/2$) as well as the total occupancy.

The extreme right-hand part in figure 9b.1 in Box 9b shows this spectrum. It is clear that only for small nucleon numbers 2, 8, 20, does a correlation with observed data show up. We shall later try to identify that part of the nuclear average field which is still missing and which was first pointed out by Mayer (1949, 1950) and Haxel, Jensen and Suess (1949). In Box 9b, we highlight this study, in particular the contribution from Mayer and present a few figures from her seminal book.

The square-well potential (with a finite depth in the interval $0 \le r \le R$ and 0 for $r > R$) can form a more realistic description of an atomic nucleus. As an example, we show the powerful properties that such a finite square-well potential

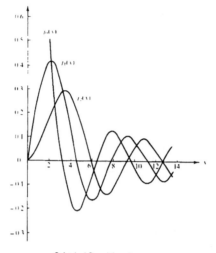

Spherical Bessel functions

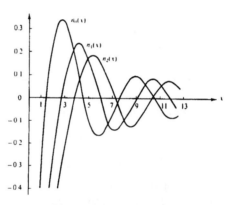

Spherical Neumann functions

Figure 9.7. Illustration of the spherical (regular) Bessel functions and (irregular) Neumann functions of order 1, 2 and 3 (taken from Arfken 1985).

has for studying simple, yet realistic situations like the deuteron. We present the more pertinent results in Box 9a.

9.4   The harmonic oscillator potential

The radial problem for a spherical, harmonic oscillator potential can be solved using the methods of section 9.2. We can start from either the Cartesian basis (i) or from the spherical basis (ii) (figure 9.8).

Table 9.1.

Orbital	X_{nl}	$N_{nl} \equiv 2(2l + 1)$	$\sum_{nl} N_{nl}$
1s	3.142	2	2
1p	4.493	6	8
1d	5.763	10	18
2s	6.283	2	20
1f	6.988	14	34
2p	7.725	6	40
1g	8.183	18	58
2d	9.095	10	68
1h	9.356	22	90
3s	9.425	2	92
2f	10.417	14	106
1i	10.513	26	132
3p	10.904	6	138
2g	11.705	18	156
⋮	⋮	⋮	⋮

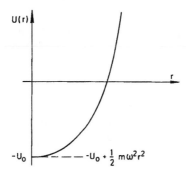

Figure 9.8. Illustration of a harmonic oscillator potential with attractive potential energy value of $-U_0$ at the origin.

(i) Using a Cartesian basis we have

$$U(r) = -U_0 + \tfrac{1}{2}m\omega^2(x^2 + y^2 + z^2). \tag{9.32}$$

The three (x, y, z) specific one-dimensional oscillator eigenvalue equations become

$$\left[\frac{d}{dx^2} + \frac{2m}{\hbar^2} \left(E_1 + \frac{U_0}{3} - \frac{1}{2}m\omega^2 x^2 \right) \right] \varphi_1(x) = 0$$

$$\left[\frac{d^2}{dy^2} + \frac{2m}{\hbar^2} \left(E_2 + \frac{U_0}{3} - \frac{1}{2}m\omega^2 y^2 \right) \right] \varphi_2(y) = 0 \qquad (9.33)$$

$$\left[\frac{d^2}{dz^2} + \frac{2m}{\hbar^2} \left(E_3 + \frac{U_0}{3} - \frac{1}{2}m\omega^2 z^2 \right) \right] \varphi_3(z) = 0,$$

with

$$E = E_1 + E_2 + E_3. \qquad (9.34)$$

The three eigenvalues are then

$$E_1 = \hbar\omega(n_1 + \tfrac{1}{2}) - U_0/3$$
$$E_2 = \hbar\omega(n_2 + \tfrac{1}{2}) - U_0/3 \qquad (9.35)$$
$$E_3 = \hbar\omega(n_3 + \tfrac{1}{2}) - U_0/3,$$

or

$$E = \hbar\omega(N + 3/2) - U_0,$$

($N = n_1 + n_2 + n_3$ with n_1, n_2, n_3 three positive integer numbers $0, 1, \ldots$). The wavefunctions for the one-dimensional oscillator are the Hermite polynomials, characterized by the radial quantum number n_i, so

$$\varphi_1(x) = N_1 \exp\left(-\frac{m\omega}{2\hbar}x^2 \right) H_{n_1}(\nu x), \qquad \left(\nu = \sqrt{\frac{m\omega}{\hbar}} \right) \qquad (9.36)$$

and

$$\varphi(x, y, z) = \varphi_1(x)\varphi_2(y)\varphi_3(z). \qquad (9.37)$$

N_1, N_2, N_3 are normalization coefficients.

(ii) We can obtain solutions that immediately separate the radial variable r from the angular (θ, φ) ones. In this case, the radial equation reduces to

$$\frac{d^2 R(r)}{dr^2} + \frac{2}{r}\frac{dR(r)}{dr} + \left(\frac{2mE}{\hbar^2} + \frac{2m}{\hbar^2}U_0 - \frac{m^2\omega^2}{\hbar^2}r^2 - \frac{l(l+1)}{r^2} \right) R(r) = 0. \qquad (9.38)$$

The normalized radial solutions are the Laguerre solutions

$$R_{nl}(r) = N_{nl}(\nu r)^l \exp\left(\frac{-\nu^2 r^2}{2} \right) \mathcal{L}_{n-1}^{l+1/2}(\nu^2 r^2), \qquad (9.39)$$

and $\nu = \sqrt{(m\omega/\hbar)}$, describing the oscillator frequency. The Laguerre polynomials (Abramowitz and Stegun 1964) are given by the series

$$\mathcal{L}_{n-1}^{l+1/2}(x) = \sum_{k=0}^{n-1} a_k^l (-1)^k x^k. \qquad (9.40)$$

Table 9.2.

N	$E_N(\hbar\omega)$	(n, l)	$\sum_{nl} 2(2l + 1)$	Total
0	3/2	1s	2	2
1	5/2	1p	6	8
2	7/2	2s, 1d	12	20
3	9/2	2p, 1f	20	40
4	11/2	3s, 2d, 1g	30	70
5	13/2	3p, 2f, 1h	42	112
6	15/2	4s, 3d, 2g, 1i	56	168

The total radial wavefunction thus is of degree $2(n - 1) + l$ in the variable r, such that $(2(n - 1) + l)$ can be identified with the major oscillator quantum number N we obtained in the Cartesian description. The total wavefunction then corresponds to energies

$$E_N = (2(n - 1) + l)\hbar\omega + \tfrac{3}{2}\hbar\omega - U_0. \qquad (9.41)$$

For a specific level (n, l) there exists a large degeneracy relative to the energy characterized by quantum number N, i.e. we have to construct all possible (n, l) values such that

$$2(n - 1) + l = N. \qquad (9.42)$$

The first few results are given in table 9.2 and in the left-hand side of figure 9b.1, and give rise to the harmonic oscillation occupancies $2, 8, 20, 40, 70, 112, 168, \ldots$ which also largely deviate from the observed values (except the small numbers at 2, 8, 20). An illustration of wavefunctions that can result from this harmonic oscillator, is presented in figure 9.9, a typical set of variations for given U as a function of l in the lower part and for given l, as a function of n, for $r R_{nl}(r)$ in the upper part.

In both the square-well potential and in the harmonic oscillator potential 2, 8, 20 are the common shell-closures and these numbers correspond to the particularly stable nuclei ^{4}He, ^{16}O, ^{40}Ca and agree with the data. The harmonic oscillator potential has the larger degeneracy of the energy eigenvalues and the corresponding wavefunctions (the degeneracy on $2(n - 1) + l$ is split for the square-well potential).

It is clear that something fundamental is missing still. We shall pay attention to the origin of the spin–orbit force, coupling the orbital and intrinsic spin, that gives rise to a solution with the correct shell-closure numbers.

9.5 The spin–orbit coupling: describing real nuclei

It was pointed out (independently by Mayer (1949, 1950) and Haxel, Jensen and Suess (1949, 1950)) that a contribution to the average field, felt by each individual

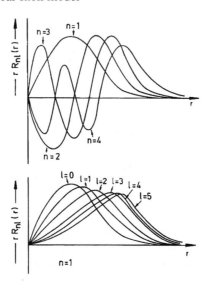

Figure 9.9. Illustration of the harmonic oscillator radial wavefunctions $r R_{nl}(r)$. The upper part shows, for given l, the variation with radial quantum number n ($n = 1$: no nodes in the interval $(0, \infty)$). The lower part shows the variation with orbital angular momentum l for the $n = 1$ value.

nucleon, should contain a spin–orbit term. The corrected potential then becomes

$$U(r) = -U_0 + \frac{1}{2}m\omega^2 r^2 - \frac{2}{\hbar^2}\alpha \hat{l} \cdot \hat{s}. \qquad (9.43)$$

The spin–orbit term, with the scalar product of the orbital angular momentum operator $\hat{l}$ and the intrinsic spin operator $\hat{s}$, can be rewritten using the total nucleon angular momentum $\hat{j}$, as

$$\hat{j} = \hat{l} + \hat{s}, \qquad (9.44)$$

and

$$\hat{l} \cdot \hat{s} = \tfrac{1}{2}(\hat{j}^2 - \hat{l}^2 - \hat{s}^2). \qquad (9.45)$$

The angular momentum operators $\hat{l}^2$, $\hat{s}^2$ and $\hat{j}^2$ form a set of commuting angular momentum operators, so they have a set of common wavefunctions. These wavefunctions, characterized by the quantum numbers corresponding to the *four* commuting operators $\hat{l}^2$, $\hat{s}^2$, $\hat{j}^2$, $\hat{j}_z$ result from angular momentum coupling the orbital ($Y_l^{m_l}(\theta, \varphi)$) and spin ($\chi_{1/2}^{m_s}(\sigma)$) wavefunctions. We denote these as

$$\varphi(\vec{r}, n(l\tfrac{1}{2})jm) \equiv R_{nl}(r) \sum_{m_l, m_s} \langle lm_l, \tfrac{1}{2}m_s | jm \rangle Y_l^{m_l}(\theta, \varphi)\chi_{1/2}^{m_s}(\sigma), \qquad (9.46)$$

(with $m = m_l + m_s$). These wavefunctions correspond to good values of $(l\frac{1}{2})jm$ and we derive the eigenvalue equation

$$\hat{l} \cdot \hat{s}\varphi(\vec{r}, n(l\tfrac{1}{2})jm) = \frac{\hbar^2}{2}[j(j+1) - l(l+1) - \tfrac{3}{4}]\varphi(\vec{r}, n(l\tfrac{1}{2})jm). \tag{9.47}$$

Since we have the two orientations

$$j = l \pm \tfrac{1}{2}, \tag{9.48}$$

we obtain

$$\hat{l} \cdot \hat{s}\varphi(\vec{r}, n(l\tfrac{1}{2})jm) = \frac{\hbar^2}{2} \left\{ \begin{matrix} l & j = l + \frac{1}{2} \\ -(l+1) & j = l - \frac{1}{2} \end{matrix} \right\} \varphi(\vec{r}, n(l\tfrac{1}{2})jm). \tag{9.49}$$

The 'effective' potential then becomes slightly different for the two orientations i.e.

$$U(r) = -U_0 + \tfrac{1}{2}m\omega^2 r^2 + \alpha \left\{ \begin{matrix} -l \\ +(l+1) \end{matrix} \right\}, \qquad j = \left\{ \begin{matrix} l + \frac{1}{2} \\ l - \frac{1}{2} \end{matrix} \right. \tag{9.50}$$

and thus the potential is more attractive for the parallel $j = l + \frac{1}{2}$ orientation, relative to the anti-parallel situation. The corresponding energy eigenvalues now become

$$\varepsilon_{n(l\frac{1}{2})j} = \hbar\omega[2(n-1) + l + \tfrac{3}{2}] - U_0 + \alpha \left\{ \begin{matrix} -l \\ l+1 \end{matrix} \right\}, \qquad j = \left\{ \begin{matrix} l + \frac{1}{2} \\ l - \frac{1}{2} \end{matrix} \right. \tag{9.51}$$

and the degeneracy in the $j = l \pm \frac{1}{2}$ coupling is broken. The final single-particle spectrum now becomes as given in figure 9.10.

We shall not discuss here the origin of this spin–orbit force: we just mention that it originates from the two-body free nucleon–nucleon two-body interaction. The spin orbit turns out to be mainly a surface effect with α, being a function of r and connected to the average potential through a relation of the form

$$\alpha(r) = U_{ls}\frac{1}{r}\frac{dU(r)}{dr}, \tag{9.52}$$

(see figure 9.11 for the $\alpha(r)$ dependence and the $j = l \pm \frac{1}{2}$ level splitting). The corresponding orbits and wavefunctions $r R_{n(l\frac{1}{2})j}(r)$ are presented in figure 9.12 for a slightly more realistic (Woods–Saxon) potential.

It is the particular lowering of the $j = l + \frac{1}{2}$ orbital of a given large N oscillator shell, which is lowered into the orbits of the $N-1$ shell, which accounts for the new shell-closure numbers at 28 ($1f_{\frac{7}{2}}$ shell), 50 ($1g_{\frac{9}{2}}$ shell), 82 ($1h_{\frac{11}{2}}$ shell), 126 ($1i_{\frac{13}{2}}$ shell),. . ..

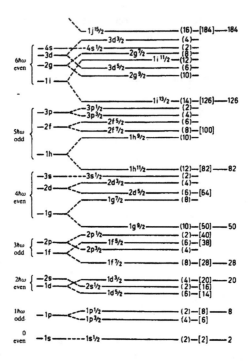

Figure 9.10. Single-particle spectrum up to $N = 6$. The various contributions to the full orbital and spin–orbit splitting are presented. Partial and accumulated nucleon numbers are drawn at the extreme right. (Taken from Mayer and Jensen, *Elementary Theory of Nuclear Shell Structure*, © 1955 John Wiley & Sons. Reprinted by permission.)

Quite often, finite potentials (in contrast to the infinite harmonic oscillator potential) such as the clipped harmonic oscillator, the finite square-well including the spin–orbit force (see table 9b.1 (Box 9b) reproduced from the *Physical Review* article of Mayer, 1950), or more realistic Woods–Saxon potentials of the form

$$U(r) = \frac{U_0}{1 + \exp((r - R_0)/a)} + \frac{U_{ls}}{r_0^2} \frac{1}{r} \frac{d}{dr} \left(\frac{1}{1 + \exp((r - R_0)/a)} \right) \hat{l} \cdot \hat{s}. \quad (9.53)$$

are used. Precise single-particle spectra can be obtained and are illustrated in figure 9.13. Here, we use $R_0 = r_0 A^{1/3}$ with $r_0 = 1.27$ fm, $a = 0.67$ fm and potentials

$$U_0 = (-51 + 33(N - Z)/A) \text{ MeV}$$
$$U_{ls} = -0.44 U_0. \quad (9.54)$$

This figure is taken from Bohr and Mottelson (1969).

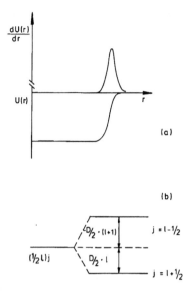

Figure 9.11. (*a*) Possible radial form of the spin–orbit strength $\alpha(r)$ as determined by the derivative of a Woods–Saxon potential. (*b*) The spin–orbit splitting between $j = l \pm \frac{1}{2}$ partners using the expression (9.51) (taken from Heyde 1991).

This single-particle model forms a starting basis for the study of more complicated cases where many nucleons, moving in independent-particle orbits, together with the residual interactions are considered. As a reference we give a detailed account of the nuclear shell model with many up-to-date applications (Heyde 1991).

9.6   Nuclear mean field: a short introduction to many-body physics in the nucleus

In the previous section we started the nucleon single-particle description from a given independent particle picture, i.e. assuming that nucleons mainly move independently from each other in an average field with a large mean-free path. The basic non-relativistic picture one has to start from, however, is one where we have A nucleons moving in the nucleus with given kinetic energy $(\vec{p}_i^2 / 2m_i)$ and interacting with the two-body force $V(i, j)$. Eventually, higher-order interactions $V(i, j, k)$ or density-dependent interactions may be considered.

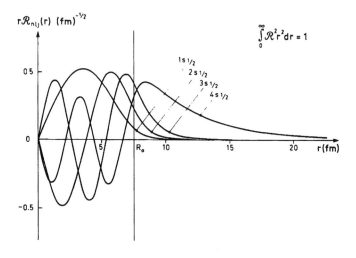

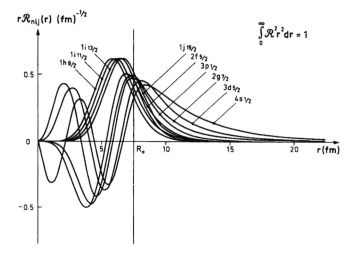

Figure 9.12. Neutron radial wavefunctions $r R_{nlj}(r)$ for $A = 208$ and $Z = 82$ based on calculations with a Woods–Saxon potential by Blomqvist and Wahlborn (1960). (Taken from Bohr and Mottelson 1969, *Nuclear Structure*, vol 1, © 1969 Addison-Wesley Publishing Company. Reprinted by permission.)

The starting point is the nuclear A-body Hamiltonian

$$H = \sum_{i=1}^{A} \frac{\vec{p}_i^{\,2}}{2m_i} + \sum_{i<j=1}^{A} V(\vec{r}_i, \vec{r}_j), \tag{9.55}$$

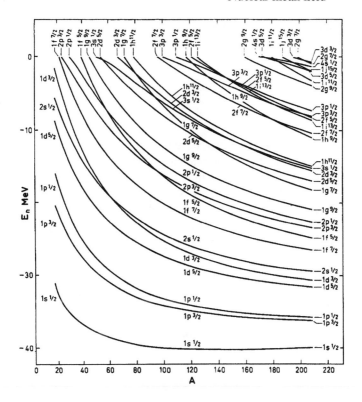

Figure 9.13. Energies of neutron orbits, calculated by Veje and quoted in Bohr and Mottelson (1969). Use was made of a Woods–Saxon potential as described in detail. (Taken from Bohr and Mottelson 1969, *Nuclear Structure*, vol 1, © 1969 Addison-Wesley Publishing Company. Reprinted by permission.)

which is at variance to the A-independent nuclear Hamiltonian

$$H_0 = \sum_{i=1}^{A} \left(\frac{\vec{p}_i^2}{2m_i} + U(|\vec{r}_i|) \right) = \sum_{i=1}^{A} h_0(i). \tag{9.56}$$

If we denote in shorthand notation all quantum numbers characterizing the single-particle motion by the notation $i \equiv n_i, l_i, j_i, m_i$; then one has

$$h_0(i)\varphi_i(\vec{r}) = \varepsilon_i \varphi_i(\vec{r}), \tag{9.57}$$

and the A-body, independent particle wavefunction becomes (not taking the anti-symmetrization into account)

$$\Psi_{1,2,\dots,A}(\vec{r}_1, \vec{r}_2, \dots, \vec{r}_A) = \prod_{i=1}^{A} \varphi_i(\vec{r}_i), \tag{9.58}$$

corresponding to the solution of H_0 with total eigenvalue E_0,

$$H_0 \Psi_{1,2,\ldots,A}(\vec{r}_1, \vec{r}_2, \ldots, \vec{r}_A) = E_0 \Psi_{1,2,\ldots,A}(\vec{r}_1, \vec{r}_2, \ldots \vec{r}_1), \qquad (9.59)$$

with

$$E_0 = \sum_{i=1}^{A} \varepsilon_i.$$

The wavefunctions $\varphi_i(\vec{r})$, however, have been obtained starting from the chosen average field $U(|\vec{r}|)$ and it is not guaranteed that this potential will give rise to these wavefunctions that minimize the total energy of the interacting system.

The Hartree–Fock method allows for such a wavefunction. In the introductory section to part D, we give a simple, but intuitive approach in order to obtain the 'best' wavefunctions $\varphi_i^{\mathrm{HF}}(\vec{r}_i)$ and corresponding energies $\varepsilon_i^{\mathrm{HF}}$ starting from the knowledge of the two-body interaction $V(\vec{r}_i, \vec{r}_j)$.

9.6.1 Hartree–Fock: a tutorial

If we consider the nucleus to consist of a number of separated point particles at positions $\vec{r}_i$ (see figure 9.14(a)) interacting through the two-body potentials $V(\vec{r}_i, \vec{r}_j)$, then a total potential acting on a given point particle at $\vec{r}_i$ is obtained by summing the various contributions, i.e.

$$U(\vec{r}_i) = \sum_j V(\vec{r}_i, \vec{r}_j). \qquad (9.60)$$

For a continuous distribution, characterized by a density distribution $\rho(\vec{r}')$, the potential energy at point $\vec{r}$ is obtained by folding the density with the two-body interaction so one obtains (figure 9.14(b))

$$U(\vec{r}) = \int \rho(\vec{r}') V(\vec{r}, \vec{r}') \, \mathrm{d}\vec{r}'. \qquad (9.61)$$

In the atomic nucleon, where the density $\rho(\vec{r}')$ is expressed using the quantum mechanical expression summed over all individually occupied orbitals $\varphi_b(\vec{r})$ ($b \equiv n_b, l_b, j_b, m_b$), we have

$$\rho(\vec{r}') = \sum_{b \in F} \rho_b(\vec{r}') = \sum_{b \in F} \varphi_b^*(\vec{r}') \varphi_b(\vec{r}'), \qquad (9.62)$$

and the one-body potential at point $\vec{r}$ becomes

$$U(\vec{r}) = \sum_{b \in F} \int \varphi_b^*(\vec{r}') V(\vec{r}, \vec{r}') \varphi_b(\vec{r}') \, \mathrm{d}\vec{r}'. \qquad (9.63)$$

The above procedure shows that if we start from a given two-body interaction, we should know the wavefunctions $\varphi_b(\vec{r}')$ in order to solve for the average field $U(\vec{r})$,

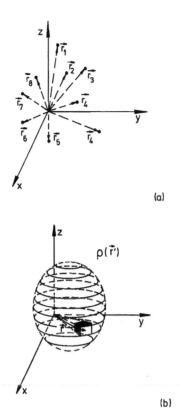

Figure 9.14. (*a*) A collection of *A* particles (with coordinates $\vec{r}_A$), representing an atomic nucleus in terms of the individual *A* nucleons, interacting via a two-body force between any two nucleons, i.e. $V(\vec{r}_i, \vec{r}_j)$. (*b*) The analogous, continuous mass distribution, described by the function $\rho(\vec{r})$. The density at a given point (in a volume element $d\vec{r}$) ρ is illustrated pictorially. This density distribution is then used to determine the average, one-body field $U(\vec{r})$.

but in order to determine $\varphi_b(\vec{r}')$ we need the average one-body field. This problem is a typical iterative problem, which we solve by starting from some initial guess for either the wavefunctions or the potential $U(\vec{r})$.

The Schrödinger equation, which we have to solve in order to accomplish the above self-consistent study, is

$$\frac{-\hbar^2}{2m}\Delta\varphi_i(\vec{r}) + \sum_{b\in F}\int \varphi_b^*(\vec{r}')V(\vec{r}, \vec{r}')\varphi_b(\vec{r}')\varphi_i(\vec{r})\,d\vec{r}' = \varepsilon_i\varphi_i(\vec{r}), \qquad (9.64)$$

which we have to solve for all values of i (i encompasses the occupied orbits b and also the other, unoccupied ones). In the Schrödinger equation we have the antisymmetrization between the two identical nucleons in orbit b (at point $\vec{r}'$) and in orbit i (at point $\vec{r}$) not yet taken care of. The correct equation follows by substituting

$$\varphi_b(\vec{r}')\varphi_i(\vec{r}) \rightarrow [\varphi_b(\vec{r}')\varphi_i(\vec{r}) - \varphi_b(\vec{r})\varphi_i(\vec{r}')], \qquad (9.65)$$

and becomes

$$-\frac{\hbar^2}{2m}\Delta\varphi_i(\vec{r}) + \sum_{b\in F}\int \varphi_b^*(\vec{r}')V(\vec{r},\vec{r}')\varphi_b(\vec{r}')\,\mathrm{d}\vec{r}' \cdot \varphi_i(\vec{r})$$

$$-\sum_{b\in F}\int \varphi_b^*(\vec{r}')V(\vec{r},\vec{r}')\varphi_b(\vec{r}) \cdot \varphi_i(\vec{r}')\,\mathrm{d}\vec{r}' = \varepsilon_i\varphi_i(\vec{r}), \qquad (9.66)$$

for $i = 1, 2, \ldots, A; A+1, \ldots$

Using the notation

$$U_{\mathrm{H}}(\vec{r}) \equiv \sum_{b\in F}\int \varphi_b^*(\vec{r}')V(\vec{r},\vec{r}')\varphi_b(\vec{r}')\,\mathrm{d}\vec{r}',$$

$$U_{\mathrm{F}}(\vec{r},\vec{r}') \equiv \sum_{b\in F}\varphi_b^*(\vec{r}')V(\vec{r},\vec{r}')\varphi_b(\vec{r}), \qquad (9.67)$$

the Hartree–Fock self-consistent set of equations to be solved becomes

$$\left\{\begin{array}{llll} -\frac{\hbar^2}{2m}\Delta\varphi_i(\vec{r}) & +U_{\mathrm{H}}(\vec{r})\varphi_i(\vec{r}) & -\int U_{\mathrm{F}}(\vec{r},\vec{r}')\varphi_i(\vec{r}')\,\mathrm{d}\vec{r}' & = \varepsilon_i\varphi_i(\vec{r}) \\ \vdots & \vdots & \vdots & \vdots \end{array}\right. . \qquad (9.68)$$

In general, solving the coupled set of integro-differential equations is complicated. In particular, the non-local part caused by $U_{\mathrm{F}}(\vec{r},\vec{r}')$ complicates the solution of these equations. For zero-range two-body interactions this integral contribution becomes local again.

The procedure to solve the coupled equations starts from initial guesses for the wavefunctions $\varphi_i^{(0)}(\vec{r})$ from which the terms $U_{\mathrm{H}}^{(0)}(\vec{r})$ and $U_{\mathrm{F}}^{(0)}(\vec{r},\vec{r}')$ could be determined, and the solution will then give rise to a first estimate of $\varepsilon_i^{(1)}$ and the corresponding wavefunctions $\varphi_i^{(1)}(\vec{r})$. We can go on with this iterative process until convergence shows up, i.e. we call these solutions $\varphi_i^{\mathrm{HF}}(\vec{r})$, $\varepsilon_i^{\mathrm{HF}}$, $U^{\mathrm{HF}}(\vec{r})$ (the local and non-local converged one- and two-body quantities (figure 9.15)).

It can be shown that using these optimum wavefunctions, the expectation value of the total Hamiltonian becomes minimal when we use the antisymmetrized A-body wavefunctions

$$\Psi_{1,2,\ldots,A}(\vec{r}_1,\vec{r}_2,\ldots,\vec{r}_A) = \mathcal{A}\Psi_{1,2,\ldots,A}(\vec{r}_1,\vec{r}_2,\ldots,\vec{r}_A), \qquad (9.69)$$

where $\mathcal{A}$ is an operator which antisymmetrizes the A-body wavefunction and Ψ is constructed from the Hartree–Fock single-particle wavefunctions.

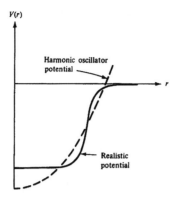

Figure 9.15. Schematic comparison of an harmonic oscillator potential with the more realistic potentials that can be derived starting from a Hartree–Fock calculation. (Taken from Frauenfelder and Henley (1991) *Subatomic Physics* © 1974. Reprinted by permission of Prentice-Hall, Englewood Cliffs, NJ.)

There are quite a number of alternative methods to derive optimal wavefunctions starting from the extremum condition that

$$\langle \Psi_{1,2,\ldots,A}(\vec{r}_1, \vec{r}_2, \ldots, \vec{r}_A) | H | \Psi_{1,2,\ldots,A}(\vec{r}_1, \vec{r}_2, \ldots, \vec{r}_A) \rangle, \qquad (9.70)$$

should be minimal E_{HF} (min). We do not concentrate here on these interesting aspects (see part D). We just mention an often used method where the 'optimal' wavefunctions are determined by expanding them on a given basis, i.e. the harmonic oscillator basis

$$\varphi_i^{HF}(\vec{r}) = \sum_k a_i^k \varphi_k^{h.o.}(\vec{r}), \qquad (9.71)$$

and minimizing E_{HF} with respect to the expansion coefficients a_i^k. It is such that in a number of cases, solutions occur that no longer conserve spherical symmetry but give rise to the lower energy of the interacting A-body system.

9.6.2 Measuring the nuclear density distributions: a test of single-particle motion

Once we have determined the Hartree–Fock single-particle orbitals, various observables characterizing the nuclear ground-state can be derived such as the nuclear binding energy E_{HF}, the various radii (mass, charge) as well as the particular charge and mass density distributions. As outlined in chapter 1, the nuclear density is given by the sum of the individual single-particle densities corresponding to the occupied orbits (if this independent-particle model is to hold

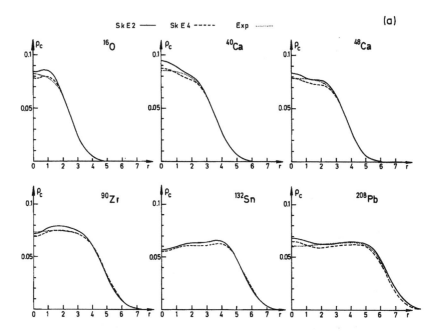

Figure 9.16. Charge densities (in units $e.fm^{-3}$) for a number of closed-shell nuclei ^{16}O, ^{40}Ca, ^{48}Ca, ^{90}Zr, ^{132}Sn and ^{208}Pb. The theoretical curves are obtained using effective forces of the Skyrme type (see chapter 10) and are compared to the data.

to a good degree)

$$\rho(\vec{r}) = \sum_{b \in F} \rho_b(\vec{r}). \tag{9.72}$$

In figure 9.16, we present a number of charge density distributions for doubly magic nuclei ^{16}O, ^{40}Ca, ^{48}Ca, ^{90}Zr, ^{132}Sn and ^{208}Pb. The theoretical curves have been determined using specific forms of nucleon–nucleon interaction that are of zero-range in coordinate space ($\delta(\vec{r}_1 - \vec{r}_2)$) but are, at the same time, density- and velocity-dependent. Such forces are called Skyrme effective interactions. We compare with the data and, in general, good agreement is observed. We also present (in figure 9.17) the various mass density distributions ρ_m (fm^{-3}) for all of the above nuclei and compare these with the nuclear matter value. This is a system which is ideal in the sense that no surface effects occur, and has a uniform density that approximates to the interior of a heavy nucleus with $N = Z = A/2$. In most cases, the electromagnetic force is neglected relative to the strong, nuclear force.

This density can be deduced from the maximum or saturation density of finite nuclei and a commonly used value is

$$\rho_{\text{nucl. matter}} = 0.16 \pm 0.02 \text{ nucleons fm}^{-3}.$$

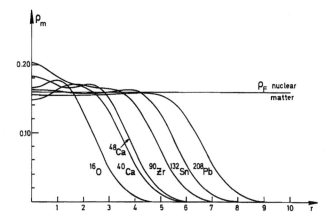

Figure 9.17. Combination of nuclear matter densities ρ_m (fm^{-3}) for the set of nuclei shown in figure 9.16. The nuclear matter density ρ_F is given for comparison.

This value is slightly higher than the average density of $3/(4\pi r_0^3) = 0.14$ nucleons fm^{-3} obtained for a finite nucleus with $r_0 = 1.2$ fm. Using a Fermi-gas model for the motion of nucleons (see chapter 8), we can derive an expression for the density

$$\rho_{\text{nucl. matter}} = \frac{2}{3\pi^2}k_F^3, \qquad (9.73)$$

where k_F is the Fermi wavevector $p_F = \hbar k_F$. Inverting, and using the above determined density, the Fermi wavevector becomes

$$k_F = \left(\frac{3\pi^2}{2}\rho_{\text{nucl. matter}}\right)^{1/3} = 1.33 \pm 0.05 \text{ fm}^{-1}. \qquad (9.74)$$

In figure 9.18, we show the binding energy per nucleon as a function of the Fermi wavevector $k_F(\text{fm}^{-1})$ in infinite nuclear matter using various nucleon–nucleon interactions. The shaded region indicates the range of values obtained by extrapolating from finite nuclei.

   A very good test of the nuclear single-particle structure in atomic nuclei would be possible if we were to know the densities in a number of adjacent nuclei such that differences determine a single contribution from a given orbital. Such a procedure was followed by Cavedon *et al* (1982) and Frois *et al* (1983). Using electron scattering experiments, the charge density distributions were determined

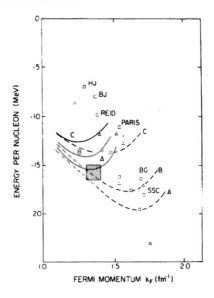

Figure 9.18. Binding energy (per nucleon) as a function of the nucleon Fermi momentum k_F in infinite nuclear matter. The shaded region represents possible values extrapolated from finite nuclei. The small squares correspond to various calculations with potentials HJ: Hamada–Johnston; BJ: Bethe–Johnson; BG: Bryan–Gersten; SSC: super-soft-core). The solid lines result from various Dirac–Brueckner calculations; dashed lines correspond to conventional Brueckner calculations. (Taken from Machleidt 1985 in *Relativistic Dynamics and Quark-Nuclear Physics*, ed M B Johnson and A Picklesimer © 1985 John Wiley & Sons. Reprinted by permission.)

in ^{206}Pb and ^{205}Tl. From the difference

$$\rho^{ch}(^{206}\text{Pb}) - \rho^{ch}(^{205}\text{Tl}) = \sum_{b \in F} |\varphi_b(\vec{r})|^2 (^{206}\text{Pb}) - \sum_{b \in F'} |\varphi_b(\vec{r})|^2 (^{205}\text{Tl})$$

$$= |\varphi_{3s1/2}(\vec{r})|^2, \tag{9.75}$$

it should be possible to obtain the structure of the 3s1/2 proton orbit. The results are shown in figure 9.19, and give an unambiguous and sound basis for the nuclear independent-particle model as a very good description of the nuclear A-body system. The method of electron scattering is, in that sense, a very powerful method to determine nuclear ground-state properties (charge distributions, charge transition and current transition densities, ...). The typical set-up at NIKHEF-K (Amsterdam), where spectrometers are installed to detect the scattered electrons and other particles emitted in the reaction process (protons, ...), is shown in figure 9.20. This set-up at Amsterdam has been working during the last few years on tests of the nuclear single-particle picture (see Box 10b).

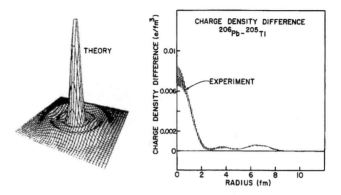

Figure 9.19. The nuclear density distribution for the least bound proton in ^{206}Pb. The shell model predicts the last (3s1/2) proton in ^{206}Pb to have a sharp maximum at the centre, as shown at the left-hand side. On the right-hand side, the nuclear charge density difference $\rho^{ch}(^{206}\text{Pb}) - \rho^{ch}(^{205}\text{Tl}) = \varphi^2_{3s1/2}(r)$ is given (taken from Frois (1983) and DOE (1983)).

9.7   Outlook: the computer versus the atomic nucleus

In the previous sections and chapters, we have seen that in order to describe fine details in the nuclear modes of motion, the required calculations become increasingly complex and one has to make use of immense computing capacities, to follow the motion of nucleons in the nucleus and to see how two nuclei interact through collisions starting from the basic two-body forces as ingredients.

The National Centre for Energy Research Computing (NERSC) provides such supercomputing capabilities to research workers whose work is supported by the Office of Energy Research (OER) of the Department of Energy (DOE). Currently (1994) there are about 4500 people who use these supercomputers with large programs for high-energy and nuclear physics. NERSC now has four multiprocessor supercomputers available: a CRAY-2 with eight processors and 134 million words of memory, a CRAY-2 with four processors and 134 million words of memory, a CRAY-2 with four processors and 67 million words of memory and a CRAY X-MP with two processors and 2 million words of memory.

The applications to nuclear physics are widespread: in particular, for high energy collisions between heavy ions the time-dependent Hartree–Fock calculations allow density variations and the various phases occurring in such processes to be followed and simulated. Large-scale shell-model calculations where the energy eigenvalue problem is solved in a basis spanning many thousands of basis vectors are typical state-of-the-art studies within the shell-model of the nucleus and will be discussed later on in part D.

Recently, at Los Alamos, Monte Carlo simulations for light nuclei have been carried out by Carlson and co-workers (Carlson *et al* 1983, 1986, 1987, 1988,

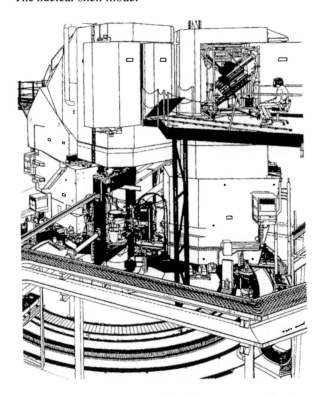

Figure 9.20. Illustration of the NIKHEF-K (Amsterdam) set-up where proton and electron spectrometers allow detailed studies on how a nucleon moves in the atomic nucleus (from Donné, private communication).

1990). Starting from accurate nucleon–nucleon interactions, one has succeeded in the study of properties of light nuclei: exact alpha-particle calculations can be performed with Green's function Monte Carlo methods. The starting point is a Hamiltonian with only nucleon degrees of freedom

$$H = -\frac{\hbar^2}{2m} \sum_i \Delta_i + \sum_{i<j} V_{i,j} + \sum_{i<j<k} V_{i,j,k} + \cdots, \quad (9.76)$$

for which we try to solve the Schrödinger equation

$$H|\Psi\rangle = E|\Psi\rangle, \quad (9.77)$$

and determine static and dynamical properties of the atomic nucleus.

Both variational Monte Carlo (VMC) and Green's function Monte Carlo (GFMC) were used to study the α-particle ground-state. The latter even provides

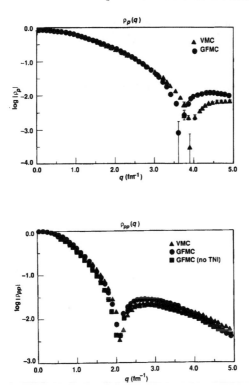

Figure 9.21. Results from *ab initio* many-body (variational Monte Carlo (VMC) and Green's function Monte Carlo (GFMC)) calculations, using supercomputer clusters, illustrate the possibility of evaluating: (top) the Fourier transform of the one-body proton density $\rho_p(q)$, and (bottom) the Fourier transform of the proton–proton distribution function $\rho_{pp}(q)$ (taken from Carlson 1990).

an exact solution, subject to the statistical errors of the method used. Without going into the details, impressive results have been obtained.

In figure 9.21, we compare for both VMC and GFMC studies the information on the one-body density that reflects the single-particle information. This is presented on a plot with the Fourier transform of the one body proton density $\rho_{proton}(\vec{r})$. There is a significant difference in the region of the second maximum indicating possible effects of exchange currents on the nucleon properties in the α-particle. In figure 9.21 there is another interesting property, i.e. the probability distribution for two protons to be separated by a distance r. Experimentally, the Fourier transform $\rho_{pp}(q)$ can be extracted from the Coulomb sum rule, which can be obtained from the many electron-scattering experiments. This quantity is characterized by a diffraction minimum due to the strong, repulsive core in the two-body force. The VMC calculations of $\rho_{pp}(q)$ produce very similar results to the much more complicated GFMC calculations.

Work is proceeding to the heavier $A = 6, 8$ systems is in progress due to the very large computing potential at various centres and the NERSC in the USA (see also chapter 12: Nuclear physics of very light nuclei).

Box 9a. Explaining the bound deuteron

The methods we have discussed can be used to explain certain properties of light, bound nucleon configurations. The deuteron is one of these very light systems for which a square-well potential can be used (figure 9a.1).

The deuteron is characterized by a $J^\pi = 1^+$ ground-state which is loosely bound ($E_B = 2.22461 \pm 0.00007$ MeV), and where proton and neutron move mainly in a relative $^3S_1$ ($^{2S+1}L_J$) state. The magnetic moment $\mu_D = +0.857406 \pm 0.000001$ μ_N is very close to the value $\mu_p + \mu_n = 0.87963\mu_N$. There is an electric quadrupole moment $Q_D = 2.875 \pm 0.002$ mb, which is small compared to the single-particle estimate.

If we express the energy, corresponding to the ground-state value $E = -E_B$, the Schrödinger equation becomes, for the one-dimensional, radial problem with zero angular momentum,

$$\frac{d^2u}{dr^2} + k^2u = 0 \qquad r < b$$
$$\frac{d^2u}{dr^2} - \alpha^2u = 0 \qquad r > b, \tag{9a.1}$$

defining

$$k^2 = \frac{M_n}{\hbar^2}(U_0 - E_B), \qquad \alpha^2 = \frac{M_n}{\hbar^2}E_B, \tag{9a.2}$$

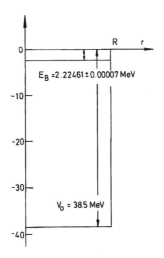

Figure 9a.1. Square-well potential, adjusted to describe correctly the binding energy E_B of the deuteron. The full depth is also given and amounts to $U_0 = 38.5$ MeV.

and using the radial solution

$$u(r) = rR(r). \tag{9a.3}$$

Approximate solutions in the two regions become

$$u(r) = A \sin kr \qquad r < b$$
$$u(r) = Be^{-\alpha(r-b)} \qquad r > b. \tag{9a.4}$$

Matching the logarithmic derivatives at $r = b$ gives

$$k \cotan kb = -\alpha, \tag{9a.5}$$

and matching the wavefunctions at $r = b$ gives

$$A \sin kb = B. \tag{9a.6}$$

These two conditions lead to the condition

$$k^2 A^2 = (k^2 + \alpha^2)B^2. \tag{9a.7}$$

The normalization of the wavefunction $4\pi \int u^2(r) \, dr = 1$ becomes

$$\frac{A^2}{2k}(2kb - \sin 2kb) + \frac{B^2}{\alpha} = \frac{1}{2\pi}. \tag{9a.8}$$

Eliminating A^2 from the last two equations, gives the value for B as

$$B = \sqrt{\frac{\alpha}{2\pi}} \left\{ 1 + \alpha b \left(1 + \frac{\alpha^2}{k^2} \right) \left(1 + \frac{\alpha}{\pi k} \right) \right\}^{-1/2}$$
$$\simeq \sqrt{\frac{\alpha}{2\pi}} \left(1 - \frac{\alpha b}{2} + \cdots \right)$$
$$\simeq \sqrt{\frac{\alpha}{2\pi}} e^{-\alpha b/2}. \tag{9a.9}$$

Knowing the binding energy E_B, we can determine the value $\alpha = 0.232 \text{ fm}^{-1}$. A best value for b can be determined from proton–neutron scattering as $b = 1.93 \text{ fm}$. This then gives $V_0 = 38.5 \text{ MeV}$. One can show that this value of U_0 and the value for b just give rise to a single, bound 1s state, all other higher-lying states 1p, 1d, 2s, being unbound. Since we also have

$$A \simeq B \simeq \sqrt{\frac{\alpha}{2\pi}} e^{-\alpha b/2}, \tag{9a.10}$$

we obtain the final wavefunctions

$$u(r) = \sqrt{\frac{\alpha}{2\pi}} e^{-\alpha b/2} \sin kr \qquad r < b$$
$$u(r) = \sqrt{\frac{\alpha}{2\pi}} e^{\alpha b/2} e^{-\alpha r} \qquad r > b, \tag{9a.11}$$

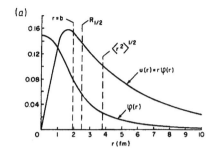

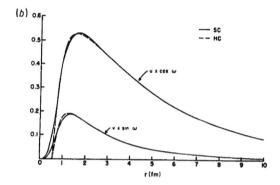

Figure 9a.2. Plot of the wavefunctions $u(r)$ and $\varphi(r)$ for the square-well potential, used to describe the deuteron. For the definition of $r = b$, $R_{1/2}$ and $\langle r^2 \rangle^{1/2}$ (see text) (a). (b) $u(r) \cos \omega$ and $v(r) \sin \omega$ for the hard-core and soft-core Reid potentials (Reid, 1968). The two functions $u(r)$ in (a) and (b) differ in their normalization by the factor $(4\pi)^{1/2}$ (taken from Hornyack 1975).

or, substituting numerical values for k, α, b;

$$u(r) = 0.160 \sin(0.938r) \ \text{fm}^{-1/2} \qquad r < b$$
$$u(r) = 0.243 \mathrm{e}^{-0.232r} \ \text{fm}^{-1/2} \qquad r > b. \qquad (9a.12)$$

In figure 9a.2, we show a plot of these wavefunctions. Here, it is clear that $R(r)$ extends far beyond $r = b$, the range of the nucleon–nucleon interaction. We also evaluate $\sqrt{\langle r^2 \rangle}$ for the deuteron wavefunction, which results in a value of 3.82 fm, as well as the value $R_{1/2}$, the radius of n–p separation where the probability of finding the deuteron up to $R_{1/2}$ is equal to the probability of finding the deuteron beyond the value of $R_{1/2}$. A value $R_{1/2} = 2.50$ fm results.

The ground-state is a pure $^3S_1$ state: this is in contrast with the small, but non-vanishing value of Q_D. This implies admixtures in the ground-state wavefunction with angular momentum $L = 2$. Using more realistic proton–neutron potentials, a mixed wavefunction containing $L = 0$ and $L = 2$ orbital components results. The radial part of the $L = 0$ component ($u(r)$ part) and of the $L = 2$ component ($v(r)$ part) are illustrated in figure 9a.2 (b) using realistic potentials: HC (hard core) and SC (soft core) Reid potential (Reid 1968).

Box 9b. Origin of the nuclear shell model

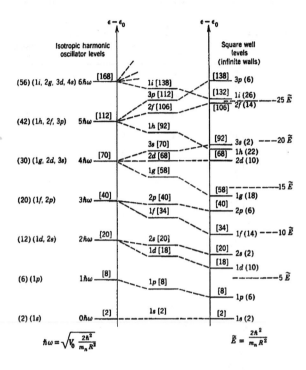

Figure 9b.1. Shell-model occupation numbers predicted for the harmonic oscillator and square-well potentials. (Taken from Mayer and Jensen © 1955 John Wiley & Sons. Reprinted by permission.)

Table 9b.1. Order of energy levels obtained from those of a square well potential by spin–orbit coupling.

Osc. no.	Square well	Spin term	No. of states	Shells	Total no.
0	1s	$1s_{1/2}$	2	2	2
1	1p	$1p_{3/2}$	4		
		$1p_{1/2}$	2	6	8
2	1d	$1d_{5/2}$	6		
		$1d_{3/2}$	4	12	
	2s	$2s_{1/2}$	2		
					20
3	1f	$1f_{7/2}$	8	8	28
		$1f_{5/2}$	6		
	2p	$2p_{3/2}$	4	22	
		$2p_{1/2}$	2		
		$1g_{9/2}$	10		
					50
4	1g	$1g_{7/2}$	8		
	2d	$2d_{5/2}$	6		
		$2d_{3/2}$	4	32	
	3s	$3s_{1/2}$	2		
		$1h_{11/2}$	12		82
5	1h	$1h_{9/2}$	10		
	2	$2f_{7/2}$	8		
		$2f_{5/2}$	6		
	3p	$3p_{3/2}$	4	44	
		$3p_{1/2}$	2		126
		$1i_{13/2}$	14		
6	1i	$1i_{11/2}$			
	2g 3d 4s				

Maria Goeppert Mayer —two-fold pioneer

Although Maria Mayer made significant contributions (leading to the Nobel Prize) starting in 1930, it was 30 years before she received a full-time faculty appointment.

Maria Goeppert Mayer

Maria Goeppert Mayer, who received the Nobel Prize in physics for her work on the shell model, died about a year and a half ago. I knew Maria almost all of my life as a "Friend of the Family"; the Mayers lived a few blocks away from us in Leonia, N.J., from 1939 to 1945, across the street in Chicago from 1945 to 1958 and then a half mile from my parents in La Jolla from 1960 on. I have memories of Maria from the Leonia period, when I was a teenager—but not physics memories. Maria stands out in my mind as completely different from the many wives I knew in Leonia, in that she had a strong commitment to a career and to science. My father, who likes to talk science to anyone all the time, loved talking to Maria, telling her his ideas and listening to her talk—as did all the other physicists and chemists, always talking about science at parties at our house. She was vital and lively, and gave a strong impression of living a satisfying life, in spite of problems of finding jobs and house-keepers. In looking back I realize how lucky I was to know her in this way. She gave the impression that being a theoretical physicist and a mother was possible, rewarding and worth a great deal. Also, as I did not know a very large number of physicists, and one of these happened to be a woman, I did not realize how few women physicists there really are.

I looked up the details of her physics career and was struck by the types of positions she had held. Until 1959, none of them are what I would term "regular" positions: voluntary associate at Johns Hopkins for nine years (she was paid approximately $100 per year); lecturer in chemistry at Columbia and part-time lecturer at Sarah Lawrence College, during the war, part-time at SAM Labs working on the atomic-bomb project; in the Chicago era, part-time senior physicist at Argonne while being voluntary professor at the University of Chicago.

According to her biographer, Joan Dash,* the unchallenging and unre-warding jobs she had in the early years were not particularly useful for her de-velopment, especially from the point of view of her image of herself. In spite of this, she managed through personal contacts, and through her husband, to continue research and to broaden her interests—it's lucky she was a theorist. She expressed great frustration with these early jobs, but never bitterness. Of course, she thought of herself as a research physicist, and my memories of the Leonia days are that she was re-garded as such by her colleagues, al-though her position as part-time lecturer of chemistry is at variance with this. In the Chicago days she was certainly considered to be a professor at Chicago although she was "voluntary" because of a nepotism rule and was not paid by the University of Chicago but by Ar-gonne. (In 1959 she was made profes-sor; this was after the Mayers had re-ceived offers from the University of California at San Diego.) I think it is a unique career, different from any other Nobel prize winner, and that only her dedication to physics, her unique abili-ties and a supportive husband sustained her through it.

In the last conversation I had with her, at the Washington APS meeting of 1971, I told her about the APS Commit-tee on Women in Physics, which was just being formed, and which interested her. She became a member of this committee.

She overcame quite subtle obstacles in the course of her life—she was a woman, a foreigner, she started her career in the days of the depression. How many of us could have achieved half the success that she did?

—EUB

* Joan Dash, *A Life of One's Own*, Harper and Row, New York (1973)

(Reprinted with permission of the American Physical Society.)

Problem set—Part C

1. Show that (neglecting both the pairing and symmetry energy terms in the liquid drop energy mass formula) the energy that is liberated in a fission process becomes maximal for an equal division in both charge and mass. Derive a value of the quantity Z^2/A where this particular fission decay mode becomes possible. Is the nucleus ^{236}U ($Z = 92$) stable against spontaneous fission?

Data given: $a_S = 13$ MeV; $a_V = 14$ MeV; $a_C = 0.6$ MeV; $a_A = 19$ MeV; $e^2/(4\pi\epsilon_0) = 1.44$ MeV fm.

2. Fusion of two identical nucleons can happen whenever the gain in binding energy between two independent touching fragments and one single nucleus (containing the two fragments) just compensates the repulsive Coulomb energy between these two fragments at the moment of touching (i.e. the energy needed to bring the two fragments from infinity until they just touch).

Determine the value of Z^2/A at which this fusion process just starts to occur.

Data given: $a_S = 13$ MeV; $a_V = 14$ MeV; $a_C = 0.6$ MeV; $a_A = 19$ MeV; $e^2/(4\pi\epsilon_0) = 1.44$ MeV fm. Hint: consider the nuclei as spherical objects and, moreover, neglect the pairing as well as the symmetry energy and in the Coulomb energy replace $Z(Z-1) \rightarrow Z^2$.

3. Show how, starting from the binding energy of three adjacent atomic nuclei $A - 1$ (even Z, even N), A (even Z, odd N) and $A + 1$ (even Z, even N), one can derive an approximate expression for the nucleon pairing energy.

Data: the binding energy for the Ca isotopes ^{40}Ca: 342 055 keV; ^{41}Ca: 350 418 keV; ^{42}Ca: 361 898 keV; ^{43}Ca: 369 831 keV; ^{44}Ca: 380 963 keV; ^{45}Ca: 388 378 keV; ^{46}Ca: 398 774 keV.

Also derive an analytic expression for the pairing energy starting from the semi-empirical mass formula. Hint: replace $Z(Z - 1)Z \rightarrow Z^2$, approximate $(1 \pm x)^n = 1 \pm n \cdot x \mp \cdots$ by its lowest order in the binomial expansion and neglect, in working out the analytic expression, all terms of the order $(1/A)^2$ relative to the order 1.

4. A nucleus can undergo fission if the force derived from the surface energy is exactly compensated by the Coulomb force taken for spherical atomic nuclei.

Determine the critical value for the ratio Z^2/A at which fission starts spontaneously.

312

Data given: $a_S = 13$ MeV; $a_C = 0.6$ MeV; $|\vec{F}| = |-dU/dR|$ with R the magnitude of the nuclear radius.

5. Consider a Fermi-gas model for atomic nuclei (containing protons and neutrons) in which the repulsive Coulomb force acting inside the nucleus gives rise to an approximate but constant Coulomb potential energy term. Thereby we consider the nucleus as a homogeneously charged sphere with radius $R = r_0 A^{1/3}$ (with $r_0 = 1.2$ fm).

Show that the following expression between the proton and neutron numbers exists

$$N = (Z^{2/3} + bZ^2 A^{1/3})^{3/2},$$

and determine the constant value of b using the Fermi-gas-model parameters only. Data given: replace $Z(Z - 1) \rightarrow Z^2$; $e^2/(4\pi\epsilon_0) = 1.44$ MeV fm; $\hbar c = 197$ MeV fm and $mc^2 = 938$ MeV.

6. More precise calculations, using the semi-empirical mass formula, make use of an additional quadratic symmetry energy term of the form $-b_A(N - Z)^4/A^3$. Derive the coefficient b_A starting from the total energy of a Fermi-gas model of an atomic nucleus containing protons and neutrons. Data given: $r_0 = 1.25$ fm; $mc^2 = 938$ MeV; $\hbar c = 197$ MeV fm.

7.

(a) Derive an approximate value for the attractive nuclear potential depth that a neutron feels starting from a Fermi-gas model. We consider the fact that the least bound neutron is still bound by approximately 8 MeV. We restrict the problem to nuclei with $N = Z = A/2$ and take as the radius parameter $r_0 = 1.3$ fm.
(b) Determine the level density $\rho(E) = dn/dE$ in the Fermi-gas model.
(c) How does the level density change if we consider the nucleons as a collection of relativistic particles?

8. In the semi-empirical mass formula, a symmetry term appears and the coefficient a_A has been determined to fit experimental masses all through the nucler mass table with as a result $a_A = 19$ MeV. Show that such a term can be deduced in a very natural way starting from the Fermi-gas model. Derive the strength that thus follows and compare with the optimal fitted value of a_A. Discuss possible explanations for the very large difference between the 'theoretical' and 'empirical' values of a_A. Data given: $r_0 = 1.2$ fm; $mc^2 = 938$ MeV; $\hbar c = 197$ MeV fm

9. The maximal kinetic energy of positrons being emitted in the β^+-decay of the nucleus ^{13}N ($Z = 7$) amounts to 1.24 MeV. After the beta-decay, no gammas are emitted. Calculate, starting from this data, the nuclear radius of the atomic nucleus with mass $A = 13$. Compare the so-derived value with the radius

calculated as the root mean square for a nucleus with $A = 13$ and constant charge density inside the nucleus.

Data given: we consider $r_0 = 1.2$ fm; $(m_n - m_p)c^2 = 1.29$ MeV and $e^2/(4\pi\epsilon_0) = 1.44$ MeV fm

10. We give the following binding energies for a number of nuclei (in units MeV)

^{14}N	104.6598	^{14}O	98.7325
^{15}N	115.4932	^{15}O	111.9569
^{16}N	117.9845	^{16}O	127.6207
^{17}N	123.8680	^{17}O	131.7650

Determine the proton and neutron separation energy in ^{16}O and in ^{17}O. What does this tell us about the properties of nuclear forces acting inside nuclei. Also bring in arguments as to which terms in the semi-empirical mass formula could be mainly responsible for the observed differences in the proton and neutron separation energy.

11.

(a) Show that the Q-value of any given nuclear reaction can be written as

$$Q = \sum \text{binding energies (final nuclei)}$$
$$- \sum \text{binding energies (initial nuclei)}.$$

(b) From which value of A can alpha-decay start to proceed in a spontaneous way? Data given: $a_V = 14$ MeV; $a_S = 13$ MeV; $a_C = 0.6$ MeV; $a_A = 19$ MeV; binding energy of the alpha-particle $= 28.3$ MeV.
Study the behaviour of $Q(A)$ in a graphical way.

(c) Under which conditions, will a given nucleus be stable against both β^-, β^+ as well as electron K-capture.

12. Show that the binding energy (per nucleon) of atomic nuclei close to the line of beta-stability can be written as

$$BE(A, Z)/A = BE \text{ (stability line)} + g \cdot (Z - Z_0)^2,$$

with Z_0 the value of Z at the stability line. In deriving this expression we have made the following approximations: replace $Z(Z-1) \rightarrow Z$; neglect the terms $(m_n - m_p)c^2 \cdot A/8a_A$ and $a_C A^{2/3}/8a_A$ in the expression for Z_0.

Determine g in as compact a form as possible. Also determine the binding energy per nucleon for nuclei that are positioned exactly on the line of beta-stability. Study the latter quantity in a graphical way and determine the maximal value starting from this graphical analysis. Data given: $a_V = 14$ MeV; $a_S = 13$ MeV; $a_C = 0.6$ MeV; $a_A = 19$ MeV.

13. Determine the particular value of Z relative to the value of Z_0 (that value of Z_0 which exactly falls on the line of beta-stability) at which a neutron is no longer bound (reaching the neutron drip-line), i.e.

$$Ae(A) \leq (A - 1)e(A - 1),$$

with $e(A) = BE(Z, N)/A$ the binding energy per nucleon.

Hint: make use of the expression of the binding energy (a parabola in $(Z - Z_0)^2$) as derived in problem 12 in which Z_0 is that value of Z that occurs at the beta-stability line itself. Also assume that one can approximately use $e_0(A) \simeq e_0(A - 1)$ and that $Z_0 \gg |Z - Z_0| \gg 1$. Also replace $Z(Z - 1) \rightarrow Z^2$ in the Coulomb energy expression.

14. Discuss and evaluate recoil energies as they appear in α-, β^-- and γ-decay.

Calculate and compare the order of magnitude for each process for an atomic nucleus with $A = 40$, $Z = 20$ and $Q_\alpha = Q_{\beta^-} = Q_\gamma = 5$ MeV. In which situation is the best approximation obtained by neglecting this recoil energy?

15. If we approximate the average nuclear potential by a square well potential

$$
\begin{aligned}
U(r) &= -U_0 & 0 \leq r \leq R_0 \\
U(r) &= \infty & r \geq R_0,
\end{aligned}
$$

which has finite depth, determine the single-particle spectrum as characterized by the quantum numbers (n, l).

Also determine both the individual and cumulative occupation of the single-particle energy spectrum. How has this to be modified in order to account for the observed stable configurations $2, 8, 20, 28, 50, 82, \ldots$?

Data given: the lowest roots of the spherical Bessel functions are given in table PC.15 and figure PC.15.

16. The three-dimensional harmonic oscillator can be studied using either Cartesian coordinates (solutions are products of Hermite functions) or spherical coordinates (Laguerre functions).

Deduce for the $N = 0$ and $N = 1$ harmonic oscillator quantum number, relations that connect the radial wavefunctions as obtained in the different bases.

17. The Hartree–Fock equations can also be derived from a variational principle for the ground-state energy

$$E = \langle \bar{\psi} | H | \bar{\psi} \rangle,$$

in which the ground-state (Hartree–Fock) energy become stationary (minimal) against small variations of the single-particle wavefunctions φ_i.

18. Harmonic oscillator wavefunctions are good approximations to the more realistic Woods–Saxon wavefunctions that are derived from a potential

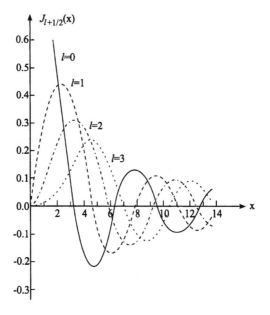

Figure PC.15.

with a finite depth which vanishes at infinity. Discuss, both in a qualitative and quantitative way for which electromagnetic operators (multipole order) describing electric transitions, differences in the radial integrals using harmonic oscillator and Woods–Saxon wavefunctions are expected to be maximal.

19. It has been shown that a strong spin–orbit coupling is necessary in nuclei in order to correctly reproduce the observed single-particle energy spectra. Also, in atomic physics, spin–orbit effects are present, coupling the orbital electron motion to its intrinsic spin (Thomas term).

Discuss the differences (and the origin of this difference) between both the sign and the magnitude of the nuclear and atomic spin–orbit interaction.

20. Determine the spin–orbit splitting energy, starting from general tensor reduction formulae in order to evaluate the corresponding matrix element

$$\langle n(\ell, 1/2) jm | \alpha(r) \hat{\ell} \cdot \hat{s} | n(\ell, 1/2) jm \rangle.$$

Hint: use the reference of de-Shalit and Talmi (1974), appendices.

21. By modifying the spherical harmonic oscillator potential into an axially deformed harmonic oscillator potential, and using the definitions

$$\omega_x = \omega_y = \omega_0(1 + 1/3\varepsilon)$$
$$\omega_z = \omega_0(1 - 2/3\varepsilon).$$

Table PC.15.

n	$l = 0, 1/2$	$l = 1, 3/2$	$l = 2, 5/2$	$l = 3, 7/2$
		$J_{l+1/2}(kr)$		
1	3.141 593	4.493 409	5.763 459	6.987 932
2	6.283 185	7.725 252	9.095 011	10.417 119
3	9.424 778	10.904 122	12.322 941	13.698 023
4	12.566 370	14.066 194	15.514 603	16.923 621
5	15.707 963	17.220 755	18.689 036	20.121 806
6	18.849 556	20.371 303	21.853 874	23.304 247
7	21.991 149	23.519 452		

n	$l = 4, 9/2$	$l = 5, 11/2$	$l = 6, 13/2$
		$J_{l+1/2}(kr)$	
1	8.182 561	9.355 812	10.512 835
2	11.704 907	12.966 530	14.207 392
3	15.039 665	16.354 710	17.647 975
4	18.301 256	19.653 152	20.983 463
5	21.525 418	22.904 551	24.262 768
6	24.727 566		

(a) determine the 'diagonal' energy corrections caused by the perturbation which is equal to the difference between the deformed and spherical harmonic oscillator terms

(b) show that now $[\hat{H}, \hat{\ell}_z] = 0$; $[\hat{H}, \hat{\ell}_x] \neq 0$; $[\hat{H}, \hat{\ell}_y] \neq 0$ and that the eigenstates no longer have a 'good' angular momentum value ℓ, or that one obtains the expansion

$$|m\rangle = \sum_m a_{l,m} |l, m\rangle.$$

22. Determine all possible states for a $(1f_{7/2})^2$ configuration, using the m-scheme and considering identical nucleons (two protons or two neutrons). How do these results change if we consider a proton and a neutron configuration to be coupled?

23. Determine all possible two-body matrix elements in the $N = 1$ harmonic oscillator shell-model space, i.e. considering the $1p_{3/2}$, $1p_{1/2}$ shell-model orbitals. Consider identical nucleons first. How will this result change if we consider the possibility of having both protons and neutrons interacting in this $N = 1$ oscillator shell-model space?

24. Determine the energy separation between a singlet $(S = 0)$ and a triplet $(S = 1)$ state of two nucleons with their individual intrinsic spins coupled, making

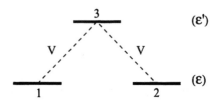

Figure PC.29.

use of a specific spin-dependent two-body interaction

$$V = -V_0(a + b\vec{\sigma}_1 \cdot \vec{\sigma}_2).$$

25. Show that the antisymmetry of the two-nucleon wavefunctions (using a harmonic oscillator model) implies that $T + S + \ell = $ odd (where ℓ is the relative orbital angular momentum).

How would this condition change for arbitrary (more general and to be expanded in a harmonic oscillator basis) single-particle wavefunctions?

26.   Determine, in detail, the (T, T_z) dependence for an isoscalar, isovector and isotensor (rank 2) energy term $E(T, T_z)$ and this for a general Hamiltonian containing kinetic energy (with slightly different proton and neutron masses), charge independent nucleon–nucleon strong interactions and a Coulomb contribution.

27. Determine the precise form of the Coulomb interaction as occurring in the nuclear many-body Hamiltonian describing A nucleons of which Z are protons, i.e. determine the precise tensorial character (tensor properties in isospin space).

28. Show that the kinetic energy term for an interacting nucleon system containing Z protons and N neutrons gives rise to a purely isovector (tensor of rank 1) interaction.

29. Study the following three-level problem where two of the states are degenerate in energy, at an energy ε, interacting with a third state at energy ε' (with $\varepsilon \neq \varepsilon'$) and with strength V. States 1 and 2 have no direct interaction (see figure PC.29).

Study the variation for changing energy difference $\varepsilon' - \varepsilon$ (numerical study).

PART D

NUCLEAR STRUCTURE: RECENT DEVELOPMENTS

Upon starting the study of the nuclear many-body system, it is a good idea to ask how we might best describe such a system. Should we consider the nucleus as a system where the single-particle degrees of freedom dominate and determine the observed nuclear structure phenomena? Can we consider the nucleus approximately as a charged liquid drop that can carry out collective vibrations and rotations? Is it, after all, more correct to speak of a collection of nucleons and nucleon resonances in a pion field or do we have to go down to the deepest level of quarks interacting with gluon fields in order to describe and understand nuclear phenomena?

Maybe the 'truth' is that the way in which the nucleus can be observed and described depends very much on the nature and energy of the probe with which we are observing and also on the way that the nucleus can show all facets, as was depicted in figure 1.1 of chapter 1. This multi-faceted aspect of the nucleus will become clear when we discuss in more detail the microscopic self-consistent and purely collective approach of describing low-lying phenomena in the nucleus.

One of the most surprising features in the behaviour of the nuclear many-body system is the existence of a large amount of orderly motion in spite of strong nucleon–nucleon interactions in the nucleus. There exists a rather large nuclear mean-free path within the nucleus and the interacting nucleon system can, in first order, be approximated by a weakly interacting system governed by a nuclear one-body mean field (chapter 10). Mean-field concepts and the related characteristics are realized to a large extent in medium-heavy and heavy nuclei. The nuclear shell model (chapter 11) proves that nucleon correlations of a limited number of protons and neutrons outside of closed-shell configurations describe a large variety of nuclear excited state properties. This is borne out most recently by many results of large-scale shell-model studies throughout the whole nuclear mass table with a variety of techniques. We also describe a number of statistical properties and discuss how the study of nuclear level distributions and level densities can give deep insight in the properties that characterize the nucleonic motion and their interactions (chapter 11). These methods leave out, to a large extent, the study of the binding in very light systems (up to mass $A \approx 10$). In recent years, *ab initio* calculations have given a good insight into the structure of these systems, starting from basic nucleon forces that exhibit the necessity for three-nucleon terms (see chapter 12). We then discuss collective properties using both geometric (using in particular the Bohr–Mottelson model with nuclear shape variables) as well as algebraic models,using symmetry concepts, the latter particularly clearly expressed using the interacting boson model (IBM) approach (chapter 13). The properties of nuclear rotation, extending even to very high angular momentum as well as the study of extreme forms of nuclear deformed shapes, represented by super- and hyper-deformation, are discussed in chapter 14.

More recently, new techniques have allowed the exploration of nuclei very far away from the region of beta-stability and now one tries to reach the extremes at the proton and neutron drip lines. Here one reaches unexplored ground where no easy extrapolations are possible. Theoretical methods common to general studies of weakly bound quantum systems will have to be invoked in understanding nuclei where almost pure neutron matter is reached or where very heavy $N = Z$ nuclei occur. We shall discuss both the experimental methods and glimpses of new physics in chapter 15. In chapter 16 we extend the discussion beyond the pure 'hadronic' proton and neutron regime by indicating the existence of a number of data in the sub-nucleonic regime (Δ-resonance, pions and mesons in the nucleus etc) that point out a connecting element towards the study of higher energy facets. We discuss recent endeavours in the search for these nuclear structure phenomena that are situated at the borderline between more traditional nuclear physics studies and the intermediate and high energy regimes. In a final epilogue (chapter 17) we underline the importance of investment in technical developments: we point out that any time new experimental methods have been extensively used to probe the atomic nucleus, unexpected physics phenomena have appeared that were most often absent in the most advanced theoretical extrapolations.

Chapter 10

The nuclear mean-field: single-particle excitations and global nuclear properties

10.1 Hartree–Fock theory: a variational approach

In many theoretical approaches, the nuclear mean field is described by means of the Hartree–Fock variational approximation. An effective interaction is introduced and the average potential is identified with the first-order contribution to the perturbation expansion of the mean field in powers of the strength of this effective interaction. The forces are determined such that a number of global nuclear properties (binding energy, nuclear charge and mass radii, density distributions, etc) are reproduced in lowest order. This Hartree–Fock approach, although clearly having its limitations, has been shown to be a very appropriate, lowest-order theory of an interacting A-body nuclear problem.

In this Hartree–Fock approach, a distinction can be made between (i) weakly-bound states, (ii) deeply-bound states as well as (iii) positive energy scattering states. A recent method to obtain both bound and scattering states from a general, optical potential, has been proposed by Mahaux and Sartor and is discussed, at length, by Mahaux (1985) (see also figure 10.1).

The starting point of Hartree–Fock theory (HF) (see also section 9.6.1 for a more intuitive discussion) is the description of the many-body wavefunction as an antisymmetrized product wavefunction

$$\Psi(1, 2, \ldots, A) = \mathcal{A}\varphi_{\alpha_1}(1)\varphi_{\alpha_2}(2)\ldots\varphi_{\alpha_A}(A), \tag{10.1}$$

where

$$\mathcal{A} \equiv \sum_P (-1)^P \hat{P},$$

with $\hat{P}$, the permutation operator, permutating the A nucleon coordinates and $(-1)^P = \pm 1$ for an even or odd permutation of the indices, $1, 2, \ldots, A$. Here, moreover $1, 2, \ldots$ is a shorthand notation for *all* coordinates describing the

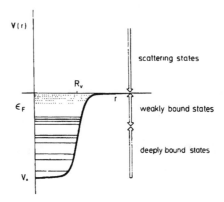

Figure 10.1. Sketch of the three important energy domains. The Fermi energy is denoted by ϵ_F. The shaded region corresponds to the occupied states in the independent particle model description of the nuclear ground state (taken from Mahaux *et al* 1985).

nucleon and $\alpha_1, \alpha_2, \ldots$ denotes *all* quantum numbers needed to characterize the basis wavefunctions. The 'best' wavefunction is determined through the variational expression

$$\delta \langle \Psi | \hat{H} | \Psi \rangle = \langle \delta \Psi | \hat{H} | \Psi \rangle = 0, \qquad (10.2)$$

for variations $\delta \Psi$ that conserve the normalization of the single-particle wavefunction i.e.

$$\int |\varphi_{\alpha_i}(\vec{r})|^2 \, d\vec{r} = 1. \qquad (10.3)$$

Starting from the Hamiltonian

$$\hat{H} = \sum_{i=1}^{A} \frac{\hat{p}_i^2}{2m_i} + \tfrac{1}{2} \sum_{i<j=1}^{A} V(\vec{r}_i, \vec{r}_j), \qquad (10.4)$$

the expectation value $\langle \Psi | \hat{H} | \Psi \rangle$ becomes

$$\langle \Psi | \hat{H} | \Psi \rangle = -\frac{\hbar^2}{2m} \sum_{i=1}^{A} \int \varphi_{\alpha_i}^*(\vec{r}) \Delta \varphi_{\alpha_i}(\vec{r}) \, d\vec{r}$$

$$+ \sum_{i<j=1}^{A} \int \varphi_{\alpha_i}^*(\vec{r}) \varphi_{\alpha_j}^*(\vec{r}') V(\vec{r}, \vec{r}') \varphi_{\alpha_i}(\vec{r}) \varphi_{\alpha_j}(\vec{r}') \, d\vec{r} \, d\vec{r}'$$

$$- \sum_{i<j=1}^{A} \int \varphi_{\alpha_i}^*(\vec{r}) \varphi_{\alpha_j}^*(\vec{r}') V(\vec{r}, \vec{r}') \varphi_{\alpha_i}(\vec{r}') \varphi_{\alpha_j}(\vec{r}) \, d\vec{r} \, d\vec{r}'. \quad (10.5)$$

Applying the variation on the $\varphi^*_{\alpha_i}(\vec{r})$, we obtain the single-particle Schrödinger equation

$$
-\frac{\hbar^2}{2m}\Delta\varphi_{\alpha_i}(\vec{r}) + \sum_{j=1}^{A}\int d\vec{r}'\, \varphi^*_{\alpha_j}(\vec{r}')V(\vec{r},\vec{r}')\varphi_{\alpha_j}(\vec{r}')\varphi_{\alpha_i}(\vec{r})
$$

$$
-\sum_{j=1}^{A}\int d\vec{r}'\, \varphi^*_{\alpha_j}(\vec{r}')V(\vec{r},\vec{r}')\varphi_{\alpha_j}(\vec{r})\varphi_{\alpha_i}(\vec{r}') = \varepsilon_{\alpha_i}\varphi_{\alpha_i}(\vec{r}), \qquad (10.6)
$$

where ε_{α_i} is the Lagrange multiplier that takes into account the single-particle wavefunction normalization. It obtains the meaning of a single-particle energy. This is apparent when restating (10.6) in the condensed form

$$
-\frac{\hbar^2}{2m}\Delta\varphi_{\alpha_i}(\vec{r}) + \int d\vec{r}'\, U(\vec{r},\vec{r}')\varphi_{\alpha_j}(\vec{r}') = \varepsilon_{\alpha_i}\varphi_{\alpha_i}(\vec{r}), \qquad (10.7)
$$

where $U(\vec{r},\vec{r}')$ describes the self-consistent field, as

$$
U(\vec{r},\vec{r}') = \delta(\vec{r}-\vec{r}')\sum_{j=1}^{A}\int d\vec{r}''\, V(\vec{r},\vec{r}'')\varphi_{\alpha_j}(\vec{r}'')\varphi^*_{\alpha_j}(\vec{r}'')
$$

$$
-\sum_{j=1}^{A}V(\vec{r},\vec{r}')\varphi_{\alpha_j}(\vec{r})\varphi^*_{\alpha_j}(\vec{r}'). \qquad (10.8)
$$

The first term is the local (direct) term corresponding to the Hartree field. The second describes the exchange term and is non-local, related directly to the two-body force $V(\vec{r},\vec{r}')$.

It could be that the two-body interaction has a more general form, i.e. it depends not only on the particle coordinates but also on their relative momenta. Then, more general methods have to be used (see Ring and Schuck 1980 and Rowe 1970).

We can introduce also the density which, for the simple product wavefunction of (10.1) becomes

$$
\rho(\vec{r}',\vec{r}) = \sum_{j=1}^{A}\varphi_{\alpha_j}(\vec{r}')\varphi^*_{\alpha_j}(\vec{r}), \qquad (10.9)
$$

with the special property that

$$
\int d\vec{r}''\, \rho(\vec{r}',\vec{r}'')\rho(\vec{r}'',\vec{r}) = \rho(\vec{r}',\vec{r}), \qquad (10.10)
$$

which also reads, in a formal way, as $\rho^2 = \rho$.

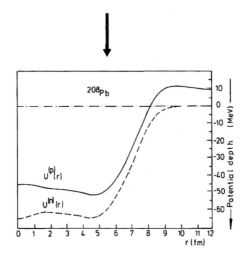

Figure 10.2. Potentials $U^{(p)}(r)$ and $U^{(n)}(r)$ in ^{208}Pb, obtained by starting from an effective Skyrme pre-parameterization (taken from Waroquier *et al* 1983).

The practical solution to the Hartree–Fock equations (10.7) with the self-consistent field given by equation (10.8) is then obtained in an iterative way as discussed in chapter 9.

In Box 10a, we discuss a force that has been used quite successfully in many recently performed Hartree–Fock calculations, which use two-body forces $V(\vec{r}_i, \vec{r}_j)$ with a zero-range structure that contains, moreover, a density dependence to which zero-range three-body terms are added. These Skyrme forces (Heyde 1991) were first used in numerical calculations by Vautherin and Brink (1972).

The local central part, determined using these SkE2 force parametrization for both protons and neutrons (Waroquier 1983) for the heavy doubly-closed shells nucleus ^{208}Pb, are illustrated in figure 10.2. The form is very much reminiscent of the phenomenological Woods–Saxon shape as discussed in chapter 9, but now the precise shape includes more details of the residual two-body interaction, including many-body effects through the density dependence $\rho((\vec{r}_1 + \vec{r}_2)/2)$ and the three-body contributions as discussed in Box 10a.

In the next section we discuss a number of observables that are used, mainly to determine the effective two-body interaction for which the major constraint is to give saturation properties in the nuclear interior at the correct equilibrium density. A number of 'global' properties so described are briefly discussed.

10.2   Hartree–Fock ground-state properties

Starting from the product Hartree–Fock wavefunction in the form of equation (10.1)

$$\Psi_{HF}(1, 2, \ldots, A) = \mathcal{A} \prod_{i=1}^{A} \varphi_{\alpha_i}^{HF}(i), \qquad (10.11)$$

a number of ground-state properties can be evaluated in a straightforward way such as

$$\langle \Psi_{HF} | \hat{H} | \Psi_{HF} \rangle = E_0,$$

$$\langle \Psi_{HF} | \sum_{i=1}^{A} \hat{r}_i^2 | \Psi_{HF} \rangle = \langle r^2 \rangle (\pi, \nu), \qquad (10.12)$$

$$\langle \Psi_{HF} | \sum_{i=1}^{A} \hat{\rho}(\vec{r}_i) | \Psi_{HF} \rangle = \rho(\vec{r})(\pi, \nu),$$

for the total binding energy, the nuclear radii for protons (π) and neutrons (ν) and more detailed density distributions.

In tables 10.1 and 10.2, we give the parametrizations for a number of Skyrme forces; SkE2, SkE4, SkIII—the original Skyrme parametrization and, correspondingly, the results of relative binding energy E_0/A, proton and neutron point rms radii r_p, r_n (fm) and charge rms radii r_c (fm). We also present a number of calculations. The single-particle energy spectra for ^{90}Zr are given in figure 10.3. Here, in particular, the large gaps at nucleon numbers 20, 28, 40 (for neutrons) and 50 for both protons and neutrons are clearly observed in this self-consistent calculation. In figures 9.16 and 9.17 of the previous chapter, a number of charge and matter density distributions have been shown for a number of double-closed shell nuclei. A comparison with the experimental data for the charge density distributions is also carried out. In figure 9.17, the various mass density distributions are plotted (for many nuclei) on a single figure to accentuate the strong similarity of the average mass density, neglecting the central density oscillatory behaviour. This figure represents the nuclear saturation property of the mass distributions in a most dramatic way.

The actual nuclear rms radii can also be evaluated as illustrated in figure 10.4, where the variation for the open neutron shell tin nuclei is presented. Here, we show the proton rms radius as a function of the neutron excess. Corrections for finite proton size and centre-of-mass motion result in an almost constant shift to a slightly higher rms value. From $A = 120$ onwards, the neutron $1h_{11/2}$ orbital starts filling, accentuating the spin-orbit contribution which results in a slight lowering in the relative increase in nuclear radius. We can distinguish two slightly different slopes in $\langle r_{charge}^2 \rangle^{1/2}$. In filling the $1g_{7/2}$ and $2d_{5/2}$ neutron orbits, we obtain an isotopic shift coefficient of 0.643, while from $A = 114$ onwards, the slope decreases further to an isotopic shift coefficient of 0.458 which is half of

Table 10.1. A list of SkE2 and SkE4 parameters which yield suitable values of nuclear matter and ground-state quantities for doubly-closed shell nuclei. The original Skyrme interaction SkIII is qiven for comparison (Beiner *et al* 1975, Waroquier *et al* 1983).

	t_0 (MeV fm^3)	t_1 (MeV fm^5)	t_2 (MeV fm^5)	t_3 (MeV fm^6)	x_0	W_0' (MeV fm^5)
SkE2	-1299.30	802.41	-67.89	19 558.96	0.270	120
SkE4	-1263.11	692.55	-83.76	19 058.78	0.358	120
SkIII	-1128.75	395.0	-95.0	14 000.0	0.45	120

	t_4 (MeV fm^8)	K (MeV)	$(E/A)_{n,m}$ (MeV)	k_F (fm^{-1})	m^*/m	a_τ (MeV)
SkE2	$-15\,808.79$	200	-16.0	1.33	0.72	29.7
SkE4	$-12\,258.97$	250	-16.0	1.31	0.75	30.0
SkIII	0.0	356	-15.87	1.29	0.76	28.2

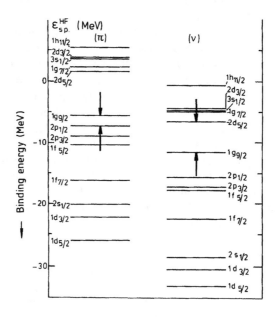

Figure 10.3. Hartree–Fock single-particle spectrum in ^{90}Zr using an effective Skyrme force (taken from Van Neck *et al* 1990).

the usual value of $\gamma = 1$, which represents the average behaviour of increase in nuclear radii. This means an approximate increase according to $A^{1/6}$. This result

Table 10.2. Binding energies per nucleon E/A (MeV), proton and neutron rms radii r_p and r_n (fm) and charge rms radii r_c (fm), corresponding to various Skyrme parametrizations in self-consistent Hartree–Fock calculations (Waroquier *et al* 1983).

	E/A	r_p	r_n	r_c
		^{16}O		
SkE2	−7.92	2.63	2.60	2.68
SkE4	−7.96	2.65	2.62	2.70
SkIII	−8.03	2.64	2.61	2.70
exp	−7.98			2.71
		^{40}Ca		
SkE2	−8.56	3.37	3.31	3.42
SkE4	−8.59	3.40	3.35	3.46
SkIII	−8.57	3.41	3.36	3.46
exp	−8.55		3.36	3.48
		^{48}Ca		
SkE2	−8.63	3.39	3.56	3.44
SkE4	−8.65	3.43	3.59	3.47
SkIII	−8.69	3.46	3.60	3.50
exp	−8.67		3.54	3.48
		^{90}Zr		
SkE2	−8.67	4.17	4.24	4.21
SkE4	−8.71	4.22	4.29	4.26
SkIII	−8.69	4.26	4.31	4.30
exp	−8.71			4.27
		^{132}Sn		
SkE2	−8.36	4.62	4.84	4.66
SkE4	−8.36	4.68	4.89	4.71
SkIII	−8.36	4.73	4.90	4.78
exp	−8.36			
		^{208}Pb		
SkE2	−7.87	5.41	5.57	5.45
SkE4	−7.87	5.47	5.62	5.50
SkIII	−7.87	5.52	5.64	5.56
exp	−7.87			5.50

is in sharp contrast to the dependence on N of the neutron rms, which increases faster than $A^{1/3}$, especially in the region of the light tin isotopes.

In calcium nuclei, which have an identical number of protons, the variation in the charge rms radius presents an interesting challenge to the existing mean-field (HF) calculations, since the charge radius in ^{48}Ca is slightly smaller than in ^{40}Ca, the spherical droplet model, indicates an increase of $\langle r^2 \rangle_{\text{charge}} \propto A^{2/3}$ that is totally at variance with the data points (see figure 10.5(a)). The

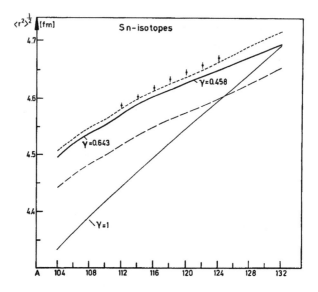

Figure 10.4. Proton rms radii in the tin isotopes. The long dashed line represents the proton point radii. The short dashed line includes the corrections for finite proton sizes and COM motion. After taking into account the electromagnetic contributions, one can derive the final charge rms, represented by the thick full line (taken from Waroquier *et al* 1979) (see text for the explanation of the γ parameter).

observed parabolic variation with N (with maximal value at the $1f_{7/2}$ mid-shell configuration at $N = 24$) cannot be explained using HF calculations only. In figure 10.5(*b*) we present the charge rms radii for the even–even calcium nuclei and only a very smooth variation with a slight increase towards ^{48}Ca is predicted. In searching for possible explanations, ground-state correlations can make up for this specific N-dependence. RPA calculations by Broglia and Barranco (1985), which include monopole, quadrupole, octupole admixtures in the 0^+ ground state can make up for the global behaviour. An excellent fit that is able to produce the odd–even staggering effect in calcium nuclei is given by explicitly treating the residual proton–neutron interaction (Talmi 1984), that uses the dependence of $\langle r^2 \rangle_{\text{charge}}$ where

$$\langle r^2 \rangle_n = \varepsilon n + \alpha n(n - 1)/2 + \beta[n/2], \qquad (10.13)$$

and $[u/2]$ denotes the largest integer not exceeding $n/2$.

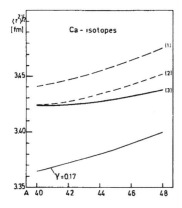
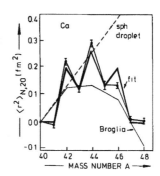

Figure 10.5. (*a*) Proton point rms radii. The full line gives the proton point rms radius. The dashed line (1) incorporates finite proton size and COM corrections, (2) includes electromagnetic neutron effects, while (3) represents the final charge rms radii (taken from Waroquier *et al*, 1979). (*b*) Comparison of measured calcium radii with the spherical droplet RPA calculations of Barranco and Broglia (1985) and a fit using the shell-model formula (10.13) as derived by Talmi (1984).

10.3 Test of single-particle motion in a mean field

10.3.1 Electromagnetic interactions with nucleons

The ideal probe to test the nucleonic motion in the nucleus is the electromagnetic interaction. This can be carried out using electron scattering off the nucleus. The electron now acts as a microscope with which to study the nucleons moving inside the nucleus. By changing the energy ($\hbar\omega$) and momentum ($\hbar\vec{q}$) of the electron, various details ranging from nuclear collective surface excitations to details of the nuclear motion can be scrutinized. In figure 10.6, we illustrate in a schematic way various dynamical processes. At the point $\hbar\omega = 0$, elastic scattering takes place and the nucleus remains in its ground state. The nuclear current consists of contributions from moving protons as well as from the nuclear magnetism, caused by the motion of intrinsic magnetic moments associated with the protons and neutrons. Charge and current distributions can be separated under certain kinematical conditions that characterize the scattering process. In the region $0 \leq \hbar\omega \leq 20$ MeV, a large number of resonances in the nuclear many-body system appear. In the region 30 MeV $\leq \hbar\omega \leq$ 150 MeV, one-nucleon emission occurs as the dominant process. This is a most interesting region, since by measuring simultaneously the energies and momenta of outgoing nucleons and scattered electrons (e'), information about the velocity distribution of a nucleon *in* a nucleus prior to the interaction can be obtained.

Figure 10.7 shows a schematic representation of the above process. We shall now adopt the most simple reaction mechanism which can be thought of: the

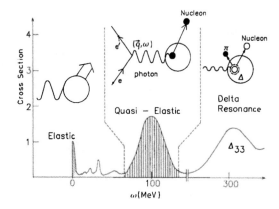

Figure 10.6. Schematic representation of the cross-section for electron scattering off the nucleus, as a function of the energy transfer $\hbar\omega$ (in MeV). A number of energy regions are indicated, each characterized by its own specific scattering mechanism as illustrated (taken from De Vries *et al* 1983).

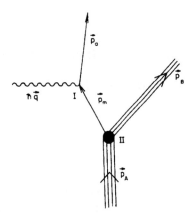

Figure 10.7.   One-photon exchange diagram for an electromagnetically induced one-particle emission process in the plane-wave impulse approximation (PWIA) (taken from Ryckebusch 1988).

detected nucleon is ejected in a one-step reaction mechanism, in which all other nucleons remain as spectators. Moreover, we adopt a plane wave description for the emitted particle. This approach is also called the 'quasi-free' approximation. Imposing momentum conservation on the vertices I and II of figure 10.7 one obtains in the laboratory system, with $|\vec{p}_A| = 0$,

$$\vec{p}_m = \vec{p}_a - \hbar\vec{q} = -\vec{p}_B. \tag{10.14}$$

The momentum $\vec{p}_m$ is customarily referred to as the missing momentum. In the quasi-free approximation, this missing momentum just represents the momentum of the detected nucleon just before it undergoes the interaction with the external electromagnetic field, in which it absorbs a momentum $\hbar\vec{q}$.

A nucleon moving in a nucleus with A interacting nucleons will be characterized by a probability distribution for its velocity. In the independent-particle model (IPM), it is useful to introduce the momentum distribution $\rho_a(\vec{p})$ for a given single-particle state, characterized by $a \equiv \{n_a, l_a, j_a; \varepsilon_a\}$. The function $\rho_a(\vec{p})$ then gives the probability to find a nucleon with momentum $\vec{p}$ in the particular state a, which becomes

$$\rho_a(\vec{p}) = \sum_{m_a} |\varphi_{a,m_a}(\vec{p})|^2 v_a^2 = \sum_{m_a} \left(\frac{1}{(2\pi\hbar)^{3/2}} \int d\vec{r} \, e^{(i\vec{p}\cdot\vec{r}/\hbar)} \varphi_{a,m_a}(\vec{r}) \right)^2 v_a^2.$$

(10.15)

Here, v_a^2 gives the occupation probability of the single-particle orbital a in the ground state. Note furthermore that

$$\int d\vec{p} \, \rho_a(\vec{p}) = (2j_a + 1)v_a^2.$$

(10.16)

Inserting the precise structure of the single-particle wavefunctions, as discussed by Ryckebusch (1988), in (10.15), one obtains the result

$$\rho_a(\vec{p}) = \frac{1}{2\pi^2\hbar^3} \left[\int dr \, r^2 j_{l_a}(pr/\hbar) \varphi_{n_a l_a j_a}(r) \right]^2 v_a^2(2j_a + 1),$$

(10.17)

with $j_{l_a}(pr/\hbar)$ the spherical Bessel function. So, determining the momentum distribution $\rho_a(\vec{p})$ immediately gives access to the nuclear single-particle wavefunctions by means of a Fourier–Bessel transformation. This momentum distribution for the case of the $3s_{1/2}$ orbital in ^{208}Pb can be plotted such that one observes a probability distribution for the nucleon moving with a certain fraction of the speed of light (see figure 10.8).

10.3.2 Hartree–Fock description of one-nucleon emission

If we intend to go beyond the most simple 'quasi-free' approach, one can start from the Hartree–Fock determination of both the bound and scattering states of the nucleon that will be ejected in the $(e, e'p)$ process. One uses a SkE force to determine *all* ingredients needed to evaluate the nuclear response to the external electromagnetic field. The full electromagnetic interaction process is considered where also the nuclear correlations (RPA correlations) inside the nucleus as well as the final-state interactions (FSI) between the outgoing nucleon and the remaining nucleus $(A - 1)$ are taken into account (figure 10.9). The original 'quasi-free' knockout is depicted in figure 10.10(a). That picture changes considerably when the RPA nuclear correlations are also taken into account.

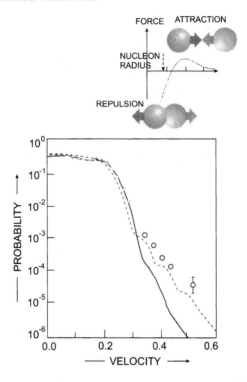

Figure 10.8. Probability distribution for a $3s_{1/2}$ proton orbital in ^{208}Pb. The full curve corresponds to a pure mean field; the broken curve contains short-range correlations (data points). The insert points out the need for a short-range repulsive part in the nucleon–nucleon forces. (Reprinted from Lapikas C and van der Steenhoven G 1996 *Nucl. Phys. News* **6** no 2, © 1996 Gordon and Breach, with permission.)

Initially, we have a similar simple picture: a nucleon becomes excited from a bound into a continuum state. The excited particle is, however, now interacting with the other nucleons, not only via the average HF field but also through the residual interaction. This is sketched in part (*b*) of figure 10.10. The residual interaction represents a means to exchange energy *and* momentum between two nucleons. More complicated processes can even be thought of. This picture finally leads to the concept of 'doorway' states, lasting long enough for a nucleon to be sent out, and the daughter ($A - 1$) nucleus to form in a definite, final state. Because of the residual interactions act in the RPA processes, the emitted nucleon is not necessarily the same as that nucleon on which the initial electromagnetic interaction took place. In this sense, the RPA picture allows a number of multi-step processes, which is in contradiction with the IPM and 'quasi-free' knockout description.

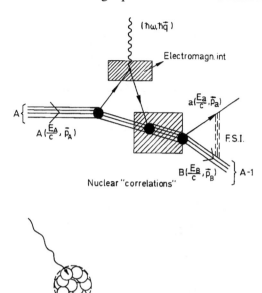

Figure 10.9. Electromagnetic interaction, exchanging energy $\hbar\omega$ and momentum $\hbar\vec{q}$ with a nucleus consisting of A nucleons (with energy E_A/c and momentum $\vec{p}_A$). In the above process, a nucleon is emitted from the nucleus (with energy E_a/c, momentum $\vec{p}_a$) leaving an $A-1$ system (energy E_B/c, momentum $\vec{p}_B$). The notation FSI denotes eventual final-state interactions between the emitted nucleon and the remaining $A-1$ nucleons).

Using the above formulation for the response of the nucleus to the electromagnetic field, cross-sections for (e,'p), (e, e'n), $(\gamma, n), \dots$ have been calculated. The large effects the RPA correlations cause in some situations are illustrated in figure 10.11 for the $^{16}O(\gamma, p_0)$ and $^{16}O(\gamma, n_1)$ reaction at $E_\gamma = 60$ MeV. The dashed line corresponds to the HF–SkE2 calculation, the solid line takes into account RPA correlations, again using the SkE2 force.

In the above discussion, it has been shown that the Hartree–Fock approximation which includes RPA residual interactions are a good description of processes where a nucleon is ejected from the nucleus. The RPA correlations, in particular, point out that one has to go beyond a single-step knockout independent-particle model description of these processes. The large amount of existing data on (γ, p), (γ, n) and $(e,e'p)$ reactions have led to a better understanding of how nucleons move in the nuclear medium.

10.3.3 Deep-lying single-hole states—fragmentation of single-hole strength

In the nuclear mean-field theory, one assumes that the description of a nucleus with closed proton and neutron shells can start with a single Slater determinant,

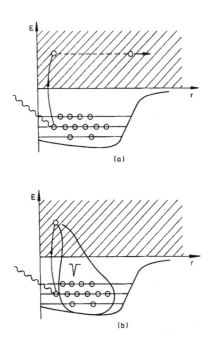

Figure 10.10. (*a*) Schematic representation of a one-nucleon emission process within an independent-particle model (IPM) and (*b*) an independent-particle model, which includes RPA correlations, represented by the 'balloon' labelled with the interaction *V* (taken from Ryckebusch 1988).

built up by single-particle states, $\varphi_{\alpha_i}(\vec{r}_i)$, that are the eigenstates of some mean field.

For such a picture, as was shown in sections 10.3.1 and 10.3.2, the removal of a nucleon from an occupied state with spherical quantum numbers n_a, l_a, j_a, can only happen with an energy transfer equal to the binding energy of that hole state. The spectral function thus reduces to a δ-peak at an energy corresponding to the single-particle energy.

The single-particle strength is now spread out by configuration mixing caused by the residual interaction. The mixing occurs between the pure hole states (1h state) and more complicated configurations such as 2h–1p, 3h–2p, components. This process is pictorially described in figure 10.12(*a*) where the high density of close-lying complicated configurations in the vicinity of a single-hole configuration is included within the dashed line region. The residual interaction will cause the single-hole strength to be fragmented in an important way. A simple and yet microscopic way to incorporate such coupling is by including effects of particle-vibration coupling. A diagrammatic representation of the Dyson equation solve the one-particle Green function, is

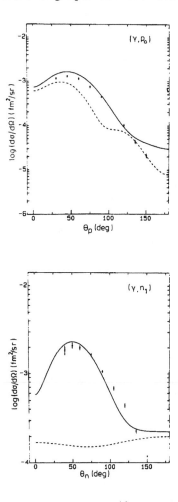

Figure 10.11. Angular distributions for the $^{16}O(\gamma, p_0)$ and (γ, n_1) reaction at $E_\gamma = 60$ MeV in a Hartree–Fock calculation using an effective Skyrme force as residual interaction (dashed line) and RPA with the same effective interaction (solid line). The figure is taken from Ryckebusch (1988) where the data points are also discussed.

given in figure 10.12(*b*). Here, both RPA polarization and correlation corrections are taken into account in a self-consistent way.

Figure 10.13 gives results for the strength distributions of deeply-bound $1d_{3/2}$ and $1d_{5/2}$ proton hole states. The theoretical results are presented using a continuous curve by folding the discrete distribution with a Lorentzian of width $2\Delta = 1$ MeV. The summed strength distribution describes the data points rather

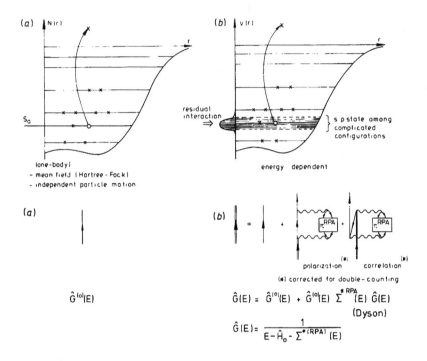

Figure 10.12. Diagrammatic representation of the Dyson equation, including RPA insertions, which describes the polarization propagator in the mass operator Σ^* (RPA). Both the free propagator (*a*) and polarization corrected propagator (*b*) are given (lower part). Also a pictorial representation of the way a nucleon is removed from its orbit in the nuclear potential well, is indicated. In (*a*) a single-particle knockout is shown whereas in (*b*) interactions with a region of more complex configurations are shown (upper part).

well in view of the complicated processes that intervene in the calculation of these hole strength distributions.

An iterative process has been built in order to handle the coupling of deep-hole states in the complex background of 2h–1p configurations using Green function techniques. The convergence of such coupling processes is depicted in figure 10.14 (Van Neck 1991). These figures, after convergence, very much resemble the observed strength distributions obtained after analysing one-nucleon knockout reactions as described by the extensive (e, e′p) data taken by the Amsterdam group (see Box 10b).

10.4 Conclusion

In the present chapter, we have presented the Hartree–Fock approximation for an interacting, non-relativistic A-body system. Starting from a variational principle,

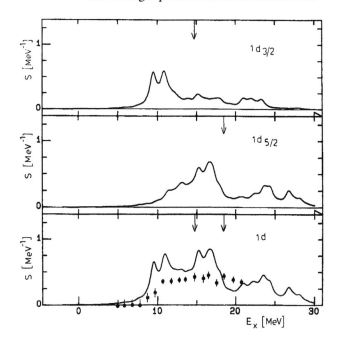

Figure 10.13. Separate strength distributions of the deeply-bound proton-hole states $1d_{3/2}$ and $1d_{5/2}$. The theoretical results are presented by a continuous curve by folding the discrete distribution with a Lorentzian of width $2\Delta = 1$ MeV (taken from Van Neck *et al* 1990).

an optimal basis and a mean-field are derived for an A-particle state's determinant wavefunction and for the energy of this A-body system.

Various tests, relating to a description of nuclear, global properties (binding energy, radii, mass- and charge-density distributions) are used to obtain effective nucleon–nucleon interactions that both saturate at the correct nuclear density and give the correct binding energy. The use of zero-range density-dependent forces (Skyrme forces and extended Skyrme forces) leads to a class of interactions which reproduce a large amount of data. We, moreover, discuss the one-nucleon knockout processes that are initiated through electromagnetic interactions with nucleons in the nucleus. Using both the quasi-free one-step nucleon knockout and more detailed Hartree–Fock and RPA calculations, information on the way that nucleons move in the nuclear interior is extracted. The large number of data on (γ, n), (γ, p) and $(e, e'p)$ reactions all point towards a basic shell-model picture, at least for those orbitals that are situated rather close to the Fermi level. A very interesting article on the nuclear mean-field theory, covering a broad range of applications ranging from the calculation of ground-state properties to properties of nuclear matter in neutron stars and concerning the dynamics of heavy-ion

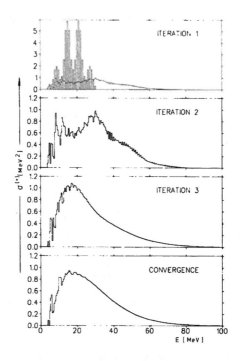

Figure 10.14. Forward part of the self-energy strength distribution in a schematic model, described in Van Neck *et al* (1991). From top to bottom we show the first, second, third and final iteration. In the top part of the figure, the second iteration result (lowest curve) is also drawn, in order to illustrate the different scaling.

collisions and spontaneous fission, has been written by Negele (1985) and serves as a good, general reference on the Hartree–Fock and mean-field description of nuclear dynamics.

Box 10a. Extended Skyrme forces in Hartree–Fock theory

In the extended Skyrme forces used, the two-body part contains an extra zero-range density-dependent term. In the three-body part, velocity-dependent terms are added. This is presented in a schematic way

$$\text{Two-body part: } V(\vec{r}_1, \vec{r}_2) = V^{(0)} + V^{(1)} + V^{(2)} + V^{(ls)} + V_{\text{Coul}} + (1 - x_3)V_0$$

SkE

$$\text{Three-body part: } W(\vec{r}_1, \vec{r}_2, \vec{r}_3) = x_3 W_0(\vec{r}_1, \vec{r}_2, \vec{r}_3) + W_1(\vec{r}_1, \vec{r}_2, \vec{r}_3, \vec{k}_1, \vec{k}_2, \vec{k}_3).$$

Here, $V^{(0)}$, $V^{(1)}$, $V^{(2)}$ and $V^{(ls)}$ have the same structure as in the original Skyrme force parametrization (Vautherin and Brink 1972) as

$$V^{(0)} = t_0(1 + x_0 P_\sigma)\delta(\vec{r}_1 - \vec{r}_2)$$
$$V^{(1)} = \tfrac{1}{2}t_1[\delta(\vec{r}_1 - \vec{r}_2)\vec{k}^2 + \vec{k}'^2\delta(\vec{r}_1 - \vec{r}_2)]$$
$$V^{(2)} = t_2\vec{k}' \cdot \delta(\vec{r}_1 - \vec{r}_2)\vec{k}'$$
$$V^{(ls)} = iW_0'(\vec{\sigma}_1 + \vec{\sigma}_2) \cdot (\vec{k}' \times \delta(\vec{r}_1 - \vec{r}_2)\vec{k}),$$

where $\vec{k}$ denotes the momentum operator, acting to the right

$$\vec{k} = \frac{1}{2i}(\vec{\nabla}_1 - \vec{\nabla}_2),$$

and

$$\vec{k}' = -\frac{1}{2i}(\overleftarrow{\nabla}_1 - \overleftarrow{\nabla}_2),$$

acting to the left. The spin-exchange operator reads

$$P_\sigma \equiv \tfrac{1}{2}(1 + \vec{\sigma}_1 \cdot \vec{\sigma}_2),$$

and the Coulomb force has its standard form. The density-dependent zero-range force V_0 reads

$$V_0 = \tfrac{1}{6}t_3(1 + P_\sigma)\rho((\vec{r}_1 + \vec{r}_2)/2)\delta(\vec{r}_1 - \vec{r}_2).$$

In the three-body part, one has the general term W_0

$$W_0 = t_3\delta(\vec{r}_1 - \vec{r}_2)\delta(\vec{r}_1 - \vec{r}_3),$$

to which a velocity-dependent zero-range term W_1 is added,

$$W_1 = \tfrac{1}{6}t_4[(\vec{k}_{12}'^2 + \vec{k}_{23}'^2 + \vec{k}_{31}'^2)\delta(\vec{r}_1 - \vec{r}_2)\delta(\vec{r}_1 - \vec{r}_3)$$
$$+ \delta(\vec{r}_1 - \vec{r}_2)\delta(\vec{r}_1 - \vec{r}_3)(\vec{k}_{12}^2 + \vec{k}_{23}^2 + \vec{k}_{31}^2)].$$

It can be shown (Waroquier 1983) that both interactions V_0 and W_0 contribute in the same way to the binding energy in even–even nuclei. The parameter x_3 has been retained so as to determine the pairing properties and thus, the properties of excited states near closed shells.

> **Box 10b. Probing how nucleons move inside the nucleus using (e, e′p) reactions**

Using electron scattering off atomic nuclei (see also chapter 15), it is possibile to eject certain particles (proton, neutron) or clusters of particles (pp, pn, nn, d, α, . . .) at the same time as the electron is being scattered. Magnetic spectrometers accurately measure the momentum of the scattered electron and of the nuclear fragments. From such measurements, the microscopic structure of nucleon single-particle motion can be reconstructed.

The cross-section for (e, e′p) reactions factorizes into an elementary electron–proton cross-section σ_{ep} and the spectral function $S(E_m, \vec{p}_m)$, which contains the nuclear structure information and can be written as

$$\frac{d^6\sigma}{de'\, d\Omega_{e'}\, dp'\, d\Omega_{p'}} = K\sigma_{ep}S(E_m, \vec{p}_m),$$

with K a kinematic factor. The spectral function represents the combined probability of finding a nucleon with momentum $\vec{p}_m$ and separation energy E_m in the nucleus. For the independent-particle model with $a \equiv \{n_a, \ell_a, j_a\}$, one can

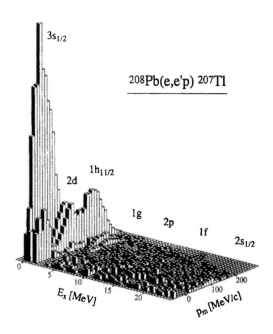

Figure 10b.1. Three-dimensional representation of the spectrum for the reaction ^{208}Pb(e, e′p) ^{207}Tl (taken from Quint 1988).

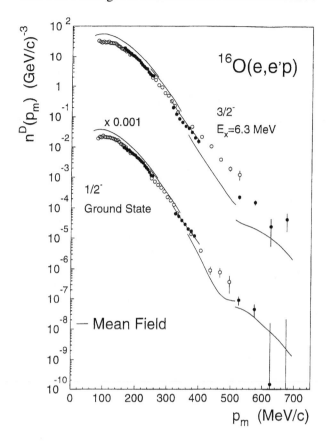

Figure 10b.2. The momentum distributions for the (e, e′p) reaction leading to the ground state of ^{15}N (scaled by 0.001) and to the first excited state at 6.3 MeV. The different data sets (filled and open circles) correspond to two different kinematical settings. The curves are mean-field predictions adjusted to the low-momentum data (reprinted from Leuschner *et al*, © 1994 with permission from Elsevier Science).

show that

$$S(E_m, \vec{p}_m) = \sum_{a(\text{occ})} |\varphi_a(\vec{p}_m)|^2 n_a \delta(E_m - \varepsilon_a),$$

where n_a is the number of protons in the occupied (occ) orbit a and $\varphi_a(\vec{p}_m)$ the single-particle wavefunction in momentum space.

In figure 10b.1 results from the ^{208}Pb(e, e′p) ^{207}Tl reaction are presented as obtained by Quint in 1988 at the 600 MeV Medium Energy Accelerator (MEA) at NIKHEF-K, Amsterdam.

More recently, the Mainz Microtron (MAMI), a continuous wave (cw) 100%-duty-cycle electron accelerator delivering an electron beam of 855 MeV in energy with a current of 100 μA, has produced high-resolution data in a previously inaccessible momentum range for reactions where a proton is knocked out of the nucleus. This electron accelerator as well as the various experimental set-ups, including a three-spectrometer facility, are described in detail by Walcher (1994).

In figure 10b.2, we illustrate the results that have been obtained probing the very high-momentum components of the nucleon single-particle wavefunctions as the nucleon moves inside the nucleus ^{16}O (Blomqvist *et al* 1995). High-momentum components mean, through the Fourier–Bessel transform, knowledge of the extreme short-range characteristics of the nuclear wavefunctions and thus tests with mean-field calculations could be extended into a region that was not accessible before. As one can observe in the figure, the cross-sections determined in this reaction $^{16}O(e, e'p)\,^{15}N$, both leading to the $\frac{1}{2}^-$ ground state at the $\frac{3}{2}^-$ excited state, cover about seven orders of magnitude. These results clearly express the high quality of the data obtained at present at electron accelerator facilities as well as the unique possibilities to probe nucleon motion inside the atomic nucleus for situations in which the nucleon is moving with a very high momentum. This allows us to compare in retrospect experimental results and theoretical concepts for the short-range nuclear wavefunction properties.

Chapter 11

The nuclear shell model: including the residual interactions

11.1 Introduction

Besides the correct reproduction of global properties of the nucleus, a microscopic self-consistent calculation should also aim at a correct description of finer details, i.e. local excitations in the nucleus. This is a highly ambitious task since the typical energy scales (binding energy $E_0 \simeq 10^3$ MeV; excitation energies $E_x \simeq$ 1–2 MeV) differ by three orders of magnitude (see figure 11.1). The calculated binding energy throughout the whole nuclear mass range can be well described by comparison with various mass table as may be seen. Now, we concentrate on whether one is also able to describe *global* and *local* properties using the same original nucleon–nucleon interaction $V(i, j)$. This is depicted in figure 11.2 which interconnects the different aspects of the problem.

One of the problems related to the study of nuclear structure at low excitation energy is the choice of the model space to be used. The starting point should be a realistic interaction, i.e. a potential which reproduces the nucleon–nucleon scattering properties in the energy region 0–500 MeV. We discuss, in section 11.2, a perturbation expansion that defines an 'effective' interaction starting from a more realistic nucleon–nucleon force.

11.2 Effective interaction and operators

It is clear, from the very beginning, that whenever a restricted space of configurations is chosen to describe the nuclear structure at low excitation energy, one will have to introduce the concepts of 'model interaction', 'model space' and 'model operators' which should account for the configuration space left out of the model space.

There are two different approaches to accomplish this goal. In a first approach, one assumes the existence of an effective interaction and determines

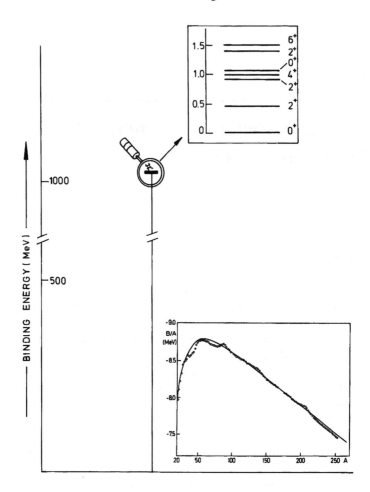

Figure 11.1. Schematic representation of the goal to describe both global (insert at lower right) aspects such as binding energies (in MeV) and local (insert at upper right) aspects such as low-lying excited states in nuclei. The energy scale for both properties differ by three orders of magnitude (taken from Heyde 1991).

this with relatively few parameters: one can either fit the specific two-body matrix elements or a simple form containing a few parameters (strength, range, etc). For a second approach one starts from a realistic force and tries to construct an effective force. This latter method requires many-body perturbation theory of which we shall give a general outline.

Before describing the more formal aspects of this method, we illustrate it with an example for ^{16}O. Having determined the average field in ^{16}O, i.e. the single-particle orbitals $\varphi_{\alpha_i}(\vec{r})$ and energies ε_{α_i}, the ground-state and excited-

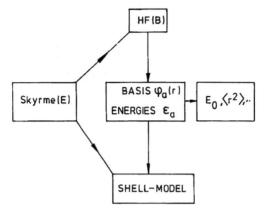

Figure 11.2. Flow diagram expressing the possibilities of carrying out, in a self-consistent way, the calculations of the average one-body field and its properties. At the same time, it allows setting up the basis to be used in more standard shell-model calculations (taken from Heyde, 1989).

state wavefunctions will, in general, be obtained via an expansion in a linear combination of a number of simple configurations. So, one obtains the expansion

$$\Psi = \sum_i a_i \Psi_i \{16 - \text{nucleon coordinates}\}. \tag{11.1}$$

Also, all one-body operators that describe measurable observables (gamma intensities, beta-decay rates, etc) will be written as a sum over the operators related to the 16 individual nucleons, or

$$\hat{O} = \sum_{i=1}^{16} \hat{O}(\vec{r}_i, \vec{\sigma}_i, \vec{\tau}_i). \tag{11.2}$$

This means that even though we have solved for the best independent-particle orbits, according to the HF theory the residual interactions amongst the nucleons imply interactions amongst *all* nucleons in the ^{16}O nucleus, or, more generally, in the A-body system. The only way to obtain a tractable problem is to define a *new* reference state instead of the real vacuum state for the nucleon excitations. So, the A-body problem simplifies into a very restricted number of 'quasi-particle' (particle–hole or more complex combinations) excitations around the Fermi level (figures 11.3(a) and (b)). At the same time, the concepts of 'model wavefunctions, effective or model interaction, effective charges, ...' are introduced to describe nuclear excitations and the related (electromagnetic) transition properties. This is illustrated, for the particular case of ^{17}O where the neutron single-particle picture with one neutron moving outside an inert ^{16}O core is compared to the

(a)

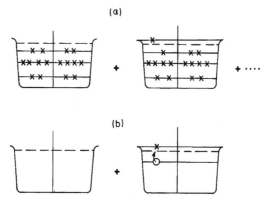

(b)

Figure 11.3. (*a*) The nucleon configurations in the case of ^{16}O, serves as a basis for solving the Schrödinger equation and is depicted relative to the real vacuum configuration. (*b*) The same configurations as in (*a*) but now shown relative to a new vacuum state which corresponds to the configuration of an unperturbed state determinant having 8 protons and 8 neutrons in the lowest Hartree–Fock $1s_{1/2}$, $1p_{3/2}$ and $1p_{1/2}$ orbits (taken from Heyde 1989).

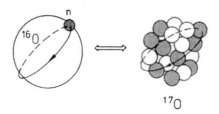

Figure 11.4. Single-particle picture of ^{17}O described by a single neutron outside a ^{16}O core and as a 17 particle picture of ^{17}O. The last neutron is not present a particular place in the latter picture (taken from Heyde 1991).

full 8 proton–9 neutron A particle problem of ^{17}O (figure 11.4). In the remaining part of this section we develop the above discussion in more detail.

In order to obtain the difference between the free n–n interaction V and the effective, model interaction V^{eff}, we consider a system described by the Hamiltonian

$$H = H_0 + V. \tag{11.3}$$

Here, H_0 denotes the unperturbed Hamiltonian with a set of unperturbed wavefunctions $\Psi_i^{(0)}$, such that

$$H_0 \Psi_i^{(0)} = E_i^{(0)} \Psi_i^{(0)} \qquad (i = 1, 2, \ldots). \tag{11.4}$$

For a particular state, the true wavefunction Ψ obeys the full Schrödinger equation;

$$(H_0 + V)\Psi = E\Psi, \tag{11.5}$$

where E represents the full energy of the many-body system.

The standard perturbation expansion then expresses the true wavefunction Ψ in the basis $\Psi_i^{(0)}$ as

$$\Psi = \sum_{i=1}^{\infty} a_i \Psi_i^{(0)}. \tag{11.6}$$

For a limited, small model space, only a small number of basis states $\Psi_i^{(0)}$ are used and one obtains a model wavefunction

$$\Psi' = \sum_{i \in M} a_i \Psi_i^{(0)}. \tag{11.7}$$

We now *impose* the condition that the effective Hamiltonian reproduces the true energy E for the corresponding model wavefunction Ψ', i.e.

$$\langle \Psi' | H^{\text{eff}} | \Psi' \rangle = E. \tag{11.8}$$

We consider, furthermore, the model wavefunction Ψ' to be normalized.

We divide the full Hilbert space into the model space M and the remaining part is given by the projection operator such that

$$\hat{P} \equiv \sum_{i \in M} |\Psi_i^{(0)}\rangle \langle \Psi_i^{(0)}|, \tag{11.9}$$

and

$$\hat{Q} = \sum_{i \notin M} |\Psi_i^{(0)}\rangle \langle \Psi_i^{(0)}|, \tag{11.10}$$

with the properties

$$\hat{P} + \hat{Q} = \hat{\mathbb{1}}, \qquad \hat{P}^2 = P, \qquad \hat{Q}^2 = Q, \qquad \hat{P}\hat{Q} = \hat{Q}\hat{P} = 0. \tag{11.11}$$

We now write the true wavefunction as follows

$$\Psi = (\hat{P} + \hat{Q})\Psi = \Psi' + \hat{Q}\Psi, \tag{11.12}$$

where we have used the result

$$\Psi' = \hat{P}\Psi,$$

and where Ψ' is the model wavefunction. We can now rewrite the Schrödinger equation (11.5), by acting on it with the $\hat{Q}$ and $\hat{P}$ operators, such that

$$(H_0 - E + \hat{Q}V\hat{Q})\hat{Q}\Psi = -\hat{Q}V(\hat{P}\Psi), \tag{11.13a}$$

and

$$(H_0 - E + \hat{P}V\hat{P})\hat{P}\Psi = -\hat{P}V(\hat{Q}\Psi). \qquad (11.13b)$$

Equation (11.13*a*) can be solved for $\hat{Q}\Psi$ with a result

$$\hat{Q}\Psi = -\hat{Q}(H_0 - E + \hat{Q}V\hat{Q})^{-1}\hat{Q}V\hat{P}(\hat{P}\Psi), \qquad (11.14)$$

which, after substitution in equation (11.13*b*), gives

$$(H_0 - E + \hat{P}V\hat{P} - \hat{P}V\hat{Q}(H_0 - E + \hat{Q}V\hat{Q})^{-1}\hat{Q}V\hat{P})\hat{P}\Psi = 0, \qquad (11.15)$$

or

$$\hat{P}\{H_0 - E + V - V\hat{Q}(H_0 - E + \hat{Q}V\hat{Q})^{-1}\hat{Q}V\}\hat{P}\Psi = 0. \qquad (11.16)$$

Introducing the model wavefunction Ψ', (11.16) can be rewritten as

$$\hat{P}(H_0 - E + V^{\text{eff}})\Psi' = 0, \qquad (11.17)$$

where the effective interaction is defined as

$$V^{\text{eff}} = V + V\hat{Q}(E - H_0 - \hat{Q}V\hat{Q})^{-1}\hat{Q}V. \qquad (11.18)$$

Equation (11.17) represents an eigenvalue problem in the model space for the true eigenvalue E at the expense of using an effective model interaction $V^{\text{eff}}(E)$. The effective interaction can be reshaped in the following way. Using the identity (operator)

$$\mathbb{1} \equiv \frac{1}{\hat{A}}(\hat{A} - \hat{B}) + \frac{1}{\hat{A}} \cdot \hat{B} \quad \text{or} \quad \frac{1}{\hat{A} - \hat{B}} \equiv \frac{1}{\hat{A}} + \frac{1}{\hat{A}}\hat{B}\frac{1}{\hat{A} - \hat{B}}, \qquad (11.19)$$

we obtain for V^{eff}, the resulting expression

$$V^{\text{eff}} = V + V\frac{\hat{Q}}{E - H_0}V^{\text{eff}}. \qquad (11.20)$$

The evaluation of the inverse operators in these expressions should be clear when they act on eigenstates. For an arbitrary function, the latter should be expanded in the eigenstates before evaluating the expressions. We can now iterate the effective interaction into a series

$$V^{\text{eff}} = V + V\frac{\hat{Q}}{E - H_0}V + V\frac{\hat{Q}}{E - H_0}V\frac{\hat{Q}}{E - H_0}V + \dots. \qquad (11.21)$$

Similarly, we obtain an expression of the true wavefunction Ψ in terms of the model wavefunction Ψ' as

$$\Psi = \Psi' + \frac{\hat{Q}}{E - H_0}V\Psi. \qquad (11.22)$$

This equation can also be iterated with the result

$$\Psi = \Psi' + \frac{\hat{Q}}{E - H_0} V\Psi' + \frac{\hat{Q}}{E - H_0} V \frac{\hat{Q}}{E - H_0} V\Psi' + \cdots. \qquad (11.23)$$

Since one has $\hat{Q}|\Psi'\rangle = 0$, the true wavefunctions are normalized according to the relations

$$\langle \Psi \mid \Psi' \rangle = \langle \Psi' \mid \Psi' \rangle = 1. \qquad (11.24)$$

Starting from the results for V^{eff} (11.21) and Ψ (11.23), we can verify the relation

$$V^{\text{eff}} \Psi' = V\Psi, \qquad (11.25)$$

showing that the action of the effective interaction on the model wavefunction gives the same result as the action of the realistic interaction on the true wavefunction.

The perturbation series for V^{eff} is of the Brillouin–Wigner type with the unknown energy E appearing in the energy denominators. It is now possible to rearrange the results such that: (i) the part of the valence nucleons and the core become separated and, (ii) the unperturbed energies E_0 come in the denominators.

Calling

$$E = E_{\text{core}} + \Delta E_{\text{core}} + E_{\text{val}} + \Delta E_{\text{c,v}}, \qquad (11.26)$$

with $E_{\text{core}} + \Delta E_{\text{core}}$ is the true core energy, E_{core} the unperturbed core energy, E_{val} the unperturbed energy of the valence nucleons and $\Delta E_{\text{c,v}}$ the remaining energy part. It has been shown (Brandow 1967) that equation (11.17) can be rephrased in the *model* space as

$$P(H_{0,\text{val.}} + V^{\text{eff}})\Psi' = (E_{\text{val.}} + \Delta E_{\text{c,v}})\Psi' \qquad (11.27)$$

with

$$H_{0,\text{val.}} \equiv H_0 - E_c. \qquad (11.28)$$

Now, equation (11.27) indeed has the form of a Schrödinger equation within the model space *only*, with eigenvalue $E_{\text{val}} + \Delta E_{\text{c,v}}$. Similarly, for the V^{eff} expression, one obtains

$$V^{\text{eff}} = V + V\frac{\hat{Q}}{E_{\text{val.}} - H_{0,\text{val.}}}V + V\frac{\hat{Q}}{E_{\text{val.}} - H_{0,\text{val.}}}V\frac{\hat{Q}}{E_{\text{val.}} - H_{0,\text{val.}}}V + \cdots. \qquad (11.29)$$

So, the Brillouin–Wigner series has been transformed into a Rayleigh–Schrödinger series where only the *unperturbed* energies E_{val} show up. Diagrammatic expansions can be used to express both V^{eff} and the wavefunction Ψ' (see Barrett and Kirson 1973, Kuo 1974, Kuo and Brown 1966).

The above methods basically link the full Hilbert space and the subsequent realistic force V and true wavefunction Ψ to a 'feasible' shell-model problem and act in a restricted model space, leading to an effective interaction V^{eff} and model

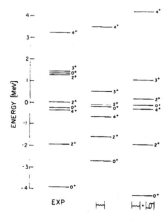

Figure 11.5. The energy spectra of ^{18}O; experimental spectrum, spectrum using the bare two-particle matrix elements of Brown and Kuo (1967) and that obtained by including the lowest-order 3p–1h intermediate states in constructing the effective interaction.

wavefunction Ψ', so that the lowest energy eigenvalues correspond to each other and that the model wavefunction Ψ' is just the projection of the true wavefunction Ψ onto the model space, spanned by the basis $\{\Psi_i^{(0)}; i = 1, 2, \ldots, M\}$.

As an example, we illustrate the low-lying two-particle spectrum in ^{18}O where the model space is the $2s_{1/2}$, $1d_{5/2}$, $1d_{3/2}$ shell-model space. The 'bare' interaction V, treated in the model space (figure 11.5), results in a spectrum that deviates strongly from the experimental spectrum. Taking into account the lowest-order corrections, due to the term

$$V \frac{\hat{Q}}{E_{\mathrm{val.}} - H_{0,\mathrm{val.}}} V, \qquad (11.30)$$

(the 'bubble' particle–hole corrections), a much improved result occurs in the case of ^{18}O. The problem of convergence of this series for V^{eff}, however, is a serious problem that has remained in a somewhat unsettled state (Schucan 1973).

The method used to determine the effective 'charges' $(e_\pi, e_\nu, g\text{-factors}, \ldots)$ goes along the same line, i.e. one asks that the true wavefunction Ψ, using the free charges in the electromagnetic operators, leads to identical matrix elements when using model wavefunctions Ψ' and effective charges, or

$$\langle \Psi' | \hat{O}^{\mathrm{eff}} | \Psi' \rangle = \langle \Psi | \hat{O} | \Psi \rangle. \qquad (11.31)$$

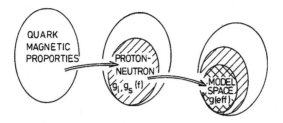

MAGNETIC PROPERTIES

Figure 11.6. Schematic illustration of the mapping procedure as explained in the text. (i) Mapping quark magnetic properties into the one-nucleon proton and neutron gyromagnetic free values. (ii) Mapping the proton and neutron free g-factors for a given nucleus into the model (or effective) gyromagnetic factors for the chosen model space (taken from Heyde 1989).

We again obtain an implicit equation for $\hat{O}^{\text{eff}}$ as a perturbation series after separating off the core and going to unperturbed valence energies

$$\hat{Q}^{\text{eff}} = \hat{O} + \hat{O}\frac{\hat{Q}}{E_{\text{val.}} - H_{0,\text{val.}}}V + \cdots. \tag{11.32}$$

Thereby $e_{\pi}^{\text{eff}} \neq e$, $e_{\nu}^{\text{eff}} \neq 0$, $g_{s,l}^{\text{eff}}(\pi, \nu) \neq g_{s,l}(\pi, \nu)$, and, moreover, the effective charges become mass dependent *and* dependent on the dimension of the model space. This process of effective or model charge construction is illustrated, for magnetic dipole g-factors, in figure 11.6. In a first step, quark degrees of freedom are included in the (renormalized) free nucleon g-factors. Subsequently, on studying moments and magnetic properties, the specific model space will again imply modifications of the earlier free g-factors into model space g-factors. In a large range of medium-heavy and heavy nuclei, one has observed a rule of thumb giving

$$g_s^{\text{eff}} \simeq 0.7g_s^{\text{free}}. \tag{11.33}$$

11.3 Two particle systems: wavefunctions and interactions

In the present section we first discuss how to construct correct two-particle wavefunctions and also give the major effects of two-nucleon interactions.

11.3.1 Two-particle wavefunctions

The two-particle angular momentum coupled wavefunctions can be constructed as

$$\Psi(j_1(1)j_2(2); JM), \tag{11.34}$$

with $j_1 \equiv \{n_1, l_1, j_1\}$ and $1, 2, \ldots$ a notation for all coordinates $\vec{r}_1, \vec{\sigma}_1, \ldots; \vec{r}_2, \vec{\sigma}_2, \ldots$. The above wavefunction is constructed via angular momentum coupling

$$\Psi(j_1(1)j_2(2); JM) = \sum_{m_1, m_2} \langle j_1 m_1, j_2 m_2 | JM \rangle \varphi_{j_1 m_1}(1) \varphi_{j_2 m_2}(2). \quad (11.35)$$

In the case of identical particles (p–p; n–n) the wavefunction should be antisymmetrized under the interchange of *all* coordinates. We explicitly construct then

(i) $j_1 \neq j_2$

$$\Psi_{as}(j_1 j_2; JM) = N \sum_{m_1, m_2} \langle j_1 m_1, j_2 m_2 | JM \rangle$$
$$[\varphi_{j_1 m_1}(1) \varphi_{j_2 m_2}(2) - \varphi_{j_1 m_1}(2) \varphi_{j_2 m_2}(1)]. \quad (11.36)$$

Rewriting in the convention of angular momentum coupling particle 1 and 2, in that order, and using the symmetry properties of Clebsch–Gordan coefficients, we obtain

$$\Psi_{nas}(j_1 j_2, JM) = \frac{1}{\sqrt{2}}[\Psi(j_1 j_2; JM) - (-1)^{j_1 + j_2 - J} \Psi(j_2 j_1; JM)]. \quad (11.37)$$

Here nas means: normalized and antisymmetrized. The nucleon coordinates need not be written any more since we use the convention of always coupling in the order (1,2).

(ii) $j_1 = j_2$
   From equation (11.37) in which we placed the restriction $j_1 = j_2 = j$, we obtain (with a different normalization though) the wavefunction

$$\Psi_{nas}(j^2; JM) = N'(1 + (-1)^J) \sum_{m_1, m_2} \langle j m_1 j m_2 | JM \rangle \varphi_{j m_1}(1) \varphi_{j m_2}(2). \quad (11.38)$$

So, one gets $N' = \frac{1}{2}$ and the restriction $J =$ even. This leads to two-particle nucleon systems with spin $J = 0, 2, 4, \ldots, 2j - 1$, in general.
   So, as an example, we can construct the two-particle configurations $(1d_{5/2}1d_{3/2})J = 1, 2, 3, 4; (1d_{5/2})^2 J = 0, 2, 4$. In evaluating the interaction energy, starting from the Hamiltonian, that describes two particles outside an inert core (which we leave out) and interacting with it by an effective interaction $V^{eff}(1, 2)$ as

$$H = H_{0, val.} + V^{eff}(1, 2), \quad (11.39)$$

with

$$H_{0, val.} = \sum_{i=1}^{2} h_0(i), \quad (11.40)$$

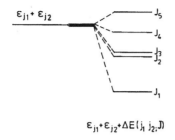

Figure 11.7. Splitting of a typical two-particle configuration $(j_1 j_2) J M$ due to the residual two-body interaction. The various states $J_1, J_2, \ldots$ are given and the energy splitting is $\Delta E(j_1 j_2; J)$.

and

$$h_0 \varphi_{j_i m_i}(\vec{r}) = \varepsilon_{j_i} \varphi_{j_i m_i}(\vec{r}), \tag{11.41}$$

we need the calculation of the two-body matrix

$$\langle j_1 j_2; J M | V^{\text{eff}}(1, 2) | j_1 j_2; J M \rangle_{\text{nas}}. \tag{11.42}$$

We so obtain the total energy of a given two-particle state (thereby lifting the degeneracy on J through the presence of $V^{\text{eff}}(1, 2)$)

$$H | j_1 j_2; J M \rangle = (E_0(j_1 j_2) + \Delta E(j_1 j_2; J)) | j_1 j_2; J M \rangle, \tag{11.43}$$

with

$$E_0(j_1 j_2) = \varepsilon_{j_1} + \varepsilon_{j_2}. \tag{11.44}$$

This process is given schematically in figure 11.7; and, for a more realistic situation, i.e. the $\langle (1d_{5/2})^2 J | V^{\text{eff}}(1, 2) | (1d_{5/2})^2 J \rangle$ matrix elements using various interactions in ^{18}O, in figure 11.8. The extended Skyrme force used, SkE4, is discussed in more detail in chapter 10 where also the more precise effect of the parameter x_3 is discussed.

Here we shall not elaborate extensively on the precise methods in order to determine the two-body matrix elements (Heyde 1991). In section 11.2, the use of realistic forces $V(1, 2)$ and the way they give rise to an effective model interaction $V^{\text{eff}}(1, 2)$ was presented. In a number of cases, the interaction itself is *not* needed: one determines the two-body matrix elements themselves in fitting the theoretical energy eigenvalues to the experimental data. This method of determining 'effective' two-body matrix elements has been used extensively. One of the best examples is the study of the sd-shell which includes nuclei between ^{16}O and ^{40}Ca and which will be discussed in section 11.5. In still another application, a general two-body form is used and a small number of parameters are fitted so as to obtain, again, good agreement between the calculated energy eigenvalues and the data.

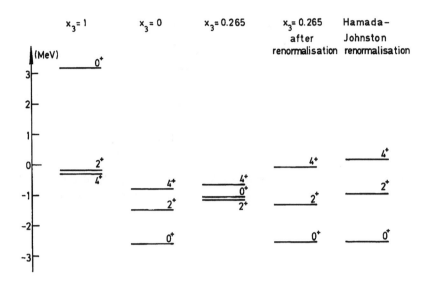

Figure 11.8. Antisymmetric and normalized effective neutron two-body matrix elements $\langle(1d_{5/2})^2 J|V|(1d_{5/2})^2 J\rangle$ in ^{18}O, evaluated with a Skyrme effective force (taken from Waroquier *et al* 1983).

This general form is dictated by general invariance properties (invariance under exchange of nucleon coordinates, translational invariance, Galilean invariance, space reflection invariance, time reversal invariance, rotational invariance in coordinate space and in charge space, etc). A general form could be

$$V^{eff}_{(1,2)} = V_0(r) + V_\sigma(r)\vec{\sigma}_1 \cdot \vec{\sigma}_2 + V_\tau(r)\vec{\tau}_1 \cdot \vec{\tau}_2 + V_{\sigma\tau}(r)\vec{\sigma}_1 \cdot \vec{\sigma}_2\vec{\tau}_1 \cdot \vec{\tau}_2, \quad (11.45)$$

and, even more extensive expressions containing tensor and spin-orbit components can be constructed. Typical radial shapes for $V_0(r)$, $V_\sigma(r)$, ... are Yukawa shapes, $e^{-\mu r}/\mu r$ with $r \equiv |\vec{r}_1 - \vec{r}_2|$. As an example, we illustrate in figure 11.9 the Hamada–Johnston potential (Hamada and Johnston, 1962). In some cases quite simple, schematic forces can be used mimicing the short-range properties of the force such as a $\delta(\vec{r}_1 - \vec{r}_2)$ or surface delta-interaction where nucleons interact only at the same place *on* the nuclear surface $\delta(\vec{r}_1 - \vec{r}_2)\delta(r_1 - R_0)$.

In many cases, the use of central radial interactions $V(|\vec{r}_1 - \vec{r}_2|)$ leads to interesting methods to evaluate the two-body matrix elements. Expanding the central interaction in the orthonormal set of Legendre polynomials, we can obtain the result

$$V(|\vec{r}_1 - \vec{r}_2|) = \sum_{k=0}^{\infty} v_k(r_1, r_2) P_k(\cos\theta_{12}), \quad (11.46)$$

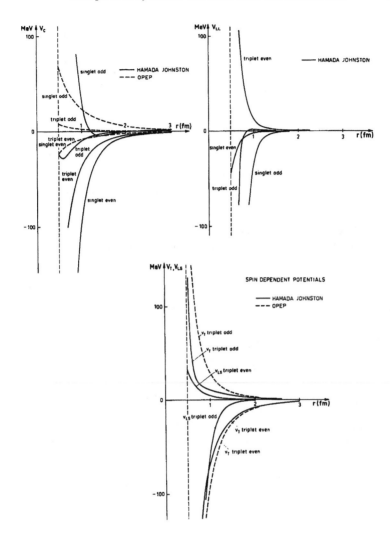

Figure 11.9. The 'realistic' nucleon–nucleon interaction potentials as obtained from the analysis of Hamada and Johnston (1962) for the central, spin-orbit, tensor and quadratic spin-orbit parts. The dotted potentials correspond to a one-pion exchange potential (OPEP) (after Bohr and Mottelson 1969).

where the index k counts the various multipole components present in the expansion. For a $\delta(\vec{r}_1 - \vec{r}_2)$ interaction, the expansion coefficients $v_k(r_1, r_2)$ become (Heyde 1991)

$$v_k(r_1, r_2) = \frac{\delta(r_1 - r_2)}{r_1 r_2} \frac{(2k+1)}{4\pi}. \tag{11.47}$$

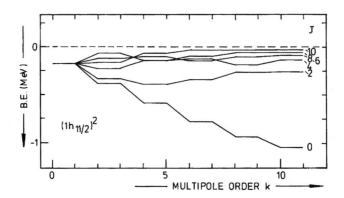

Figure 11.10. Various multipole (k = 0, ..., 11) contributions to the $\langle(1h_{11/2})^2 J|V|(1h_{11/2})^2 J\rangle$ two-body matrix elements using a pure δ-force interaction. Only even multipoles ($k = 0, 2, \ldots$) give an attractive contribution.

The final result concerning the two-body matrix elements results in the expression

$$\langle j_1 j_2; JM|\delta(\vec{r}_1 - \vec{r}_2)|j_1 j_2, JM\rangle_{\text{nas}}$$

$$= F^0(2j_1 + 1)(2j_2 + 1)\begin{pmatrix} j_1 & j_2 & J \\ \frac{1}{2} & -\frac{1}{2} & 0 \end{pmatrix}^2 (1 + (-1)^{l_1+l_2+J})/2, \quad (11.48)$$

where F^0 is a Slater integral, expressing the strength of the interaction. The J-dependence, however, only rests in the Wigner $3j$-symbol and the phase factor. For the $(1f_{7/2})^2 J = 0, \ldots 6$ configurations, the relative energy shifts become (in units $4F^0$)

$$\Delta E(1f_{7/2})^2 \quad \begin{aligned} J &= 0 : 1 \\ J &= 2 : 5/21 = 0.238 \\ J &= 4 : 9/77 = 0.117 \\ J &= 6 : 25/429 = 0.058. \end{aligned}$$

In figure 11.10, we present the case of the $(1h_{11/2})^2 J = 0, 2, \ldots, 10$ two-body matrix elements. It is clear that it is mainly the 0 spin coupling that gives rise to a large binding energy. This expresses the strong pairing correlation energy between identical nucleons in the nucleus. All other $J = 2, 4, \ldots, 10$ states remain close to the unperturbed energy of $2\varepsilon(1h_{11/2})$ (the dashed line). We also give the contributions of the various multipole k components, where $k = 0, 1, 2, \ldots, 11$. Only the even k values bring in extra binding energy. A striking result is the steady increase in binding energy for the 0 spin state, with increasing multipole order and thus, it is the high multipoles that are responsible for the attractive pairing part. The low multipoles give a totally different spectrum. A number of examples will be discussed in more detail in section 11.3.

11.3.2 Configuration mixing: model space and model interaction

In many cases when we consider nuclei with just two valence nucleons outside closed shells, it is not possible to single out one orbit j. Usually a number of valence shells are present in which the two nucleons can, in principle, move.

Let us consider the case of ^{18}O with two neutrons outside the ^{16}O core. In the simplest approach the two neutrons move in the energetically most favoured orbit, i.e. in the $1d_{5/2}$ orbit (figure 11.11). Thus we can only form the $(1d_{5/2})^2 0^+, 2^+, 4^+$ configurations and then determine the strength of the residual interaction V_{12} so that the theoretical $0^+ - 2^+ - 4^+$ spacing reproduces the experimental spacing as well as possible. The next step is to consider the full sd model space with many more configurations for each J^π value. For the $J^\pi = 0^+$ state, we have three configurations, i.e. the $(1d_{5/2})^2 0^+, (2s_{1/2})^2 0^+$ and the $(1d_{3/2})^2 0^+$ configurations. In the latter situation, the strength of the residual interaction V_{12} will be different from the model space where only the $(1d_{5/2})^2 0^+$ state is considered. Thus one generally concludes that the strength of the residual interaction depends on the model space chosen or $V_{12} = V_{12}$ (model space) such that, the larger model space is, the smaller V_{12} will become in order to get a similar overall agreement. In the larger model spaces, one will, in general, be able to describe the observed properties the nucleus better than with the smaller model spaces. This argument only relates to effective forces using a given form, i.e. a Gaussian interaction, an (M)SDI interaction, etc for which only the strength parameter determines the overall magnitude of the two-body matrix elements in a given finite dimensional model space. Thus one should not extrapolate to the full (infinite dimensional) configuration space in which the bare nucleon–nucleon force would be acting.

Thus, in the case of ^{18}O where the $2s_{1/2}$ and $1d_{5/2}$ orbits separate from the higher-lying $1d_{3/2}$ orbit, for the model spaces one has

$$J^\pi = 0^+ \rightarrow (1d_{5/2})^2_{0^+} (2s_{1/2})^2_{0^+}$$
$$J^\pi = 2^+ \rightarrow (1d_{5/2})^2_{2^+} (1d_{5/2}2s_{1/2})_{2^+}$$
$$J^\pi = 3^+ \rightarrow (1d_{5/2}2s_{1/2})_{3^+}$$
$$J^\pi = 4^+ \rightarrow (1d_{5/2})^2_{4^+}.$$

The energy eigenvalues for the $J^\pi = 0^+$ states, for example, will be the corresponding eigenvalues for the eigenstates of the Hamiltonian

$$H = H_0 + H_{\text{res}},$$
$$= \sum_{i=1}^{2} h_0(i) + V_{12}, \tag{11.49}$$

where the core energy corresponding to the closed shell system E_0 is taken as the reference value.

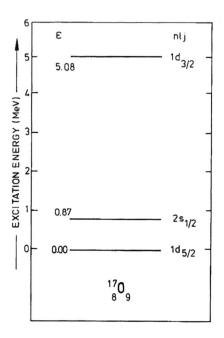

Figure 11.11. The neutron single-particle energies in $^{17}_{8}O_9$ (relative to the $1d_{5/2}$ orbit) for the $2s_{1/2}$ and $1d_{3/2}$ orbits. Energies are taken from the experimental spectrum in ^{17}O (taken from Heyde 1991).

The wavefunctions will, in general, be linear combinations of the possible basis functions. This means that for $J^\pi = 0^+$ we will get two eigenfunctions

$$|\Psi_{0^+;1}\rangle = \sum_{k=1}^{n} a_{k,1}|\psi_k^{(0)}; 0^+\rangle,$$

$$|\Psi_{0^+;2}\rangle = \sum_{k=1}^{n} a_{k,2}|\psi_k^{(0)}; 0^+\rangle, \qquad (11.50)$$

where for the particular case of ^{18}O we define

$$|\psi_1^{(0)}; 0^+\rangle \equiv |(1d_{5/2})^2; 0^+\rangle$$
$$|\psi_2^{(0)}; 0^+\rangle \equiv |(2s_{1/2})^2; 0^+\rangle.$$

Before turning back to the particular case of ^{18}O, we make the method more general. If the basis set is denoted by $|\psi_k^{(0)}\rangle$ ($k = 1, 2, \ldots, n$), the total wavefunction can be expanded as

$$|\Psi_p\rangle = \sum_{k=1}^{n} a_{kp}|\psi_k^{(0)}\rangle. \tag{11.51}$$

The coefficients a_{kp} have to be determined by solving the Schrödinger equation for $|\Psi_p\rangle$, or

$$H|\Psi_p\rangle = E_p|\Psi_p\rangle. \tag{11.52}$$

In explicit form this becomes (using the Hamiltonian of (11.49))

$$(H_0 + H_{\text{res}}) \sum_{k=1}^{n} a_{kp}|\psi_k^{(0)}\rangle = E_p \sum_{k=1}^{n} a_{kp}|\psi_k^{(0)}\rangle, \tag{11.53}$$

or

$$\sum_{k=1}^{n} \langle\psi_l^{(0)}|H_0 + H_{\text{res}}|\psi_k^{(0)}\rangle a_{kp} = E_p a_{lp}. \tag{11.54}$$

Since the basis function $|\psi_k^{(0)}\rangle$ corresponds to eigenfunctions of H_0 with eigenvalues (unperturbed energies) $E_k^{(0)}$, we can rewrite (11.54) in shorthand form as

$$\sum_{k=1}^{n} H_{lk} a_{kp} = E_p a_{lp}, \tag{11.55}$$

with

$$H_{lk} \equiv E_k^{(0)} \delta_{lk} + \langle\psi_l^{(0)}|H_{\text{res}}|\psi_k^{(0)}\rangle. \tag{11.56}$$

The eigenvalue equation becomes a matrix equation

$$[H][A] = [E][A]. \tag{11.57}$$

This forms a secular equation for the eigenvalues E_p which are determined from

$$\begin{vmatrix} H_{11} - E_p & H_{12} & \dots H_{1n} \\ H_{21} & H_{22} - E_p & \dots H_{2n} \\ \vdots & \vdots & \vdots \\ H_{n1} & \dots & H_{nn} - E_p \end{vmatrix} = 0. \tag{11.58}$$

This is a nth degree equation for the n-roots E_p ($p = 1, 2, \dots, n$). Substitution of each value of E_p separately in (11.55) gives a set of linear equations that can be solved for the coefficients a_{kp}. The wavefunctions $|\Psi_p\rangle$ can be orthonormalized since

$$\sum_{k=1}^{n} a_{kp} a_{kp'} = \delta_{pp'}. \tag{11.59}$$

From (11.59) it now follows that

$$\sum_{l,k=1}^{n} a_{lp'} H_{lk} a_{kp} = E_p \delta_{pp'}, \tag{11.60}$$

or, in matrix form

$$[\tilde{A}][H][A] = [E], \tag{11.61}$$

with $[\tilde{A}] = [A]^{-1}$. Equation (11.61) indicates a similarity transformation to a new basis that makes $[H]$ diagonal and thus produces the n energy eigenvalues. In practical situations, with n large, this process needs high-speed computers. A number of algorithms exist for $[H]$ (Hermitian, real in most cases) matrix diagonalization which we do not discuss here (Wilkinson 1965): the Jacobi method (small n or $n \lesssim 50$), the Householder method ($50 \lesssim n \leq 200$), the Lanczos algorithm ($n \gtrsim 1000$, requiring the calculation of only a small number of eigenvalues, normally the lowest lying ones). In cases where the non-diagonal matrix elements $|H_{ij}|$ are of the order of the unperturbed energy differences $|E_i^{(0)} - E_j^{(0)}|$, large configuration mixing will result and the final energy eigenvalues E_p can be very different from the unperturbed spectrum of eigenvalues $E_p^{(0)}$. If, on the other hand, the $|H_{ij}|$ are small compared to $|E_i^{(0)} - E_j^{(0)}|$, energy shifts will be small and even perturbation theory might be applied.

Now to make these general considerations more specific, we discuss the case of $J^\pi = 0^+$ levels in ^{18}O for the $(1d_{5/2}2s_{1/2})$ model space. As shown before, the model space reduces to a two-dimensional space $n = 2$ (11.50) and the 2×2 energy matrix can be written as

$$H = \begin{bmatrix} 2\varepsilon_{1d_{5/2}} + \langle(1d_{5/2})^2; 0^+|V_{12}|(1d_{5/2})^2; 0^+\rangle & \langle(1d_{5/2})^2; 0^+|V_{12}|(2s_{1/2})^2; 0^+\rangle \\ \langle(2s_{1/2})^2; 0^+|V_{12}|(1d_{5/2})^2; 0^+\rangle & 2\varepsilon_{2s_{1/2}} + \langle(2s_{1/2})^2; 0^+|V_{12}|(2s_{1/2})^2; 0^+\rangle \end{bmatrix}. \tag{11.62}$$

These diagonal elements yield the first correction to the unperturbed single-particle energies $2\varepsilon_{1d_{5/2}}$ and $2\varepsilon_{2s_{1/2}}$, respectively (the diagonal two-body interaction matrix elements H_{11} and H_{22}, figure 11.12). The energy matrix is Hermitian which for a real matrix means symmetric with $H_{12} = H_{21}$ and, in shorthand notation gives the secular equation

$$\begin{bmatrix} H_{11} - \lambda & H_{12} \\ H_{12} & H_{22} - \lambda \end{bmatrix} = 0. \tag{11.63}$$

We can solve this easily since we get a quadratic equation in λ

$$\lambda^2 - \lambda(H_{11} + H_{22}) - H_{12}^2 + H_{11}H_{22} = 0, \tag{11.64}$$

with the roots

$$\lambda_\pm = \frac{H_{11} + H_{22}}{2} \pm \frac{1}{2}[(H_{11} - H_{22})^2 + 4H_{12}^2]^{1/2}. \tag{11.65}$$

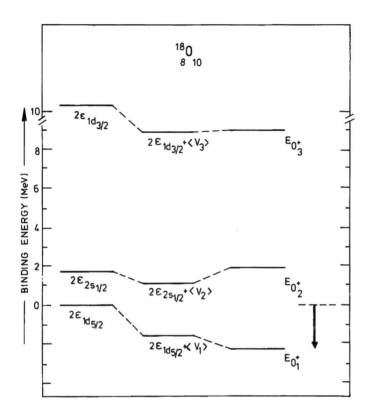

Figure 11.12. Various energy contributions to solving the secular equation for ^{18}O. On the extreme left, unperturbed energies are given. In the middle part, diagonal two-body matrix elements are added. On the extreme right, the final resulting energy eigenvalues 0_i^+ ($i = 1, 2, 3$) from diagonalizing the *full* energy matrix are given (taken from Heyde 1991).

The difference $\Delta\lambda \equiv \lambda_+ - \lambda_-$ then becomes

$$\Delta\lambda = [(H_{11} - H_{22})^2 + 4H_{12}^2]^{1/2}, \tag{11.66}$$

and is shown in figure 11.13. Even for $H_{11} = H_{22}$, the degenerate situation for the two basis states, a difference of $\Delta\lambda = 2H_{12}$ results. It is as if the two levels are repelled over a distance of H_{12}. Thus $2H_{12}$ is the minimal energy difference. In the limit of $|H_{11}-H_{22}| \gg |H_{12}|$, the energy difference $\Delta\lambda$ becomes asymptotically equal to $\Delta H \equiv H_{11} - H_{22}$.

The equation for $\lambda_\pm$ given in (11.65) is interesting with respect to perturbation theory.

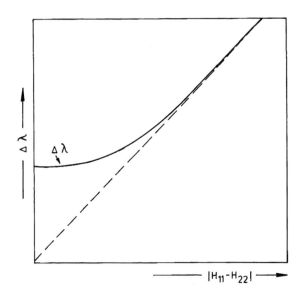

Figure 11.13. The variation of the eigenvalue difference $\Delta\lambda \equiv \lambda_+ - \lambda_-$ for the two-level model of equations (11.63)–(11.66), as a function of unperturbed energy difference $\Delta H \equiv |H_1 - H_2|$ (taken from Heyde 1991).

(i) If we consider the case $|H_{11} - H_{22}| \gg |H_{12}|$, then we see that one can obtain by expanding the square root around $H_{11} = H_{22}$

$$\lambda_1 = H_{11} + \frac{H_{12}^2}{H_{11} - H_{22}} + \cdots,$$

$$\lambda_2 = H_{22} + \frac{H_{12}^2}{H_{22} - H_{11}} + \cdots, \tag{11.67}$$

where we use the expansion $(1 + x)^{1/2} \cong 1 + \frac{1}{2}x + \cdots$.

(ii) One can show that if the perturbation expansion does not converge easily, one has to sum the full perturbation series to infinity. The final result of this sum can be shown to be equal to the square root expression in (11.65).

It is also interesting to study (11.65) with a constant interaction matrix element H_{12} when the unperturbed energies H_{11}, H_{22} vary linearly, i.e. $H_{11} = E_1^{(0)} + \chi a$ and $H_{22} = E_2^{(0)} - \chi b$ (figure 11.14). There will be a crossing point for the unperturbed energies at a certain value of $\chi = \chi_{\text{crossing}}$. However, the eigenvalues E_1, E_2 will first approach the crossing but then change directions (no-crossing rule). The wavefunctions are also interesting. First of all, we study the wavefunctions analytically, i.e. the coefficients a_{kp}.

We get

$$|\Psi_1\rangle = a_{11}|\psi_1^{(0)}\rangle + a_{21}|\psi_2^{(0)}\rangle$$
$$|\Psi_2\rangle = a_{12}|\psi_1^{(0)}\rangle + a_{22}|\psi_2^{(0)}\rangle. \qquad (11.68)$$

If we use one of the eigenvalues, say λ_1, the coefficients follow from

$$(H_{11} - \lambda_1)a_{11} + H_{12}a_{21} = 0, \qquad (11.69)$$

or

$$\frac{a_{11}}{a_{21}} = \frac{-H_{12}}{H_{11} - \lambda_1}. \qquad (11.70)$$

The normalizing condition $a_{11}^2 + a_{21}^2 = 1$ then gives

$$a_{21} = (1/(1 + (H_{12}/(H_{11} - \lambda)^2)))^{1/2}, \qquad (11.71)$$

and similar results for the other coefficients. In the situation that $H_{11} = H_{22}$, the absolute values of the coefficients a_{11}, a_{12}, a_{21} and a_{22} are *all* equal to $1/\sqrt{2}$. These coefficients then also determine the wavefunctions of figure 11.14 at the crossing point. One can see in figure 11.14 that for the case of $\chi = 0$, one has

$$E_1 \simeq E_1^{(0)} \quad \text{and} \quad |\Psi_1\rangle \simeq |\psi_1^{(0)}\rangle,$$
$$E_2 \simeq E_2^{(0)} \quad \text{and} \quad |\Psi_2\rangle \simeq |\psi_2^{(0)}\rangle.$$

On the other hand, after the level crossing and in the region where again $|H_{11} - H_{22}| \gg |H_{12}|$, one has

$$E_1 \simeq H_{22} \quad \text{and} \quad |\Psi_1\rangle \simeq |\psi_2^{(0)}\rangle,$$
$$E_2 \simeq H_{11} \quad \text{and} \quad |\Psi_2\rangle \simeq |\psi_1^{(0)}\rangle,$$

so that one can conclude that the 'character' of the states has been interchanged in the crossing region, although the levels never actually cross!

The full result for ^{18}O is then depicted in figure 11.15, where respectively, the unperturbed two-particle spectrum, the spectrum adding the diagonal matrix elements as well as the results after diagonalizing the full energy matrix, are discussed. This example explains the main effects that show up on setting up the study of energy spectra near closed shells.

In the next two sections (11.4 and 11.5) we shall discuss in somewhat more detail examples of studies in nuclei containing just two valence nucleons, in doubly-closed shell nuclei *and* when carrying out large-scale shell-model calculations according to state-of-the-art calculations.

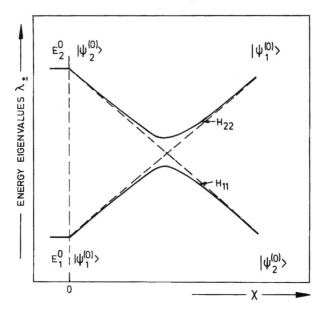

Figure 11.14. Variation of the eigenvalues $\lambda_\pm$ obtained from a two-level model as a function of a parameter χ which describes the variation with unperturbed energies H_{11}, H_{22} assuming a linear variation, i.e. $H_{11} = E_1^{(0)} + \chi a$; $H_{22} = E_2^{(0)} - \chi b$ (with H_{12} as interaction matrix element). The wavefunction variation is also shown (taken from Heyde 1991).

11.4   Energy spectra near and at closed shells

11.4.1   Two-particle spectra

We use the methods discussed in section 11.3 in order to perform a full diagonalization of the energy matrix of (11.39) within a two-particle basis.   The basic procedure in setting up the energy matrix consists in determining *all* partitions of two-particles over the available single-particle space $\varphi_{j1}, \varphi_{j2}, \ldots, \varphi_{jk}$ (figure 11.16(*a*)).

As discussed in section 11.2, it is possible to study the excitation modes for two-particle (or two-hole) energy spectra using a closed shell configuration and determining the effective forces such that particle–hole excitations in this core are implicitly taken into account. Starting from realistic forces: Hamada–Johnston, Tabakin, Reid potential, Skyrme forces, one has to evaluate therefore the 'renormalized' interaction that acts in the two-particle or two-hole model space. Using schematic, delta, surface-delta, etc forces, the parameters such as strength, range, eventual spin, isospin dependences are determined so as to obtain a good reproducibility within a given small two-nucleon valence space.

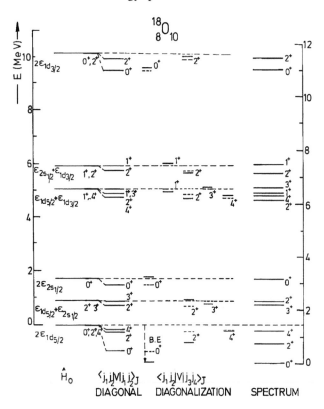

Figure 11.15. Full description of the construction of the two-neutron energy spectrum in the case of ^{18}O. At the extreme left, the unperturbed (degenerate in J^π) two-particle spectrum is given. Next (proceeding to the right) the addition of the diagonal matrix elements is indicated. Then, the result of diagonalizing the total Hamiltonian in the various J^π subspaces is given. Finally, at the extreme right, the total spectrum is given. The energy scale refers to the single-neutron energy spectrum of figure 11.11.

Therefore, the latter forces are only used in a particular mass region whereas the realistic forces can be used throughout the nuclear mass table; the renormalization effects are the aspects needed to bring the realistic force in line with the study of a particular mass region.

In figure 11.17, we compare, as an illustration, the typical results occurring in two-particle spectra, the experimental and theoretical energy spectra for ^{18}O, ^{42}Ca, ^{50}Ca, ^{58}Ni and ^{134}Te. The force used is the SkE2 force ($x_3 = 0.43$). The typical result is the large energy gap between the ground state 0^+ and the first excited 2^+ level. The overall understanding of two-particle spectra is quite good and is mainly a balancing effect between the pairing and quadrupole components in the force.

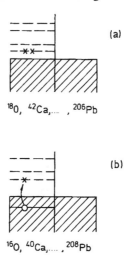

Figure 11.16. (*a*) Schematic representation of the model space for two valence nucleons (particles or holes) outside closed shells. (*b*) Similar representation of the model space but now for one particle-one hole (1p–1h) excitations *in* a closed-shell nucleus (taken from Heyde 1989).

11.4.2 Closed-shell nuclei: 1p–1h excitations

For doubly-closed shell nuclei, the low-lying excited states will be mainly within the 1p–1h configurations (figure 11.16(*b*)), where, due to charge independence of the nucleon–nucleon interaction and the almost identical character of the nuclear average proton and neutron fields (taking the Coulomb part aside), the proton 1p–1h and neutron 1p–1h excitations are very nearly degenerate in the unperturbed energy (figure 11.18). It is the small difference in mass between proton and neutron and the Coulomb interaction that induce slight perturbations on the isospin symmetry (charge symmetry) in this picture. Although the isospin formalism can be used to describe the 1p–1h excitation in doubly-closed shell nuclei, we consider the explicit difference between the proton and neutron 1p–1h configurations and, only later, check on the isospin purity of the eigenstates.

Since we know how to couple the proton 1p–1h and neutron 1p–1h excitations to a definite isospin, e.g.

$$|ph^{-1}; JM\rangle_\pi = \frac{1}{\sqrt{2}}[|ph^{-1}, JM, T = 1\rangle + |ph^{-1}, JM, T = 0\rangle]$$

$$|ph^{-1}; JM\rangle_\nu = \frac{1}{\sqrt{2}}[|ph^{-1}; JM, T = 1\rangle - |ph^{-1}, JM, T = 0\rangle], \quad (11.72)$$

we can, in the linear combination of the wavefunctions, expressed in the proton-neutron basis, substitute the expression (11.72) so as to obtain an expansion of the

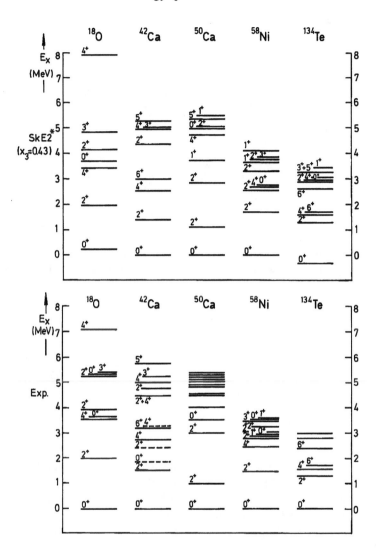

Figure 11.17. Two-particle energy spectra of some (doubly-closed shell+two nucleons) nuclei using a Skyrme effective force (SkE2*) (Waroquier *et al* 1983). Only positive parity states are retained. The full set of data is discussed in Heyde (1991).

wavefunctions in the isospin (J, T) coupled basis. Formally, the wavefunctions

$$|i, JM\rangle = \sum_{ph,\rho} c(ph^{-1}(\rho); iJ)|ph^{-1}, JM\rangle, \qquad (11.73)$$

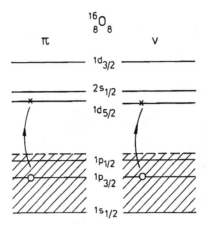

Figure 11.18. The model space used to carry out a 1p–1h study of ^{16}O. The full (sd) space for unoccupied and (sp) space for occupied configurations is considered (taken from Heyde 1991).

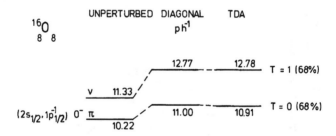

Figure 11.19. Using a Skyrme force (SkE4*) (see text) to determine both the Hartree–Fock single-particle energies and the two-body matrix elements, we present (i) unperturbed proton (π) and neutron (ν) 0^- (1p–1h) energies; (ii) energies, including the diagonal matrix elements (repulsive) and, (iii) the final results obtained after diagonalizing the interaction in the $J^\pi = 0^-$ space. The isospin purity is also given as T (in %) (taken from Heyde 1991).

where i denotes the number of eigenstates for given J and ρ the charge quantum number ($\rho \equiv \pi, \nu$), can be rewritten in the isospin coupled basis as

$$|i, JM\rangle = \sum_{ph,T} c'(ph^{-1}; iJT)|ph^{-1}; JM, T\rangle, \qquad (11.74)$$

where here, for each $(p, h)^{-1}$ combination, we sum over the $T = 0$ and $T = 1$ states. From the actual numerical studies, it follows that the low-lying states are mainly $T = 0$ in character and the higher-lying are the $T = 1$ states.

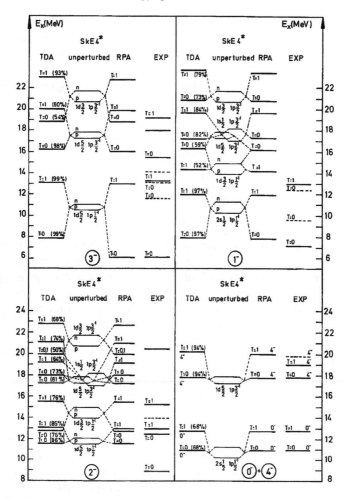

Figure 11.20. Negative parity states in ^{16}O. A comparison between the TDA (Tamm–Dancoff approximation), RPA (random-phase approximation) using the SkE4* force and the data is made. The isospin purity is given in all cases (T in %). Experimental levels, drawn with dashed lines have mainly a 3p–3h character (taken from Waroquier 1983).

The Hartree–Fock field in ^{16}O has been determined using the methods of chapter 10 for various Skyrme forces. We discuss here the results for the 1p–1h spectrum in ^{16}O using the same effective Skyrme force (SkE4*) that was used in order to determine the average Hartree–Fock properties. We illustrate, in figure 11.19, as an example, the proton and neutron 1p–1h 0^- configurations $|(2s_{1/2}1p_{1/2}^{-1}); 0^-\rangle$. A net energy difference of $\Delta\epsilon = 1.11$ MeV results. The diagonal ph^{-1} matrix elements are slightly repulsive and finally, the non-diagonal matrix element is small. So, in the final answer, rather pure proton ph^{-1} and neutron ph^{-1} states result. This is clearly illustrated by expressing the wavefunctions on the basis of equation (11.74) where the maximal purity is 68%.

Considering to the 1^- state, within the $1\hbar\omega$ 1p–1h configuration space, we can only form the following basis states

$$|2s_{1/2}(1p_{1/2})^{-1}; 1^-\rangle$$
$$|2s_{1/2}(1p_{3/2})^{-1}; 1^-\rangle$$
$$|1d_{3/2}(1p_{1/2})^{-1}; 1^-\rangle$$
$$|1d_{3/2}(1p_{3/2})^{-1}; 1^-\rangle$$
$$|1d_{5/2}(1p_{3/2})^{-1}; 1^-\rangle.$$

Taking both the proton and neutron 1p–1h configurations, a 10-dimensional model space results. Using the methods discussed in the present chapter, one can set up the eigenvalue equation within this model space using a given effective force (TDA approximation or the RPA approximation if a more complicated ground-state wavefunction is used). The resulting spectra for the $0^-, \ldots, 4^-$ states are presented in figure 11.20 where we illustrate in all cases:

(i)   the unperturbed energy, giving both the charge character of the particular ph^{-1} excitation;
(ii) the TDA diagonalization results (with isospin purity);
(iii) the RPA diagonalization results;
(iv) the experimental data.

One observes, in particular, for the lowest $1^-, 3^-$ levels the very pure $T = 0$ isospin character. Since the unperturbed proton and neutron 1p–1h configurations are almost degenerate, the diagonalization implies a definite symmetry character ($T = 0$ and $T = 1$) according to the lower- and high-lying states.

We illustrate a study of the 3^-, collective isoscalar state in ^{208}Pb (Gillet 1964) (figure 11.21). Here, as a function of the dimension of the 1p–1h configuration model space, the convergence properties in the TDA and RPA approach are illustrated. One notices that: (i) the RPA eigenvalue is always lower than the corresponding TDA eigenvalue and that (ii) for a given strength of the residual interaction, the dimension of the 3^- configuration space affects the final excitation energy in a major way. This aspect of convergence should be tested in all cases.

As a final example (figure 11.22), we give a state-of-the-art self-consistent shell-model calculation for ^{40}Ca within a 1p–1h configuration space and thereby use the extended Skyrme force parametrization as an effective interaction.

It has become clear that the nuclear shell model is able to give a good description of the possible excitation modes in nuclei where just a few nucleons are interacting outside a closed shell (two-particle and two-hole nuclei) and that

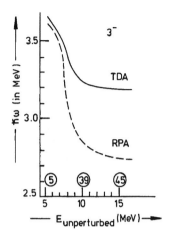

Figure 11.21. The octupole 3^- state in ^{208}Pb using the TDA and RPA methods as a function of the 1p–1h model space dimension. The number of 1p–1h components as well as its unperturbed energy are indicated on the abscissa. The force used and the results are discussed by Gillet (1966).

it is even possible to describe the lowest-lying excitations in the doubly-closed shell nuclei as mainly 1p–1h excitations. In order to get a better insight into the possibilities and predictive power of the nuclear shell model *and* of its limitations, we should look to examples of large-scale shell-model calculations. In the next section we discuss a few such examples.

11.5 Large-scale shell-model calculations

Models have come in very naturally: in particular, the shell model has been a robust guide for more than 50 years now. From the most simple hand-by-hand approach up to the present large-scale shell-model diagonalizations of the nuclear eigenvalue problem (reaching dimensions for a basis of $\approx 10^9$ in the fp shell-model space (Caurier *et al* 1997)), this has been used as a benchmark to explain many observed properties.

The lightest nuclei, taking the 'p' shell, were studied a long time ago by Cohen and Kurath (Cohen and Kurath 1965). The 'sd' model space has been covered by Brown and Wildenthal (Brown and Wildenthal 1988), in great detail.

Modern large-scale shell-model configurations try to treat many or all of the possible ways in which nucleons can be distributed over the available single-particle orbits that are important in a particular mass region. It is conceivable that all nuclei situated between ^{16}O and ^{40}Ca might be studied using the full $(2s_{1/2}, 1d_{5/2}, 1d_{3/2} - \pi, \nu)$ model space. It is clear that nuclei in the mid-shell

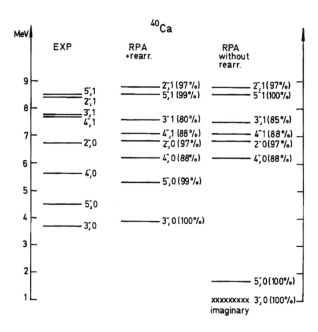

Figure 11.22. Low-lying negative parity (J^π, T) states in ^{40}Ca. The theoretical levels correspond to a fully self-consistent RPA calculation using the SkE2 force. Isospin purity (T in %) is also indicated (taken from Waroquier *et al* 1987).

region ^{28}Si, ... one will obtain a very large model space and extensive numerical computations will be needed. The increase in computing facilities has made the implementation of large model spaces and configuration mixing possible.

The current research aims at a theoretical understanding of energy levels and also of all other observables (decay rates) that give test possibilities for the model spaces used, missing configurations and/or deficiencies in the nucleon–nucleon interactions (Brown and Wildenthal 1988).

In the present section, we discuss the achievements of the (sd) shell and also its shortcomings.

The sd shell contains 24 active m-states, i.e. states characterized by the quantum numbers n, l, j, m, t_z. As an example, in ^{28}Si for $M = 0$ and $T_z = 0$ one has 93 710 states that form the model space and, in the JT scheme, the $J = 3$, $T = 1$ space contains 6706 states. We illustrate the distribution of unperturbed states in ^{26}Mg for the 2^+ states constructed in this sd space, with the unperturbed basis configuration denoted by

$$\Psi_{(0)}(2^+) = |(1d_{5/2})^{n_1}_{J_1} (2s_{1/2})^{n_2}_{J_2} (1d_{3/2})^{n_3}_{J_3}; 2^+\rangle. \qquad (11.75)$$

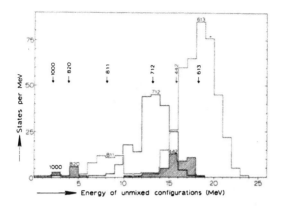

Figure 11.23. The number of unmixed configurations with $J^\pi = 2^+$ in ^{26}Mg for a number of different particle distributions over the $1d_{5/2}$, $2s_{1/2}$ and $1d_{3/2}$ orbits. The energy is relative to the energy of the lowest-lying configurations. A surface delta two-body interaction was used (Brussaard and Glaudemans 1977) and experimental single-particle energies. The particle distributions are denoted by (n_1, n_2, n_3), the number of particles in the $1d_{5/2}$, $2s_{1/2}$ and $1d_{3/2}$ orbits, respectively. The arrows indicate the unperturbed energies $n_2(\varepsilon_{2s_{1/2}} - \varepsilon_{2d_{5/2}}) + n_3(\varepsilon_{1d_{3/2}} - \varepsilon_{1d_{5/2}})$ (taken from Brussaard and Glaudemans 1977).

Here, the partitions (n_1, n_2, n_3) determine the relative, unperturbed energy in this model space as

$$E_0 = n_2(\varepsilon_{1s_{1/2}} - \varepsilon_{1d_{5/2}}) + n_3(\varepsilon_{1d_{3/2}} - \varepsilon_{1d_{5/2}}) + \Delta E. \qquad (11.76)$$

The distribution in E_0 for the basis states $\Psi_{(0)}(2^+)$ is then given in figure 11.23, where also the diagonal interaction matrix element $(A' = A - A_{\text{core}})$ energy

$$\Delta E \equiv \langle \Psi_0(2^+)| \sum_{i<j=1}^{A'} V(1,2)|\Psi_0(2^+)\rangle, \qquad (11.77)$$

has been incorporated. The arrows give the energy E_0, ignoring ΔE, for the various (n_1, n_2, n_3) distributions. In each case, these basis configurations need to be constructed and used to construct the energy matrix. Fast methods are used to determine the energy eigenvalues, (Lanczos algorithm—Brussaard and Glaudemans 1977) since in most cases one is only interested in the energy eigenvalues corresponding to the lowest eigenstates.

Current large-scale sd shell-model calculations have been performed without truncating, even when evaluating the nuclear properties at the mid-shell region (^{28}Si). Brown and Wildenthal (1988) have pioneered the study in this particular mass region. As an example, we show the slow convergence related to the choice

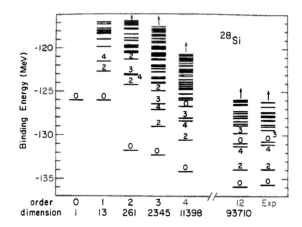

Figure 11.24. The various results for the energy levels in $^{28}_{14}\text{Si}_{14}$, calculated within the full sd model space. Starting from the extreme left with a $(1d_{5/2})^{12}(2s_{1/2})^0(1d_{3/2})^0$ configuration we increase the model space by lifting $1, 2, \ldots$ up to 12 particles out of the $1d_{5/2}$ orbit up to the extreme general $(1d_{5/2})^{n_1}(2s_{1/2})^{n_2}(1d_{3/2})^{n_3}$ ($n_1 + n_2 + n_3 = 12$) situation. The dimension of the full model space is given in each case. The experimental spectrum is given for comparison. (Taken from Brown and Wildenthal. Reproduced with permission form the Annual Review of Nuclear Sciences **38** © 1988 by Annual Reviews Inc.)

of the model space. Starting from the extreme 'closed subshell' configuration $(1d_{5/2})^{12}(2s_{1/2})^0(1d_{3/2})^0$ (order 0 in figure 11.24), one observes the evolution in the energy spectrum when enlarging the model space in successive breaking the $(1d_{5/2})^{12}$ configuration and ending with the 12 particles being distributed in *all* possible ways over the $1d_{5/2}, 2s_{1/2}$ and $1d_{3/2}$ orbits without any further constraints. The same effective force has been used in the various calculations (order $0 \rightarrow 12$) and, at order $= 12$, a good reproduction of the ^{28}Si experimental energy spectrum, exhibiting a number of collective features, is reached. Further details, in particular relating to the evaluation of electromagnetic properties in the sd shell nuclei, are discussed by Brown and Wildenthal (1988).

More recently, large-scale shell-model calculations have been performed incorporating the full fp shell thereby accessing a large number of nuclei where the alleged doubly-closed shell nucleus ^{56}Ni is incorporated in a consistent way. The Strasboug–Madrid group, using the code ANTOINE (using the m-scheme) has used computer developments as well as two decades of computer development on shell-model methods to accomplish an important step. At present, issues relating to a shell closure at $Z = N = 28$ can now be studied in a broader perspective. Moreover, properties that are most often analysed within a deformed basis and subsequent features like deformed bands, backbending etc are now accessible within a purely shell-model context. Hope is clearly present to bridge the gap between spherical large-scale shell-model studies on one side, and phenomenological studies starting from an intitial deformed mean-field on the other side. A number of interesting references for further study of fp-shell model calculations are given: Caurier 1994, Caurier 1995, Zuker 1995, Caurier 1996, Dufour 1996, Martinez-Pinedo 1996, Retamosa 1997, Martinez-Pinedo 1997 (see Box 11a). Conferences and workshops concentrating on shell-model developments have been intensively pursued in recent years and we refer to the proceedings of some of these workshops in order to bring the reader in contact with recent developments: Wyss 1995, Covello 1996.

It is very clear that efforts are going on to extend straightforward shell-model calculations within large model spaces. At present, the evaluation of eigenvalues in model spaces approaching 10^9 are within reach.

In figure 11.25, we indicate how the CPU speed has become faster as a function of the year with a good number of examples. At the same time, we also illustrate, in figure 11.25, the possibility of handling increasingly large model spaces, as indicated on the figure.

Impressive though these calculations are, one may ask the question of whether the essential degrees of freedom active in certain mass regions could also be derived from clever truncations to the huge shell-model space. There now exist many such truncation methods which would bring me outside the main scope of the present discussion but I refer the reader to (Heyde 1999, Heyde 1994, Heyde 1998) for extensive reviews.

One can still wonder if those rather complex wavefunctions will ultimately give answers as to why certain nuclei are very much like spherical systems while others exhibit properties that are close to the dynamical deformed and collective vibrational systems. Work has to be done in order to find appropriate truncation schemes that guide shell-model studies, not just in a 'linear' way (exhausting all model configurations within a certain spherical model boundary like the sd or fp-shell model studies Kar 1997, Yokoyama 1997, Nakada 1996, Nakada 1997a, b), but exploit other symmetries that may show up in order to cope with more exotic nuclear properties.

A particular elegant approach, the shell-model Monte Carlo method, has been worked out by Otsuka and co-workers (Honma *et al* 1995, Mizusaki *et al* 1996, Honma *et al* 1996, Ostuka *et al* 1998, Mizusaki *et al* 1999, Otsuka 2002, Otsuka 2002). Instead of just extending the model spaces by counting all possible partitions of nucleons outside some inert core, the new shell-model Monte Carlo method aims to construct a selected number of basis states that are optimized to the mass region and the nucleon–nucleon force using stochastic methods (Monte Carlo basis generation or importance truncation scheme). In this case, diagonalizations in the fpg shell turn out to be restricted to no more than 30–50 basis states (Honma 1995). The early method has been improved over the years (see (Otsuka 2002)), resulting in a powerful method that even allows nuclei

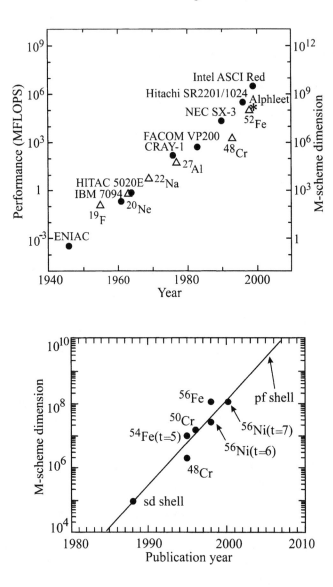

Figure 11.25. Evolution of the CPU speed of computers and shell-model dimensions. A number of typical examples are given for the corresponding years (taken from (Arima 2002)) (upper part), and, the M-scheme dimensions for the $M = 0$ model space (mainly the fp shell) as a function of the publication year (the index t denotes the number of particles that become excited from the $1f_{7/2}$ orbital into the rest of the fp shell and indicates the amount of truncation). (Taken from (Mizusaki 2002) (lower part).) (Reprinted from 2002 *Nucl. Phys.* A **704** 1c and 190c © 2002 with permission from Elsevier).

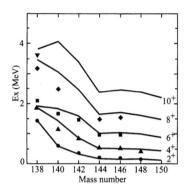

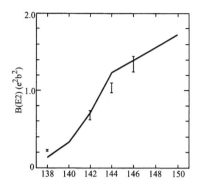

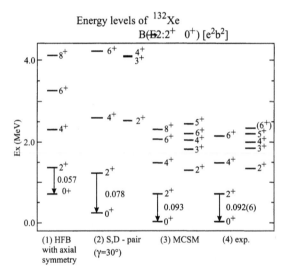

Energy levels of ^{132}Xe
$B(E2:2^+\ 0^+)\ [e^2b^2]$

Figure 11.26. The energy levels of the ground band in the even–even Ba nuclei up to spin $J = 10$ from shell-model Monte Carlo (SMMC) calculations, as compared to the data (upper left-hand part); the electromagnetic E2 reduced transition probabilities $B(E2)$ for the corresponding Ba nuclei (upper right-hand part), and the detailed results in ^{132}Xe on energies and $B(E2, 2^+ \rightarrow 0^+)$ values. Comparison between results from Hartree–Fock–Bogoliubov and specially truncated model spaces (S–D pairs) with the SMMC and the data is carried out (lower part of figure). (Taken from (Shimizu *et al* 2002).) (Reprinted from 2002 *Nucl. Phys.* A **704** 244c © 2002 with permission from Elsevier.)

in the rare-earth mass region to be treated (see figure 11.26 for the Xe and Ba nuclei).

One of the strong points is that one is able to describe e.g. spherical structures, deformed rotational bands and almost superdeformed bands using a single Hamiltonian and using the same model space. Therefore, the method allows evolving nuclear structure phenomena along the axis of increasing excitation energy to be described. The other strong point is the feasibility to handle many valence particles outside an inert core. In regular, even large-scale shell-model calculations, quite often one has to bring in truncations in the number of ways in which particles in the valence space become 'frozen'. Here now, one can handle long series of isotopes or isotones using the same techniques. This feature is essential in order to be able to treat collective phenomena and the way they evolve from vibrational towards rotational. Again, the Hamiltonian and model space remain untouched and it is only the number of interacting nucleons that generates the necessary variations in the nuclear structure. As we shall see, in chapter 15, this method can handle dramatic changes in the underlying shell structures, i.e. well-known magic numbers may well change and the shell-model Monte Carlo method is able to handle those important modifications in the mean field.

In spite of the enormous model spaces, a number of problems have been noticed, in particular regarding the nuclear binding energy or, related to that quantity, regarding the separation energies of various quantities, e.g. the two neutron separation energies $S_{2n}(Z, A)$. In figure 11.27, the $S_{2n}(Z, A)$ values are presented where the diameter of the black circles is a measure, at each point (Z, A), of the deviation

$$\Delta S_{2n} \equiv |S_{2n}(Z, A)_{\text{exp.}} - S_{2n}(Z, A)_{\text{theory}}|. \tag{11.78}$$

It can be seen, in particular for nuclei near $Z = 11, 12$ and with neutron number at or very near to the shell closure $N = 20$, that in the independent shell-model picture, large deviations from zero appear for ΔS_{2n}. In these nuclei, the nuclei appear to be *more* strongly bound, indicating, in general, a lack of convergence in the model spaces considered. Explanations for these deficiencies which, at the same time signal the presence of new physics, have been offered by Wood *et al* (1992) by allowing neutron particle–hole excitations to occur at the $N = 20$ closed shell. The creation of components in the wavefunction where, besides the regular sd model space, neutron 2p–2h components are present, gives rise to increased binding energy *and* the onset of a small zone of deformation (chapter 13). The wavefunctions become

$$\Psi(J^\pi M) = \sum_{j(\text{sd})} a((\text{sd})^n; i J)\Psi_i((\text{sd})^n, JM)$$

$$+ \sum_{k(\text{sd,fp})} b((\text{sd})^{n-2}(\text{fp})^2; k J)\bar\Psi_k((\text{sd})^{n-2}(\text{fp})^2; JM),$$

where the extra configurations are denoted by the b coefficients and basis states $\bar\Psi_k((\text{sd})^{n-2}(\text{fp})^2; JM)$. The increasased binding energy arising from these

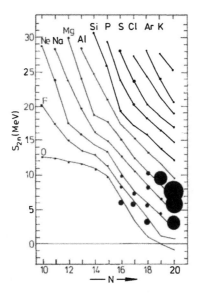

Figure 11.27. Two-neutron separation energies S_{2n} along the sd-isotopic chains. The lines connect the theoretical points. The data are indicated by dots, or solid circles. Their diameter is a measure of the deviation between the theoretical and experimental S_{2n} values (taken from Wood *et al* 1992).

$(sd)^n \rightarrow (sd)^{n-2}(fp)^2$-excitations (called intruder excitations) as well as the appearance of low-lying, deformed states in this mass region is discussed by Wood *et al* (1992).

The presence of these intruder excitations near to closed shells is a more general feature imposing severe limitations on the use of limited model spaces. On the other hand, because of the rather weak coupling between the model spaces containing the regular states and the one containing the intruder excitations, various other approaches designed to handle these intruder excitations have been put forward (Heyde *et al* 1983, Wood *et al* 1992). An overview of the various regions, in the nuclear mass table, where intruder excitations have been observed is given in figure 11.28.

Approximations to the large-scale shell-model calculation, when many nucleons are partitioned over many single-particle orbits, have been put forward and accentuate the importance of the strong pairing energy correlations in the interacting many-nucleon system. The approximation starts from the fact that we have, until now, considered a sharp Fermi level in the distribution of valence nucleons over the 'open' single-particle orbits. The short-range strong pairing force is then able to scatter pairs of particles across the sharp Fermi level leading to 2p–2h, 4p–4h, ... correlations into the ground state. When, now, many valence nucleons are in the open shells as occurs under certain conditions, one can obtain

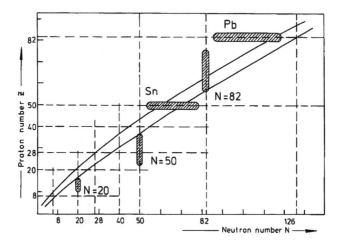

Figure 11.28. A number of regions (shaded zones) are drawn on the mass map where clear evidence for low-lying 0^+ intruder excitations across major closed shells has been observed. Here, we only present the $N = 20$, $N = 50$, $Z = 50$ and $Z = 82$ mass regions.

a smooth probability distribution for the occupation of the single-particle orbits. So, pairing correlations become important and modify the nuclear ground state nucleon distribution in a major way (see figure 11.29). The consequences of these pairing correlations, which are most easily handled within the Bardeen–Cooper–Schrieffer (or BCS) approximation, are presented in Heyde (1991) and references therein. The concepts of quasi-particle excitations, BCS theory, quasi-spin, seniority scheme are introduced in order to study the nuclear pairing correlations in a mathematically consistent way. We shall, in this presentation, not go into much detail and refer to the literature for extensive discussions (Ring and Schuck 1980, Eisenberg and Greiner (1976), Rowe (1970)).

11.6 A new approach to the nuclear many-body problem: shell-model Monte Carlo methods

In the discussion of the nuclear shell model in the preceding sections (sections 11.3, 11.4 and 11.5), it has been made clear that one aims at solving the nuclear many-body problem as precisely as possible. The basic philosophy used is one of separating out of the full many-body Hamiltonian

$$\hat{H} = \sum_i t_i + \tfrac{1}{2} \sum_{i,j} V_{i,j}, \tag{11.79}$$

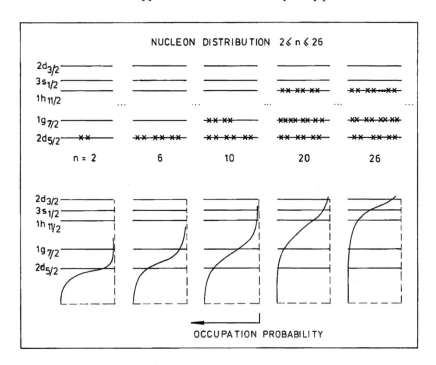

Figure 11.29. Distribution of a number of nucleons n ($2 \leq n \leq 26$) over the five orbits $2d_{5/2}$, $1g_{7/2}$, $1h_{11/2}$, $3s_{1/2}$, $2d_{3/2}$ in the 50–82 region. In the upper part we depict one (for each n) of the various possible ways in which the n particles could be distributed over the appropriate single-particle levels. In the lower part we indicate, for the corresponding n situation, the optimal pair distribution according to the pairing or BCS presumption (see Heyde 1991).

a one-body Hamiltonian, containing an average field U_i, and a correction Hamiltonian $\hat{H}_{\text{res}}$, describing the residual interactions

$$\hat{H} = \sum_i (t_i + U_i) + \tfrac{1}{2} \sum_{i,j} V_{i,j} - \sum_i U_i = \hat{H}_0 + \hat{H}_{\text{res}}. \qquad (11.80)$$

One then constructs a many-body basis $|\psi_i\rangle$ with which one computes the many-body Hamiltonian matrix elements $H_{i,j} = \langle \psi_i | H | \psi_j \rangle$, constructs the Hamiltonian energy matrix and solves for the lowest-energy eigenvalues and corresponding eigenvectors. Highly efficient diagonalizaton algorithms have been worked out over the years (see also sections 11.3, 11.4 and 11.5) allowing typical cases with dimensions of 100 000 in the m-scheme and 3000 to 5000 in angular momentum and isospin (J, T) coupled form. The largest cases that can be handled, although they make it much more difficult to exploit the symmetries of the matrix and finding an optimal 'loading' balance of the energy matrix, at present go up to 10^7–10^8. For some typical codes used, we refer to Schmid *et al* (1997) with the code VAMPIR, Brown and Wildenthal (1988), McRae *et al* (1988) with the OXBASH code, Caurier (1989) with the code ANTOINE, Nakada *et al* (1994). Otsuka and coworkers (Honma *et al* (1996), Mizusaki *et al* (1996)) devised a special technique in which a truncated model space is generated stochastically using an ingenious way of generating and selecting the basis states. Diagonalization in this truncated space can lead to very close upper limits for the lowest-energy eigenvalues and this for systems with dimension up to 10^{12}. At present, one can solve the full sd space and the fp shell partly, but for heavier nuclei one needs to take into account the solution of an approximation to the full large-scale shell-model method as the model spaces for rare-earth nuclei reach values of the order of 10^{16}–10^{20}.

All of the above methods are restricted to the description of ground-state and low-lying excitation modes and are within a $T = 0$ (zero-temperature) limit to the many-body problem. To circumvent the various difficulties and principal restrictions in handling the large-scale shell-model space diagonalization, an alternative treatment of the shell-model problem has been suggested by the group of Koonin and coworkers (Johnson *et al* (1992), Lang *et al* (1993), Ormand *et al* (1994), Alhassid *et al* (1994), Koonin *et al* (1997a, b)). This method is based on a path-integral formulation of the so-called Boltzmann operator $\exp(-\beta \hat{H})$, where $\hat{H}$ is the Hamiltonian as depicted in (11.79) and (11.80), and β is the reciprocal temperature $1/T$. Starting from this expression, one can determine the nuclear many-body thermodynamic properties through the partition function

$$Z_\beta = \hat{\text{Tr}}(e^{-\beta \hat{H}}), \tag{11.81}$$

and derived quantities like the internal energy U, the entropy S, etc. So, instead of finding the detailed spectroscopic information at the $T = 0$ limit, in this new approach one starts from the other limit: trying to extrapolate from the higher-temperature regime into the low-energy region of the atomic nucleus.

In a first step, one can separate the one-body Hamiltonian $\hat{H}_0$ from the more difficult two-body part $\hat{H}_{\text{res}}$ when evaluating the exponential expression. If the two parts of $\hat{H}$ would commute, this would become

$$e^{-\beta(\hat{H}_0 + \hat{H}\text{res})} = e^{-\beta \hat{H}_0} e^{-\beta \hat{H}_{\text{res}}}. \tag{11.82}$$

In general, this is not the case though. In the regime of small β values (high temperature T), the above separation is still possible with an error of third-order in the parameter β. One can reduce this error by using the Suzuki–Trotter formula (von der Linden 1992) resulting in the expression

$$e^{-\beta(\hat{H}_0 + \hat{H}\text{res})} = e^{-\frac{\beta}{2} \hat{H}_0} e^{-\beta \hat{H}_{\text{res}}} e^{-\frac{\beta}{2} \hat{H}_0} + \theta(\beta^3). \tag{11.83}$$

For use at higher β values (low temperature T), one needs to split the inverse temperature into a number of inverse temperature intervals N_t. This then leads to the Suzuki–Trotter formula (von der Linden 1992)

$$
\begin{aligned}
e^{-\beta(\hat{H}_0+\hat{H}\text{res})} &= [e^{-\frac{\beta}{N_t}(\hat{H}_0+\hat{H}\text{res})}]^{N_t} \\
&= e^{-\frac{\beta}{2N_t}\hat{H}_0} e^{-\frac{\beta}{N_t}\hat{H}\text{res}} e^{-\frac{\beta}{N_t}\hat{H}_0} e^{-\frac{\beta}{N_t}\hat{H}\text{res}} \ldots e^{-\frac{\beta}{N_t}\hat{H}_0} \\
&\quad \times e^{-\frac{\beta}{N_t}\hat{H}\text{res}} \times e^{-\frac{\beta}{2N_t}\hat{H}_0} + \theta(\beta^3).
\end{aligned}
\tag{11.84}
$$

The more difficult part comes from the exponential parts in (11.84) but there exist methods to decompose the exponential of a two-body operator into a sum of exponentials of one-body operators known as the Hubbard–Stratonovich transform (Stratonovich 1957, Hubbard 1959). In the simplest case where the two-body residual interaction Hamiltonian $\hat{H}_\text{res}$ can be written as the square of a one-body operator

$$
\hat{H}_\text{res} = -\hat{A}^2,
\tag{11.85}
$$

one arrives at the decomposition

$$
e^{-\beta\hat{H}_\text{res}} = e^{\beta\hat{A}^2} = \frac{1}{\sqrt{2\pi}} \int_{-\infty}^{+\infty} e^{-\frac{\sigma^2}{2}} e^{\sigma\sqrt{2\beta}\hat{A}} \, d\sigma.
\tag{11.86}
$$

This is an exact decomposition and transforms the exponential of a two-body Hamiltonian into a continuous sum of exponentials of one-body operators. Instead of evaluating the continuous sum of the auxiliary field σ, one can use a discrete decomposition by replacing the exponential weight integral by a Gaussian quadrature formula. This method remains essentially the same for a Hamiltonian that is a sum of squares of commuting observables, and even non-commuting observables (Rombouts 1997), but leads to multi-dimensional integrals over many auxiliary fields $(\sigma_1, \sigma_2, \sigma_3, \ldots, \sigma_m)$. Here again, replacement in terms of three- to four-point Gaussian quadrature formulae leads to a sum over discrete auxiliary-field configurations. The extension of the Hubbard–Stratonovich transform for a general and realistic Hamiltonian (equation (11.80)) remains possible and is discussed in detail by Rombouts (1997) and Koonin *et al* (1997a, b).

Coming back to equations (11.85) and (11.86), one can rewrite the exponential over the two-body Hamiltonian in a specified inverse-temperature slice in the form

$$
e^{-\frac{\beta}{N_t}\hat{H}_\text{res}} = \sum_\sigma e^{\hat{A}_\sigma}.
\tag{11.87}
$$

If we then apply this decomposition to every two-body term of the expansion (11.84) one finally obtains the full expression

$$
\begin{aligned}
e^{-\beta\hat{H}} &\cong \sum_{\sigma_1,\sigma_2,\ldots\sigma_{N_t}} e^{-\frac{\beta}{2N_t}\hat{H}_0} e^{\hat{A}_{\sigma_1}} e^{-\frac{\beta}{N_t}\hat{H}_0} e^{\hat{A}_{\sigma_2}} \ldots e^{-\frac{\beta}{N_t}\hat{H}_0} e^{\hat{A}_{\sigma_{N_t}}} e^{-\frac{\beta}{2N_t}\hat{H}_0} \\
&= \sum_\sigma e^{-\hat{S}_\sigma(\beta)} = \sum_\sigma \hat{U}_\sigma.
\end{aligned}
\tag{11.88}
$$

Here, the operator $\hat{U}_\sigma$ is a product of exponentials of one-body operators and, as such, an exponential of a one-body operator $\hat{S}_\sigma(\beta)$ itself. An exponential of a one-body operator can be seen as the Boltzmann operator for a (non-Hermitian) mean-field operator. Therefore, it can be represented by an $N_S \times N_S$ matrix U_σ with N_S the dimension of the one-body space. The calculation of the partition function then amounts to algebraic manipulations of small matrices, even for very large regular shell-model valence spaces. Expression (11.88) thus reduces the correlated many-body problem to a sum over systems of independent particles.

By now circumventing the diagonalization problem of matrices of huge dimension that were intractable (dimensions 10^{20} and higher), solving the many-body problem (or the partition function and derived quantitities) reduces to carrying out the summations shown in equation (11.88) which looks almost as intractable as the original problem because of the huge number of terms. The solution here comes from taking samples when carrying out the summation following Monte Carlo sampling methods. The aim is in general to compute the ratio of two sums, given by

$$E(f) = \frac{\sum_x f(x)w(x)}{\sum_x w(x')}, \tag{11.89}$$

where the number of states is very large. We assume that the function $w(x)$ is always positive; for negative $w(x)$ values a serious problem arises called the 'sign' problem discussed in detail by Koonin *et al* (1997a, b) and Rombouts (1997). The trick of performing this sum ((11.88) and (11.89)) consists in approximating the sum with a limited sample $S = \{x^{[1]}, x^{[2]}, \ldots, x^{[m]}\}$ with the $x^{[i]}$ values distributed according to the weight function $w(x)$. Then the central limit theorem assures that for large enough M, the sample average

$$E_S(f) = \frac{1}{M} \sum_{i=1}^{M} f(x^{[i]}) \cong E(f), \tag{11.90}$$

converges to the average value $E(f)$. The statistical error is proportional to $1/\sqrt{M}$ (see figure 11.30 for a schematic illustration of this method). The problem in generating a sample that is modulated according to the weight function $w(x)$ is solved through Markov-chain Monte Carlo sampling with, as a sacrifice, the fact that the $x^{[i]}$ values are no longer independent. It has been shown that the results obtained using this particular sampling technique converge to the exact results if large enough samples are drawn (Rombouts 1997). One of the most efficient methods was introduced by Metropolis *et al* (1953) and was originally suggested for the calculation of thermodynamic properties of molecules. The Metropolis algorithm can be explained using a simple example. Starting from a state x^i, a trial state x' is drawn randomly in the interval $[x^{i-d}, x^{i+d}]$. If $w(x') > w(x^i)$ then one sets $x^{i+1} = x'$, otherwise one sets $x^{i+1} = x'$ with probability $w(x')/w(x^i)$ and $x^{i+1} = x^i$ with probability $1 - w(x')/w(x^i)$. This procedure guarantees

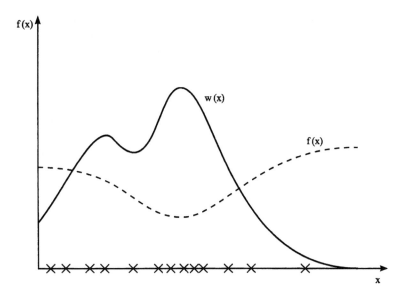

Figure 11.30. Schematic illustration of a function $f(x)$ sampled with a weight function $w(x)$. The sampling points are also given.

that in the long run the states $x^i, x^{x+1}, x^{i+2}, \ldots, x^M$ are distributed according to $w(x)$. This specific Monte Carlo sampling technique is widely used in many different fields such as statistical physics, statistics, econometrics, biostatistics, etc. Detailed discussions of the mathematical details are given in the thesis of Rombouts (1997).

This all holds if $w(x)$ plays the role of a weight function, i.e. if $w(x) > 0$ for all x. It can happen for certain choices of the Hamiltonian that $w(x) < 0$. If the average sign of the function tends to zero, it follows that the statistical error tends to infinity. This sign problem can be avoided by making a special choice for the Hamiltonian, e.g. using a pairing-plus-quadrupole force in the study of even–even nuclei. For more general interactions the sign problem is present and is particularly important at low temperatures. Possible ways of coping with this problem have been suggested by Koonin *et al* (1997, 1998) but this issue remains, at present, a topic of debate.

We are now in a position to combine all of the above elements, i.e. (i) the fact that the exponential of a one-body operator can be expressed by an $N_S \times N_S$ matrix with N_S the dimension of the space of one-particle states, (ii) the Hubbard–Stratonovich decomposition of the exponential of a two-body operator in a sum of exponentials over one-body operators introducing auxiliary fields σ and (iii) the Markov-chain Monte Carlo sampling of very large sums to evaluate thermodynamic properties of the nuclear many-body system. So one is mainly interested in the evaluation of expectation values of quantum-mechanical observables $\hat{A}$ over a given ensemble (canonical, grand-canonical or microcanonical). In, for example, the canonical ensemble, where the system has a fixed number of particles while the energy of the system can still fluctuate, one obtains the result

$$\langle \hat{A} \rangle_C = \frac{1}{Z_\beta} \hat{\text{Tr}}_N (\hat{A} e^{-\beta \hat{H}}). \qquad (11.91)$$

In the expression (11.91), the trace is taken with respect to the particular space of N-particle states (Tr_N) whereas the more general trace symbol ($\hat{\text{Tr}}$) is used for the diagonal summation over the full many-body space. The above expression can be cast in the form (making use of equations (11.88) and (11.89))

$$\langle \hat{A} \rangle_C = \frac{\sum_\sigma f_A(\sigma) w(\sigma)}{\sum_{\sigma'} w(\sigma')}, \qquad (11.92)$$

with

$$w(\sigma) = \hat{\text{Tr}}_N (\hat{U}_\sigma), \qquad (11.93)$$

$$f_A(\sigma) = \hat{\text{Tr}}_N (\hat{A} \hat{U}_\sigma) / w(\sigma), \qquad (11.94)$$

in which the weight function and the function we wish to sample are defined. One then has to apply the Monte Carlo methods discussed before in evaluating the quantum-mechanical averages. Technical details on the optimal way to evaluate the various traces are discussed in Rombouts (1997) and Koonin (1997a, b). In those references, the techniques to evaluate averages within the grand-canonical and the microcanonical ensemble are also discussed. Of particular interest is the study of the internal energy, the specific heat, the entropy and the free energy. All these properties can also be derived once the partition function is known.

Combining all of the discussions given above it should become clear that a new avenue has been opened recently allowing for an exact solution of the nuclear many-body problem within a given statistical error. The starting point is the evaluation of the partition function or the Boltzmann operator. Making use of the Hubbard–Stratonovich decomposition of the two-body part of the Hamiltonian, it becomes possible to rearrange the Boltzmann operator as a huge sum solely over products of exponentials of one-body operators. This sum is subsequently sampled using Markov-chain Monte Carlo methods. The whole formalism is a finite-temperature method. The calculational efforts become more severe the lower the energy of the many-body system becomes. One has also to note the presence of a sign problem for general Hamiltonians and, in particular, at low temperatures. Finally, one has to stress that thermodynamic nuclear properties are derived and the transformation into spectroscopic information (using an inverse Laplace transform) is far from trivial.

As a first, albeit schematic illustration, we present the study of the internal energy and the specific heat for a pairing Hamiltonian using a $(1h_{11/2})^6$ model space and constant pairing strength. In figure 11.31 we compare the results for U (internal energy) and specific heat C as a function of the temperature. It becomes

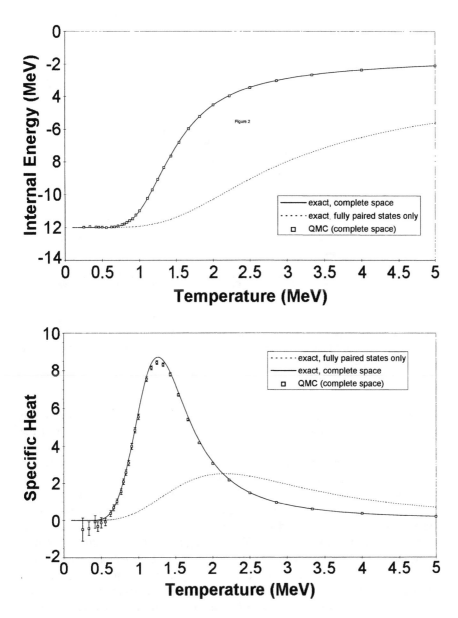

Figure 11.31. The internal energy (U) (upper part) and the specific heat (C) (lower part) as a function of temperature T. The calculation uses a $(1h_{11/2})^6$ model space. We give (a) the exact shell-model results (full line), (b) the exact result using a fully paired model space (dashed line) and (c) the quantum Monte Carlo results (reprinted from Rombouts *et al* © 1998 by the American Physical Society).

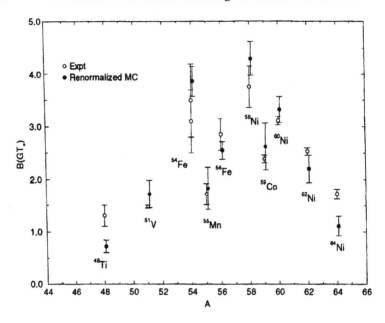

Figure 11.32. Comparison of the renormalized Gamow–Teller strength, as calculated with the shell-model Monte Carlo approach (reprinted from Koonin *et al* © 1997a with permission from Elsevier Science).

very clear that the exact results for this pairing problem and the Monte Carlo approximation, using the complete space, coincide within the precision of the drawing of the curves. The peak around 1.25 MeV in the specific heat corresponds to the break-up of the pairing structure. In figure 11.32 we give, as a benchmark example, the Gamow–Teller strength, calculated using shell-model Monte Carlo methods and compared with the experimental numbers. More realistic studies have been carried out and are discussed by Koonin *et al* (1997a, b) and references therein. Applications to ground-state properties of medium-mass nuclei (fp model space) with attention to the Gamow–Teller strength have been performed. Studies of the thermal influence on pairing correlations and on the rotational motion have also been carried out by these same authors. They discuss more exotic properties like double β-decay and the study of a fully microscopic structure of collective, γ-soft nuclei and they give an outlook for the study of giant resonances and multi-major shell studies.

The Monte Carlo method as described earlier can also be successfully used in the study of nuclear level densities (Nakada and Alhassid 1997). Accurate knowledge of level densities is very important in the study of nuclear reactions of various types, in particular for neutron and proton capture rates. The nucleosynthesis of many of the heavy elements known in nature proceed exactly via these capture processes (s and r for neutrons, rp for protons). Most theoretical approaches take as a starting point the Fermi gas model (see chapter 8), which leads to the Bethe formula (Bethe 1936) which describes the many-particle level densities starting from the single-particle level-density parameter a. Shell corrections and the most important two-body correlations are taken into account in a highly empirical way. In the backshifted Bethe formula (BBF) for example, the ground-state energy is shifted by an amount Δ (see chapter 7). This particular formulation gives a fairly good description of level densities in many nuclei if both the quantities a and Δ are fitted for each individual nucleus. Therefore, the use of this method is of limited scope in order to try to reach a firm understanding of level densities.

The nuclear shell model leads to an ideal framework for determining the level densities by explicit calculations (see chapter 11) but present day shell-model capabilities cannot cope with the very large model spaces that would be needed to describe the level densities in the region of neutron resonances through conventional diagonalization methods. It is here that the Monte Carlo shell-model methods come in since they provide the necessary framework to describe the thermal properties of nuclei in a highly excited state. Studies have been carried out by Alhassid and Nakada (Nakada Alhassid 1997, Nakada and Alhassid 1998, Alhassid *et al* 1999, Alhassid 2000, Langanke 1998) in the Fe mass region with success, even making a specific study of the parity dependence in the level density expressions.

Recent results on the density of accessible levels in rare-earth nuclei by Melby and Guttormsen (Melby *et al* 1999, Guttormsen *et al* 2000) have led to the possibility of extracting the nuclear temperature as a function of excitation energy in the region 0–6 MeV. The behaviour in the nuclear temperature gives first indications of a possible onset of pair breaking signaling a possible phase transition in the pairing model in nuclei.

11.7 Spectral properties of the nuclear shell-model many-body system

In the study of the simplified situation in which just two configurations are considered as a basis to treat residual interactions amongst those basis configurations, we have discussed in section 11.3.2 (see, in particular figure 11.14), a no-crossing rule results such that for a crossing of the energies for these unperturbed configurations, the energy eigenvalues are characterized by a closest approach of $2|V|$ with V the value of the interaction matrix element. This simple two-level model can, of course, be generalized to the situation of large-scale configuration mixing problems. Moreover, coming back to the simple case of just two interacting configurations of section 11.3.2, it may still be that the two levels with equal spin and parity J^{π} cross after diagonalization if an extra quantum number would characterize the wavefunction. A typical example is the case of the isospin quantum number T. For the case of a Hamiltonian which conserves the isospin symmetry exactly, no isospin mixing will show up and the isospin quantum number T characterizing the wavefunctions will remain a good quantum number, besides the spin and parity. So, one can have the situation that $E(J^{\pi}, T) = E(J^{\pi}, T')$. Here, one clearly notices that a study of the spectral properties, e.g. the distribution of the nearest-neighbor level distances denoted by $s_{ij} \equiv E_i - E_j$, carries interesting information on the characteristics of the residual interactions in the many-body system.

A property that is especially sensitive to the interactions amongst the nucleons in the atomic nucleus is the nuclear level spacing distribution. Starting from a simple 2×2 model Hamiltonian in which the elements of the matrix

$$\begin{pmatrix} H_{11} & H_{12} \\ H_{21} & H_{22} \end{pmatrix}, \tag{11.95}$$

are randomly distributed (uncorrelated), we can then consider an ensemble of 2×2 matrices in which the probability for a given matrix is specified by some function $P(H)$ and which is given by the relation

$$P(H_{11}) = P(H_{22}) = P(H_{12}) = P(H_{21}). \tag{11.96}$$

One can then study the statistical properties of the eigenvalues and eigenvectors of this ensemble (Bohr and Mottelson 1969, Guhr *et al* 1998). If we call E_1 and E_2 the two eigenvalues, one can obtain (in an unnormalized form) a probability distribution for the relative energy difference s, between the eigenvalues (with s defined as $s \equiv E_1 - E_2$)

$$P(s) = s \cdot \exp(-s^2), \tag{11.97}$$

which is referred to as the Wigner distribution (Bohr and Mottelson 1969, Guhr *et al* 1998) (see figure 11.33).

This simple 2×2 matrix analysis illustrates the types of results that can be obtained starting from random matrix ensembles. However, in order to study the more general properties of the distributions of levels, one should extend these methods to the properties of matrices with large dimensions. Essentially the same Wigner distribution results. It is also said to correspond to the Gaussion Orthogonal Ensemble of matrices (GOE) (extensions to Gaussian Unitary Ensembles or GUE is easy) (Bohr and Mottelson 1969, Guhr *et al* 1998).

One can similarly show that in non-interacting systems (systems where, because of the presence of certain symmetries, for example isospin symmetry (Guhr and Weidenmüller 1990), that govern the Hamiltonian describing the system, a number of vanishing H-matrix elements appear), the energy spacing is described by a Poisson distribution (see figure 11.33) and is given by the expression

$$P(s) = \exp(-s). \tag{11.98}$$

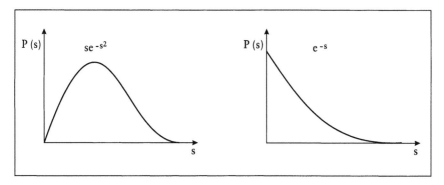

Figure 11.33. Illustration of the two extreme distributions: the Wigner probability distribution $s \exp(-s^2)$ corresponding to randomly distributed levels and the Poisson distribution $\exp(-s)$ corresponding to a non-interacting system. (Reprinted from 'From Nucleus to the Atomic Nucleus' (K Heyde) © 1998 with permission of Springer Verlag.)

It is now possible to connect these level spacing distributions via a semiclassical argument to classical ideas of integrable and non-integrable systems (Bohigas and Weidenmüller 1988, Weidenmüller 1986, Gutzwiller 1990, Gutzwiller 1992, Stöckmann 1990). In considering a certain bounded region where classical motion is periodic (of integrable), e.g. the rectangular domain or the domain spanned by two concentric circles (see figure 11.34), or non-integrable (chaotic or partially chaotic), e.g. a stadium domain or in the case of the well-known Sinai billiard boundary conditions (Weidenmuller 1986, Bohigas *et al* 1984) (see figure 11.35), one can then regard the domain walls in a quantum-mechanical two-dimensional potential problem such that the wavefunctions have to vanish at the boundaries imposed by the particular potential domain. Solving the corresponding two-dimensional Schrödinger equation with a potential that is constant inside the domain and takes an infinite value outside of the domain, one can solve for the eigenvalues and corresponding wavefunctions with appropriate boundary conditions. If we now study the distribution of the energy eigenvalues as $P(s)$, where s is a measure of the distance between adjacent levels in the domain, a remarkable result shows up. For the classically integrable systems, the resulting $P(s)$ distribution follows a Poisson law whereas for the non-integrable systems in which chaotic motion can result, a Wigner GOE distribution law is obtained. Thus, the conjecture, made by Bohigas, expresses the fact that a one-to-one correspondence exists between a classical system and the related quantum system (Bohigas and Weidenmüller 1988). Closely related is the periodic orbit theory (POT) (Gutzwiller 1990, Gutzwiller 1992), in which a description of quantum spectra is given in terms of classical closed orbitals. What comes out of this discussion is the fact that after the extraction of the (non-generic) bouncing-ball orbitals (or classical closed orbitals) of the classical system, the quantum

Rectangle   Circle

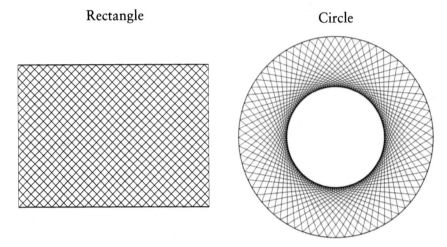

Stadium

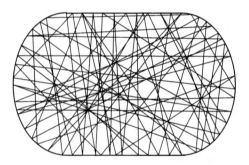

Figure 11.34. Some illustrative examples of domains (rectangle, region between two concentric circles) in which periodic, regular integrable motion can result compared to a potential domain (stadium) in which random, non-integrable motion results. (Reprinted from 'From Nucleus to the Atomic Nucleus' (K Heyde) © 1998 with permission of Springer Verlag.)

counterpart behaves like a typical chaotic system. The study of the generic system then relates to a study of level statistics and also width statistics.

A very interesting approach is connected to the experimental study of microwave frequencies in a two-dimensional (2D) microwave resonator. It builds on the observation that the Schrödinger equation $(\Delta + k^2)\psi = 0$ with $k^2 = \sqrt{(2mE/\hbar^2)}$ is identical to the Helmholtz equation for the z-component of the electric field E_z in the 2D resonator $(\Delta + k^2)E_z = 0$ with $k = 2\pi f/c$ with a potential in the Schrödinger equation conform with the geometrical shape

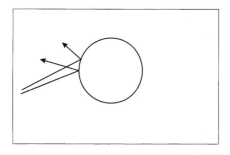 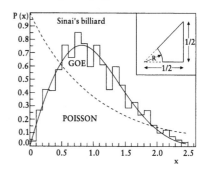

Figure 11.35. The Sinai billiard domain in which two, initially close, trajectories diverge in time. The right-hand part of the figure shows the distribution of level spacings (nearest-neighbor distribution $P(x)$) corresponding to the energy eigenvalues for the Sinai potential domain. This latter distribution is very well described by the GOE or Wigner distribution. (Reprinted from Weidenmüller H 1986 *Comm. Part. Nucl. Phys.* **16** no 4 199 © 1986 Gordon and Breach with permission.)

of the resonator. Therefore, the motion of the quantum particle in a given potential can be simulated by means of the electromagnetic waves inside the 2D superconducting microwave resonator in an experimental way. Such experiments have recently been performed at the Technical Universität Darmstadt by Richter's group (see (Gräf *et al* 1992, Alt *et al* 1995, Richter 1999, Dembowski *et al* 1990) for an overview of work in this field), using various model configurations (one-quarter stadium, ...).

This discussion justifies the interest in a study of various spectral properties of atomic nuclei. One can certainly find experimentally interesting cases. The difficulty is that it is necessary to sample a sufficiently large ensemble of levels of the same spin and parity in a given nucleus and study its nearest-neighbour level distributions. Until recently, the only useful information came from neutron and proton resonance studies. More recently, the situation has improved through the availability of the 'Nuclear Data Ensemble (NDE)' which consists of all spin $\frac{1}{2}$ s-wave neutron resonances measured by the group at TUNL (Haq *et al* 1982) (see figure 11.36), the completely known spectrum of ^{26}Al between the ground state and the proton threshold and spectroscopic data obtained via extensive (n, γ) experiments over many nuclei.

Recently, an interesting observation was made that was not expected at all related to the way standard nuclear structure regular features are produced. The general thought was that this is due to the special properties of the nuclear two-body forces. It was shown by Johnson *et al* (Johnson *et al* 1998) that when random two-body interactions are used in the study of the low-energy properties of many-body systems, just by specifying an invariance under a particle–hole conjugation, patterns appeared that are very similar to the results of standard two-

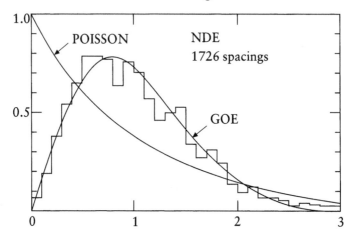

Figure 11.36. Nearest-neighbor spacing distribution $P(x)$ corresponding to the Nuclear Data Ensemble (NDE) consisting of 1726 level spacings. The NDE distribution follows a GOE distribution very closely. (Reprinted from Richter A 1993 *Nucl. Phys.* A **553** 417c © 1993 with permission from Elsevier.)

body interactions: observation of 0^+ ground states, separated by a gap from the higher excited states, evidence for phonon vibrations at low energy, etc. The interplay of the regular characteristics generated by the shell model and the random choice of the original interactions has lead to a study of this topic over the last 5–6 years. We refer to (Velazques *et al* 2003, Bijker and Frank 2000, Zhao and Arima 2001, Horoi *et al* 2001) for more details on this issue and for a good entry for further study of this topic.

Level statistics within the nuclear shell model has been investigated starting from given model Hamiltonians in a number of theoretical studies: we cite a number of papers that can be used as a starting point for further studies in this field ((Volya *et al* 2002, Horoi *et al* 1995, Moline *et al* 2000, Caurier *et al* 1996) and references therein). In these situations, the generated sets of levels are consistent mainly with GOE statistics, as expected. Collective models have also been studied (see discussion in chapters 13 and 14 concentrating on vibrational and rotational motion) with most interesting results. These studies cover a large class of collective degrees of freedom such as transitions between certain symmetries within the Interacting Boson Model (IBM—see chapter 13, section 13.3) (Pato *et al* 1994), properties in deformed nuclei related to the K-quantum number (Bosetti *et al* 1996) and to axial deformed rotors (Heiss *et al* 1995) and references therein and particle–rotor systems (Kruppa *et al* 1995). Recent studies in this respect have concentrated on three different properties and their spectral properties. Detailed studies of the level spacings for the scissors mode (see Box 13b) have been performed. The ratio of the energy for the first 4^+ to 2^+ states, called $R_{4/2}$, covering 1306 2^+ levels in 169 nuclei has been used to study the level statistics for groups of states with similar collective properties (expressed by this $R_{4/2}$ ratio). Most distributions are consistent with GOE but for certain ratios, near 2, 2.5 and 3.3, those ratios that appear for pure harmonic vibrators, γ-soft nuclei and axial rotors, respectively, distributions consistent with a Poisson law show up (Abul-Magd *et al* 2004). Finally, a study of the properties of so-called pygmy dipole resonances (PDM) has been carried out by the Darmstadt group of Richter (Richter 2002), covering an ensemble of 154 1^- levels in ^{138}Ba, ^{140}Ce, ^{144}Sm and ^{208}Pb and the level distributions are quite consistent with a Poisson distribution. Large-scale quasi-particle calculations (QPM) generating a set of 841 1^- states, by the same group, favor a GOE distribution. This difference is not fully understood yet but might be due to missing levels in the experimental studies.

This field of the study of the spectral properties that appear within the atomic nucleus has been richly documented in recent years and has been the subject of a number of very well-written overview papers and review talks. Since a thorough discussion of this subject goes outside the scope of the present textbook, I refer the reader to a list of seminal papers from which a good overview can be gained for further study.

Box 11a. Large-scale shell-model calculations: making contact with deformed bands

---

We have presented, in this chapter, a large number of results produced by the nuclear shell model provided a good effective interaction, adapted to the model space, is used. Starting from these robust features, one can try to examine a number of key questions, for example:

- Is it possible to describe rotational motion starting from a spherical shell-model approach, applied to nuclei beyond the sd-model space?
- How can one truncate the huge model space but still retain the essential physics?.
- Does the possibility exist to derive objects that resemble the concept of an intrinsic state by studying the wavefunctions obtained?
- How do the spherical shell model and deformed models (see chapter 13) compare?

The pf shell is clearly a good region to study these questions because the model space is large enough (it consists of the $1f_{7/2}$, $2p_{3/2}$, $1f_{5/2}$ and $2p_{1/2}$ orbitals that can contain, in the mid-shell point, 20 particles outside a ^{40}Ca core). It also contains orbitals that are strongly connected through the quadrupole operator r^2Y_2, an essential criterion in order to be able to have collective quadrupole properties develop, as shown previously in the sd shell by Elliott (Elliott 1958). If one considers the nucleus ^{48}Cr with four valence protons and four valence neutrons in the fp shell, the experimental spectrum of the lowest collective band shows a clear collective structure. The shell-model calculations have been carried out using the code ANTOINE (see section 11.5 and references in that section as well as (Poves 1997, Poves 1999, Langanke and Poves 2000) for details and further references on the technical details that appear in constructing and diagonalizing the energy matrices).

Here, we anticipate the discussion of collective excitations, which will be discussed in chapter 13, but it is clear that the experimental spectrum, when plotted in a special way, i.e. plotting the angular momentum J of the given state *versus* the energy difference of the corresponding states $E(J) - E(J-2)$, is very consistent with that of a rotor at the bottom part of the band (see figure 11a.1) (Lenzi *et al* 1996, Poves 2003). It does not fit at all the energy relations that are fixed when considering a vibrational structure. Moreover, some strange 'backbending' in the curve is observed when passing through a given interval in angular momentum (between $J = 8$ and 14) when again a rotor-like structure seems to emerge. This particular phenomenon will be discussed at some length in chapter 14 when discussing rotational motion at high angular momentum. What becomes clear though is that the interactions amongst the four protons and four neutrons moving in all possible ways within the fp shell-model space do produce collective properties that are characteristic for deformed systems that carry out

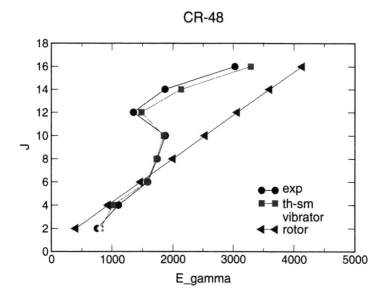

Figure 11a.1. A comparison of the experimental data on the ground-state band in ^{48}Cr (Lenzi *et al* 1996), plotted in a particular way, i.e. plotting the total angular momentum *J versus* the energy difference $E(J) - E(J - 2)$ (also called the backbending plot) with the results from a large-scale shell-model study (Poves 1997, Poves 1999) and references therein) and with the collective-model predictions for a pure rotor and vibrational ground band. (Courtesy of A Poves (Poves 2003) with reference to Caurier *et al* 1994 *Phys. Rev. C* **50** 225.)

rotational motion. The agreement between the results of the large-scale shell-model calculations and the experimental data is impressive.

One can also make a more quantitative analysis of the shell-model results by re-analysing the shell-model results using collective rotational expressions for the band member energies and the electromagnetic E2 decay properties (decay rates and quadrupole moments). These results indicate that, up to spin $J = 10$, the shell-model results are consistent with a rotor-like description for both the energies and the E2 decay rates.

As a final point, it is most interesting to note that when reducing the large model space to just the $1f_{7/2}$, $2p_{3/2}$ subspace, the results do not change in an essential way. This then gives an interesting hint towards a way to truncate model spaces but still keeping the essential physics intact (in the present case, the rotational collective motion).

So it seems that the spherical shell model is able, through the correlations that result amongst the many nucleon partitions by means of the specific effective force, to generate collective properties that would normally only be ascribed to collective phenomena in atomic nuclei.

Chapter 12

Nuclear physics of very light nuclei

12.1 Introduction: Physics of nuclear binding and stability

Most of our present-day understanding of nuclear structure and the ways in which protons and neutrons 'stick' together under the influence of the strong nucleon–nucleon force derives from a study of a small 'patch' in the (Z, N) plane of atomic nuclei, at rather low excitation energy and small rotation (or spin) and from study of a limited number of open decay channels, be it natural decay or induced via nuclear reactions. Descriptions of this interacting system of A nucleons using mean-field techniques (see chapter 10) or extensive shell-model studies (see chapter 11) have been fairly successful. The interactions used in these methods, however, have been fixed, in particular, by using information from nuclei that appear at or very near to the region of β stable nuclei. Therefore, in order to extend these methods to very light nuclei or nuclei very far away from the valley of stability, other methods will have to be used.

Talking about the 'physics' for nuclei far from stability gives, at least, the impression that the basic rules of binding many-body systems under the influence of the strong nucleon–nucleon force would be totally different from what one notes near the valley of β-stability. Starting from the two-body nucleon–nucleon interaction, as determined from the scattering between nucleons, all kinds of correlations caused by the presence of more nucleons, the so-called 'medium' effects, will show up and will quickly make the picture more complicated to follow starting from the two-body forces. We will come back to this issue later, and in more detail, in section 12.2.

One can illustrate just how quickly the picture gets complicated, by looking at the most simple combinations of protons and neutrons, i.e. the deuteron (see also chapter 9, Box 9a). We know that the binding energy amounts to $E_{\mathrm{B}} = 2.224\,61 \pm 0.000\,07$ MeV and has spin-parity $J^{\pi} = 1^{+}$. The motion essentially corresponds to the two nucleons moving in a relative $L = 0$ orbital angular momentum state and have spins coupled to $S = 1$, forming mainly an $^{3}S_{1}$ state. One can easily solve for the wavefunctions when simplifying the potential

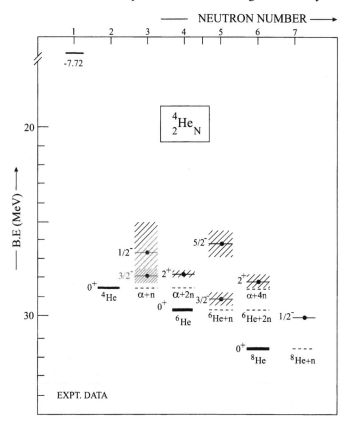

Figure 12.1. Binding energy for the various He isotopes (mass $A = 3$–9). Particle unbound states are represented as hatched regions. Thresholds are given as short-broken lines.

to a finite square-well with depth V_0 and range R. What one notes is that if we consider a potential that vanishes at $R = 1.93$ fm (determined from p–n scattering data), the value of $\sqrt{\langle r^2 \rangle}$ for the deuteron wavefunction becomes 3.82 fm and the probability of finding the two nucleons outside the range of the potential amounts to 66%. These results, which have been confirmed by much more complicated calculations, show that the simplest two-neutron system is very loosely bound.

Another interesting example for learning about binding and stability is to study very light nuclei (Pieper *et al* 2001, Pieper 2002, Pieper *et al* 2002, Wiringa and Pieper 2002). So, we consider the series of He nuclei by adding more and more neutrons: ^{2}He obviously does not exist but then we go to ^{3}He, ^{4}He, ^{6}He, ^{8}He and that is it. Even though the nuclei ^{5}He and ^{7}He have a positive binding energy, they are unstable against particle emission (see figure 12.1).

So, one has to be cautious when talking about 'bound' systems and 'stable' systems. *The first is an absolute statement (the total mass (or energy) of the*

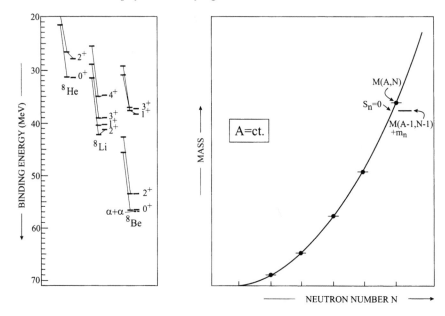

Figure 12.2. A typical example for the relative stability of ^{8}He, ^{8}Li and ^{8}Be (connected through a chain of β^--decays), ending in the instable configuration against the formation of 2α particles (^{8}Be $\rightarrow \alpha + \alpha$) (left-hand side). In the right-hand part, the relative masses in a β^--decay chain for fixed A (n $\rightarrow$ p + e$^-$ + $\bar{\nu}_e$) are given. In moving towards the neutron-rich side, a situation will occur in which the mass of the nucleus $M(A, N)$ will no longer be stable and this will lead to neutron drip-off (the situation of a nucleus $A-1(N-1)$ with one neutron less plus a neutron such that $M(A, N) \geq M(A - 1, N - 1) + m_n$).

system is less than the sum of the masses of the individual constituents) whereas the other is a relative statement (one compares the binding energy, for all possible partitions of a given number of protons Z and neutrons and the actual nucleus $^A_Z X_N$ itself to decide on particular 'particle' stability, or, can study the instability against transformations for fixed A like β^-, β^+ and EC decay (see also figure 12.2 as an illustration of both types of unstable nuclei). One can follow this in a plot (the ground-state binding energy relative to the first particle threshold) and one notes some interesting staggering pointing towards correlations of a pairing type. So, the He nuclei with three or five neutrons are just not stable against particle emission. One can do this for much heavier nuclei and one generally notes that these staggering effects are a genuine property of all nuclei. This is what I mean by 'correlations' or 'medium' effects which are governed by the fact that (a) the nucleon–nucleon force inside the nucleus does not show specific charge dependences and (b) the nucleon–nucleon force 'saturates', so in a system with A nucleons, the binding energy essentially scales with the number of nucleons A, not the number of all possible two-body interaction energies $A(A - 1)/2$.

In general, when S_p and S_n become zero, one can no longer add a single proton or neutron and the stability of the nuclei comes to an end at the 'so-called' drip-lines. They mark out a region in the (N, Z) plane within which all known nuclei are situated. Experimentally, at present, one has come quite close to mapping out the nuclei up to the proton drip-line. This is not at all the case for neutron-rich nuclei (except for very light nuclei). Again, even though nothing about the basic physics of binding nucleons (protons,neutrons) is changed, one will surely get unexpected properties because of important medium effects. Nuclei in the Sn region near the drip-line can reach a ratio $N/Z \propto 2.4$–2.5 and one expects differences relative to the structure near the valley of stability. Can one go to such extremes as nuclei consisting of neutrons or protons only? The latter, again, is obviously impossible because of the extra Coulomb forces in the system. The point I would like come back to is that this early exercise we made on the He nuclei, cannot be repeated for the Sn nuclei and one will have to use different methods to study the stability of these nuclei.

The most simple approach to discuss nuclear stability is to use the liquid-drop model (LDM) which contains a volume, a surface and a Coulomb term, the latter two counteracting the volume binding-energy term (see chapter 7). One also considers a symmetry-energy term favouring $N = Z$ nuclei, making all other combinations less bound and a pairing term (which we have already noted to be present in the He nuclei). The full expression looks like

$$BE(A, Z) = a_V A + a_S A^{2/3} - a_C Z^2/A^{1/3} - a_A (A - 2Z)^2/A \pm a_P/A^{1/2}. \quad (12.1)$$

For light nuclei with $Z = 5$—the B nuclei—the binding energy relative to the first particle threshold (so this is actually the neutron or proton separation energy depending whether one is moving into the neutron-rich or proton-rich direction) is presented in figure 12.3.

Similarly, if one inspects the separation energy for two neutrons, S_{2n}, one also notices a general drop moving away from the β-stability line (now without the staggering) and, if one moves through a series of isotopes, one notes that it is essentially the symmetry term which makes up for this effect approaching the drip-line at $S_{2n} = 0$ (see chapter 7). This means that there are various drip-lines, depending on which particle or group of particles one is looking at (it is clear that because of the pairing effect, slight differences in the last stable nucleus will show up). There are, however, in the LDM a number of assumptions which consider protons and neutrons to be contained within a given potential (in the same volume in space) with a constant total density ρ_0 and with the density ratio $\rho_n = N/Z\rho_p$. This will surely change far from the region near β-stability and we will discuss examples of experiments pointing this out.

Coming back to the question of extreme forms of exotic nuclei, one can ask how far one could extrapolate the simple liquid-drop concept. An extreme but at the same time straightforward extension leads to the choice of putting $Z = 0$, hence considering a pure neutron system. This gives rise to a binding energy per

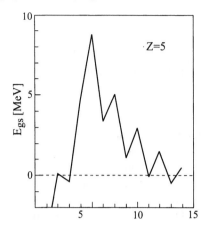

Figure 12.3. Binding energy, relative to the first particle threshold for the system with $Z = 5$ protons—this quantity is called E_{gs}—as a function of the neutron number N. (Figure taken from Marques Moreno (2003) and see references therein.) (Reprinted from 'Halos, molecules and multineutrons', *Proc. Ecole Joliet-Curie de Physique Nucléaire*, September 2002, with permission.)

nucleon of

$$BE(A, Z = 0) \approx (a_{\mathrm{V}} - a_{\mathrm{A}}) - \frac{a_{\mathrm{S}}}{A^{1/3}}. \tag{12.2}$$

This asymptotic expression only depends on the difference between the volume and symmetry constants and for typical choices ($a_{\mathrm{V}} \approx 15$ MeV and $a_{\mathrm{A}} \approx 23$ MeV) the neutron system is unbound by about 8 MeV/nucleon. This all depends on a good knowledge of these coefficients, in particular the symmetry strength, which can be derived from a Fermi gas model and expanding the total energy around the symmetric point $N = Z$ (see chapter 8). Applying the LDM to light systems (Marques Moreno 2003), using a standard parametrization, one notes that the symmetry term is overestimated (see figure 12.4).

In some versions, the surface-to-volume ratio is also considered when evaluating the symmetry term and this has the general effect of weakening the parameter a_{A} (see broken curves). For light nuclei, these effects are still rather small but, for a neutron nucleus, this amounts to an effect of almost 20 MeV/neutron. When the system becomes very neutron-rich, the hypothesis behind a single liquid drop of protons and neutrons, distributed over the same volume, does not hold any longer. Even a simple estimate can be made considering a Fermi gas model in which the density remains ρ_0 all over the nucleus but one with two phases: a central core with $N = Z$ with the extra neutrons outside the core forming a neutron skin. The resulting symmetry term remains linear in A for a nucleus with all neutrons but the strength a_{A}, needed to describe known nuclei, is much smaller and is about 6 MeV. This leads to the dotted lines. It is certainly too simplistic but figure 12.3 displays how little we know about these systems: calculations that describe known nuclei within a

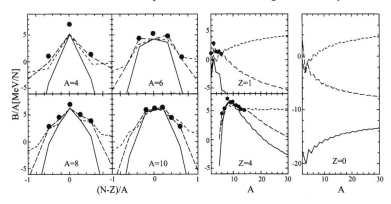

Figure 12.4. Binding energy per nucleon for different isobars (left), H and Be isotopes (middle) and pure neutron systems ($Z = 0$) (right). The symbols are the data, the lines are results of the liquid drop model using various asymmetry terms: standard (filled), surface corrected (broken) and one derived from a neutron-skin density model. (Figure taken from Marques Moreno (2003). (For permission, see caption of figure 12.3.))

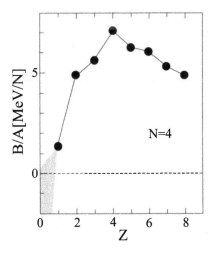

Figure 12.5. Binding energy per nucleon for nuclei containing four neutrons and a varying number of protons (up to $Z = 8$) and extrapolating towards a system with just four neutrons. (Figure taken from Marques Moreno (2003). (For permission, see caption of figure 12.3.))

few MeV/nucleon could lead to diverging effects of up to 20 MeV/nucleon. Of course, in discussing an approximate model for a neutron 'nucleus' (star?), one would also have to incorporate a gravitational term into equation (12.2) which then can lead to a stable configuration (see chapter 7, Box 7a).

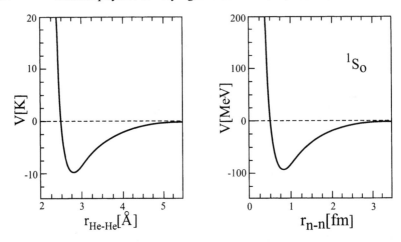

Figure 12.6. Comparison between the ^{3}He interatomic potential and the neutron–neutron potential acting in the $^1S_0$ singlet state (with $S = 0$ and $T = 1$). (Figure taken from Marques Moreno (2003). (For permission, see caption of figure 12.3.))

A last extreme example, when extrapolating very light and very neutron-rich systems, one notes a most intriguing feature (see figure 12.5) for systems with just four neutrons: it turns out that ^{8}Be is the most strongly bound (energy/nucleon). In progressing towards lower and lower Z values and going through the sequence ^{8}Be $\rightarrow$ ^{7}Li $\rightarrow$ ^{6}He $\rightarrow$ ^{5}H, there is a steady drop albeit with a staggering which comes from pairing effects. The next 'nucleus' is ^{4}n which probably goes on decreasing but with the pairing part, this is not yet clear (as an aside: even if the four-neutron system should be barely bound, there are no bound subsystems to decay to, so this would lead to a bound system!). The fact that two neutrons do not form a bound state is often used as an argument against the formation of multi-neutron systems. However, there exists an analogy with atomic physics if we consider the system of ^{3}He atoms. Indeed, ^{3}He atoms act like fermions and their interaction, although attractive cannot form a dimer (see figure 12.6 where the potentials for the nucleon case is compared to the case of two ^{3}He atoms). Since one knows that, for a large number of atoms, the ^{3}He atoms form a liquid drop, a number of theoretical studies have been carried out in order to find a good and realistic estimate of the critical number of atoms needed in order to be able to form a bound system, leading to the result $N \approx 40$ (Baranco *et al* 1997, Guardida and Navarro 2000). Very recent calculations, using modern nucleon–nucleon potentials and incorporating both three- and four-body effects ((Pieper 2003) and the references therein) cannot reconcile a bound four-neutron system with the experimental properties of binding in very light nuclei unless one destroys some benchmark results in the study of these light nuclei (see section 12.2).

12.2   The theoretical study of very light systems

One may ask the question, at this point whether one could not start from *ab initio* methods and start for the lightest nuclei, using a Hamiltonian containing both two- and three-body terms such as

$$H = \sum_i^A T_i + \sum_{i<j} V_{i,j} + \sum_{i<j<k} V_{i,j,k}. \qquad (12.3)$$

Because there is no natural central point in an atomic nucleus, this is a very difficult job and the group of Pandharipande and co-workers (Pieper *et al* 2001, Pieper 2002, Pieper *et al* 2002, Wiringa and Pieper 2002, Carlson and Sciavilla 1998) has come as far as mass $A = 10$. The two-nucleon interactions used are parametrized through some 60 parameters that fit all known nucleon–nucleon scattering data with a χ^2/data of ≈ 1. The Argonne v_{18} potential used can be written as a sum of electromagnetic and one-pion exchange terms and a shorter-range phenomenological part. The electromagnetic terms include one-and two-photon exchange Coulomb interactions, vacuum polarization, Darwin–Foldy and magnetic moment terms with appropriate proton and neutron form factors. The one-pion exchange part contains the usual Yukawa and tensor radial functions with a short-range cutoff. This whole part plus the phenomenological part can be written as a sum of 18 terms. The first eight terms contain the operators $\{1, \vec{\sigma}_i \cdot \vec{\sigma}_j, \hat{S}_{ij}, \vec{L} \cdot \vec{S}\} \otimes \{1, \vec{\tau}_i \cdot \vec{\tau}_j\}$, six extra terms contain quadratic orbital terms like $\{L^2, L^2 \vec{\sigma}_i \cdot \vec{\sigma}_j, (\vec{L} \cdot \vec{S})^2\} \otimes \{1, \vec{\tau}_i \cdot \vec{\tau}_j\}$ and four extra terms that explicitly break charge independence. The radial forms associated with each of these operators are determining by fitting data on nucleon–nucleon scattering containing 1787 pp and 2514 np data in the energy range 0–350 MeV. Of course, both the binding energy of the deuteron and the nn scattering length have been included in this fit.

In order to go beyond two-body systems, a three-body term needs to be included: such three-body terms are written as sums of two-pion-exchange and shorter-range phenomenological contributions. In the more recent studies, in order to describe the isospin dependence for neutron-rich systems correctly, i.e. to have repulsive contributions in isospin $T = 1/2$ triples and attractive in $T = 3/2$ triples, a number of modifications with respect to the original three-body Urbana potentials was necessary (leading to the so-called 'Illinois' three-body potentials).

In light nuclei, it turns out that these three-body terms can contribute from 15% up to almost 50% in the nuclear binding energy (increasing mostly with increasing neutron excess).

The methods used start from a trial wavefunction $\Psi_T^{J,\pi}$ which is first constructed for the given nucleus and optimized and contains information about the way nucleons are distributed over the lowest $1s_{1/2}$, $1p_{3/2}$, $1p_{1/2}$ orbitals. This trial wavefunction is then used as the starting point for a Green-function Monte Carlo calculation (GFMC) which projects the exact lowest-energy state

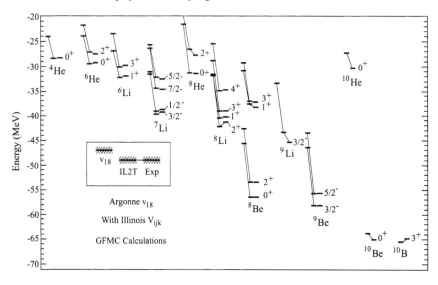

Figure 12.7. Energy levels for the light nuclei using a Quantum Monte Carlo (QMC) calculation using two-and three-body forces (Pieper 2002). (Reprinted from Pieper S C 2002 *Eur. Phys. J.* A **13** 75 © 2002 with permission from Springer-Verlag.)

with the same quantum numbers by propagating it in imaginary time or evaluating $\Psi_0 = \lim_{\tau \to 0} \exp(-(H - E_0)\tau)\Psi(\text{trial})$ (Koonin *et al* 1997).

These are very complex calculations (see figure 12.7). The number of spin–isospin components in a given trial wavefunction $\Psi_T^{J,\pi}$ rapidly grows with the number of participating nucleons in the nucleus: a calculation for a state in ^{8}Be e.g. involves about 30 times more floating-point operations compared to one for ^{6}Li and ^{10}Be requires more than 50 times what is needed for ^{8}Be.

A most interesting result arises in the study of the mass $A = 5$ and $A = 8$ chains for which mass numbers it is known that no stable elements appear in nature. By studying in detail the effect of the various terms in the full two-plus-three-body force on the binding energy and relative stability of the various elements, it became clear that a pure central potential cannot be reconciled with the experimental data. It seems that the relative strength of the tensor force compared to the spin–orbit term is crucial in explaining these peculiar properties of the instability of elements with $A = 5$ and $A = 8$. This particular tensor force resembles the force that couples two magnetic moments. This same force component, which is known to be instrumental in bringing in the necessary d-wave correlations in the deuteron s-wave ground state in order to explain the observed electric quadrupole moment (see figure 12.8), is also needed, it seems, to provide the key in understanding the instability for mass $A = 5$ and $A = 8$.

Calculations of this type are feasible up to mass $A = 10, 12$ so it is clear that other methods will be needed to extend full-scale shell-model studies for such

Figure 12.8. Constant density surfaces for a polarized deuteron in the $M_d = \pm 1$ (left-hand part) and $M_d = 0$ (right-hand part) states. The deuteron has a total angular momentum of spin-1 and can be oriented in a specific direction, for example by using an external magnetic field, with possible spin projections $M_d = +1$ (parallel), $M_d = -1$ (anti-parallel), or $M_d = 0$ (perpendicular). The force which can be attributed to the exchange of a π meson at long range has a strong tensor character which leads to these unusual shapes. The length of the dumb-bell and the diameter of the doughnut are both about 1.5 fm (10^{-15} m) (Carlsson 1998). (Reprinted from Carlsson *et al* 1998 *Rev. Mod. Phys.* **70** 743 © 1998 by the American Physical Society.)

light nuclei. Many results can be consulted in the literature and I cite here some of the most recent publications in this field ((Pieper 2002) and references therein).

Also no-core shell-model (NCSM) studies have been carried out recently, moving as far as $A = 10$ and 12 compatible with present-day computer 'technology' (see Barrett *et al* (2002, 2003), Navratil and Ormand (2003) for more details). Here, too, one treats all A nucleons as active particles and one starts with a Hamiltonian that is translationally invariant in the A nucleon coordinates but adds a harmonic oscillator centre-of-mass potential. This confining potential is needed in order to compute the effective interaction and also provides a basis in order to carry out detailed calculations. The strong nucleon–nucleon interactions inside the nucleus lead, in general, to very slowly converging results in the harmonic oscillator basis. Therefore, in contrast to this *ab initio* method, the bare nucleon–nucleon interaction is of limited help and so we start by utilizing an 'effective' force. Because, in this NCSM approach, all nucleons remain active, there is a systematic procedure in order to derive this effective force, starting from the bare nucleon–nucleon interaction and the three-body forces. This is a particular strength of this NCSM approach which has been applied to p-shell nuclei and may well be applicable to some sd-shell nuclei.

Nuclear structure—transition section

After studying the nuclear many-body problem in part C, in which we have discussed both the simplest collective static properties (chapter 7), described using a liquid-drop model, as well as the simplest independent-particle model approach i.e. the Fermi gas model (chapter 8), followed by a systematic exposition of the independent-particle shell model of nucleons moving in a simplified average potential (chapter 9), we have treated the interactions of the nucleons forming nuclei in the first three chapters of part D, extensively (chapters 10, 11 and 12). Here, we subsequently discussed nuclear mean-field methods starting from the Hartree–Fock method and pointed out the experimental facts that form the backbone of support for the use of mean-field concepts (chapter 10), the nuclear shell model incorporating residual interactions and the most recent accomplishments using large-scale shell-model studies, Monte Carlo shell-model calculations, level densities and the study of nuclear interactions derived from the study of the spectral properties of the nuclear many-body system (chapter 11) as well as the state-of-the-art *ab initio* approach to model and describe light nuclei (chapter 12).

Before entering the study of the dynamics in collective models (chapters 13 and 14), we present in figure CD.1, the way in which the methods described in chapters 7–12 fit into an as-complete as-possible description of nuclei and nuclear properties. It has become clear that, for masses up to $A \approx 10$, *ab initio* few-body calculations starting from two- and three-body forces, fitted to nucleon–nucleon scattering, and using Green function methods after variational Monte Carlo calculations to determine the starting trial wavefunction, form a solid starting point. Using no-core shell-model studies, one is, at present, able to extend this range in A a little (drawn in figure CD.1 with the second arrow). It then turns out that large-scale shell-model calculations have provided a confident and robust basis for describing nuclei with increasing mass number covering the p-shell, the full sd-shell and, more recently, treat the fp-shell. These calculations are essentially $0\hbar\omega$ calculations although, in certain regions (near closed shells at $N = 8$, 20, 28 and 40), excitations across a shell closure have been considered in an approximate way. More recently, a new algorithm—the Monte Carlo shell-model approach—has allowed the study of medium–heavy nuclei. This is schematically shown in figure CD.1 by the third arrow. At present, the study

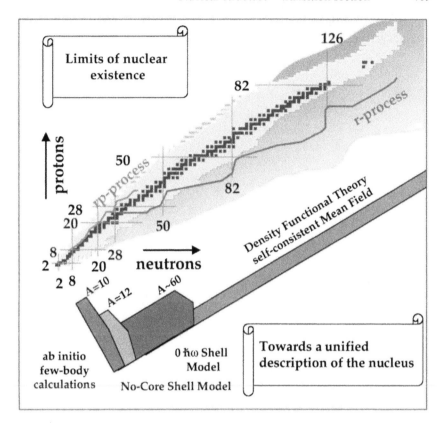

Figure CD.1. Schematic view of the various regions of applicability for modelling the atomic nucleus. (Courtesy of W Nazarewicz.)

of still heavier nuclei beyond masses $A \approx$ 60–70 can only be realized by using particular truncations to the nuclear shell model or, more generically, by making use of mean field methods. Thereby, the mean-field can be allowed to develop the essential degrees of freedom (quadrupole collectivity, pairing modes, octupole and/or higher-multipole modes of collectivity or a combination of these modes). At present, efforts are being made to build in the necessary correlations and thereby lift mean-field methods beyond the lowest-order adiabatic, non-interacting intrinsic structures. The region possibly treated by such methods is denoted by the larger line. In this context, figure CD.1 gives a pictorial overview of the interrelations between these earlier chapters.

In the next chapters, 13 and 14, we describe other collective models that allow for the vibrational and rotational dynamics that is superposed on the collective statics of the intrinsic mean-field solutions. Moreover, we also underline the power of symmetries to describe a large class of nuclear collective excitations, making use of an interacting system of $l = 0$ and $l = 2$ bosons.

Chapter 13

Collective modes of motion

In its most simple form, the nuclear shell model only describes the motion of nucleons as independent particles in an average field. Even with the residual interactions, as discussed extensively in chapter 11, the nuclear shell model with a spherical average field is not always a very appropriate starting point from which to describe the coherent motion of the many valence nucleons, except in some schematic models. Still, the nuclear shell model, especially as used before, contains a serious predictive power when discussing nuclei not too far removed from the closed-shell configurations.

Using specific nuclear reactions, certain excitation modes have been observed in many nuclei where both the proton and neutron number are far away from the closed-shell configurations. A number of regular, *collective* features related to the lowering of the first excited 2^+ state far below the energy needed to break nucleon 0^+ coupled pairs are very clear for the mass region $Z \geq 50$, $N \leq 82$ and are illustrated in figure 13.1. The multipole excitations of the nuclear charge and mass distributions can be used to obtain a quantitative measure of the collectivity of e.g. the lowest 2^+ level. An illustration of the $B(E2; 0_1^+ \rightarrow 2_1^+)$ values in nuclei with $N \geq 82$, $Z \leq 98$, with values in Weisskopf units, clearly illustrates the zone of large quadrupole collectivity near $A \simeq 160$–180 and in the actinide region with $A \geq 230$ (figure 13.2).

In the present chapter, we discuss the major elementary, collective building blocks or degrees of freedom for the understanding of open-shell nuclei. The most important feature in that respect is the indication of multi-quanta structures. We shall therefore discuss the collective *vibrational* and collective *rotational* characteristics in the atomic nucleus. These model approaches are very effective in describing the variety of collective features and its variation with valence numbers throughout the nuclear mass table. The dominant mode we shall concentrate on is the quadrupole mode. The residual proton–neutron interaction is particularly important in describing the low excitation energy and its variation in large mass regions. The low value of $E_x(2_1^+)$ and the correspondingly large

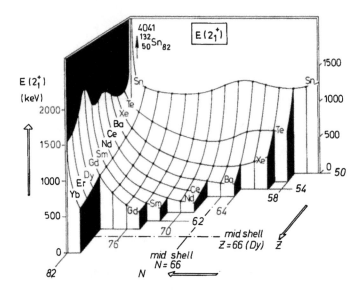

Figure 13.1. Landscape plot of the energy of the first excited 2^+ state $E_x(2_1^+)$ in the region $50 \leq Z \leq 82$ and $50 \leq N \leq 82$. The lines connect the $E_x(2_1^+)$ values in isotope chains (taken from Wood 1992).

increase in the related $B(E2; 0_1^+ \rightarrow 2_1^+)$ reduced transition probability are the clear indicators of nuclear, collective motion.

In section 13.1 we discuss vibrational (low-lying isoscalar and high-lying giant resonances) excitations. In section 13.2, we address the salient features relating to nuclear, collective rotational motion. In section 13.3, algebraic methods emphasizing the nuclear symmetries are presented with a number of applications relating to the interacting boson model. In section 13.4, we discuss recent results on nuclear shape coexistence.

13.1 Nuclear vibrations

A very interesting method for describing nuclear coherent excitations starts from a multipole expansion of the fluctuations in the nuclear density distribution around a spherical (or deformed) equilibrium shape, i.e.

$$V(\vec{r}_i, \vec{r}_j) = \sum_\lambda u_\lambda(r_i, r_j) \hat{Y}_\lambda(\hat{r}_i) \hat{Y}_\lambda(\hat{r}_j). \tag{13.1}$$

The above expression then leads to the corresponding multipole components in the average field

$$U(\vec{r}_i) = \sum_\lambda u_\lambda(r_i) Y_\lambda(\hat{r}_i), \tag{13.2}$$

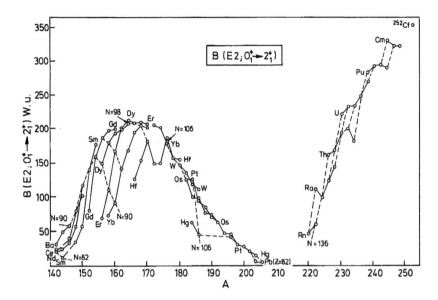

Figure 13.2. Systematics of the $B(E2; 0_1^+ \rightarrow 2_1^+)$ values for the even–even nuclei with $N \geq 82$, $Z \leq 98$. The $B(E2)$ values are expressed in Weisskopf units (WU) (taken from Wood 1992).

when averaging over the particle with the coordinate $\vec{r}_j$ in the nucleon–nucleon interaction itself. It is now such that, in particular, the low multipoles determine the average field properties in a dynamical way, e.g. $\lambda = 0$ results in a spherical field; $\lambda = 2$ determines the quadrupole field, $\lambda = 3$ the octupole deformed field component, etc. As an illustration, we show the way nucleons are moving within the nucleus in performing the $\lambda = 2$ quadrupole variations in the average field, and then also in the corresponding density distributions (figure 13.3).

13.1.1 Isoscalar vibrations

The description of the density variations of a liquid drop has been developed by Bohr and Mottelson, Nilsson and many others in the Copenhagen school. Here, one starts with a nuclear shape, for which the radius vector in the direction θ, φ, where θ, φ are the polar angles of the point on the nuclear surface, is written as

$$R(\theta, \varphi) = R_0 \left(1 + \sum_{\lambda\mu} \alpha_{\lambda\mu} Y_{\lambda\mu}^*(\theta, \varphi) \right),$$ (13.3)

where λ describes the multipolarity of the shape (see figure 13.4). The dynamics of this object in the small amplitude limit of harmonic oscillations in the $\alpha_{\lambda\mu}(r)$ coordinates around a spherical equilibrium shape is described by starting from the

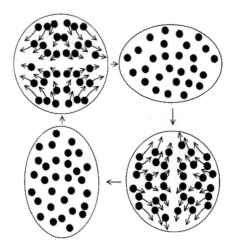

Figure 13.3. Giant quadrupole excitations indicate the collective vibrational excitation of protons against neutrons. The various phases in the oscillation are pictorially indicated. Here, a distortion from the spherical shape to an ellipsoidal shape occurs. The various flow patterns redistributing protons and neutrons out of the spherical equilibrium shape are given by the arrows. (Taken from Bertsch 1983. Reprinted with permission of *Scientific American*.)

linear $(2\lambda + 1)$-dimensional oscillator Hamiltonian

$$H_{\text{vibr}} = \sum_{\lambda,\mu} \frac{B_\lambda}{2} |\dot{\alpha}_{\lambda\mu}|^2 + \sum_{\lambda,\mu} \frac{C_\lambda}{2} |\alpha_{\lambda\mu}|^2. \qquad (13.4)$$

Defining the momentum $\pi_{\lambda,\mu} = B_\lambda \dot{\alpha}_{\lambda\mu}^*$ or $-i\hbar\partial/\partial\dot{\alpha}_{\lambda\mu}$, one can quantize the Hamiltonian of equation (13.4). It is, therefore, most convenient to define creation and annihilation operators for the oscillator quanta of a given multipolarity via the relations

$$b_{\lambda\mu}^+ = \sqrt{\frac{\omega_\lambda B_\lambda}{2\hbar}} \left(\alpha_{\lambda\mu} - \frac{i}{\omega_\lambda B_\lambda} (-1)^\mu \pi_{\lambda-\mu} \right)$$

$$b_{\lambda\mu} = \sqrt{\frac{\omega_\lambda B_\lambda}{2\hbar}} \left((-1)^\mu \alpha_{\lambda-\mu} + \frac{i}{\omega_\lambda B_\lambda} \pi_{\lambda\mu} \right). \qquad (13.5)$$

Using the standard boson commutation relations

$$[b_{\lambda'\mu'}, b_{\lambda\mu}^+] = \delta_{\lambda\lambda'}\delta_{\mu\mu'}, \qquad (13.6)$$

the oscillator Hamiltonian can be rewritten in the compact form

$$H_{\text{vibr}} = \sum_\lambda \hbar\omega_\lambda \sum_\mu (b_{\lambda\mu}^+ b_{\lambda\mu} + \tfrac{1}{2}). \qquad (13.7)$$

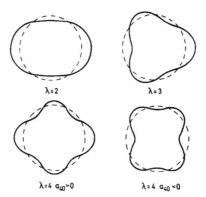

Figure 13.4. Nuclear shape changes corresponding to quadrupole ($\lambda = 2$), octupole ($\lambda = 3$) and hexadecupole ($\lambda = 4$) deformations (taken from Ring and Schuck 1980).

In the above, the frequencies are given as $\omega_\lambda = \sqrt{C_\lambda/B_\lambda}$. The ground state has no phonons $b_{\lambda\mu}|0\rangle = 0$ and the many phonon states can be obtained by acting with the $b^+_{\lambda\mu}$ operators in the ground state. A multiphonon (normalized) state can then be obtained as

$$\prod_{\lambda\mu} \frac{(b^+_{\lambda\mu})^{n_{\lambda\mu}}}{\sqrt{n_{\lambda\mu}!}}|0\rangle. \qquad (13.8)$$

A harmonic multi-phonon spectrum is obtained. Since each phonon carries an angular momentum λ, one has to handle angular momentum coupling with care. For quadrupole phonons, the two-phonon states, containing angular momentum $J = 0, 2, 4$ are constructed as (figure 13.5)

$$|n_2 = 2; JM\rangle = \frac{1}{\sqrt{2}} \sum_{\mu_1,\mu_2} \langle 2\mu_1, 2\mu_2|JM\rangle b^+_{2\mu_1} b^+_{2\mu_2}|0\rangle. \qquad (13.9)$$

Many ($n_2 = 2, 3, 4, \ldots$) quadrupole phonon states can be constructed using angular momentum coupling. Other methods exist, however, to classify the many phonon states using group theoretical methods (see section 13.3).

In the above system, the general method of constructing the many-phonon states can be used for the various multipoles. What is interesting, besides the energy spectra, are the electromagnetic decay properties in such a nucleus described by harmonic density oscillatory motion. The λ-pole transition can be described, using a collective approach, by the operator

$$\mathcal{M}(E\lambda, \mu) = \frac{Ze}{A} \int d^3r\, r^\lambda Y'_{\lambda\mu}(\hat{r})\rho(\vec{r}). \qquad (13.10)$$

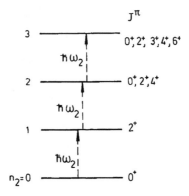

Figure 13.5. Multi-phonon quadrupole spectrum. On the left side, the number of quadrupole phonons n_2 is given. On the right side, the various possible angular momenta (J^π) are given.

In expanding the density around the equilibrium value ρ_0 but always using a constant density for $r \leq R_0$ and vanishing density outside ($r > R_0$), we obtain

$$\rho(\vec{r}) \cong \rho_0(r) - R_0 \frac{\partial \rho_0}{\partial r} \sum_{\lambda\mu} \alpha_{\lambda\mu} Y^*_{\lambda\mu}(\hat{r}) + \theta(\alpha^2), \qquad (13.11)$$

leading to the operator

$$\mathcal{M}(E\lambda, \mu) = \frac{3}{4\pi} Z e R_0^\lambda \left(\frac{\hbar}{2\omega_\lambda B_\lambda} \right)^{1/2} (b^+_{\lambda\mu} + (-1)^\mu b_{\lambda-\mu}). \qquad (13.12)$$

So, the $E\lambda$ radiation follows the selection rule $\Delta n_\lambda = \pm 1$ and the $E\lambda$ moments of all states vanish. The $B(E\lambda; \lambda \to 0^+)$ for the $n_\lambda = 1 \to n_\lambda = 0$ transition becomes

$$B(E\lambda; \lambda \to 0^+) = \left(\frac{3R_0^\lambda Z e}{4\pi} \right)^2 \frac{\hbar}{2\omega_\lambda B_\lambda}, \qquad (13.13)$$

and one obtains the ratio in $B(E2)$ values for the quadrupole vibrational nucleus

$$B(E2; 4_1^+ \to 2_1^+) = 2B(E2; 2_1^+ \to 0_1^+). \qquad (13.14)$$

Typical values for the $B(E2; 2_1^+ \to 0_1^+)$ values are of the order of 10–50 Weisskopf units (wu). A number of these ratios are fully independent of the precise values of the C_λ, B_λ coefficients and are from a purely geometric origin.

In figures 13.6 and 13.7 some examples of nuclei, exhibiting many characteristics of quadrupole vibrational spectra, are presented for cadmium nuclei. In the first one, the smooth variation in detailed spectra is given for

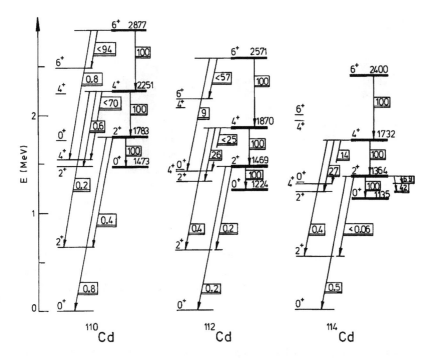

Figure 13.6. Relative $B(E2)$ values for the deformed bands in ^{110,112,114}Cd (thick lines). The other excitations present part of the multi-phonon quadrupole vibrational spectrum (taken from Wood *et al* 1992).

$^{110-114}$Cd. The thickened levels correspond to a class of states related to excitations across the $Z = 50$ closed shell and fall outside of the quadrupole vibrational model space: they are the 'intruder' bands in this $Z = 50$ mass region. In figure 13.7, detailed $B(E2)$ values are given with the thickness being proportional to the relative $B(E2)$ values.

The parameters B_λ and C_λ can then be derived using the assumption of irrotational flow for the inertial quantity B_λ and considering both the surface energy and Coulomb energy to derive the restoring force parameter C_λ. The results of the derivation (see Ring and Schuck 1980), are

$$B_\lambda = \frac{1}{\lambda} \frac{3 A m R_0^2}{4\pi}, \tag{13.15}$$

with m the nucleon mass and $R_0 = r_0 A^{1/3}$

$$C_\lambda = (\lambda - 1)(\lambda + 2) R_0^2 a_s - \frac{3(\lambda - 1)}{2\pi(2\lambda + 1)} \frac{Z^2 e^2}{R_0}, \tag{13.16}$$

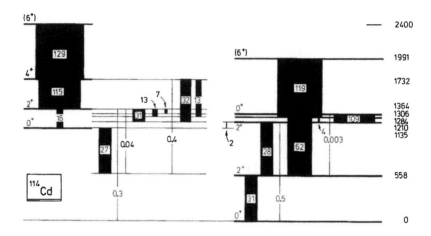

Figure 13.7. The complete low-lying energy spectrum in ^{114}Cd where the quadrupole vibrational structure is very much apparent. The thickness of the arrows is proportional to the $B(E2)$ values. The absolute $B(E2)$ values are given in Weisskopf units (WU). The data are taken from Nuclear Data Sheets (Blachot and Marguier 1990).

where a_s denotes the surface energy constant with a value of 18.56 MeV. A comparison of the liquid drop C_λ value with the corresponding value, deduced from those cases where both the $B(E2; 2_1^+ \rightarrow 0_1^+)$ and $\hbar\omega_2(\simeq E_x(2_1^+))$ are determined experimentally (using the harmonic expressions given before), is carried out in figure 13.8 for the whole mass region $20 \le A \le 240$. Very strong deviations between the liquid drop C_2 and 'experimental' C_2 values occur. The small values in the rare-earth ($A \simeq 150$–180) and actinide region ($A \simeq 240$) indicate large deviations from the vibrational picture. This is the region where energy spectra express more of a rotational characteristic compared with the mass regions near closed shells ($Z \cong 50$, $Z \cong 82$) where values rather close to C_2 (liquid drop) occur.

13.1.2 Sum rules in the vibrational model

In general, many 2^+ states appear in the even–even nuclei so it can become difficult to decide which state corresponds to the phonon excitation or to understand how much the phonon state is fragmented. Model independent estimates are obtained using sum-rule methods.

Summing in an energy-weighted way, one can evaluate the expression

$$S(E\lambda) = \sum_f (E_f - E_0)B(E\lambda; 0 \rightarrow \lambda_f), \qquad (13.17)$$

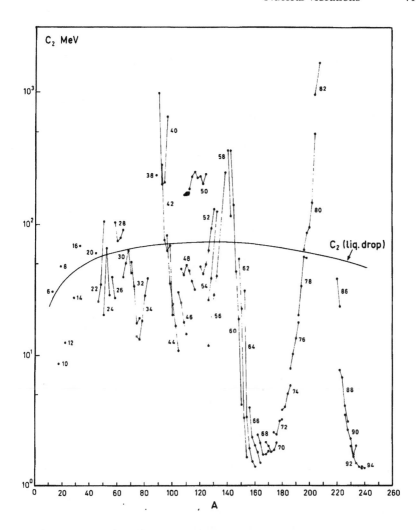

Figure 13.8. Systematics of the restoring force parameter C_2 for the low-frequency quadrupole mode. This quantity is calculated using the expression $C_2 = \frac{5}{2}\hbar\omega_2(\frac{3}{4\pi}ZeR^2)^2/B(E2; 0_1^+ \rightarrow 2_1^+)$ and $R = 1.2A^{1/3}$ fm. The frequencies $\hbar\omega_2$ and the $B(E2)$ values are taken from the compilation of Stelson and Grodzins (1965). This result is only appropriate in as much as a harmonic approximation holds. The liquid-drop value of C_2 is also given. (Taken from Bohr and Mottelson, *Nuclear Structure*, vol 2, © 1982 Addison-Wesley Publishing Company. Reprinted by permission.)

where we sum over all possible λ states ($f = 1, 2, \ldots$). A possibility exists to evaluate this sum rule without having to know all the intermediate state λ_f in using an equality of the right-hand side in equation (13.17) with a double commutator expression

$$S(E\lambda) = \frac{1}{2}\sum_\mu \langle 0|[[\mathcal{M}(E\lambda, \mu), H], \mathcal{M}^*(E\lambda, \mu)]|0\rangle. \qquad (13.18)$$

For a Hamiltonian, containing only kinetic energy and velocity independent two-body interactions, the result becomes

$$\mathcal{M}(E\lambda, \mu) = e\sum_p r_p^\lambda Y_{\lambda\mu}(\hat{r}_p), \qquad (13.19)$$

giving

$$[\mathcal{M}(E\lambda, \mu), H] = \frac{\hbar^2}{m}e\sum \vec{\nabla}(r^\lambda Y_{\lambda\mu}(\hat{r})) \cdot \vec{\nabla}$$

$$[[\mathcal{M}(E\lambda, \mu), H], \mathcal{M}^*(E\lambda, \mu)] = \frac{\hbar^2}{m}e^2\sum \vec{\nabla}(r^\lambda Y_{\lambda\mu}(\hat{r})) \cdot \vec{\nabla}(r^\lambda Y_{\lambda\mu}^*(\hat{r})), \quad (13.20)$$

leading to the sum-rule value

$$S(E\lambda) = \frac{Ze^2\hbar^2}{2m}\sum_\mu \langle 0|\vec{\nabla}(r^\lambda Y_{\lambda\mu}(\hat{r})) \cdot \vec{\nabla}(r^\lambda Y_{\lambda\mu}^*(\hat{r}))|0\rangle$$

$$= \frac{Ze^2\hbar^2}{2m}\frac{\lambda(2\lambda + 1)^2}{4\pi}\langle r^{2\lambda-2}\rangle. \qquad (13.21)$$

Evaluating the above expression for a constant density nucleus one obtains

$$S(E\lambda)_{T=0} = \frac{3}{4\pi}\lambda(2\lambda + 1)\frac{Z^2 e^2\hbar^2}{2mA}R_0^{2\lambda-2}, \qquad (13.22)$$

if isospin is included, because in the ntermediate sum the $\Delta T = 0$ transitions contribute Z/A of this sum and the $\Delta T = 1$ transitions contribute N/A. Typically, a low-lying collective state exhausts only about 10% of this $T = 0$ sum rule.

In the purely harmonic oscillator model, using the irrotational value of B_λ and the harmonic $B(E\lambda; 0 \to \lambda)$ value, it shows that the product $\hbar\omega_\lambda B(E\lambda; 0 \to \lambda)$ exactly exhausts the sum rule for $T = 0$ transitions, as derived in equation (13.22).

So, as one observes, the evaluation of $S(E\lambda)$ is a very good measure of how much of the 'collective oscillator phonon mode λ' is found in the nucleus. It also gives a good idea on the percentage of this strength which is fragmented in the atomic nucleus.

13.1.3   Giant resonances

As discussed in section 13.1.2, it is shown that the main contribution to the energy-weighted sum rule is situated in the higher-lying states. The strength

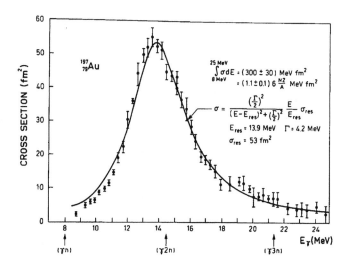

Figure 13.9. The total photoabsorption cross section for ^{197}Au, illustrating the absorption of photons on a giant resonating electric dipole state.   The solid curve shows a Breit–Wigner shape. (Taken from Bohr and Mottelson, *Nuclear Structure*, vol. 2, © 1982 Addison-Wesley Publishing Company. Reprinted by permission.)

is mainly peaked in the unbound nuclear region and shows up in the form of a wide resonance.   The electric dipole ($E1$) isovector mode, corresponding to neutrons oscillating against protons, can be excited particularly strongly via photon absorption.   A typical example for the odd-mass nucleus ^{197}Au is illustrated in figure 13.9. Here, the thresholds for the various (γ, n), $(\gamma, 2n)$ and $(\gamma, 3n)$ reactions are indicated. The very wide, 'giant' resonance stands out very strongly. One can note the peaking near to $E_x \simeq 14$ MeV. The width is mainly built from a decay width (Γ_{decay}) caused by particle emission and a spreading width (Γ_{spread}) caused by coupling of the resonance to non-coherent modes of motion which results in a damping of the collective motion. In the light nuclei, an important fine structure in the giant resonance region is observed indicating the underlying microscopic structure of the giant dipole resonance which, in a nucleus like ^{16}O, is built from a coherent superposition of 1p–1h excitations, coupled to $J^{\pi} = 1^-$, where one sums over proton and neutron 1p–1h excitations in an equivalent way.

The resonance energy varies smoothly with mass number A according to a law that can be well approximated by $E_{\text{res}}(1^-) \propto 79 A^{-1/3}$ MeV for heavy nuclei (figure 13.10). The resonance exhausts the energy-weighted $E1$, $S(E1, T = 1)$ sum rule almost to 100% with a steady decrease towards the light nuclei with only a 50% strength.

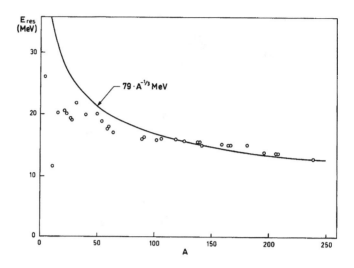

Figure 13.10. Systematics of the dipole resonance frequency. The experimental data are taken from the review article by Hayward (1965) except for ^{4}He. In the case of deformed nuclei, where two resonance maxima appear, a weighted mean of the two resonance energies is given. The solid curve results from the liquid-drop model. (Taken from Bohr and Mottelson, *Nuclear Structure*, vol. 2, © 1982 Addison-Wesley Publishing Company. Reprinted by permission.)

In a macroscopic approach, this electric dipole resonance can be understood as an oscillation of a proton fluid versus the neutron fluid around the equilibrium density $\rho_p = \rho_n = \rho_0/2$. Using hydrodynamical methods, a wave equation expresses the density variations in time and space as

$$\Delta\delta\rho_p - \frac{1}{c'^2}\frac{\partial^2\delta\rho_p}{\partial t^2} = 0, \qquad (\delta\rho_p \equiv \rho_p - \rho_0/2) \qquad (13.23)$$

can be derived (Eisenberg and Greiner 1970) with the velocity c' determined by the symmetry energy and the equilibrium density and results in a value of $c' \simeq 73$ fm/10^{-21} s. For oscillations in a spherical nucleus, the equation (13.23) becomes, with $k = \omega/c'$,

$$\Delta\delta\rho_p + k^2\delta\rho_p = 0, \qquad (13.24)$$

and leads to solutions for $\lambda = 1$ given by

$$\delta\rho_p \propto j_1(kr)Y_{1\mu}(\hat{r}). \qquad (13.25)$$

The value of k and thus of the corresponding energy eigenvalue results from the constraint that no net current is outgoing at the nuclear surface. This is expressed mathematically by

$$\frac{d}{dr} j_1(kR) = 0, \qquad \text{or} \qquad kR = 2.08. \tag{13.26}$$

Using the above values for the lowest root, the previous value of c' and, $R = r_0 A^{1/3}$ (with $r_0 = 1.2$ fm), one obtains the value of $E_{res}(1^-) \simeq 82 A^{-1/3}$ MeV, very close to the experimental number.

Using the results from section 13.1.2 in determining the sum rule $S(E1, T = 1)$ one derives easily the value of

$$S(E1; T = 1) = \frac{9}{4\pi} \frac{\hbar^2 e^2}{2m} \frac{NZ}{A}, \tag{13.27}$$

indeed, exhausting the full $T = 1$ sum rule, according to a purely collective model description. The compatibility of this approach with the single-particle picture has been elucidated by Brink (1957) who showed that the $E1$ dipole operator, separating out the unphysical centre-of-mass motion results in an operator

$$\mathcal{M}(E1) = e \frac{NZ}{A} \vec{r}, \tag{13.28}$$

with $\vec{r}$ the relative coordinate of the proton and neutron centres of mass. So clearly, the $E1$ operator acts with a resulting oscillation in this relative coordinate with strength eNZ/A.

Many more giant resonances corresponding to the various other multipoles have been discovered subsequently. The electric quadrupole and octupole resonances have been well studied in the past. Of particular interest is the giant monopole (a compression) mode for which the energy of the $E0$ centroid gives a possibility of studying the nuclear compressibility parameter; a value which is very important in describing nuclear matter at high density in order to map out the nuclear equation of state. An extensive review on giant resonances has been given by Speth and Van der Woude (1981) and Goeke and Speth (1982).

Various giant resonances have been studied, where besides variations in the nuclear proton and neutron densities, the spin orientation is also changed in a cooperative way. The Gamow–Teller resonance is the best studied in that respect; a mode strongly excited in (p, n) nuclear reactions. The knowledge of these spin-flip resonances, in retrospect, gives information on the particular spin (isospin) components in the nucleon–nucleon interaction, depicted via the $\vec{\sigma}_1 \cdot \vec{\sigma}_2 \vec{\tau}_1 \cdot \vec{\tau}_2$ component.

In the study of giant resonance excitations, the recent observation of double giant resonances was a breakthrough in the study of nuclear phenomena at high excitation energy. The observation could only be carried out in an unambiguous way using pion double-charge exchange at Los Alamos. This clear-cut evidence is presented in Box 13a.

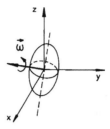

Figure 13.11. Rotational motion of a deformed nucleus (characterized by the rotational vector $\vec{\omega}$). The internal degrees of freedom are described by β and γ (see equation (13.30) and the external, rotational degrees of freedom are denoted by the Euler angles Ω (taken from Iachello 1985).

13.2 Rotational motion of deformed shapes

13.2.1 The Bohr Hamiltonian

In contrast to the discussion of section 13.1; other small amplitude harmonic vibrations can occur around a non-spherical equilibrium shape. So, the potential $U(\alpha_{2\mu})$ (see equation (13.4)) in the collective Hamiltonian can eventually show a minimum at a non-zero set of values $(\alpha_{2\mu})_0$. In this case, a stable deformed shape can result and thus, collective rotations described by the collective variables $\alpha_{2\mu}$, in the laboratory frame. For axially symmetric objects, rotation around an axis, perpendicular to the symmetry axis (see figure 13.11) can indeed occur. Such modes of motion are called collective rotations. We shall also concentrate on the quadrupole degree of freedom since it is this particular multipolarity ($\lambda = 2$) which plays a major role in describing low-lying, nuclear collective excitations.

Starting from a stable, deformed nucleus and a set of intrinsic axes, connected to the rotating motion of the nucleus as depicted in the lab frame, one can, in general, relate the transformed collective variables $a_{\lambda\mu}$ to the laboratory $\alpha_{\lambda\mu}$ values. The transformation is described, according to (appendix B):

$$Y_{\lambda\mu}(\text{rotated}) = \sum_{\mu'} D^{\lambda}_{\mu'\mu}(\Omega) Y_{\lambda\mu'}(\text{lab})$$

$$a_{\lambda\mu} = \sum_{\mu'} D^{\lambda}_{\mu'\mu}(\Omega) \alpha_{\lambda\mu'}. \qquad (13.29)$$

Thereby, the nuclear radius $R(\theta, \varphi)$ remains invariant under a rotation of the coordinate system.

For axially symmetric deformations with the z-axis as symmetry axis, all $\alpha_{\lambda\mu}$ vanish except for $\mu = 0$. These variables $\alpha_{\lambda 0}$ are usually called β_{λ}. For a quadrupole deformation, we have five variables $\alpha_{2,\mu}$ ($\mu = -2, \ldots, +2$). Three of them determine the orientation of the liquid drop in the lab frame (corresponding to the Euler angles Ω). Using the transformation from the lab into the rotated, body-fixed axis system, the five $\alpha_{2\mu}$ reduce to *two* real independent variables a_{20} and $a_{22} = a_{2-2}$ (with $a_{21} = a_{2-1} = 0$). One can define the more standard parameters

$$a_{20} = \beta \cos \gamma$$
$$a_{22} = \frac{1}{\sqrt{2}} \beta \sin \gamma. \tag{13.30}$$

Using the Y_{20} and $Y_{2\pm2}$ spherical harmonics in the intrinsic system, we can rewrite $R(\theta, \varphi)$ as

$$R(\theta, \varphi) = R_0 \left\{ 1 + \beta \sqrt{\frac{5}{16\pi}} (\cos \gamma (3 \cos^2 \theta - 1) + \sqrt{3} \sin \gamma \sin^2 \theta \cos 2\varphi) \right\}. \tag{13.31}$$

In figure 13.12, we present these nuclear shapes for $\lambda = 2$ using the polar angles.

(a) γ values of $0°$, $120°$ and $240°$ yield prolate spheroids with the 3, 1 and 2 axes as symmetry axes;
(b) $\gamma = 180°$, $300°$ and $60°$ give oblate shapes;
(c) with γ not a multiple of $60°$, triaxial shapes result;
(d) the interval $0° \leq \gamma \leq 60°$ is sufficient to describe all possible quadrupole deformed shapes;
(e) the increments along the three semi-axes in the body-fixed systems are evaluated as

$$\delta R_1 = R\left(\frac{\pi}{2}, 0\right) - R_0 = R_0 \sqrt{\frac{5}{4\pi}} \beta \cos\left(\gamma - \frac{2\pi}{3}\right)$$
$$\delta R_2 = R\left(\frac{\pi}{2}, \frac{\pi}{2}\right) - R_0 = R_0 \sqrt{\frac{5}{4\pi}} \beta \cos\left(\gamma + \frac{2\pi}{3}\right)$$
$$\delta R_3 = R(0, 0) - R_0 = R_0 \sqrt{\frac{5}{4\pi}} \beta \cos \gamma, \tag{13.32}$$

or, when taken together,

$$\delta R_k = R_0 \sqrt{\frac{5}{4\pi}} \beta \cos\left(\gamma - \frac{2\pi}{3} k\right) \qquad k = 1, 2, 3. \tag{13.33}$$

• In deriving the Hamiltonian describing the collective modes of motion, we start again from the collective Hamiltonian of equation (13.4) but with the potential energy changed into an expression of the type

$$U(\beta, \gamma) = \tfrac{1}{2} C_{20}(a_{20}(\beta, \gamma) - a_{20}°)^2 + C_{22}(a_{22}(\beta, \gamma) - a_{22}°)^2, \tag{13.34}$$

corresponding to a quadratic small amplitude oscillation but now around the equilibrium point $(a_{20}°, a_{22}°, a_{2-2}°)$. At the same time, collective rotations can

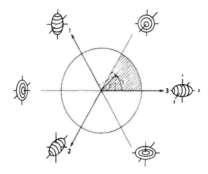

Figure 13.12. Various nuclear shapes in the (β, γ) plane. The projections on the three axes are proportional to the various increments δR_1, δR_2 and δR_3 (see equations (13.32)) (taken from Ring and Schuck 1980).

occur. Even more general expressions of $U(\beta, \gamma)$ can be used and, as will be pointed out in chapter 14, microscopic shell-model calculations of $U(\beta, \gamma)$ can even be carried out. Some typical $U(\beta, \gamma)$ plots are depicted in figure 13.13 corresponding to: (i) a vibrator β^2 variation; (ii) a prolate equilibrium shape; (iii) a γ-soft vibrator nucleus and (iv) a triaxial rotor system. The more realistic example for $^{124}_{56}\text{Te}_{68}$, as derived from the dynamical deformation theory (DDT), is also given in figure 13.14(*a*). Here, the full complexity of possible $U(\beta, \gamma)$ surfaces becomes clear: in ^{124}Te a γ-soft ridge becomes clear giving rise, after quantization, to the energy spectrum as given in figure 13.14(*b*). A fuller discussion on how to obtain the energy spectra after quantizing the Bohr Hamiltonian will be presented.

• The next step, the most difficult one, is the transformation of the kinetic energy term in equation (13.4). The derivation is lengthy and given in detail by Eisenberg and Greiner (1987). The resulting Bohr Hamiltonian becomes

$$H = T(\beta, \gamma) + U(\beta, \gamma), \tag{13.35}$$

with $U(\beta, \gamma)$ as given in equation (13.34) and

$$T = T_{\text{rot}} + \tfrac{1}{2} B_2(\dot{\beta}^2 + \beta^2 \dot{\gamma}^2), \tag{13.36}$$

where

$$T_{\text{rot}} = \frac{1}{2} \sum_{k=1}^{3} \mathcal{J}_k \omega_k^2. \tag{13.37}$$

Here, ω_k describes the angular velocity around the body-fixed axis k and $\mathcal{J}_k$ are functions of β, γ given as

$$\mathcal{J}_k = 4 B_2 \beta^2 \sin^2 \left(\gamma - \frac{2\pi}{3} k \right) \qquad k = 1, 2, 3. \tag{13.38}$$

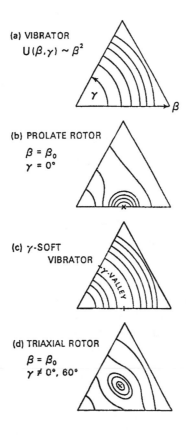

Figure 13.13. Different potential energy shapes $U(\beta, \gamma)$ in the β, ($\gamma = 0° \rightarrow \gamma = 60°$) sector corresponding to a spherical vibrator, a prolate rotor, a γ-soft vibrator and a triaxial rotor respectively (taken from Heyde 1989).

For fixed values of β and γ, T_{rot} is the collective rotational kinetic energy with moments of inertia $\mathcal{J}_k$. With β, γ changing, the collective rotational and β, γ vibrational energy become coupled in a complicated way. Using the irrotational value for B_2 (section 13.1), these irrotational moments of inertia become

$$\mathcal{J}_k^{\text{irrot}} = \frac{3}{2\pi} m A R_0^2 \beta^2 \sin^2 \left(\gamma - \frac{2\pi}{3} k \right) \qquad k = 1, 2, 3, \qquad (13.39)$$

whereas for *rigid* body inertial moments, one derives

$$\mathcal{J}_k^{\text{rigid}} = \frac{2}{5} m A R_0^2 \left(1 - \sqrt{\frac{5}{4\pi}} \beta \cos \left(\gamma - \frac{2\pi}{3} k \right) \right) \qquad k = 1, 2, 3. \qquad (13.40)$$

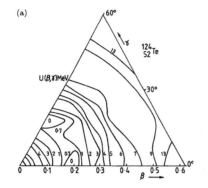
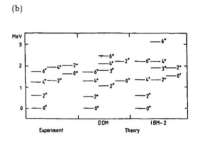

Figure 13.14. The contour plots of $U(\beta, \gamma)$ for $^{124}_{52}$Te obtained from the dynamical deformation model (DDM) calculations of Kumar (1984). The corresponding collective spectra in ^{124}Te are also given (taken from Heyde 1989).

We can remark that: (i) $\mathcal{J}^{\text{irrot}}$ vanishes around the symmetry axes; (ii) $\mathcal{J}^{\text{irrot}}$ shows a stronger β-dependence ($\sim \beta^2$) compared to a β-dependence only in $\mathcal{J}^{\text{rigid}}$; (iii) the experimental moments of inertia $\mathcal{J}^{\text{exp}}$ can, in a first step be derived from the 2^+_1 excitation energy assuming a pure rotational $J(J+1)$ spin dependence. A relation with the deformation variable β can be obtained with the result

$$\mathcal{J}^{\text{exp}} \simeq \frac{\hbar^2 \beta^2 A^{7/3}}{400} \ (\text{MeV}^{-1}). \tag{13.41}$$

A systematic compilation of $\mathcal{J}^{\text{exp}}$ for nuclei in the mass region ($150 \le A \le 190$) are presented in figure 13.15 where, besides the data for various types of nuclei, the rigid rotor values are also drawn. In general, one obtains the ordering with

$$\mathcal{J}^{\text{irrot}} < \mathcal{J}^{\text{exp}} < \mathcal{J}^{\text{rigid}}. \tag{13.42}$$

An extensive analysis of moments of inertia, in the framework of the variable moment of inertia (VMI) model (Scharff-Goldhaber *et al* 1976, Davidson 1965, Mariscotti *et al* 1969) gives the reference $\mathcal{J}_0$. This value gives a very pictorial, overall picture of the softness of the nuclear collective rotational motion (figure 13.16).

• The next step now is the quantization of the classical Hamiltonian as given in equation (13.35). There exists no unique prescription in order to quantize the motion relating to the β and γ variables. Commonly, one adopts the Pauli prescription, giving rise to the Bohr Hamiltonian

$$\hat{H}_{\text{coll}} = \frac{-\hbar^2}{2B_2} \left[\beta^{-4} \frac{\partial}{\partial \beta} \left(\beta^4 \frac{\partial}{\partial \beta} \right) + \frac{1}{\beta^2 \sin 3\gamma} \frac{\partial}{\partial \gamma} \left(\sin 3\gamma \frac{\partial}{\partial \gamma} \right) \right] + \hat{T}_{\text{rot}} + U(\beta, \gamma), \tag{13.43}$$

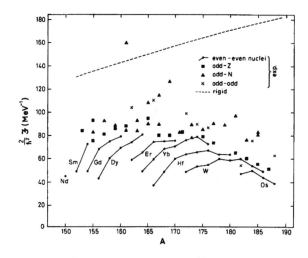

Figure 13.15. Systematics of moments of inertia for nuclei in the mass region $150 \leq A \leq 190$. These moments of inertia are derived from the empirical levels given in the *Tables of Isotopes* (Lederer *et al* 1967). (Taken from Bohr and Mottelson, *Nuclear Structure*, vol. 2, © 1982 Addison-Wesley Publishing Company. Reprinted by permission.)

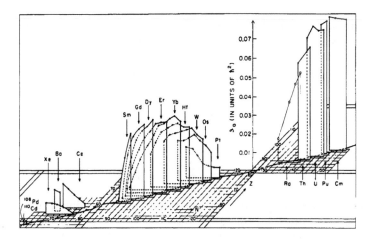

Figure 13.16. Calculated ground-state moments of inertia $\mathcal{J}_0$ in even–even nuclei as a function of N and Z. (Taken from Scharff-Goldhaber *et al* 1976. Reproduced with permission from the *Annual Review of Nuclear Sciences* **26**, © 1976 by Annual Reviews Inc.)

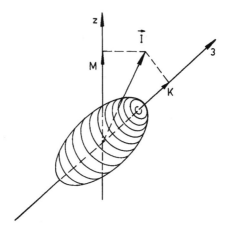

Figure 13.17. Relationships between the total angular momentum $\vec{J}$ and its projections M and K on the laboratory and intrinsic 3-axis, respectively.

with

$$\hat{T}_{\text{rot}} = \frac{\hat{I}_1^2}{2\mathcal{J}_1} + \frac{\hat{I}_2^2}{2\mathcal{J}_2} + \frac{\hat{I}_3^2}{2\mathcal{J}_3}. \tag{13.44}$$

Here, the operators $\hat{I}_k$ describe the total angular momentum projections onto the body-fixed axes. The orientation and the projection quantum numbers are indicated in figure 13.17.

A general expression for the collective wavefunction is derived as

$$|\psi_M^J\rangle = \sum_K g_K(\beta, \gamma)|JMK\rangle, \tag{13.45}$$

with

$$|JMK\rangle = \sqrt{\frac{2J+1}{8\pi^2}} D_{MK}^J(\Omega). \tag{13.46}$$

A number of symmetry operations are related to the axially symmetric rotor and these symmetries put constraints on the collective wavefunctions. We do not, at present, go into a detailed discussion as given by Eisenberg and Greiner (1987). At present, we only discuss the salient features related to the most simple modes of motion and also related to the collective Bohr Hamiltonian of an axially symmetric case, equation (13.43).

Singling out a deep minimun in $U(\beta, \gamma)$ at the deformation $\beta = \beta_0$ and $\gamma = 0°$, we expect rotations on which small amplitude vibrations become superimposed. In this situation, one obtains, to a good approximation, the Hamiltonian of an axial rotor with moments of inertia $\mathcal{J}_0 = \mathcal{J}_1(\beta_0, 0) =$

$\mathcal{J}_2(\beta_0, 0)$ which is written as

$$\hat{T}'_{\text{rot}} = \frac{\hat{I}^2 - \hat{I}_3^2}{2\mathcal{J}_0}. \tag{13.47}$$

We here distinguish

(i) $K = 0$ *bands* $(J_3 = 0)$. The wavefunction, since rotational axial vibrational motions decouple, now becomes

$$|\psi^J_{M, K=0}\rangle = g_0(\beta, \gamma)|J, M, K = 0\rangle. \tag{13.48}$$

A spin sequence $J = 0, 2, 4, 6, \ldots$ appears and describes the collective, rotational motion. For the vibrational motion, one can approximately also decouple the a_{20} (β-vibrations) from the a_{22} (γ-vibrations) oscillations. Superimposed on each vibrational (n_β, n_γ) state, a rotational band is constructed, according to the energy eigenvalue

$$E_{n_\beta, n_\gamma}(J) = \hbar\omega_\beta(n_\beta + \tfrac{1}{2}) + \hbar\omega_\gamma(2n_\gamma + 1) + \frac{\hbar^2}{2\mathcal{J}_0} J(J + 1), \tag{13.49}$$

with $n_\beta = 0, 1, 2, \ldots$; $n_\gamma = 0, 1, 2, \ldots$ and with ω_β and ω_γ the β and γ vibrational frequencies. These bands, in particular for $n_\beta = 1, n_\gamma = 0$; $n_\beta = 0$, $n_\gamma = 1$ have been observed in many nuclei. A schematic picture of various combined vibrational modes is given in figure 13.18. The most dramatic example of a rotational band structure, expressing the $J(J + 1)$ regular motion, as well as the interconnecting $E2$ gamma transition, is depicted in the ^{242,244}Pu nuclei (figure 13.19).

(ii) $K \neq 0$ *bands*. Here, symmetrized, rotational wavefunctions are needed to give good parity, with the form

$$|\psi^J_{M, K}\rangle = g_K(\beta, \gamma)\frac{1}{\sqrt{2}}[|JMK\rangle + (-1)^J|JM - K\rangle], \tag{13.50}$$

and with even K values. For such $K \neq 0$ bands, the spin sequence results with $J = |K|, |K|+1, |K|+2, \ldots$. Here now, the γ-vibration couples to the rotational motion, with the resulting energy spectrum

$$E_{K, n_\beta, n_\gamma}(J) = \hbar\omega_\beta(\eta_\beta + \tfrac{1}{2}) + \hbar\omega_\gamma\left(2n_\gamma + 1 + \frac{|K|}{2}\right) + \frac{\hbar^2}{2\mathcal{J}_0}[J(J + 1) - K^2]. \tag{13.51}$$

13.2.2 Realistic situations

The above discussion of a standard Bohr Hamiltonian, encompassing regular, collective rotational bands following a $J(J + 1)$ spin law with, at the same time, vibrational β and γ bands, is an idealized situation. In many nuclei, transitional

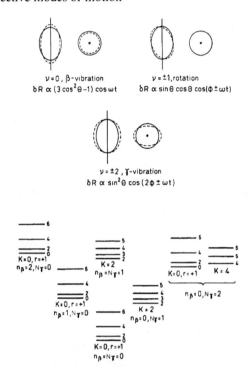

Figure 13.18. Various quadrupole shape oscillations in a spheroidal nucleus. The upper part shows projections of the nuclear shape in directions perpendicular and parallel to the symmetry axis. The lower part shows the spectra associated with excitations of one or two quanta, including the specific values of the various oscillation energies $\hbar\omega_\beta$, $\hbar\omega_\gamma$. The value of N_γ used in the figure that classifies the gamma vibration is defined as $N_\gamma = 2n_\gamma + |K|/2$. The rotational energy is assumed to be given by $J(J+1) - K^2$ and harmonic oscillatory spectra are considered. (Taken from Bohr and Mottelson, *Nuclear Structure*, vol. 2, © 1982 Addison-Wesley Publishing Company. Reprinted by permission.).

spectra lying between the purely harmonic quadrupole vibrational system and the rotational limit can result. A good indicator to 'locate' collective spectra in the energy ratio $E_{4_1^+}/E_{2_1^+}$ which is 2 for pure, harmonic vibrators and 3.33 for a pure, rigid rotor spectrum. Near closed shells (see chapter 11), values for $E_{4_1^+}/E_{2_1^+}$, quite close to 1 can show up, indicating the presence of closed shells and the extreme rigidity, relating to $(j)^2 J$ shell-model configurations. In figure 13.20, we present this ratio for most even–even nuclei, giving a very clear view of the actual 'placement' of nuclei. Only a few nuclei (in the deformed regions) approach the rigid rotor $J(J+1)$ limit. The transition from the vibrational model, on

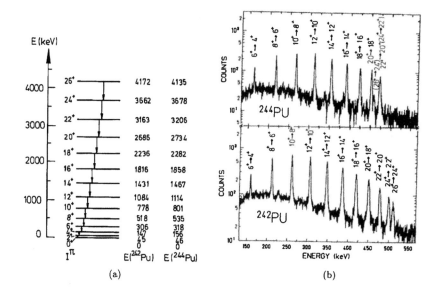

Figure 13.19. (*a*) Band structure observed in ^{242,244}Pu. The $E2$ transitions connecting the members of the band are shown by arrows. The gamma lines corresponding to these transitions are marked, e.g., $6^+ \rightarrow 4^+$ (at ~ 160 keV) in part (*b*). The discontinuity in transition energy at high spin is discussed in chapter 13. (*b*) Coulomb excitation with 5.6–5.8 MeV/u ^{208}Pu. Doppler shift-corrected gamma spectra for ^{242}Pu and ^{244}Pu are shown (Spreng *et al* 1983) (taken from Wood 1992).

one side, and the rotational (with β, γ vibrations included), on the other side, is schematically illustrated in figure 13.21.

In addition, detailed studies of triaxial even–even nuclei have been carried out (Meyer-ter-Vehn 1975), with a classification of the 'measure' of triaxiality: here, the rotational spectra are modified because of the asymmetric character. Here too, K is no longer a good quantum number.

Another topic is the influence of the odd-particle, coupling in various ways (strong, weak, intermediate) to the nuclear collective motion. We do not discuss these extensions of the nuclear collective motion and refer to the very well documented second volume of Bohr and Mottelson (1975).

13.2.3 Electromagnetic quadrupole properties

Besides the regular spacing in the collective, rotational band structure, other observables indicate the presence of coherence in the nuclear motion such as quadrupole moments and electric quadrupole enhanced $E2$ transition probabilities. The specific behaviour of quadrupole moments has already been discussed in chapter 1 (sections 1.6, 1.7 and Box 1e).

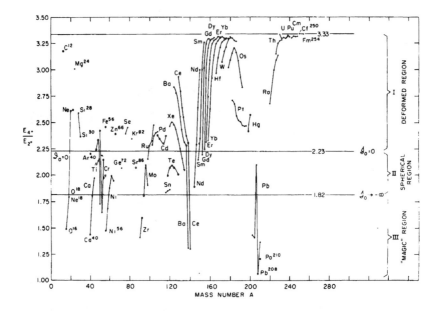

Figure 13.20. Experimental values of the $R \equiv E_{4+}/E_{2+}$ ratio in even–even nuclei. The top horizontal line corresponds to the ideal rotor ratio (3.33). The interval $2.67 \leq R \leq 3.33$ corresponds to the prediction of the asymmetric rotor. Ratios in the interval $2.23 \leq R \leq 3.33$ fit most nuclei in the transitional region (at the point $R = 2.23$ the moment of inertia $\mathcal{J}_0$ vanishes). The nuclei with an even smaller ratio of R correspond to nuclei having just a few nucleons (two) outside a closed shell configuration. (Taken from Scharff-Goldhaber Reproduced with permission from the *Annual Review of Nuclear Science* **26** © 1976 by Annual Reviews Inc.)

In the ground-state band for a fixed β value and $\gamma = 0°$, the value of the intrinsic quadrupole moment is obtained, starting from the collective $\hat{Q}_{\lambda\mu}$ operator which, in lowest order, is

$$\hat{Q}_{\lambda\mu} = \frac{3e}{4\pi} Z R_0^2 \alpha_{\lambda\mu}, \qquad (13.52)$$

and transformed into the intrinsic body-fixed system results in the quadrupole moment is

$$Q_0 = \sqrt{\frac{16\pi}{5}} \frac{3}{4\pi} Z e R_0^2 \beta. \qquad (13.53)$$

For the $E2$ transition rates, the basic strength does not change and is related to the intrinsic ground-state band structure, expressed via Q_0 (and β). The general spin dependence is expressed via the Clebsch–Gordan coefficient taking

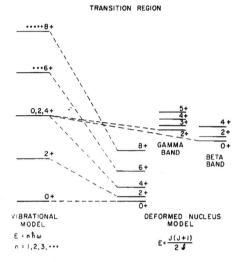

Figure 13.21. Schematic level scheme connecting the vibrational multi-phonon limit to the spectrum resulting in a deformed nucleus exhibiting at the same time gamma- and beta-vibrational excitations. The dashed lines are drawn to guide the eye. (Taken from Scharff-Goldhaber Reproduced with permission from the *Annual Review of Nuclear Science* **26** © 1976 by Annual Reviews Inc.)

the angular momentum coupling and selection rules into account, as

$$B(E2; J_i \rightarrow J_f) = \frac{5}{16\pi} Q_0^2 \langle J_i K, 20 | J_f K \rangle^2. \tag{13.54}$$

For the $K = 0$ band, and using the explicit form of the Clebsch–Gordan coefficient, one obtains

$$B(E2; J_i + 2 \rightarrow J_i) = \frac{5}{16\pi} Q_0^2 \frac{3}{2} \frac{(J_i + 1)(J_i + 2)}{(2J_i + 3)(2J_i + 5)}. \tag{13.55}$$

The spectroscopic quadrupole moment becomes

$$Q = Q_0 \frac{3K^2 - J(J+1)}{(2J+3)(J+1)}. \tag{13.56}$$

So, in general, the band-head with $J = K$ has a non-vanishing spectroscopic quadrupole moment, except for the $K = 0$ ($J = 0$) band. In the latter case, we can only get information on Q_0 by studying excited states. A way to determine Q_0 is by studying the $B(E2)$ values in the ground-state band. It can thereby be tested in a way, which can extend up to high-spin states in the collective ground-state band and shows that the Q_0 value and thus deformation β remains a constant, even in those cases where the energy spacings deviate from the $J(J+1)$ law. The

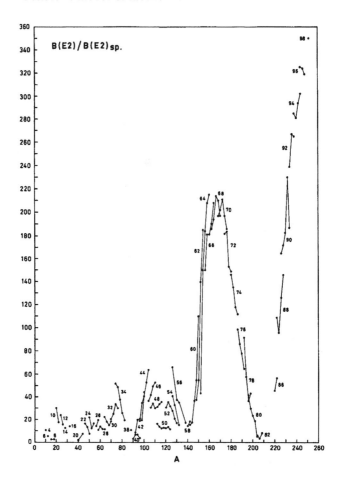

Figure 13.22. The $B(E2; 0_1^+ \rightarrow 2_1^+)$ values in even–even nuclei, expressed in single-particle units, taken as $B(E2)_{\mathrm{sp}} = 0.30A^{4/3}e^2$ fm^4 ($R = 1.2A^{1/3}$ fm). The larger part of $B(E2)$ values are from the compilation of Stelson and Grodzins (1965). (Taken from Bohr and Mottelson, *Nuclear Structure*, vol. 2, © 1982 Addison-Wesley Publishing Company. Reprinted by permission.)

collectivity is easily observed by plotting the $B(E2)$ value of ground-state bands for various mass regions, in single-particle $B(E2)$ units. Values up to $\simeq 200$ (in the region near $A \simeq 160$–180) and even extending up to $\simeq 350$ (in the region at $A \simeq 250$) indicate that enhanced and coherent modes of motion, have been observed (figure 13.22).

13.3 Algebraic description of nuclear, collective motion

13.3.1 Symmetry concepts in nuclear physics

Besides the use of collective shape variables from the early studies in nuclear collective motion (the Bohr–Mottelson model), there exist also possibilities to describe the well-ordered coherent motion of nucleons, by the concept of symmetries related to algebraic descriptions of the nuclear many-body system.

The simplest illustration, which contrasts the standard coordinate quantum mechanical representation with algebraic descriptions, can be obtained in the study of the harmonic oscillator. One way is to start from the one-dimensional problem, described by the equation

$$-\frac{\hbar^2}{2m}\frac{d^2\varphi(x)}{dx^2} - \frac{1}{2}m\omega^2 x^2 \varphi(x) = E\varphi(x), \tag{13.57}$$

with, as solutions, the well-known Hermite functions. Alternatively, one can define an algebraic structure, defining creation and annihilation operators

$$\hat{b}^+ \equiv \frac{1}{\sqrt{2}}\left(\lambda x - \frac{1}{\lambda}\frac{d}{dx}\right)$$

$$\hat{b} \equiv \frac{1}{\sqrt{2}}\left(\lambda x + \frac{1}{\lambda}\frac{d}{dx}\right), \tag{13.58}$$

with $\lambda = \sqrt{(m\omega/\hbar)}$, such that the Hamiltonian reads

$$\hat{H} = \hbar\omega(\hat{b}^+\hat{b} + \tfrac{1}{2}). \tag{13.59}$$

This forms a one-dimensional group structure $U(1)$ for which the algebraic properties of many boson excitations can be studied, according to the n-boson states

$$|n\rangle = \frac{1}{\sqrt{n!}}(\hat{b}^+)^n|0\rangle, \tag{13.60}$$

that are solutions to the eigenvalue equation

$$\hat{H}|n\rangle = E_n|n\rangle = \hbar\omega(n + \tfrac{1}{2})|n\rangle. \tag{13.61}$$

It is the extension of the above method to more complex group structures that has led to deep insights in describing many facets of the atomic nucleus. We first discuss briefly a few of the major accomplishments (see figure 13.23).

1932: The concept of isospin symmetry, describing the charge independence of the nuclear forces by means of the isospin concept with the $SU(2)$ group as the underlying mathematical group (Heisenberg 1932). This is the simplest of all dynamical symmetries and expresses the invariance of the Hamiltonian against the exchange of all proton and neutron coordinates.

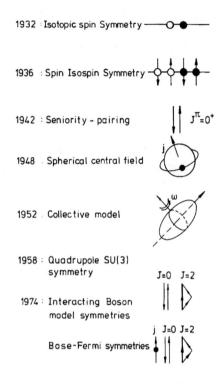

1932 : Isotopic spin Symmetry

1936 : Spin Isospin Symmetry

1942 : Seniority - pairing

1948 . Spherical central field

1952 . Collective model

1958 : Quadrupole SU(3)
symmetry

1974 : Interacting Boson
model symmetries

Bose-Fermi symmetries

Figure 13.23. Pictorial representation of some of the most important nuclear symmetries developed over the years (taken from Heyde 1989).

1936: Spin and isospin were combined by Wigner into the $SU(4)$ supermultiplet scheme with $SU(4)$ as the group structure (Wigner 1937). This concept has been extensively used in the description of light α-like nuclei ($A = 4 \times n$).

1948: The spherical symmetry of the nuclear mean field and the realization of its major importance for describing the nucleon motion in the nucleus were put forward by Mayer (Mayer 1949), Haxel, Jensen and Suess (Haxel *et al* 1949).

1958: Elliott remarked that in some cases, the average nuclear potential could be depicted by a deformed, harmonic oscillator containing the $SU(3)$ dynamical symmetry (Elliott 1958, Elliott and Harvey 1963). This work opened the first possible connection between the macroscopic collective motion and its microscopic description.

1942: The nucleon residual interaction amongst identical nucleons is particularly strong in $J^{\pi} = 0^+$ and 2^+ coupled pair states. This 'pairing' property is a cornerstone in accounting for the nuclear structure of many spherical nuclei near closed shells in particular. Pairing is at the origin of seniority, which is related with the quasi-spin classification and group as used first by Racah in describing the properties of many-electron configurations in atomic physics (Racah 1943).

1952: The nuclear deformed field is a typical example of the concept of spontaneous symmetry breaking. The restoration of the rotational symmetry, present in the Hamiltonian, leads to the formation of nuclear rotational spectra. These properties were discussed earlier in a more phenomenological way by Bohr and Mottelson (Bohr 1951, 1952; Bohr and Mottelson 1953).

1974: The introduction of dynamical symmetries in order to describe the nuclear collective motion, starting from a many-boson system, with only s ($L = 0$) and d ($L = 2$) bosons was introduced by Arima and Iachello (Arima and Iachello 1975, 1976, 1978, 1979). The relation to the nuclear shell model and its underlying shell-structure has been studied extensively (Otsuka *et al* 1978b). These boson models have given rise to a new momentum in nuclear physics research.

In the next paragraphs, we discuss some of the basic ingredients behind the interacting boson model description of nuclear, collective (mainly quadrupole) motion. This interacting boson model (IBM) relies heavily on group theory; in particular the $U(6)$ group structure of interacting s and d bosons.

13.3.2   Symmetries of the IBM

In many problems in physics, exact solutions can be obtained if the Hamiltonian has certain symmetries. Rotational invariance leads in general to the possibility of characterizing the angular eigenfunctions by quantum numbers l, m. These quantum numbers relate to the representation of the 0(3) and 0(2) rotation groups in three and two dimensions, respectively.

The group structure of the IBM can be discussed in a six-dimensional Hilbert space, spanned by the s^+ and d_μ^+ ($-2 \le \mu \le 2$) bosons. The column vector

$$b_\mu^+ \equiv \begin{pmatrix} s^+ \\ d_{+2}^+ \\ \vdots \\ d_{-2}^+ \end{pmatrix}, \tag{13.62}$$

transforms according to the group $U(6)$ and the s and d boson states

$$\underbrace{b_\mu^+ b_\nu^+ b_\kappa^+ \ldots}_{N} |0\rangle, \tag{13.63}$$

form the totally symmetric representations of the group $U(6)$, which is characterized by the number of bosons N. Here, no distinction is made between proton and neutron bosons, so we call N the total number of bosons. The 36 bilinear combinations

$$\hat{G}_{\mu\nu} \equiv b_\mu^+ b_\nu \qquad (\mu, \nu = 1, 2, \dots, 6), \tag{13.64}$$

with $b^+ \equiv s^+, d_\mu^+$, form a $U(6)$ Lie algebra and are the generators that close under commutation. This is the case, if the generators, for a general group, X_a fulfil the relation

$$[X_a, X_b] = \sum_c c_{ab}^c X_c. \tag{13.65}$$

The general two-body Hamiltonian within this $U(6)$ Lie algebra consists of terms that are linear (single-boson energies) and quadratic (two-body interactions) in the generators $G_{\mu\nu}$, or

$$\hat{H} = E_0 \sum_{\mu\nu} \varepsilon_{\mu\nu} \hat{G}_{\mu\nu} + \frac{1}{2} \sum_{\substack{\mu\nu \\ \kappa\sigma}} U_{\mu\nu,\kappa\sigma} \hat{G}_{\mu\nu} \hat{G}_{\kappa\sigma}$$

$$= E_0 + \hat{H}'. \tag{13.66}$$

This means that, in the absence of the interaction $\hat{H}'$, *all* possible states (13.63) for a given N will be degenerate in energy. The interaction $\hat{H}'$ will then split the different possible states for a given N and one needs to solve the eigenvalue equation to obtain energies and eigenvectors in the most general situation.

There are now situations where the Hamiltonian $\hat{H}'$ can be rewritten exactly as the sum of Casimir (invariant) operators of a complete chain of subgroups of the largest one, i.e. $G \supset G' \supset G' \supset \dots$ with

$$\hat{H}' = \alpha C(G) + \alpha' C(G') + \alpha' C(G') + \cdots. \tag{13.67}$$

The eigenvalue problem for (13.67) can be solved in closed form and leads to energy formulae in which the eigenvalues are given in terms of the various quantum numbers that label the irreducible representation (irrep) of $G \supset G' \supset G' \dots$

$$E = \alpha \langle C(G) \rangle + \alpha' \langle C(G') \rangle + \alpha' \langle C(G') \rangle + \cdots, \tag{13.68}$$

where $\langle \cdots \rangle$ denotes the expectation value. An illustration of this process, where the original, fully degenerate large multiplet of states contained in the symmetric irrep $[N]$ is split under the influence of the various terms contributing to (13.68), is given in figure 13.24.

The general Lie group theoretical problem to be solved thus becomes:

(i)   to identify all possible subgroups of $U(6)$;
(ii) to find the irrep for these various group chains;
(iii) evaluate the expectation values of the various, invariant, Casimir operators.

Problems (ii) and (iii) are well-defined group-theoretical problems but (i) depends on the physics of the particular nuclei to be studied.

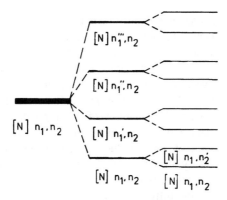

$[N]\, n_1''',n_2$

$[N]\, n_1'',n_2$

$[N]\, n_1,n_2$

$[N]\, n_1',n_2$

$[N]\, n_1,n_2'$

$[N]\, n_1,n_2$ $[N]\, n_1,n_2$

Figure 13.24. Degeneracy splitting for the case of a Hamiltonian, expressed as a sum of Casimir invariant operators (see equation (13.68)) which consecutively splits the energy spectrum. In the figure, a splitting of the level with the quantum numbers n_1, n_2 is caused by the perturbation $\hat{H}'$ given in equation (13.66) (taken from Heyde 1989).

U(6) subgroup chains

There are three group reductions that can be constructed, called the $U(5)$, $SU(3)$ and $O(6)$ chains corresponding to three different illustrations of nuclear collective quadrupole motion.

(i)   The 25 generators $(d^+\tilde{d})_m^{(l)}$ close under commutation and form the $U(5)$ subalgebra of $U(6)$. Furthermore, the 10 components $(d^+\tilde{d})_m^{(1)}$ and $(d^+\tilde{d})_m^{(3)}$ close again under commutation and are the generators of the $O(5)$ algebra. Here, the $(d^+\tilde{d})_m^{(1)} \propto \hat{L}$ are the generators of $O(3)$ and a full subgroup reduction

$$U(6) \supset U(5) \supset O(5) \supset O(3) \supset O(2), \qquad (13.69)$$

is obtained.

(ii) The $SU(3)$ chain is obtained according to the reduction

$$U(6) \supset SU(3) \supset O(3) \supset O(2), \qquad (13.70)$$

and,

(iii) the $O(6)$ group chain according to the group reduction

$$U(6) \supset O(6) \supset O(5) \supset O(3) \supset O(2). \qquad (13.71)$$

The Casimir invariant operators for a given group are found by the condition that they commute with all the generators of the group. Precise methods have been obtained in order to derive and construct the various Casimir invariant operators as well as the corresponding eigenvalues (Arima and Iachello 1988).

Each group is characterized by its irreducible representations (irrep) that carry the necessary quantum numbers (or representation labels) to define a basis in which the IBM Hamiltonian has to be diagonalized. The major question is: what representations of a subgroup belong to a given representation of the larger group? One situation we know well is that the representations of $O(3)$ are labelled by l and those of its $O(2)$ subgroup by m_l with condition $-l \leq m_l \leq +l$. This procedure is known as the reduction of a group with respect to its various subgroups. The fact that e.g. two quadrupole phonons couple to $J^\pi = 0^+, 2^+, 4^+$ states only, is an example of such a reduction. The technique of Young tableaux then supplies the necessary book-keeping device in group reduction.

In the IBM, making no distinction between proton and neutron bosons (the IBM-1) one only needs the fully symmetric irrep of $U(6)$ so that only a single label is needed e.g. $[N, 0, 0, 0, 0, 0] \equiv [N]$.

Here, we shall not derive in detail the various reduction schemes for the $U(5)$, $SU(3)$ nor of the $O(6)$ group chains and just give the main results,

$$
\begin{array}{llllll}
\text{(i)} & U(6) \supset & U(5) \supset & O(5) & \supset & O(3) \supset & O(2) \\
& | & | & | & & | & | \\
& [N] & [n_d] & [v] & \ldots n_\Delta \ldots & L & M
\end{array} \tag{13.72}
$$

$$
\begin{array}{llllll}
\text{(ii)} & U(6) \supset & SU(3) & \supset & O(3) \supset & O(2) \\
& | & & & \\
& [N] & (\lambda, \mu) & \ldots K \ldots & L & M
\end{array} \tag{13.73}
$$

$$
\begin{array}{llllll}
\text{(iii)} & U(6) \supset & O(6) \supset & O(5) & \supset & O(3) \supset & O(2) \\
& | & | & | & & | & | \\
& [N] & [\sigma] & [v] & \ldots n_\Delta \ldots & L & M.
\end{array} \tag{13.74}
$$

Here the labels n_Δ, in (i) and (iii), and K, in (ii), are needed to classify all states in the reduction from one group to the next lower one, since these reductions are not fully reducible.

Dynamical symmetries of IBM-1

In the situation where the Hamiltonian is written in terms of the Casimir invariants of a single chain, the problem is solvable in an analytic, transparent way. These limiting cases are called the dynamical symmetries contained within the IBM-1.

A most general Hamiltonian (containing up to quadratic Casimir operators for the subgroups, i.e. the C_{1U5}, C_{2U5}, C_{2O6}, C_{2O5}, C_{2O3}, C_{SU3}) is of the form

$$
H = \varepsilon C_{1U5} + \alpha C_{2U5} + \beta C_{2O5} + \gamma C_{2O3} + \delta C_{SU3} + \eta C_{2O6}. \tag{13.75}
$$

The parameters are simply related to those of the more phenomenological parametrization of the IBM-1 Hamiltonian and are discussed by Arima and Iachello (1984). The above Hamiltonian contains Casimir operators of *all* subgroups and eigenvalues can only be obtained in a numerical way.

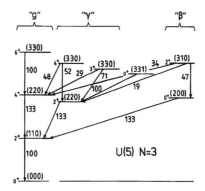

Figure 13.25. The $U(5)$ spectrum for the $N = 3$ boson case. The quantum numbers (n_d, v, n_s) denote the d-boson number, the d boson seniority and the number of boson triplets coupled to 0^+, respectively. The $B(E2)$ values are given, normalized to $B(E2; 2_1^+ \rightarrow 0_1^+)$ (taken from Lipas 1984).

Table 13.1. $U(5)$-chain quantum numbers for $N = 3$.

n_d	v	n_Δ	λ	L
0	0	0	0	0
1	1	0	1	2
2	2	0	2	4, 2
	0	0	0	0
3	3	0	3	6, 4, 3
		1	0	0
	1	0	1	2

Particular choices, however, lead to the dynamical symmetries. For $\delta = \eta = 0$, we find the dynamical $U(5)$ symmetry by using the eigenvalues of the various Casimir invariants ($\langle C_{1U5} \rangle = n_d$, $\langle C_{2U5} \rangle = n_d(n_d + 4)$, $\langle C_{2O5} \rangle = v(v + 3)$, $\langle C_{2O3} \rangle = L(L + 1)$), and the eigenvalue for (13.75) (with $\delta = \eta = 0$) becomes

$$E_{U(5)} = \varepsilon n_d + \alpha n_d(n_d + 4) + \beta v(v + 3) + \gamma L(L + 1), \qquad (13.76)$$

and we discuss this limit in more detail (figure 13.25).

We consider $N = 3$ with the reduction of its irrep labels as given in table 13.1. States with the same L, distinguished by the v-quantum number, occur first at $n_d = 4$, only and states of the same L, v to be distinguished by n_Δ only occur from $n_d = 6$ onwards. The energy spectrum of figure 13.25 is very reminiscent of an anharmonic quadrupole vibrator. The important difference, though, is the cut-off in the IBM-1 at values of $n_d = N$. Only for $N \rightarrow \infty$ does the $U(5)$ limit correspond to the vibrator, encountered in the geometrical shape description of section 13.1. This cut-off at fixed N is a more general feature of the IBM.

Besides the energy spectrum for $N = 3$, relative $E2$ reduced transition probabilities can also be derived using the same group-theoretical methods. They are normalized to the value $B(E2; 2_1^+ \rightarrow 0_1^+) = 100$ (arbitrary units; relative values). Here, the cut-off shows up in a simple way and is given by the expression

$$B(E2)_N = \frac{N - n_d}{N} B(E2)_{N \rightarrow \infty}, \qquad (13.77)$$

where n_d refers to the final state and $B(E2)_{N \rightarrow \infty}$ is the corresponding phonon model value of section 13.1. So, the IBM predicts a general fall-off in the $B(E2)$ values when compared to the geometrical models. The N-dependence will, however, be slightly different according to the dynamical symmetry one considers, and outside of these symmetries this effect only appears after numerical studies. Extensive studies of finite N effects in the IBM have been discussed by Casten and Warner (1988).

Besides the $N = 3$ schematic example just discussed, we compare in figure 13.26(a) the vibrational limit for a large boson number and give a comparison for the ^{110}Cd spectrum (figure 13.26(b)). Similar studies in the $SU(3)$ and $O(6)$ limit have been discussed by Arima and Iachello (1978, 1979).

In many cases, though, a more general situation than a single dynamical symmetry will result. Transitions between the various dynamical symmetries can be clearly depicted using a triangle representation with one of the symmetries at each corner. In these cases, using a single parameter, the continuous change between any two limits can be followed. In figure 13.27(a); this triangle is shown and the three legs contain a few, illustrative cases for particular transition: (i) $U(5) \rightarrow O(6)$ in the Ru,Pd nuclei; (ii) $U(5) \rightarrow SU(3)$ in the Sm,Nd nuclei and, (iii) $O(6) \rightarrow SU(3)$ in the Pt,Os nuclei. The corresponding change is related to the change in the $B(E2; 2_1^+ \rightarrow 0_1^+)$ value which, expressed in single-particle units (spu), equal to Weisskopf units, gives a particular variation that can be obtained from the analytical expressions for $B(E2; 2_1^+ \rightarrow 0_1^+)$ in the dynamical symmetries (figure 13.27(b)). Finally, in figure 13.27(c), we give the transitional spectra in between the quadrupole vibrational (or $U(5)$) limit ($\xi = 0$) and the $O(6)(\xi = 1.0)$ limit. In particular for the $0_1^+, 2_1^+, 4_1^+, 6_1^+$ structure (the yrast band structure) a steady but smooth variation is clearly shown.

A very clear, but not highly technical review on the IBM-1 has been given by Casten and Warner (1988) and expands on most of the topics discussed in the present section.

13.3.3 The proton–neutron interacting boson model: IBM-2

Even though the IBM-1, which has a strong root in group-theoretical methods, has been highly successful in describing many facets of nuclear, collective quadrupole

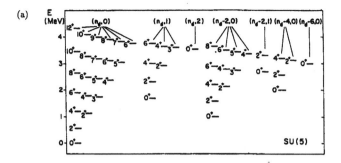

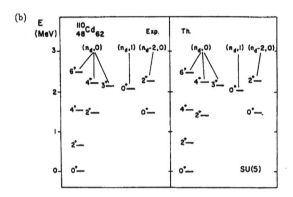

Figure 13.26. (*a*) Theoretical $U(5)$ spectrum with $N = 6$ bosons. The notation (v, n_Δ) is given at the top and explained in the text. (*b*) Experimental energy spectrum for ^{110}Cd ($N = 7$) and a $U(5)$ fit (taken from Arima and Iachello 1976, 1988).

motion, no distinction is made between the proton and neutron variables. Only in as much as the many-boson states are realized as fully symmetric representations of the $U(6)$ group structure, does the IBM-1 exhaust the full model space.

Besides the fact that, from a group-theoretical approach, the basis is now spanned by the irrep of the product group $U_\pi(6) \otimes U_\nu(6)$, a shell-model basis to the IBM-2 can be given.

In the present section we discuss the shell-model origin (i) and, the possibility of forming mixed-symmetry states in the proton and neutron charge label, which characterizes the enlarged boson space.

Shell-model truncation

In describing a given nucleus $^A_Z X_N$ containing a large number of valence nucleons outside of a closed shell, the pairing properties and, to a lesser extent, the quadrupole component give rise to strongly bound $J^\pi = 0^+$ and 2^+ pairs. It

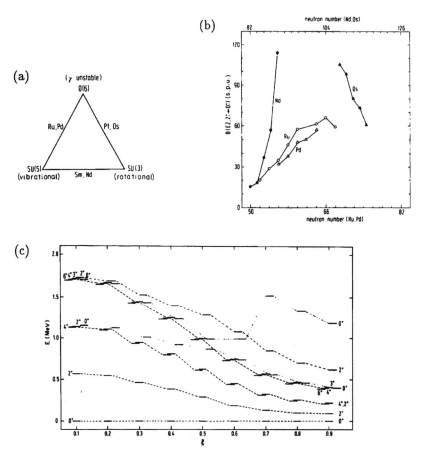

Figure 13.27. (*a*) Symmetry triangle illustrating the three symmetries of the IBM-1, i.e. the $U(5)$, $SU(3)$ and $O(6)$ limits, corresponding to the vibrational, rotational and γ-unstable geometrical limit and the three connecting links. (*b*) Experimental $B(E2; 2_1^+ \rightarrow 0_1^+)$ values (in spu) for the Ru (open circles), Pd (open triangles), Nd (filled circles) and Os (filled triangles). (*c*) Calculated excitation energies for the low-lying excited states in the $U(5) \rightarrow U(6)$ transitional region as a function of the transition parameter ξ and for $N = 14$ bosons (taken from Stachel *et al* 1982).

is this very fact that will subsequently give the possibility of relating nucleon 0^+ and 2^+ coupled pairs to s and d boson configurations.

The 0^+, 2^+ pair truncation (called S, D pair truncation) allows us to use a restricted model space in order to describe the major, low-lying collective excitations. The corresponding, fermion basis states can be depicted as

$$\{[(S_\pi^+)^{N_{s\pi}} (D_\pi^+)^{N_{d\pi}}_{\gamma_\pi J_\pi}] \otimes [(S_\nu^+)^{N_{s\nu}} (D_\nu^+)^{N_{d\nu}}_{\gamma_\nu J_\nu}]\}^{(J)}_M |0\rangle, \qquad (13.78)$$

Table 13.2. Corresponding lowest seniority SD fermion and sd-boson states.

Fermion space (F)			Boson space (B)		
$n = 0,$	$v = 0$	$\|0\rangle$	$N = 0,$	$nd = 0$	$\|0\rangle$
$n = 2,$	$v = 0$	$S^+\|0\rangle$	$N = 1,$	$nd = 0$	$s^+\|0\rangle$
	$v = 2$	$D^+\|0\rangle$		$nd = 1$	$d^+\|0\rangle$
$n = 4,$	$v = 0$	$(S^+)^2\|0\rangle$	$N = 2,$	$nd = 0$	$(s^+)^2\|0\rangle$
	$v = 2$	$S^+D^+\|0\rangle$		$nd = 1$	$s^+d^+\|0\rangle$
	$v = 4$	$(D^+)^2\|0\rangle$		$nd = 2$	$(d^+)^2\|0\rangle$
$\cdots$	$\cdots$		$\cdots$	$\cdots$	

with

$$N_{s_\pi} + N_{d_\pi} = \frac{n_\pi}{2} = N_\pi$$

$$N_{s_\nu} + N_{d_\nu} = \frac{n_\nu}{2} = N_\nu, \tag{13.79}$$

where n_π (n_ν) denotes the number of valence protons (neutrons).

Using the Otsuka–Arima–Iachello (OAI) mapping procedure (Otsuka *et al* 1978) the fermion many-pair configurations (13.78) are mapped onto real boson states where now $S_\rho^+ \to s_\rho^+$, $D_\rho^+ \to d_\rho^+$ ($\rho \equiv \pi, \nu$) carries out the transformation into boson configurations. This fermion basis $\to$ boson basis mapping uses the methods outlined in chapter 11 when constructing an effective interaction, that starts from a realistic interaction. The constraint in the mapping is that matrix elements remain equal, or,

$$\langle \psi_F | \hat{H}_F | \psi_F \rangle = \langle \phi_B | \hat{H}_B | \phi_B \rangle, \tag{13.80}$$

for the lowest fermion and boson configuration. A correspondence table illustrates this mapping in a very instructive way (see table 13.2).

For a simple fermion Hamiltonian, containing pairing and quadrupole–quadrupole interactions, the general form of the boson Hamiltonian, using the OAI mapping of equation (13.80), now becomes

$$\hat{H}_B = E_0 + \varepsilon_{d_\pi} \hat{n}_{d_\pi} + \varepsilon_{d_\nu} \hat{n}_{d_\nu} + \kappa \hat{Q}_\pi^{(2)} \cdot \hat{Q}_\nu^{(2)} + \hat{V}_{\pi\nu} + \hat{V}_{\nu\nu} + \hat{M}_{\pi\nu}, \tag{13.81}$$

where the various operators are:

$$\hat{n}_{d_\rho} \equiv d_\rho^+ \cdot \tilde{d}_\rho \quad \text{(the boson number operator)} \tag{13.82}$$

$$\hat{Q}_\rho^{(2)} \equiv (d^+ s + s^+ \tilde{d})_\rho^{(2)} + \chi_\rho (d^+ \tilde{d})_\rho^{(2)}. \tag{13.83}$$

The various parameters in the Hamiltonian (13.81) (ε_{d_ρ}, κ, χ_π, χ_ν, $E_0, \ldots$) are related to the underlying nuclear shell structure. The quantity ε_{d_π} (ε_{d_ν}) denotes the d-boson proton (neutron) energy; κ the quadrupole interaction strength and the remaining terms $\hat{V}_{\pi\pi}$, $\hat{V}_{\nu\nu}$, $\hat{M}_{\pi\nu}$ describe remaining interactions amongst identical ($\pi\pi$, $\nu\nu$) bosons. The last term has a particular significance, is known as the Majorana interaction and will be discussed later.

The whole mapping process is schematically shown in figure 13.28 where we start from the full fermion space and end in the boson (sd) model space. So, we have indicated that, starting from a general large-scale shell-model approach, and using residual pairing- and quadruopole forces in the fermion space, a rather interesting approximation is obtained, namely the (s, d) OAI mapping in the boson model space. With this Hamiltonian, one is able to describe a large class of collective excitations in medium-heavy and heavy nuclei. One can also, from the Hamiltonian (13.81) taken as an independent starting point, consider the parameters ε_{d_ρ}, κ, χ_π, χ_ν, ... as *free* parameters which are determined so as to describe the observed nuclear properties as well as possible. The general strategy to carry out such an IBM-2 calculation in a given nucleus e.g. $^{118}_{54}\text{Xe}_{64}$ is explained in figure 13.29 where the corresponding fermion and boson model spaces are depicted. IBM-2 calculations for the even–even Xe nuclei for the ground-state $(0^+, 2^+, 4^+, 6^+, 8^+)$ band; as well as for the γ-band $(2_2^+, 3_1^+, 4_2^+, 5_1^+)$ and for the β-band $(0_2^+, 2_3^+, 4_3^+)$ are given in figure 13.30. These results are typical examples of the type of those obtained in the many other mass regions of the nuclear mass table (Arima and Iachello 1984). A computer program NPBOS has been written by Otsuka which performs this task is a general Hamiltonian (Otsuka 1985).

Mixed-symmetry excitations

Incorporating the proton and neutron boson degrees of freedom in the extended boson model space of the IBM-2, the group structure becomes

$$U_\pi(6) \otimes U_\nu(6) \supset U_{\pi+\nu}(6) \supset \cdots, \qquad (13.84)$$

where the irrep are formed by the product irrep

$$[N_\pi] \otimes [N_\nu] \supset [N_\pi+N_\nu, 0, \ldots] \oplus [N_\pi+N_\nu-1, 1, 0, \ldots] \oplus [N_\pi+N_\nu-2, 2, 0, \ldots]. \qquad (13.85)$$

Now, the one-row irrep, two-row (and even more complicated and less symmetric) irrep can be constructed. These states are now called mixed-symmetry states, which correspond to non-symmetric couplings of the basic proton and neutron bosons.

The most simple illustration is obtained in the vibrational limit when considering just one boson of each type (figure 13.31). The s_π^+, d_π^+ and s_ν^+, d_ν^+ can be combined in the *symmetric* states that relate to the

$$0^+(s_\pi^+ s_\nu^+), \; 2^+ \left(\frac{1}{\sqrt{2}} (s_\pi^+ d_\nu^+ + s_\nu^+ d_\pi^+) \right), \; 4^+(d_\pi^+ d_\nu^+), \qquad (13.86)$$

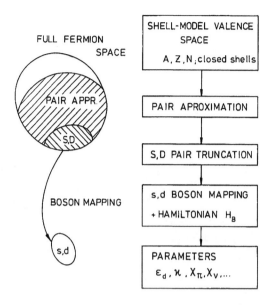

Figure 13.28. Schematic representations of the various approximations underlying the interacting boson model (IBM) when starting from a full shell-model calculation (taken from Heyde 1989).

harmonic quadrupole spectrum, coinciding with the simpler IBM-1 picture. The antisymmetric states now are $2^+(1/\sqrt{2}(s_\pi^+ d_\nu^+ - s_\nu^+ d_\pi^+))$, 1^+, $3^+(d_\pi^+ d_\nu^+)$. The energy of the lowest 2^+ state, relative to the 0^+ symmetric state, is governed by the strength of the $\hat{M}_{\pi\nu}$ Majorana operator, which shifts the classes of states having a different symmetry character.

In the rotational, or $SU(3)$ limit, similar studies can be made but now the lowest mixed-symmetry state becomes the 1^+ level, the band-head of a $1^+, 3^+, \ldots$ band. The search for such states has been an interesting success story and the early experiments are discussed in Box 13b. These mixed-symmetry states, when taking the classical limit of the corresponding algebraic model, are related to certain classes of proton and neutron motion where the two nuclear fluids move in anti-phase, isovector modes. In the vibrational example of figure 13.31, low-lying isovector modes are generated whereas the rotational out-of-phase motion of protons versus neutrons corresponds to a 'scissor'-like mode. This scissor mode, for which the 1^+ state is the lowest-lying in deformed nuclei, can be strongly excited via $M1$ transitions starting from the ground state 0^+ configuration. By now, an extensive amount of data on these 1^+ states has been obtained. Various reactions were used: (e, e'), (γ, γ'), (p, p') to excite low-lying orbital-like 1^+ excitations in deformed nuclei which occur near $E_x \simeq 3$ MeV

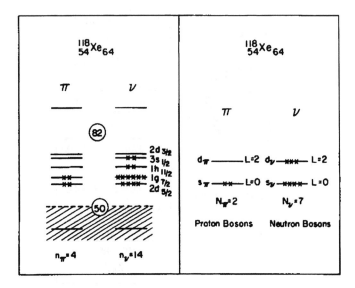

Figure 13.29. Schematic outline of how to perform a realistic IBM-2 calculation for a given nucleus. Here, the example of $^{118}_{54}\text{Xe}_{64}$ is used. We can (i) determine the nearest closed shells $(50, 50)$, (ii) determine the number of bosons ('valence' particle number divided by two) and, (iii) make an estimate of the important parameters ε, κ, χ_π and χ_ν (taken from Iachello 1984).

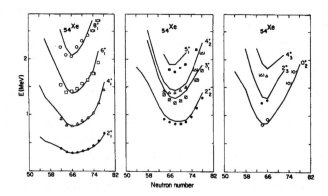

Figure 13.30. Comparison of IBM-2 calculations and experimental energy spectra for the even–even Xe nuclei (taken from Scholten 1980).

while the spin–flip strength is situated at higher energies ($5\,\text{MeV} \le E_x \le 9\,\text{MeV}$) in a double-hump-like structure. In Box 13b, the (e, e') experiments at Darmstadt,

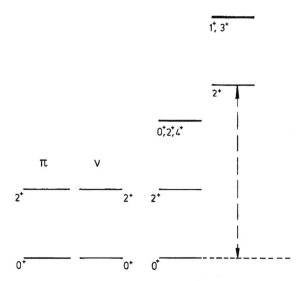

Figure 13.31. Schematic representations of a coupled proton–neutron system where one boson of each type ($N_\pi = 1, N_\nu = 1$) is present. The symmetric states ($0^+; 2^+; 0^+, 2^+, 4^+, \ldots$) and the mixed-symmetry ($2^+; 1^+, 3^+, \ldots$) ones are drawn in the right-hand part.

which showed the first evidence for the existence of mixed-symmetry 1^+ states, is presented.

13.3.4 Extension of the interacting boson model

The algebraic techniques that form the basic structure of the interacting boson as discussed in the previous sections have been extended to cover a number of interesting topics. Without aiming at a full treatment of every type of application that might have been studied we here mention a number of major extensions

(i) It has been observed over the years that with the use of s and d bosons only it is very difficult, if not impossible, to address the problem of high-spin states. The systematic treatment of bosons with higher angular momentum, in particular the inclusion of a hexadecapole or g-boson, has been worked out both in group-theoretical studies as well as by numerical treatments of much larger model spaces including a number of such g-bosons (see, for example, the extensive discussion in the book edited by Casten 1993). A systematic treatment, using the program package 'Mathematica' has been carried out by Kuyucak (1995) and Li (1996).

(ii) The extension of the general framework to encompass the coupling of an odd particle to the underlying even–even core, described by a regular interacting boson model, is referred to as the interacting boson–fermion model (IBFM). This seems at first to be rather similar to well known particle–core coupling. In the IBFM, there are, however, a number of elements that may give deeper insight into the physics of particle–core coupling. Firstly, there exists the possibility to relate the collective Hamiltonian to the specific single-particle structure present in a given mass region and, secondly, the symmetries that are within the even–even core system can, for given mass regions, be combined with the single-particle space into larger groups. The full structure of the treatment of odd-mass nuclei is covererd in a monograph (Iachello and Van Isacker 1992). A number of these IBFM extensions as well as more detailed reference to the literature is discussed by Van Isacker and Warner (1994).

(iii) Using supersymmetric structures, in which both the fermion degrees of freedom (the odd protons and/or neutrons) and the bosons (nucleon pairs) are put together, both even–even, odd–mass and odd–odd nuclei put in given 'quartets' could be described in a unified way

It had been suggested that it might be possible to observe the effects of supersymmetric relations in atomic nuclei in the mass region of heavy Pt and Au nuclei, in particular that the odd–odd nucleus ^{196}Au would form the ultimate nucleus for a stringent test. Starting from the known nuclear structure data on the other nearby nuclei ^{194}Pt, ^{195}Pt and ^{195}Au, the supersymmetric relations then fix the nuclear level structure and its properties in ^{196}Au in an unambiguous way as these four nuclei form such a 'quartet'. So it was clear that state-of-the-art experiments concentrating on that odd–odd Au nucleus could fill in the missing element for constructing the full quartet of nuclei. The boson–fermion transformations implied by the supersymmetry then 'lock' the four nuclei up in a fixed structure (see left-hand part of figure 13.32).

Metz *et al* (1999) have found solid experimental evidence that supersymmetry relationships do hold for these four particular nuclei. They used the Tandem accelerator in München to study one-neutron transfer reactions leading to the odd–odd Au nucleus. This provides information about the excited energy states in the nucleus of ^{196}Au. The full experimental program also consisted of the study of in-beam gamma-ray spectroscopy, conversion electron spectroscopy, studied at the PSI (Villigen, Switzerland) and on studying proton–deuteron pick-up reactions leading to the same odd–odd Au nucleus. The nucleus ^{195}Pt was measured at the same time to obtain a reference data set (Metz *et al* 2000, Gröger 2000). As a result, it was observed that, using the supersymmetric relationships between energy levels in the other three nuclei of the 'quartet', the theoretical and predicted results were in excellent agreement. Even the detailed results of the neutron transfer reaction were very well explained. A popular account of these most interesting Bose–Fermi interrelations are discussed by J Jolie (see Jolie 2002).

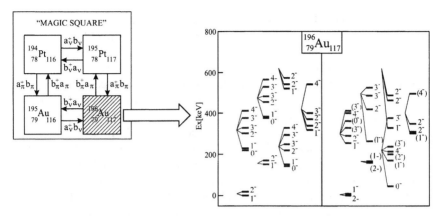

Figure 13.32. Through supersymmetry, bosonic and fermionic systems with a constant number $\mathcal{N}$ (sum of the number of bosons (paired nucleons) and the number of fermions (unpaired nucleons)), are related through given operators (the $b_\pi, b_\nu, b_\pi^\dagger, b_\nu^\dagger$ operators annihilate or create proton or neutron bosons, whereas the $a_\pi, a_\nu, a_\pi^\dagger, a_\nu^\dagger$ operators annihilate or create proton or neutron fermions) (left-hand part of the figure). The predictive power of this supersymmetry classification is shown by comparing the lowest part of the energy spectra with the data for ^{196}Au. (Right-hand part: adapted from Gröger J *et al* 2000 *Phys. Rev.* C **62** 064304 © 2000 by the American Physical Society.)

(iv) In extending the region of applicability to light nuclei, one cannot go on by just handling proton–proton and neutron–neutron pairs; one needs to complement this by considering proton–neutron pairs on an equal footing. This is because the building blocks of the proton–neutron boson model (IBM-2) do not form an isospin invariant model. Extensions into an IBM-3 model by including the so-called δ-boson with isospin $T = 1$ and projection $M_T = 0$ and, even further, by also taking both the $T = 1$ and the $T = 0$ (with projection $M_T = 0$) proton–neutron bosons into account in the IBM-4 model have been constructed and studied in detail. We again refer to Van Isacker and Warner (1994) for a detailed discussion of this isospin extension of the original interacting boson model. In that review article, which concentrates on various extensions, references to the more specific literature on this topic are presented.

13.4 Shape coexistence and phase transitions

13.4.1 Shape coexistence: introduction and experimental facts

The concept of deformed shapes and the appearance of different shapes in a given nucleus was introduced in nuclear physics as early as 1937 by the work of N Bohr and F Kalckar (Bohr and Kalckar 1937). In those early days, little did one expect how fruitful these ideas would turn out to be. The discovery of the first excited state with spin 0^+ in the doubly-magic nucleus ^{16}O and its subsequent interpretation starting from rearranging four particles from occupied orbitals into empty orbitals above the Fermi level, resulting in a cooperative strong binding energy effect and a subsequent highly deformed shape coexisting with the spherical ground state (Morinaga 1956) opened up a new field in nuclear physics research, devoted to the investigation and understanding of shape coexistence. Soon after, it was realized that the atomic nucleus, on its way to fission, had to undergo a number of shape changes in which a specific shape could be trapped as an isomeric state in a secondary potential minimum, called fission isomers (Björnholm and Lynn 1980). Shape coexistence, invoking multiple shapes, was predicted and also observed in many spherical nuclei near magic shells and these particular phases could be linked to the occupation of very specific up- and/or down-sloping orbitals, coined 'intruder orbitals', which allowed for a simple understanding of the phenomenon of shape coexistence (Heyde *et al* 1983, Wood *et al* 1992). The method put forward in these papers could be used to predict shape coexistence e.g. in the Sn nuclei, around mass number $A = 116$ and in the Pb nuclei from mass number $A = 196$ and below. Once fast rotation was employed as a new tool to study nuclear shapes spinning up nuclei very fast, as in the case of ^{152}Dy, a 'superdeformed' shape was discovered with axis ratios for the prolate deformed ellipsoid of 2:1, coexisting with single-particle excitations corresponding to oblate shapes (Nolan and Twin 1988). This research field has 'exploded' in recent years due to the highly increased technical capabilities in detecting gamma-radiation emitted during various types of nuclear reactions (Gammasphere, Euroball, AGATA and GRETA as exponents of the next-generation gamma-ray spectrometers). The variety of shapes occurring in atomic nuclei continues to be a topic of active and rapidly evolving research as exemplified in the recent paper by Andreyev *et al* (Andreyev *et al* 2000) (see Box 13c).

Excellent examples of complete structures representing shape coexistence have been observed in the last decade in the Cd isotopes. These isotopes form an excellent test of shape coexistence as here the intruder states have their lowest excitation energy (at mid shell) exactly at the line of stability. This gives a unique possibility for studying the six stable even–even Cd isotopes in an as complete way as possible. Studies like those of the structure of ^{112}Cd (Deleze *et al* 1993) revealed that these isotopes exhibit complete three-phonon spherical structures together with complete more deformed O(6)-like intruder excitations. Experiments are planned to study the very neutron-rich $^{124,126,128}Cd$ isotopes in order to explore the behaviour of the family of intruder states (Jokinen *et al*).

Another more dramatic example of shape coexistence shows up in the data for the Pb region when removing neutrons from the closed shell at $N = 126$. These data point towards the appearance of specific particle–hole (p–h) excitations across the closed shell at $Z = 82$. It is precisely the energy gap at the $Z = 82$, $N = 126$ closed shell of only approximately 3.5 MeV, combined with a

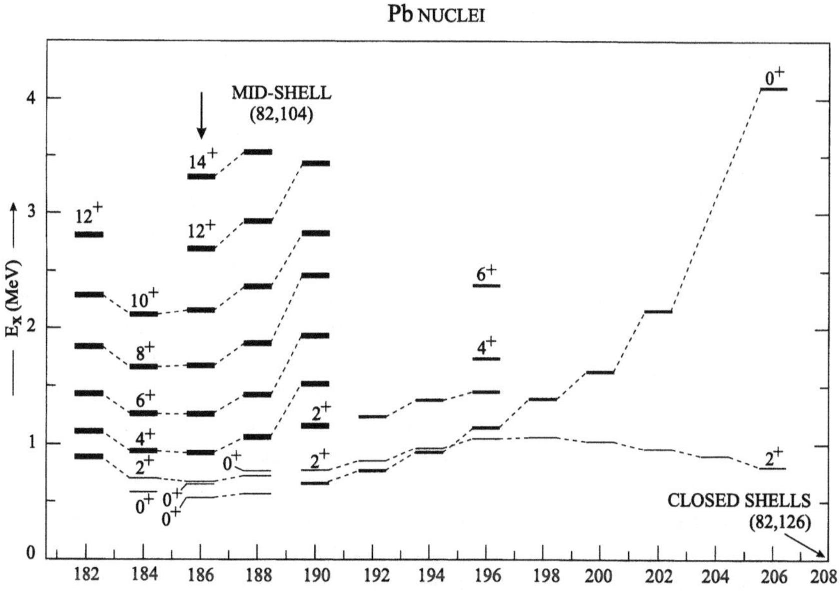

Pb NUCLEI

Figure 13.33. Systematics of the lowest 0^+ states in the even–even Pb nuclei. The first excited 2^+ state is also given for reference. The band members of the yrast structure are given in the mass region $182 \leq A \leq 190$. (Reprinted from Fossion R *et al* 2003 *Phys. Rev. C* **67** 024306 © 2003 by the American Physical Society.)

very large open neutron shell (filling the 82–126 orbitals) that enables the proton–neutron quadrupole–quadrupole force to lower the excitation energy of 2p–2h,4p–4h, etc configurations so as to approach the ground state (for the Pb and Hg nuclei) and even cross it (for the Pt and possibly the Po nuclei too) (Heyde *et al* 1983, Wood *et al* 1992). Because of the increased quadrupole collectivity associated with these p–h excitations, collective bands are observed on top of the low-lying 0^+ intruder excitations and so indicates the presence of shape coexistence (see figure 13.33 for the most recent systematics and also Julin (2001)).

An extensive overview of shape coexistence in both odd-mass and even–even nuclei has been given in (Heyde *et al* 1983, Wood *et al* 1992) and the references in these papers give an extensive overview of many of the data that have firmly pointed to the coexistence of various different nuclear shapes.

13.4.2 Shape phase transitions in atomic nuclei

In a large number of many-body systems, such as atomic nuclei, molecules, atomic clusters, polymers, etc, one has observed the appearance of phase transitions whenever a drastic change in the geometry of the systems occurs. A typical example is the situation in which a molecular phase transtion results from the change between a linear chain and a planar configuration. Such phase transitions are called 'shape' phase transitions.

One of the foremost studied types of phase transitions is associated with atomic nuclei in which the nuclear shape changes from spherical into a quadrupole deformed shape (Bohr and Mottelson 1975). Such phase transitions are characterized by a rather sudden variation in a number of physical observables that are also called 'order'-parameters and this is a function of an externally changing 'control'-parameter. In the case of shape changes, the 'order'-parameter is the quadrupole deformation, expressed by the quantity β. For the earlier phase transitions in atomic nuclei, the 'control'-parameter is given by the dimensionless coupling strength which characterizes the quadrupole–quadrupole forces in the nucleus and makes the nucleus change from a spherical shape to a deformed shape. Even though the atomic nucleus is characterized by a rather restricted number of interacting nucleons (typically of the order $\approx$50–200) and no phase transitions are possible in the strict thermodynamical sense (Gross 2001), it is still possible to describe phase transitions, making use of the classical limit in having the number of interacting objects N go to infinity. Numerical studies making use of the concept of dynamical symmetries in various systems (atomic nuclei (Arima and Iachello 1975, Iachello and Arima 1988), molecules (Iachello 1981, Iachello and Levine 1995)), indicate that, in relatively 'small' systems, with a restricted number of particles, large variations in the 'order'-parameters do show up for small variations in the 'control'-parameters, albeit these are more smeared out compared to the discontinuous changes in the 'order'-parameters in infinite systems.

In order to model these phase transitions in atomic nuclei, in which many properties result from the interplay of protons and neutrons interacting by means of an effective force inside the atomic nucleus, one has to separate out the essential building block inside the atomic nucleus. We have discussed the presence of an average potential in chapters 9–11. In this chapter, 13, it has been shown that, to a very good approximation, nucleons coupling into 0^+ coupled pairs can be treated as bosons. A condensate of such pairs represents the ground state of many nuclei to a large extent (Talmi 1993). Studying the symmetry properties of this interacting boson system (Interacting Boson Model or IBM for short (Arima and Iachello 1975, Iachello and Arima 1987) forms a very good starting point for addressing these questions. The three symmetries—the $U(5)$, $SU(3)$ and $O(6)$ limits (corresponding to anharmonic vibrations, rotational motion and triaxial-unstable nuclei)—are the benchmarks for studying phase transtions when moving from one limiting case to another. The ground-state binding energy is a sensitive probe and the study of the variation in the binding energy (and of other observables, which then play the role of an 'order'-parameter) as well as its lowest derivatives towards the 'control'-parameter will give a tool for characterizing the order of the phase transition. Recent work by Iachello (2000, 2001, 2003) has studied these questions, starting from the differential equation that describes nuclear quadrupole collective motion in the collective space of the β and γ variables.

These results have given a strong impulse to the study of phase transitions in (finite) atomic nuclei, in particular those related to the quadrupole degree of freedom. Studies to find experimental indications of the actual appearance of these benchmarks (called E(5), X(5), Y(5)) have been vigorously pursued recently. Besides Iachello's studies, groups have concentrated on the study of finite (albeit large N) studies of the IBM over the available 'control'-parameter space (Casten 1981) and on the classical limit of the IBM (carrying out the limit for $N \to \infty$).

Box 13a. Double giant resonances in nuclei

In many nuclei, double and multiple phonon excitations have been observed for the low-lying isoscalar vibrations. Although there is no reason that multiple giant resonances could not be observed, it has taken until recently for the experimental observation of such excitations. Their properties confirm general aspects of nuclear structure theory.

The possibility of a double giant resonance was first pointed out by Auerbach (1987) as a new kind of nuclear collective excitation. The questions then relate to understanding the excitation energy, width, integrated strength and other characteristics that arise from the properties of the single giant resonance. In nuclear structure theory, such double excitations should be understandable both from a macroscopic (coherent, collective two-phonon excitations) and a microscopic (a coherent picture of 2p–2h excitations) picture.

Double resonances have not been observed until very recently, mainly because of experimental difficulties of detecting such excitations at the very high energy of the lower resonance due to the high underlying background. Work at the Los Alamos Meson Physics Facility (LAMPF) has given the first clear indication for such double giant resonances. At LAMPF, one has one of the most intense pion beams available with π^0, $\pi^{\pm}$ components and the possibility of a rich spectrum of nuclear reactions such as single-charge exchange (SCX) and double-charge exchange (DCX).

Since the GDR (Giant Dipole Resonance, section 13.1.3) has been strongly excited in SCX reactions, it was suggested that DCX reactions, viewed as a

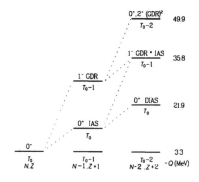

Figure 13a.1. Schematic diagram of single and double resonances expected in pion single-charge-exchange (π^+, π^0) and pion double-charge-exchange (π^+, π^-) reactions. The numbers to the right are the Q values for the ground state and the three double resonances observed in DCX experiments on ^{93}Nb. (Taken from Mordechai and Moore 1991. Reprinted with permission of *Nature* © 1991 MacMillan Magazines Ltd.)

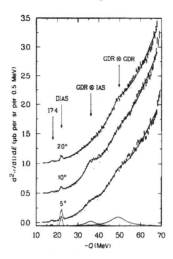

Figure 13a.2. Double differential cross-section for the (π^+, π^-) reaction on ^{93}Nb at $T_\pi = 295$ MeV, the laboratory kinetic energy of the incoming pion, and $\theta_{lab} = 5°$, 10° and 20°. The arrows indicate the three double resonances in pion DCX: the DIAS; the GDR⊗IAS and the GDR⊗GDR (see text for a more extensive explanation of the notation). (Taken from Mordechai and Moore 1991. Reprinted with permission of *Nature* © 1991 MacMillan Magazines Ltd.)

sequence of two SCX reactions, would excite, for nuclei with $N - Z \gg 1$, the double giant resonances rather strongly. In figure 13a.1, the schematic energy-level diagram of 'single' and 'double' resonances anticipated in pion SCX (π^+, π^0) and pion DCX (π^+, π^-) is given. The numbers are the Q-values for the ground state and the three double resonances observed in the DCX experiment on ^{93}Nb. In this figure, also the double isobaric analogue state (DIAS) is given: it corresponds to acting twice with the $\hat{t}_-$ operator, changing two of the excess neutrons into protons.

The spectra were taken at the energetic pion channel at Los Alamos using a special set-up for pion DCX (Greene 1979). Outgoing pions can be detected with a kinetic energy range 100–300 MeV and an energy resolution of $\simeq 140$ keV.

We present, in figures 13a.2 and 13a.3 the double-differential cross-sections for the (π^+, π^-) reaction on ^{93}Nb and ^{40}Ca, respectively. The arrows indicate the various double resonances: the DIAS, GDR⊗IAS and GDR⊗GDR for ^{93}Nb and the double-giant resonance GDR⊗GDR in ^{40}Ca. The double dipole resonance has a width of ~ 8–10 MeV which is larger than the width of the single dipole resonance by a factor 1.5–2.0 in agreement with theoretical estimates for the width of such a double giant resonance (Auerbach 1990).

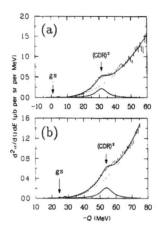

Figure 13a.3. (*a*) Double differential cross-section for the (π^-, π^+) reaction on ^{40}Ca at $T_\pi = 295$ MeV and $\theta_{\text{lab}} = 5°$. The arrows indicate the fitted location of the ground state (gs) and the giant resonance $(\text{GDR})^2$. The dashed line gives the background and the solid line is a fit to the spectrum. (*b*) The same as in (*a*) but now for the inverse reaction ^{40}Ca(π^+, π^-) ^{40}Ti. (Taken from Mordechai and Moore 1991. Reprinted with permission of *Nature* © 1991 MacMillan Magazines Ltd.) See figure 13a.1.

There exists a unique feature about the DCX reactions in the simplicity with which one can measure (π^+, π^-) and (π^-, π^+) reactions on the same target. This is illustrated for ^{40}Ca in figure 13a.3.

Concluding, unambiguous evidence for the appearance of double giant resonances in atomic nuclei has now been obtained. The identification, based on the energies at which they appear, the angular distributions and cross-section, is strong. The results indicate that the collective interpretation of two giant resonance excitations remains valid, in particular for nuclei with $N - Z \gg 1$, and that these excitations appear as a general feature in nuclear structure properties. Remaining questions relate to the missing 0^+ strength of the GDR $\otimes$ GDR excitation and the fast-increasing width of the GDR $\otimes$ IAS excitation with mass number.

Box 13b. Magnetic electron scattering at Darmstadt: probing the nuclear currents in deformed nuclei

---

The first experimental evidence for a low-lying 1^+ collective state in heavy deformed nuclei, predicted some time ago, was obtained by Bohle *et al* (1984). The excitation can be looked upon in a two-rotor model (Lo Iudice and Palumbo 1978, 1979) as being due to a contra-rotational motion of deformed proton and neutron distributions. In vibrational nuclei, isovector vibrational motion can be constructed with a 2^+ level as a low-lying excited state (figure 13b.1).

The initial impulse to study such 1^+ excitations in deformed nuclei came from the IBM-2 theoretical studies predicting a new class of states, called mixed-symmetry states (section 13.3). In this IBM-2 approach, the $K^\pi = 1^+$ excitation mode corresponds to a band head of a $K^\pi = 1^+$ band and is related to the motion of the valence nucleons (protons versus neutrons), in contrast to the purely collective models where, initially *all* nucleons were considered to participate in the two-rotor collective mode of motion. Realistic estimates of both the 1^+ excitation energy and the $M1$ strength determined in the IBM-2 represented the start of an intensive search for such excitations in deformed nuclei.

The experiment at the Darmstadt Electron Linear Accelerator (DALINAC) concentrated on the nucleus ^{156}Gd. The inelastically scattered electrons were detected using a $169°$ double focusing magnetic spectrometer at a scattering angle of $\theta = 165°$, at bombarding energies $E_{e^-} = 25, 30, 36, 42, 45, 50$ and 56 MeV and $E_{e^-} = 42$ MeV at $\theta = 117°$ and $E_{e^-} = 45$ MeV at $\theta = 105°$. The spectrum taken at $E_{e^-} = 30$ MeV reveals a rich fine structure but the only strong transition is to a state at $E_x = 3.075$ MeV. This state is almost absent in the $E_{e^-} = 50$ MeV spectrum, in which, however, the collective 3^- level at

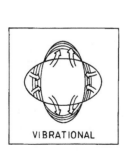
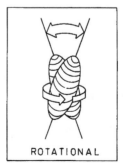

VIBRATIONAL   ROTATIONAL

Figure 13b.1. Schematic drawing of the non-symmetric modes of relative motion of the proton and neutron degrees of freedom. In the left part, vibrational anti-phase oscillatory motion of protons against neutrons is given. In the right part the collective rotational anti-phase 'scissor' mode is presented.

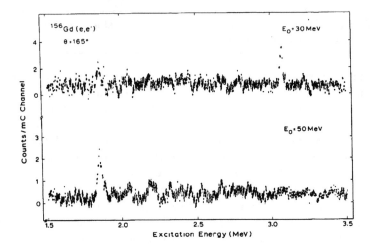

Figure 13b.2. High-resolution electron inelastic scattering spectra for ^{156}Gd at electron endpoint energies of $E_0 = 30$ MeV and 50 MeV. The strongly excited state at $E_x = 3.075$ MeV as a $J^\pi = 1^+$ state, corresponding to the rotational 'scissor'-like excitation (taken from Bohle *et al* 1984).

$E_x = 1.852$ MeV is strongly excited (figure 13b.2). The state at $E_x = 3.075$ MeV dominates all spectra at low incident electron energies and has a form factor behaviour consistent with a 1^+ assignment.

A detailed discussion of a number of features: excitation energy, $M1$ transition strength and form factors all point towards the 1^+ mixed-symmetry character. In particular, the $B(M1, 0_1^+ \rightarrow 1_1^+) = 1.3 \pm 0.2\mu_N^2$ value is of the right order of magnitude as given by the IBM-2 prediction for the $SU(3)$ limit

$$B(M1; 0_1^+ \rightarrow 1_1^+) = \frac{3}{4\pi} \frac{8N_\pi N_\nu}{2N - 1}(g_\pi - g_\nu)^2 \mu_N^2, \qquad (13b.1)$$

(with $N = N_\pi + N_\nu$) which, using $N_\pi = 7$, $N_\nu = 5$ and $g_\pi = 0.9$, $g_\nu = -0.05$, gives a value of $2.5\mu_N^2$. The purely collective two-rotor models result in $B(M1)$ values that are one order of magnitude bigger than the observed value.

Other interesting properties of this 1^+ mode are hinted at by comparing (p, p') and (e, e') data. The (p, p') scattering at $E_p = 25$ MeV and $\theta = 35°$ is mainly sensitive to the spin–flip component and so, the combined results point out the mainly orbital character of the 1^+ 3.075 MeV state in ^{156}Gd, as illustrated in figure 13b.3.

Since the original study, electro- and photo-nuclear experiments have revealed the 1^+ mode in nuclei ranging from medium-heavy fp-shell nuclei to a large number of deformed nuclei and to thorium and uranium. The transition strength is quite often concentrated in just a few 'collective' states with little

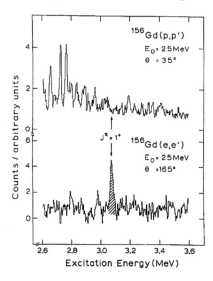

Figure 13b.3. High-resolution inelastic proton and electron scattering spectrums of ^{156}Gd. The 'scissor' mode, leading to a strongly excited $J^\pi = 1^+$ state (hatched) is missing in the proton spectrum (taken from Richter 1988).

Figure 13b.4. Photograph of the new superconducting S-Dalinac (electron accelerator at the Institüt für Kernphysik der Technische Hochschule, Darmstadt). The re-circulating part and a scattering chamber (in the centre foreground) are clearly visible (courtesy of Richter 1991).

spreading. Finally, in figure 13b.4, we give an illustration of the newly constructed DALINAC set-up.

Box 13c. A triplet of different states in ^{186}Pb

In most atomic nuclei with an even number of protons and neutrons, the low-lying excitation modes are mainly built from nucleon pair breaking of the otherwise superfluid phase or from vibrational or rotational motion. It has been shown, however, that in certain parts of the nuclear mass table, if one type of nucleons (e.g. protons) forms a closed-shell configuration, whereas the other type of nucleon (e.g. neutrons) has a maximal number of 'valence' particles (situated mid-way between closed shells), such as is the case for ^{116}Sn or ^{186}Pb, a subtle rearrangement of a few nucleons near the Fermi level may well result in a totally new macroscopic shape for that nucleus.

The recent development of recoil-decay-tagging techniques (RDT) and heavy-ion-induced fusion–evaporation reactions have led to a wealth of new data in nuclei that have a neutron number near $N = 104$ and are situated in the Pb region. The large body of experiments carried out in recent years have given ample evidence for unexpected new collective band structures in both the single closed shell $Z = 82$ Pb nuclei as well as in the adjacent Hg, Pt and Po nuclei below and above the proton shell closure respectively (Julin *et al* 2001) (see also section 13.4).

The nucleus ^{186}Pb can be studied in detail in order to find out about the critical position in which this nucleus is situated, i.e. having $Z = 82$ and $N = 104$. The best way to study this nucleus is through α- decay but then the nucleus ^{190}Po needs to be formed. This can be accomplished through a four-neutron emission which results after the fusion reaction of 255 MeV ^{52}Cr ions with ^{142}Nd target nuclei. The cross section is very small (only about 300 nb), which corresponds to the production of about 300 atoms of ^{190}Po per hour but with a background that is much higher than the interesting phenomena for which one is searching. The separator for heavy-ion production (SHIP) velocity filter at GSI, Darmstadt, developed for the study of super heavy elements (see chapter 1) (Hofmann 1998), however, can separate the correct events from the background and implant them in an efficient detection system. The study of the α-decay then unambiguously identifies two very low-lying excited 0^+ states below 650 keV which is a very special situation (Andreyev *et al* 2000).

Calculations, making use of a deformed mean-field approach or using a deformed Woods–Saxon potential in order to study the possible equilibrium states (Nazarewicz 1993, Andreyev *et al* 2000), have indicated the possibilities of producing rather close-lying oblate and prolate minima in the total energy surface for the Pb nuclei while approaching the neutron mid-shell region at $N = 104$, next to the spherical ground-state configuration as shown in figure 13c.1.

It seems that, as for the Pb region, the conditions that allow the presence of such very low-lying intruding 0^+ states i.e. a moderate energy gap at the $Z = 82$ proton shell closure combined with a very large open neutron shell at $N = 104$ (having 22 valence neutrons) match precisely.

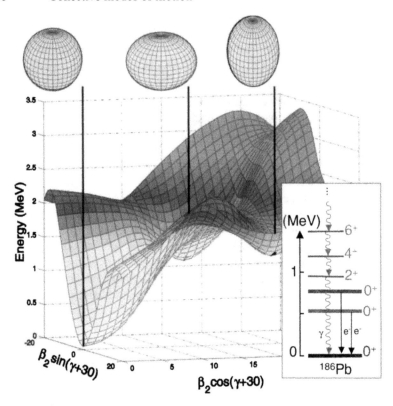

Figure 13c.1. Calculated potential energy surface for ^{186}Pb. The spherical, oblate and prolate energy minima are indicated by arrows and a pictoral figure of the corresponding shape is drawn (see (Andreyev 2000) for more details). (Reprinted with permission from Andreyev *et al* 2000 *Nature* **405** 430 © 2000 MacMillan Magazines Ltd.)

An extensive overview of shape coexistence in other mass regions for both odd-mass and even–even nuclei has been given in (Heyde *et al* 1983, Wood *et al* 1992) and the references in these papers give an extensive overview of many of the data that have firmly pointed out the coexistence of various different nuclear shapes.

Chapter 14

Deformation in nuclei: shapes and rapid rotation

In the discussion of chapter 13, it became clear that in a number of mass regions, coherence in the nuclear single-particle motion results in collective effects. The most dramatic illustrations of these collective excitations is the observation of many rotational bands that extend to very high angular momentum states. At the same time, one could prove that in the bands, the nucleus seems to acquire a strongly deformed shape that is given, in a quantitative way, by the large intrinsic quadrupole moment. The concept of shape, shape changes and rotations of shapes at high frequencies will be the subject of the present chapter.

In section 14.1 we shall concentrate on the various manifestations in which the deformed shape, and thus the average field, influences the nuclear single-particle motion. The clearest development is made using the Nilsson potential, which gives rise to all of the salient features characteristic of deformed single-particle motion. It is also pointed out how the total energy of a static, deformed nucleus can be evaluated and that a minimization of this total energy expression will give rise to stable, deformed minima. In section 14.2 we study the effect of rotation on this single-particle motion and discuss the cranking model as a means of providing a microscopic underpinning of the rotational nuclear structure. Finally, in section 14.3, we apply the above description to attempt to understand nuclear rotation at high and very-high spins and the phenomenon of superdeformation as an illustration of these somewhat unexpected, exotic nuclear structure features.

14.1 The harmonic anisotropic oscillator: the Nilsson model

Starting from a spheroidal distribution or deformed oscillatory potential with frequencies ω_x, ω_y and ω_z in the directions x, y, z, the Hamiltonian governing the nuclear single-particle motion becomes

$$H_{\text{def}} = -\frac{\hbar^2}{2m}\Delta + \frac{1}{2}m(\omega_x^2 x^2 + \omega_y^2 y^2 + \omega_z^2 z^2). \tag{14.1}$$

The three frequencies are then chosen to be proportional to the inverse of the half-axis of the spheroid, i.e.

$$\omega_x = \overset{\circ}{\omega}\frac{R_0}{a_x}; \dots, \tag{14.2}$$

with a necessary condition of volume conservation

$$\omega_x \omega_y \omega_z = \overset{\circ}{\omega}_0{}^3. \tag{14.3}$$

This Hamiltonian is separable in the three directions and the energy eigenvalues and eigenfunctions can easily be constructed using the results for the one-dimensional harmonic oscillator model.

For axially symmetric shapes, one can introduce a deformation variable δ, using the following prescription

$$\omega_\perp (= \omega_x = \omega_y) = \omega_0(\delta)(1 + \tfrac{2}{3}\delta)^{1/2}$$
$$\omega_z = \omega_0(\delta)(1 - \tfrac{4}{3}\delta)^{1/2}, \tag{14.4}$$

where volume conservation is guaranteed up to second order in δ giving the deformation dependence $\omega_0(\delta)$ as

$$\omega_0(\delta) = \overset{\circ}{\omega}_0(1 + \tfrac{2}{3}\delta^2). \tag{14.5}$$

According to Nilsson (1955), one introduces a deformation dependent oscillator length $b(\delta) = (\hbar/m\omega_0(\delta))^{1/2}$ so that dimensionless coordinates, expressed by a prime $\vec{r}', \theta', \varphi', \dots$, can be used and the Hamiltonian of equation (14.1) now becomes

$$H_{\text{def}} = \hbar\omega_0(\delta)\left(-\frac{1}{2}\Delta' + \frac{r'^2}{2} - \frac{1}{3}\sqrt{\frac{16\pi}{5}}\delta r'^2 Y_{20}(\hat{r}')\right). \tag{14.6}$$

This Hamiltonian reduces to the spherical, isotropic oscillator potential to which a quadrupole deformation or perturbation term has been added. It is clear that this particular term will split the m degeneracy of the spherical solutions (j, m) and the amount of splitting will be described by the magnitude of the 'deformation' variable δ. In the expression (14.6), we can identify the quantity $\frac{1}{3}\sqrt{16\pi/5}\delta$ with the deformation parameter β as used in chapter 13 (section 13.1).

Since spherical symmetry is broken but axial symmetry remains, the solutions to the Hamiltonian (14.6) can be obtained using cylindrical coordinates

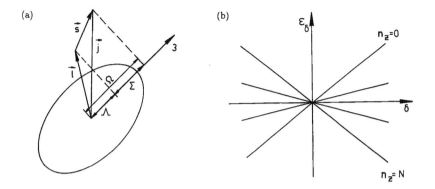

Figure 14.1. (*a*) Coupling scheme indicating the various angular momenta $(\vec{l}, \vec{s})\vec{j}$ and their projection $(\Lambda, \Sigma)\Omega$, on the symmetry axis (3-axis) respective for a particle moving in a deformed axialy symmetric potential. (*b*) Energy levels, corresponding to an anisotropic harmonic oscillator potential, as a function of deformation (δ) and for varying number of oscillator quanta along the 3-axis, denoted by n_z ($n_z = 0, \ldots, N$).

with the associated quantum numbers n_z, n_ρ, m_l (with m_l the projection of the orbital angular momentum on the symmetry axis), also called $\Lambda = m_l$.

With the relations $N = n_x + n_y + n_z = n_z + 2n_\rho + m_l$, the eigenvalues become

$$\varepsilon_\delta(N, n_z) \cong \hbar\overset{\circ}{\omega}_0 \left[(N + 3/2) + \delta\left(\frac{N}{3} - n_z\right) \right], \qquad (14.7)$$

which implies a splitting, linear in the number of oscillator quanta in the z direction, as a function of the deformation variable δ.

Including also intrinsic spin, with projection $\Sigma = \pm\frac{1}{2}$, the total projection $\Omega = \Lambda + \Sigma$ of spin on the symmetry axis remains a good quantum number and characterizes the eigenstates in a deformed potential with axial symmetry. The coupling scheme is drawn in figure 14.1(*a*) as well as the energy splitting of equation (14.7), in a schematic way in figure 14.1(*b*). One can then characterize a deformed eigenstate with the quantum numbers

$$\Omega^\pi [N, n_z, \Lambda], \qquad (14.8)$$

with π the parity of the orbit, defined as $\pi = (-1)^l$.

Large degeneracies remain in the axially symmetric harmonic oscillator if the frequencies $\omega_\perp$ and ω_z are in the ratio of integers, i.e. $\omega_z/\omega_\perp = p/q$. For $p : q = 1 : 1$ one regains the spherical oscillator; for $p : q = 1 : 2$ one obtains 'superdeformed' prolate or, if 2:1, oblate shapes, etc. In figure 14.2, a number of important ratios are indicated on the axis of deformation.

Even though this level scheme contains the major effects relating to nuclear deformation, the strong spin–orbit force needs to be added to transform it to a

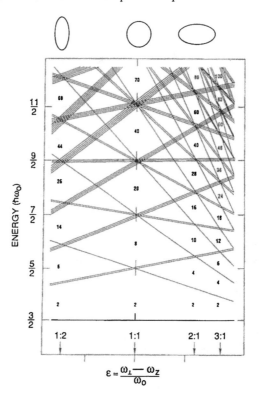

Figure 14.2. Single-particle level spectrum of the axially symmetric harmonic oscillator, as a function of deformation (ε). Here, $\omega_0 = \frac{1}{3}(2\omega_\perp + \omega_z)$. The orbit degeneracy is $n_\perp + 1$ which is illustrated by artificially splitting the lines. The arrows indicate the characteristic deformation corresponding to the ratio of $\omega_\perp / \omega_z = 1/2, 1/1, 2/1$ and $3/1$ (taken from Wood *et al* 1992).

realistic deformed single-particle spectrum. Moreover, in the original Nilsson parametrization, a $\hat{l}^2$ term has also been added to simulate a potential which appears to be more flat in the nuclear interior region, compared to the oscillator potential. So, the Nilsson Hamiltonian, descibing the potential is

$$H_{\text{Nilsson}} = \hbar\omega_0(\delta)\left(-\frac{1}{2}\Delta' + \frac{r'^2}{2} - \beta r'^2 Y_{20}(\hat{r}')\right) - \kappa \hbar \overset{\circ}{\omega}_0 (2\hat{l}\cdot\hat{s} + \mu(\hat{l}^2 - \langle \hat{l}^2 \rangle_N),$$

(14.9)

where 2κ describes the spin-orbit strength and $\kappa\mu$ the $\hat{l}^2$ orbit energy shift. The new terms, however, are no longer diagonal in the basis $|Nn_z, \Lambda\Sigma(\Omega)\rangle$, or in the equivalent $|Nlj, \Omega\rangle$ spherical basis. It is also easily shown that $[H_{\text{Nilsson}}, \hat{j}^2] \neq 0$ and only Ω^π are the remaining correct quantum numbers. The Hamiltonian (14.9)

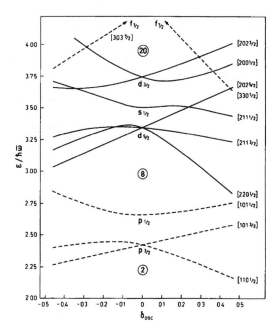

Figure 14.3. Spectrum of a single particle moving in a spheroidal potential ($N, Z < 20$). The spectrum is taken from Mottelson and Nilsson (1959). The orbits are labelled by the asymptotic quantum numbers $[N, n_z, \Lambda, \Omega]$ and refer to prolate deformation. A difference in parity is indicated by the type of lines (full lines: positive parity; dashed lines: negative parity). (Taken from Bohr and Mottelson, *Nuclear Structure*, vol. 2, © 1982 Addison-Wesley Publishing Company. Reprinted by permission.)

has to be diagonalized in a basis. The original Nilsson article considered the basis $|Nn_z\Lambda\Omega\rangle$ to construct the energy matrix. The results for such a calculation for light nuclei is presented in figure 14.3 and is applicable to the deformed nuclei that are situated in the p-shell ($A \simeq 10$) and in the mid-shell s, d region ($A \simeq 26, 28$).

In the region of small deformation, the quadrupole term $\alpha r'^2 Y_{20}(\hat{r}')$ can be used as a perturbation and be evaluated in the basis in which the spin-orbit and $\hat{l}^2$ terms become diagonal i.e. in the $|Nlj, \Omega\rangle$ basis. The matrix element

$$\langle Nlj\Omega|r'^2 Y_{20}(\hat{r}')|Nlj\Omega\rangle \propto \frac{3\Omega^2 - j(j+1)}{(2j-1)(2j+3)}, \tag{14.10}$$

is obtained and describes the (Ω, j) dependence near zero deformation.

For very large deformations, on the other hand, the $\hat{l}, \hat{s}$ and $\hat{l}^2$ term can be neglected relative to the quadrupole deformation effect. In this limit, the quantum numbers of the anisotropic harmonic oscillator become good quantum numbers.

They are also called the asymptotic quantum numbers $\Omega^\pi [N, n_z, \Lambda, \Sigma]$ and are discussed in detail by Nilsson (1955).

The original Nilsson potential, with the $r'^2 Y_{20}(\hat{r}')$ quadrupole term has the serious drawback of non-vanishing matrix elements, connecting the major oscillator quantum number N to $N \pm 2$. Using a new coordinate system gives rise to 'stretched' coordinates with a corresponding deformation parameter ε_2 ($\equiv \varepsilon$), the $|\Delta N| = 2$ couplings are diagonalized and form a more convenient basis and representation to study deformed single-particle states. Using a natural extension to deformations, other than just quadrupole, a 'modified' harmonic oscillator potential is given by

$$U_{\mathrm{MHO}} = \tfrac{1}{2} \hbar \omega_0(\varepsilon) \rho_t^2 [1 + 2\varepsilon_1 P_1(\cos \theta_t) - \tfrac{2}{3} \varepsilon_2 P_2(\cos \theta_t) + 2 \sum_{\lambda=3}^{\lambda_{\max}} \varepsilon_\lambda P_\lambda(\cos \theta_t)],$$

(14.11)

where the stretched coordinates ρ_t, θ_t are defined by

$$\cos \theta_t = \cos \theta \left[\frac{1 - \tfrac{2}{3} \varepsilon_2}{1 + \varepsilon_2(\tfrac{1}{3} - \cos^2 \theta)} \right]^{1/2},$$

(14.12)

and

$$\rho_t^2 = \xi^2 + \eta^2 + \zeta^2,$$

(14.13)

with (ξ, η, ζ) the 'stretched' coordinates.

The parameters κ, μ determine the basic that orders and splits the various single-particle orbits and their values are such that at zero deformation the single-particle and single-hole energy spectra are well reproduced. The first evaluation was carried out in the article by Nilsson *et al* but various adjustments, that depend on the specific nuclear mass regions under study, have been carried out.

The Nilsson model has been highly successful in describing a large amount of nuclear data. Even though, at first sight, a Nilsson level scheme can look quite complicated, a number of general features result (see figure 14.4).

(i) Each spherical (n, l, j) level is now split into $j + \tfrac{1}{2}$ double-generate states, according to the $\pm \Omega$ degeneracy. The Nilsson states are most often still characterized by the $[Nn_z \Lambda \Omega]$ quantum numbers even though these are *not* good quantum numbers, in particular for small deformations.

(ii) According to equation (14.10), orbits with the lower Ω values are shifted downwards for positive (prolate) deformations and upwards for negative (oblate) deformations. This can be understood in a qualitative way by looking at the orientations of these different Ω orbits relative to the z-axis.

(iii) For large deformations we see that levels with the same n_z value are moving in almost parallel lines (see figure 14.4). This peculiar feature is related to a further, underlying symmetry that appears in the deformed shell model (pseudo-spin symmetries).

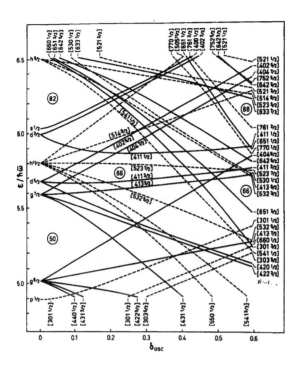

Figure 14.4. As for figure 14.3 but now for $(50 < Z < 82)$. The figure is taken from Gustafson *et al* (1967).

(iv) Using the spherical basis $|Nlj\Omega\rangle$ to expand the actual Nilsson orbits,

$$|\Omega_i\rangle = \sum c^i_{lj}|Nlj, \Omega\rangle, \qquad (14.14)$$

that are near to zero-deformation, the coefficients c^i_{lj} lead to only small admixtures, in particular for the highest (nlj) spherical orbit in each N shell, i.e. $1g_{9/2}$ orbit in the $N = 4$ oscillator shell.

(v) The slope of the Nilsson orbits, characterized by $|\Omega_i\rangle$ and ε_{Ω_i}, is related to the quadrupole single-particle matrix element, or (see equations (14.9) and (14.10))

$$\frac{d\varepsilon_{\Omega_i}}{d\beta} = -\hbar\omega_0(\delta)\langle\Omega_i|r'^2 Y_{20}(\hat{r}')|\Omega_i\rangle. \qquad (14.15)$$

The matrix elements of the quadrupole operator $r'^2 Y_{20}(\hat{r}')$ have been evaluated using the $|Nlj\Omega\rangle$ basis, and are presented in equation (14.10).

It is now possible to determine the total energy of the nucleus as a function of nuclear deformation. This calculation will allow us to determine the stable, nuclear shapes in a very natural way. At first sight one might think of adding the various deformed single-particle energies; however, one should minimize the total many-body Hamiltonian given by

$$\hat{H} = \sum_{i=1}^{A} t_i + \frac{1}{2} \sum_{i,j} V(i,j). \tag{14.16}$$

The average, one-body field, determined in a self-consistent way by starting from the two-body interaction (chapter 10), is known to be given by

$$U(i) = \sum_{j=1(j \neq i)}^{A} V(i,j), \tag{14.17}$$

and the full Hamiltonian is be rewritten as

$$\hat{H} = \frac{1}{2} \sum_{i=1}^{A} h_i + \frac{1}{2} \sum_{i=1}^{A} t_i, \tag{14.18}$$

with

$$h_i = t_i + U(i). \tag{14.19}$$

In the case of the oscillator potential $U(i)$ and, because of the well-known virial theorem $\langle t_i \rangle = \langle U_i \rangle = \frac{1}{2} \langle h_i \rangle$, we obtain for the total ground-state energy, if deformation is taken into account

$$E_0(\delta) \equiv \langle \hat{H} \rangle = \frac{3}{4} \sum_{i=1}^{A} \langle h_i \rangle = \frac{3}{4} \sum_{i=1}^{A} \varepsilon_i(\delta), \tag{14.20}$$

where $\varepsilon_i(\delta)$ denotes the eigenvalues of the Nilsson potential. This procedure allows an approximate determination of the ground-state equilibrium values δ_{eq} but absolute values are not so well determined since residual interactions are not well treated in the above procedure. A typical variation of $E_0(\delta)$ as a function of deformation is depicted in figure 14.5 where, besides the total energy, the liquid-drop model variation of the total energy (dashed line) is also presented (chapter 7).

The basic reason for difficulties in producing the correct total ground-state energy resides in the fact that this property is a bulk property and even small shifts in the single-particle energies can give rise to large errors in the binding energy. To obtain *both* the global (liquid-drop model variations) and local (shell-model effects) variations in the correct way as a function of nuclear deformation, Strutinsky developed a method to combine the best properties of both extreme nuclear model approximations. We do not discuss this Strutinsky procedure (Strutinsky 1967, 1968) which results in adding a shell correction to the liquid-drop model energy

$$E = E_{LDM} + E_{SHELL}, \tag{14.21}$$

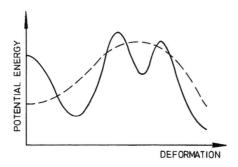

Figure 14.5. Schematic variation of the energy with deformation for a nucleus with a second minimum. The dashed line corresponds to the liquid-drop model barrier.

with

$$E_{\text{SHELL}} = \sum_{i=1}^{A} \varepsilon_i(\delta) - \tilde{E}_{\text{SHELL}}, \qquad (14.22)$$

where $\tilde{E}_{\text{SHELL}}$ subtracts that part of the total energy already contained in the liquid-drop part E_{LDM}, but leaves the shell-model energy fluctuations. It can be shown that the shell-correction energy is largely correlated to the level density distribution near the Fermi energy. The nucleus is expected to be more strongly bound if the level density is small since nucleons can then occupy more strongly bound single-particle orbits (figure 14.6). As a general rule, in quantum systems, large degeneracies lead to a *reduced* stability. So, a new definition of a magic or closed-shell nucleus is one that is the least degenerate compared to its neighbours. To illustrate the above procedure in more realistic cases, we show in figure 14.7 the E_{SHELL} shell-correction energy, using the modified harmonic oscillator, in which quadrupole, hexadecupole and 6-pole deformations have been included. The presence of the strongly bound (negative E_{SHELL} values) spherical shells at 20, 28, 50, 82, 126 and 184 clearly shows up. Many other deformed shells show up at the same time with the corresponding values of $\varepsilon_2(\varepsilon_4, \varepsilon_6)$ indicated.

The above shell-correction description can also be obtained from Hartree–Fock theory (chapter 10) when the density ρ can be decomposed in a smoothly varying part, ρ_0, and a fluctuating part, $\tilde{\rho}$, that take into account the shell corrections near the Fermi level (Ring and Schuck 1980).

The Strutinsky and/or Hartree–Fock total energy calculations are *the* methods to study ground-state properties (binding energy, deformation at equilibrium shape) in many regions of the nuclear mass table.

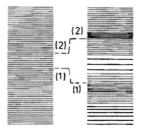

Figure 14.6. Comparison of an equally spaced level density distribution to a schematic shell-model level density. The binding energy of the Fermi level (1), in the right-hand case is stronger than in the left-hand case, whereas for the situation (2), the opposite result is true (taken from Ring and Schuck, 1980).

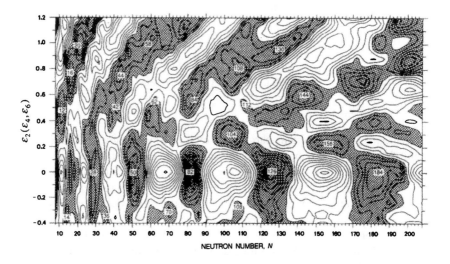

Figure 14.7. Shell-energy correction diagram in the (ε_2, N) plane. The resulting figure is taken from Ragnarsson and Sheline (1984).

14.2   Rotational motion: the cranking model

Up until now, collective rotational motion was considered to be a purely macroscopic feature related to a bulk property of the nucleus. However, nuclear collective motion is built from a microscopic underlying structure which is necessary to determine the collective variables and parameters.

The cranking model allows the inertial parameters to be determined and has many advantages. It provides a fully microscopic description of nuclear rotation; it handles collective and single-particle excitations on an equal footing and it

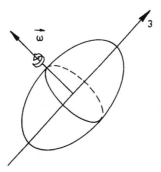

Figure 14.8. Pictorial representation of the rotational axis ($\vec{\omega}$) and the rotational motion about it at angular frequency ($\vec{\omega}$). The rotational axis is perpendicular to the intrinsic symmetry axis (3).

extends even to very high-spin states. The drawbacks are that it is a non-linear theory and that angular momentum is not conserved.

The first discussion of cranking was given by Inglis (Inglis 1954, 1956) in a semi-classical context. Here, we briefly present the major steps in the derivation of the cranking model: the model of independent particles moving in an average potential which is rotating with the coordinate frame fixed to that potential.

We consider a single-particle potential U with a fixed shape rotating with respect to the rotational axis $\vec{\omega}$ (see figure 14.8). Thus, we can express the time dependence (choosing the axes as shown in figure 14.8) as

$$U(\vec{r}; t) = U(r, \theta, \varphi - \omega t; 0), \qquad (14.23)$$

and the time dependence only shows up when a φ-dependence on U occurs which implies axial asymmetry around the rotational axis $\vec{\omega}$.

By means of the unitary transformation $\hat{U} = \exp(\mathrm{i}(\omega/\hbar)Jt)$, with $\vec{\omega} \cdot \vec{J} = -\mathrm{i}\hbar\omega(\partial/\partial\varphi)$, one induces a transformation of an angle $\varphi = \omega t$ around the rotational axis. We can define the transformed wavefunction as

$$\psi_r = \hat{U}\psi, \qquad (14.24)$$

with

$$\mathrm{i}\hbar\frac{\partial\psi_r}{\partial t} = \mathrm{i}\hbar\hat{U}\frac{\partial\psi}{\partial t} + \mathrm{i}\hbar\frac{\partial\hat{U}}{\partial t}\psi, \qquad (14.25)$$

or

$$\mathrm{i}\hbar\frac{\partial\psi_r}{\partial t} = (h(t = 0) - \vec{\omega} \cdot \vec{J})\psi_r. \qquad (14.26)$$

The latter equation (14.26) contains an explicit time-independent Hamiltonian $h_\omega \equiv h(t = 0) - \vec{\omega} \cdot \vec{J}$ and can be solved in the standard way, leading to

$$h_\omega\psi_r = (h(t = 0) - \vec{\omega} \cdot \vec{J})\psi_r = \varepsilon'_\omega\psi_r. \qquad (14.27)$$

We can obtain the eigenvalues of the original Hamiltonian $h(t)$ as

$$\varepsilon_\omega = \langle \psi | h(t) | \psi \rangle = \langle \psi_r | h(t = 0) | \psi_r \rangle = \varepsilon'_\omega + \omega \langle \psi_r | \vec{J} | \psi_r \rangle, \qquad (14.28)$$

with the Coriolis interaction $\vec{\omega} \cdot \vec{J}$.

For systems with spin, the operator that generates rotations is $\vec{j} = \vec{l} + \vec{s}$. The orientation of the rotational axis is conventionally chosen as parallel to the x-axis, and perpendicular to the symmetry axis. In more complicated situations, we require $\vec{\omega}$ to be parallel to a principal axis of the potential. So, the general many-body Hamiltonian of the cranking model is

$$\hat{H}_\omega = \hat{H} - \omega \hat{J}_x, \qquad \left(\hat{J}_x = \sum_{i=1}^{A} \hat{j}_x(i) \right),$$

$$= \sum_{i=1}^{A} h_\omega(i), \qquad (14.29)$$

where $\hat{H}$ is a sum of the individual deformed potentials. The energy in the lab system, $E(\omega)$ is

$$E(\omega) = \langle \Psi_\omega | \hat{H} | \Psi_\omega \rangle = \langle \Psi_\omega | \hat{H}_\omega | \Psi_\omega \rangle + \omega \langle \Psi_\omega | \hat{J}_x | \Psi_\omega \rangle, \qquad (14.30)$$

where $|\Psi(\omega)\rangle$ describes the ground-state Slater determinant. We can now expand

$$E(\omega) = E(\omega = 0) + \tfrac{1}{2} \mathcal{J}_1 \omega^2 + \cdots, \qquad (14.31)$$

and, since for $\omega = 0$ we have $\langle \Psi_0 | \hat{J}_x | \Psi_0 \rangle = 0$, and thus

$$J(\omega) = \langle \Psi_\omega | \hat{J}_x | \Psi_\omega \rangle = \mathcal{J}_2 \omega + \cdots. \qquad (14.32)$$

We can point out that $\mathcal{J}_1 = \mathcal{J}_2$ and also obtain that

$$\omega = \frac{\mathrm{d}E}{\mathrm{d}J}. \qquad (14.33)$$

Since the angular frequency is not an observable, we have to find some means to determine it from the actual energies and spins. According to Inglis we can do somewhat better, by including zero-point oscillations, using

$$J = \langle \Psi_\omega | J_x | \Psi_\omega \rangle = \hbar \sqrt{J(J+1)}, \qquad (14.34)$$

and, in first order, we obtain

$$\omega = \frac{\hbar \sqrt{J(J+1)}}{\mathcal{J}_1}. \qquad (14.35)$$

The energy expression finally is

$$E(J) = E(0) + \frac{\hbar^2}{2\mathcal{J}_1} J(J+1).$$ (14.36)

As the deformed potential of the unperturbed system is filled up to the Fermi level, the perturbation term $\omega\hat{J}_x$ can excite one-particle one-hole excitations. The perturbed wavefunction becomes in lowest-order perturbation theory

$$|\Psi\rangle = |\Psi_0\rangle + \omega \sum_{p,h} \frac{\langle ph|\hat{J}_x|\Psi_0\rangle}{\varepsilon_p - \varepsilon_h} a_p^+ a_h |\Psi_0\rangle,$$ (14.37)

where the Nilsson energies, $\varepsilon_p, \varepsilon_h$, are the single-particle energies of the deformed Hamiltonian $\hat{H}_{\text{def}}$. The expectation value of $\hat{J}_x$, up to first order in ω, then becomes

$$J = \langle\Psi|\hat{J}_x|\Psi\rangle = 2\omega \sum_{p,h} \frac{|\langle ph|\hat{J}_x|\Psi_0\rangle|^2}{\varepsilon_p - \varepsilon_h},$$ (14.38)

and leads to the Inglis cranking expression for the moment of inertia $\mathcal{J}_{\text{Inglis}}$ as

$$\mathcal{J}_{\text{Inglis}} = 2 \cdot \sum_{p,h} \frac{|\langle p|\hat{J}_x|h\rangle|^2}{\varepsilon_p - \varepsilon_h}.$$ (14.39)

This Inglis cranking formula leads to moments quite close to the rigid-body moment of inertia[1].

It was pointed out, however, in chapter 13 (section 13.2) that the experimental moments of inertia are somewhat smaller than these rigid-body values. It is the residual interaction, which was not considered in the above discussion and, the pairing correlations in particular, that give rise to an important quenching in equation (14.39). Not only are the energy denominators increased to the unperturbed $2qp$ energy but also pairing reduction factors become very small when away from the Fermi level that determine the major effects to the reduction.

For realistic, heavy nuclei, the Nilsson deformed potential discussed in section 14.1 will be used with the modifications to the Nilsson single-particle energies due to the term $-\omega\hat{j}_x$. So, the cranked Nilsson model starts from the new, extended single-particle Hamiltonian

$$h'(\omega) = h_{\text{Nilsson}} - \omega\hat{j}_x.$$ (14.40)

We discuss, and illustrate in figure 14.9, the salient features implied by the cranking term, for the sd-shell orbits.

---

[1] In a number of books and/or references on collective motion, certain expressions are slightly different with respect to the factors $\hbar$ and $\hbar^2$. Here, the angular momentuum operator $\hat{J}_x$ contains the factor $\hbar$ implicitly. If not, *all* expressions (equations (14.23) through (14.39)), where the angular momentum operator $\hat{J}$ (or $\hat{J}_x$) occurs, need an extra factor $\hbar$ for each power of $\hat{J}$ (or $\hat{J}_x$).

(i) The Coriolis and centrifugal forces affect the intrinsic structure of a rotating nucleus. Depending on whether a nucleon is moving clockwise or anticlockwise, the Coriolis interaction gives rise to forces that have opposite sign, thus breaking the time-reversal invariance. The nuclear rotation specifies a preferential direction in the nucleus. At $\omega = 0$, the usual Nilsson scheme is recovered.

(ii) The cranked Hamiltonian of equation (14.40) is still invariant under a rotation of π around the x-axis and the two levels correspond to eigenstates of the 'signature' operator $R_x = e^{i\pi \hat{j}_x}$ (with eigenvalues $r_x = \pm i$).

(iii) Some levels show a very strong dependence on the rotational frequency ω, i.e. orbits corresponding to large j and small Ω values. They show strong Ω-mixing and alignment along the x-axis.

(iv) For even–even nuclei at not too large angular velocities, pairing correlations have to be included. These counteract the alignment and try to keep nucleons coupled to form 0^+ pairs (figure 14.10). For the odd-mass nuclei, the high frequency can bring high-lying orbits down to very low energies and even modify the ground-state structure of the intrinsic ground band. The same can also happen in even–even nuclei where, due to the large rotational energy at these high frequencies, the two quasi-particle configuration based on such a pair of highly-aligned particles becomes the yrast configuration (see figure 14.11). This band crossing then results in a number of dramatic changes in the collective regular band structure, known as 'backbending' which is discussed in Box 14a. The superconducting ground-state pair correlations break up at a critical frequency, ω_{crit}, which may be compared to the critical magnetic field break-up in the superconducting phase of a material at low temperature. The similarity is shown in a schematic, but illustrative, way in figure 14.12.

14.3   Rotational motion at very high spin

14.3.1   Backbending phenomenon

In the present chapter it has been shown that a large amount of angular momenta can be obtained by collective motion (i.e. a coherent contribution of many nucleons to the rotational motion). It is important that the nucleus exhibits a stable, deformed shape. Subsequently, rigid rotation will contribute angular momentum and energy according to the expression

$$E(I) = E_0 + \frac{\hbar^2}{2\mathcal{J}} J(J+1). \qquad (14.41)$$

This collective band structure does not give the most favourable excitation energy for a particular spin. In section 14.2, it was shown that for high rotational frequencies around an axis perpendicular to the nuclear symmetry axis,

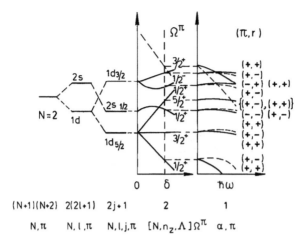

Figure 14.9. The full $s - d$ $(N = 2)$ energy spectrum. The various terms contributing to the splitting of the $(N + 1)(N + 2)$ degeneracy are given in a schematic way. Subsequently the orbit $\hat{l}^2$, spin-orbit $\hat{l} \cdot \hat{s}$, axially deformed field and cranking term gives an additional breaking contribution to the degeneracy. These degeneracies, as well as the corresponding good quantum numbers are indicated in each case (taken from Garrett, 1987).

alignment can result and the strong Coriolis force $\vec{\omega} \cdot \vec{J}$ gradually breaks up the pairing correlations. Thus, other bands can become energetically lower than the original ground-state intrinsic band. This crossing phenomenon is associated with 'backscattering' properties (see later).

Besides the first, collective rotational motion, angular momentum can be acquired by non-collective motion. Here, the alignment of the individual nuclear orbits along the nuclear symmetry axis contributes to the total nuclear spin. The system does not have large deformed shapes but remains basically spherical or weakly deformed. The two processes are illustrated in figure 14.13.

A large variety of band structures have been observed in deformed nuclei. In general, quite important deviations appear relative to the simplest $J(J + 1)$ spin-dependence given in equation (14.41). An expansion of the type

$$E(J) = E_0 + AJ(J + 1) + B(J(J + 1))^2 + C(J(J + 1))^4 + \cdots, \quad (14.42)$$

is quite often used. The drawback is the very slow convergence. The cranking formula for the energy, now expressed in terms of the rotational frequency, and higher-order series have been used with a lot of success. The parametrization of Harris (1965) gives the expansion

$$E(J) = \alpha \omega^2 + \beta \omega^4 + \gamma \omega^6 + \cdots. \quad (14.43)$$

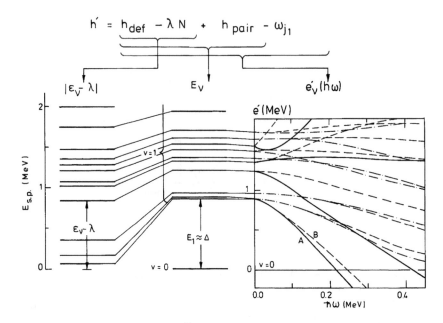

Figure 14.10. Spectra of Nilsson states (extreme left), one quasi-particle energies E_ν and routhians $e'_\nu(\hbar\omega)$ (right) illustrating the effect of pairing and rotation on the single-particle motion $(\pi, \alpha) \equiv (+, \frac{1}{2}), (+, -\frac{1}{2}), (-, \frac{1}{2})(-, -\frac{1}{2})$ are given by the solid, short-dashed, dot-dashed and long-dashed lines. The Hamiltonian with its various terms and the specific effects the various contributions cause, are indicated with the accolades and arrows. The spectrum is appropriate for single-neutron motion in ^{165}Yb (taken from Garrett 1987).

The relations connecting E, ω and J are given in the cranking equation and, needing an equation giving J as a function of ω (and higher powers), we use

$$\frac{\mathrm{d}E}{\mathrm{d}\omega} = \frac{\mathrm{d}E}{\mathrm{d}J}\frac{\mathrm{d}J}{\mathrm{d}\omega} = \hbar\omega\frac{\mathrm{d}J}{\mathrm{d}\omega}, \tag{14.44}$$

and

$$J(\omega) = 2\alpha\omega + \tfrac{4}{3}\beta\omega^3 + \tfrac{6}{5}\gamma\omega^5 + \cdots, \tag{14.45}$$

$$\mathcal{J}(\omega) = 2\alpha + \tfrac{4}{3}\beta\omega^2 + \tfrac{6}{5}\gamma\omega^4 + \cdots. \tag{14.46}$$

So, a plot of $\mathcal{J}$, as a function of ω^2, will give a mainly linear dependence, if just a two-parameter Harris formula is used (figure 14.14).

Nuclei can be obtained in these very high angular momentum states, mainly through heavy-ion induced reactions (HI, xn) (Morinaga 1963). The states that are populated subsequently, decay, through a series of statistical low-spin transitions, into the high-spin lower energies yrast structure (figure 14.15).

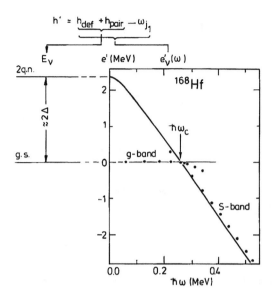

Figure 14.11. In the extreme left-hand part, the pairing effect for two quasi-particle excitations is given. To the right, the effect of Coriolis plus centrifugal terms is illustrated for a highly-aligned high-j low-Ω configuration. The experimental points correspond to the yrast sequence in ^{168}Hf and are compared with the cranking model calculations (lines). The various separate contributions from the nuclear Hamiltonian are clearly given. It is shown that at the crossing frequency ($\hbar\omega_c$) the rotational 'correlation' energy counteracts the nuclear pairing correlation energy (taken from Garrett 1987).

A most interesting way to study the high-spin physics can be obtained by plotting the moment of inertia against the rotational frequency squared $(\omega)^2$, since, using the Harris parametrization, in lowest order (equation (14.46)), a quadratic relationship results. For low-spin values one indeed observes straight lines that follow the data points to a good approximation (see the examples of ^{158}Dy and ^{162}Er in figure 14.16). In these two nuclei, however, a very steep, almost vertical, change shows up: the ω value remains constant while the moment of inertia increases rapidly. This picture presents a serious breakdown of the classic rotational picture and has been called 'backbending'. The first examples were studied by Johnson *et al* in ^{160}Dy (Johnson *et al* 1971) and it is striking that the energy difference between adjacent levels in the ground-state band,

$$\Delta E_{J,J-2} = \frac{\hbar^2}{2\mathcal{J}}(4J - 2), \qquad (14.47)$$

exhibits a decrease for certain spin values.

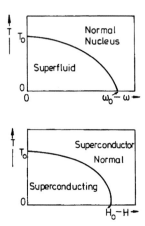

Figure 14.12. Comparison of rotational and temperature-dependent quenching of nuclear pair correlations and the magnetic- and temperature-dependent quenching of 'superconductivity'.

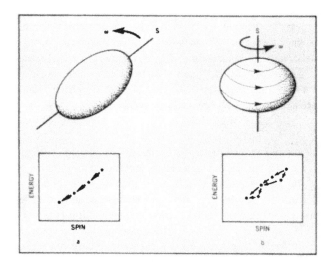

Figure 14.13. Two different ways of describing angular momentum for a given nucleus (upper part). In the case (*a*), the nucleus acquires angular momentum by a collective rotation, resulting in a simple yrast spectrum. In case (*b*), the nucleus changes its spin by rearranging individual nucleon orbits resulting in complex energy spectra (Mottelson 1979).

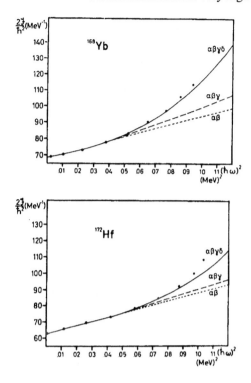

Figure 14.14. A plot of the moment of inertia $2\mathcal{J}/\hbar^2$ (MeV^{-1}) versus the square of the angular frequency $(\hbar\omega)^2$ (MeV)^2 (see text for ways how to extract these quantities from a given rotational band) (taken from Johnson and Szymanski 1973).

In constructing the $\mathcal{J} = \mathcal{J}(\omega^2)$ plots, we need to have a good value for the nuclear, rotational frequency ω, as deduced from the observed energies and spin-values, since the classical value gives

$$\omega = \frac{\mathrm{d}E}{\mathrm{d}J},\tag{14.48}$$

one should, of course, in a quantum mechanical treatment, plot the derivative against the value of $\sqrt{J(J+1)}$, to give

$$\hbar\omega = \frac{\mathrm{d}E}{\mathrm{d}\sqrt{J(J+1)}}.\tag{14.49}$$

and if we use energy differences between J and $J - 2$ levels, the result is

$$(\hbar\omega)_2 \simeq \frac{\Delta E_{J,J-2}}{\sqrt{J(J+1)} - \sqrt{(J-1)(J-2)}},\tag{14.50}$$

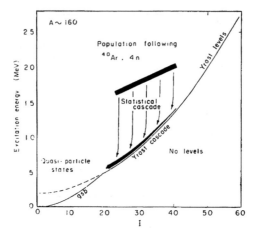

Figure 14.15. Excitation energy plotted against angular momentum in a nucleus with mass $A \simeq 160$ that is produced in an $(^{40}\text{Ar}, 4n)$ reaction. The range of angular momentum and energy populated in such a reaction is shown together with the decaying, statistical gamma decay cascades (taken from Newton 1970).

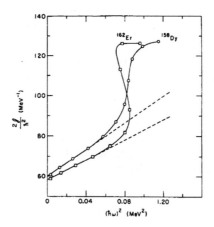

Figure 14.16. Illustration of the moment of inertia $(2\mathcal{J}/\hbar^2)$ against the angular frequency $(\hbar\omega)^2$ for ^{162}Er and ^{158}Dy.

in contrast to the simpler expression

$$(\hbar\omega)_1 \simeq \frac{\Delta E_{J,J-2}}{2}. \tag{14.51}$$

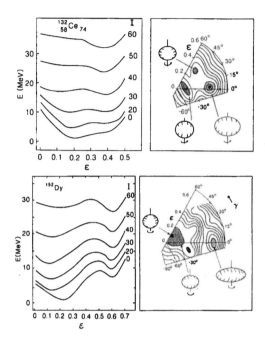

Figure 14.17. Total potential energy curves (as a function of the angular momentum) for the two even–even nuclei $^{132}_{58}$Ce$_{74}$ and $^{152}_{66}$Dy$_{86}$, as a function of quadrupole deformation. The various minima in the potential (ε, γ) energy surface are shown at fixed spin values of $40\hbar$ in ^{132}Ce and ^{152}Dy (right-hand part). (Taken from Nolan and Twin 1988. Reproduced with permission from the *Annual Review of Nuclear Science* **38** © 1988 by Annual Reviews Inc.)

An even better expression of equation (14.50) takes the 'curvature' correction in the derivative versus $J(J+1)$ into account, such that

$$\Delta(\sqrt{J(J+1)}) = \frac{d\sqrt{J(J+1)}}{dJ(J+1)}\Delta J(J+1)$$

$$= (2J-1)/\sqrt{J^2 - J + 1}, \qquad (14.52)$$

and gives the value

$$(\hbar\omega)_3 = \frac{\Delta E_{J,J-2}}{2J-1}\sqrt{J^2 - J + 1}. \qquad (14.53)$$

The higher band could (i) correspond to a larger deformation compared to the ground-state band, (ii) correspond to the non-superfluid state, through the Coriolis anti-pairing effect (Mottelson and Valatin 1960) (CAP) or (iii) be a particular two-quasi-particle band with large angular momentum alignment along the rotational

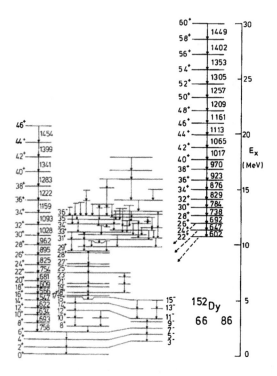

Figure 14.18. The level scheme of ^{152}Dy. All energies are given in keV. The coexistence of a number of quite distinct structures: normal deformed bounds (left), non-collective particle–hole structure (middle) and superdeformed band (right), are shown (taken from Wood *et al* 1992).

axis. The backbending then results in the sudden aligning of a pair of nucleons, as discussed in section 14.2.

14.3.2 Deformation energy surfaces at very high spin: super- and hyperdeformation

As was pointed out in section 14.1, the total energy of the deformed atomic nucleus cannot be obtained in a reliable way at large deformations because the bulk part of the total energy is not properly accounted for in the Nilsson model. With the Strutinsky prescription this becomes feasible.

Similarly, total energy surfaces can be calculated as a function of angular momentum J *and* deformation, using the same methods but now we have to determine the energy of a rotating, liquid drop as the reference energy E_{LDM} (rotating). The total energy is

$$E(\beta, \gamma, J) = E_{\text{LDM}}(\text{rotating}; \beta, \gamma, J) + E_{\text{shell}}(\beta, \gamma, J), \tag{14.54}$$

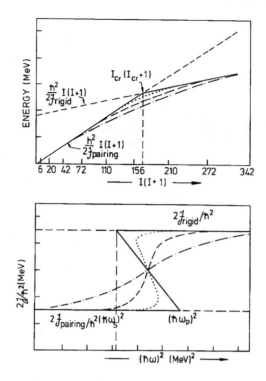

Figure 14.19. Schematic illustration of the backbending mechanism. In the upper part the crossing bands: the ground-state band and the excited band, corresponding to the moments of inertia, $\mathcal{J}_{\text{pairing}}$ and $\mathcal{J}_{\text{rigid}}$, respectively, are given as a function of $J(J+1)$. In the lower part, the $2\mathcal{J}/\hbar^2$ versus $(\hbar\omega)^2$ figure, deduced from the upper part, is constructed. The various curves: full line, dotted line, dashed line and dot-dashed lines correspond to various situations where the mixing between the two bands at and near the crossing zone varies from zero to a substantial value.

and can be evaluated, either at constant frequency ω or at constant angular momentum J. One then has to diagonalize the Nilsson, or more generally, deformed potential in the rotating frame ω which is called solving for the 'Routhian' eigenvalues and eigenfunctions.

Many such calculations have been carried out in deformed nuclei but also in a number of light p and sd-shell nuclei: (Åberg *et al* 1990).

It was shown in these studies that the total energy surfaces, as a function of angular momentum, gave rise to particularly stable shapes with axes ratio 2:1:1 (axially symmetric). The existence of these strongly elongated shapes was known from fission isomeric configurations in the $A \simeq 220$ region. There was some evidence for such shapes in the region near $A \simeq 150$. The essential idea,

Figure 14.20. A diagram showing the gammasphere set-up with a honeycomb of detectors surrounding the interaction region (centre). Not less than 110 high resolution Ge detectors and 55 BGO scintillation counters can be placed around the target (taken from Goldhaber 1991).

emphasized by the cranking calculations, was that a precise value of rotational motion was necessary to stabilize the very strongly deformed nuclear shape (figure 14.17).

Since its discovery in 1985, superdeformation has become, both experimentally and theoretically a very active subfield in nuclear physics research. By now, many examples in both the $N = 86$ region but also in the much heavier region around ^{194}Hg have been observed. The dramatic level spectrum of ^{152}Dy is shown in figure 14.18 where a superdeformed band extends up to spin $J = 60\hbar$ (Wood *et al* 1992).

Detailed review articles have recently been written by Åberg *et al* (1990), Nolan and Twin (1988), Janssen and Khoo (1992) and we refer the reader to these very instructive articles for more details.

The physics described in this backbending phenomenon can be understood in terms of the crossing of two bands (in a number of cases, higher backbending points have been observed). We illustrate this in figure 14.19 where we consider two bands with slightly different moments of inertia: a ground-state with moment of inertia ($\mathcal{J}_{\text{pairing}}$) and a rigid rotor moment of inertia ($\mathcal{J}_{\text{rigid}}$) where the latter is the bigger one. Since the remaining residual interaction is a function of the coupling strength a crossing (no-crossing rule) does not occur and we obtain a region where

$$\frac{\mathrm{d}\omega}{\mathrm{d}J} = \frac{\mathrm{d}^2 E}{\mathrm{d}J^2} < 0, \tag{14.55}$$

becomes negative. The ideal rotational spectra then correspond to horizontal lines in the $\mathcal{J} = \mathcal{J}(\omega^2)$ plots and it is the intersections which cause the transitions (in discrete or smooth transitions from the ground-state band into the upper band with the higher moment of inertia. If we could stay in the ground-state band, a smooth behaviour would be observed. The data, however, normally go across the levels of the yrast band structure.

In the above discussion, not very much was said about the physics of the higher band, which crosses the ground-state band at a critical frequency ω_{crit}.

It is important to use high-resolution germanium detectors with the largest possible fraction of detected events in the full energy peak in order to be able to observe the very weak γ-transitions in the superdeformed band. Various constraints (γ–γ coincidences, energy sum restrictions, etc) emphasize the need for large detector arrays. It was the TESSA-3 array at Daresbury, UK, that allowed for the first identification of superdeformed structures. In the USA, a super-array, 'Gammasphere' has been constructed for high-multiplicity gamma ray detection and is presented in figure 14.20 (Goldhaber 1991) in a schematic way.

While many superdeformed bands have been observed and catalogued (Firestone and Singh 1994), the precise excitation energy as well as the exact spin values have also been a point of discussion and serious experimental search. This is partly due to the fact that at these high excitation energies between 5–8 MeV, the density of states with more 'standard' deformation is very high and so the very superdeformed band levels are embedded in a background of states that will 'drain' out intensity in the γ-decay, resulting in an almost continuous low-energy γ-ray bump in the observed spectra. There is still a chance of direct single-step transitions that may well be detected by imposing severe coincidence constraints between known γ-transitions within the superdeformed band and transitions within the low-energy bands (Janssens and Stephens 1996, Beausang *et al* 1996, Garrett *et al* 1984). Such experiments have now been carried out with positive results. The first one at Gammasphere, detecting unambiguous single-gamma transitions in the nucleus ^{194}Hg (see figure 14.21) was performed by Khoo *et al* (1996). This method allows for a unique determination of excitation energy and, eventually, of the precise spin values in the superdeformed bands.

Speculations have been made recently about even more elongated shapes that seem to be allowed by the cranked, very-high-spin deformed shell-model calculations (Phillips 1993, Garrett 1988). These hyperdeformed states correspond to elongated nuclei with an ellipsoidal shape at a major-to-minor axis ratio of 3:1 and may indeed occur at very high spin. It is not obvious,

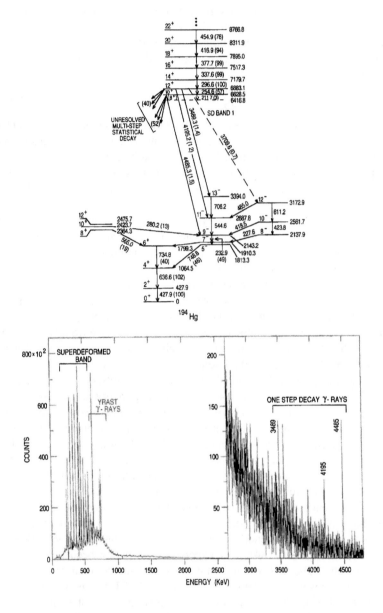

Figure 14.21. Spectrum of γ-rays, in coincidence with the superdeformed band in ^{194}Hg. The high-energy portion of the spectrum shows some of the transitions associated with the decay out of the superdeformed band. The study resulted in a partial level scheme as shown in the upper part of the figure. (Reprinted from Janssens and Stephens ©1996 Gordon and Breach.)

however, whether such states of extreme deformation can indeed be formed and give rise to a unique sequence of γ-transitions, deexciting while staying within the hyperdeformed band structure. It seems that at these very elongated shapes, the potential barrier preventing proton emission is reduced at the tips of the nucleus and so cooling via proton emission of the highly excited compound nucleus may result.

It was a team headed by Galindo-Uribarri (Galindo-Uribarri *et al* 1993) that carried out experiments at Chalk River using a beam of ^{37}Cl bombarding a target of ^{120}Sn thereby forming the ^{157}Ho compound nucleus. The emission of a single proton and a number of neutrons meant that the final nucleus was a dysprosium (Dy) isotope. Using the emitted proton as one element in detecting coincident gamma-rays in the final Dy nucleus, the final intensity observed could be enhanced to an observable level. From the nuclear reaction kinematics and other hints it seemed as if the most probable elements studied were ^{152}Dy and ^{153}Dy. The final analyses suggested the observation of a rotational band with a moment of inertia much larger than those of corresponding superdeformed bands in these Dy isotopes. Translated into axis ratios, the observed result was consistent with a ratio of almost 3:1.

A number of years ago, a generation of powerful gamma-ray spectrometers, spanning a full 4π geometry, came into use—Gammasphere in the USA (Goldhaber 1991), Eurogam, a UK/France undertaking, GASP in Italy—thereby expanding the efficiency and possibilities for weak transitions and multiple coincidences to be studied. More recently, besides an upgrade of Gammasphere, the EUROBALL collaboration, which is a common project by Denmark, France, Germany, Italy, Sweden and the UK, is providing a pioneering experimental facility for the study of nuclear structure. Using composite Ge counters, EUROBALL is a prototype of a new generation of gamma-ray detector arrays. It consists of 239 Ge crystals geometrically arranged so as to cover 45% of the total solid angle possible. It has been installed at LNL (Legnaro, Italy) and also at IReS-Vivitron (Strasbourg, France) and a good number of new results have come out of the use of this array (Simpson 1997, de Angelis *et al* 2003) We would like to mention that the use of beams of unstable ions will allow nuclear structure studies for very exotic nuclei. Thereby, EUROBALL will be used at the GSI (Darmstadt) in the Rising project. A full report describing both the physics and the technical lay-out has just been published and can be consulted at http://www-dapnia.cea.fr/Sphn/Deformes/EB/eb-report-final.pdf.

The combined use of such gamma-arrays with magnetic spectrometers has proven to be a highly successful method for reaction channel selection and has been exploited recently in the study of nuclei very far away from the region of β-stability (as an example we quote the RITU gas-filled spectrometer used at the Jyväskylä cyclotron)

As a result of the research and development connected to building the most advanced detection systems for nuclear physics research, the concept of γ-ray tracking has resulted in projects like GRETA (Deleplanque *et al* 1999).

We should mention that gamma-ray arrays, using segmented Ge detectors, at radioactive ion beam facilities such as EXOGAM at Spiral (Ganil,France) (Simpson *et al* 2000), the Mini-Ball at Rex-Isolde at CERN (Eberth *et al* 1997) and the VEGA project (Gerl *et al* 1998) at GSI,Darmstadt, have started. It is, however, with the Advanced GAmma Tracking Array (AGATA), that an array will be built with a two to three order of magnitude increase in sensitivity. The detector will be realized within a European collaboration and is intended to be used in experimental campaigns at radioactive and stable beam facilities throughout Europe. Details can be found on the web-page http://www-dapnia.cea.fr/Sphn/Deformes/Agata/index.shtml.

In conclusion, it is appropriate to point out once again the stabilizing effect created by rapid nuclear rotation. This can even result in particular stable and stabilized nuclear shapes with large elongations for specific proton and neutron numbers. This result has been one of the most surprising to appear in the dynamics of the nuclear A-body problem.

> **Box 14a. Evidence for a 'singularity' in the nuclear rotational band structure**

It has been shown that the rotational band structure in doubly-even deformed nuclei implies moments of inertia that, in some cases, strongly deviate from the simple picture of a rigid rotor. Nucleons are strongly coupled into 0^+ pairs in the lowest, intrinsic ground-state structure. The rotational motion and the corresponding strong Coriolis force, which try to break up the pairing correlations, lead to a regular, smooth increase of the moment of inertia with increasing angular momentum in the nucleus. A critical point was suggested to appear around $J = 20$, where a transition occurred into a non-pairing mode (in units $\hbar$). So, one has to study the properties of nuclear, collective bands at high spin.

A large amount of spin can be brought into the nucleus using (HI, xn) or (α, xn) reactions. The ground-band in ^{160}Dy was studied by the (α, 4n) reaction on ^{160}Gd with 43 MeV α-particles at the Stockholm cyclotron (Johnson *et al* 1971). The transitions following this reaction are presented in figure 14a.1 and are clearly identified up to spin 18.

Measurements of the excitation function for the gamma transitions and angular distribution coefficients of the various transitions, combined with coincidence data prove that all of the above gamma transitions belong to ^{160}Dy.

It can be concluded that a cascade of fast $E2$ transitions is observed, up to spin 18. The regularity in the band structure is clearly broken when the $18^+ \rightarrow 16^+$ and $16^+ \rightarrow 14^+$ transitions are reached. We discussed in section 14.3 that,

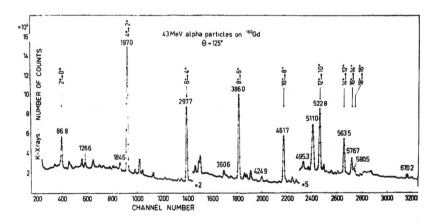

Figure 14a.1. The γ-ray spectrum recorded at an angle of $\theta = 125°$ relative to the beam. The indicated peaks are assigned to the (α, 4n) reaction leading to final states in ^{160}Dy (taken from Johnson *et al* 1971).

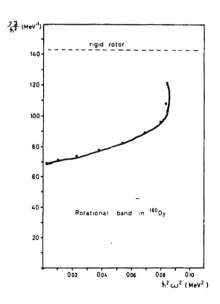

Figure 14a.2. The moment of inertia $(2\mathcal{J}/\hbar^2)$ in ^{160}Dy as a function of rotational frequency. The horizontal dashed line represents the moment of inertia corresponding to rigid rotational motion (taken from Johnson *et al* 1971).

for an axial rotor, the angular rotational frequency and moment of inertia are defined according to the expressions

$$\hbar\omega = \frac{dE}{d\sqrt{J(J+1)}}, \qquad (14a.1)$$

$$2\mathcal{J}/\hbar^2 = \left(\frac{dE}{dJ(J+1)}\right)^{-1}, \qquad (14a.2)$$

and the way to evaluate the derivatives, using finite differences, has been outlined in section 14.3. The drastic change in the moment of inertia, plotted as a function of ω^2 when the 16^+ state is reached, can be interpreted as due to a phase transition between a superfluid state and the normal state. When the pairing correlations disappear, the moment of inertia should rapidly reach the rigid rotor value (see figure 14a.2). The rotational frequency remains almost constant in ^{160}Dy with a steady increase in the moment of inertia with increasing angular momentum.

These observations represented the first, direct evidence for large deviations from the regular behaviour of a ground-state rotational band and clearly indicate the presence of other, nearby bands which modify the yrast structure.

Box 14b. The superdeformed band in ^{152}Dy

In section 14.3, it was pointed out that rapid rotation can stabilize the nuclear shape at extreme elongations for certain proton and neutron numbers. Thereby, shell and energy corrections produce local maxima that remain over a rather large span of angular momenta. The axes ratio 2:1:1 was suggested in the mass region $A \simeq 150$ near $N = 86$ to give rise to very large deformed shapes, called 'superdeformation'.

The first superdeformed band was observed in ^{152}Dy by Twin *et al* (1986) in ^{108}Pd $(^{48}$Ca$, x$n$)^{156-x\mathrm{n}}$Dy reaction, carried out at the TESSA-3 spectrometer in Daresbury. In the reaction, the projectile and target fuse to form a compound nucleus with an excitation energy of 70–90 MeV and maximum spin of $70\hbar$. The compound nucleus boils off a number of nucleons (neutrons mainly) followed by gamma-ray emission. The reaction used to populate ^{152}Dy and the resulting,

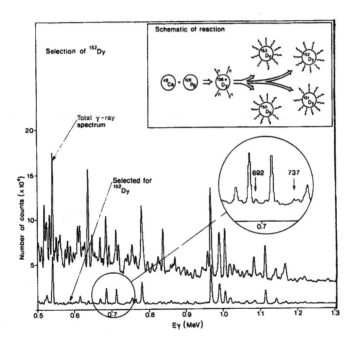

Figure 14b.1. The top inset shows schematically the reaction ^{108}Pd $(^{48}$Ca$, x$n$)^{156-x\mathrm{n}}$Dy. The total gamma-ray spectrum is composed of transitions from *all* the final products from the reaction. One can select decays in ^{152}Dy by a coincidence condition so that γ-transitions, associated with the superdeformed band, are clearly distinguished. (Taken from Nolan and Twin 1988. Reproduced with permission from the *Annual Review of Nuclear Science* **38** © 1988 by Annual Reviews Inc.)

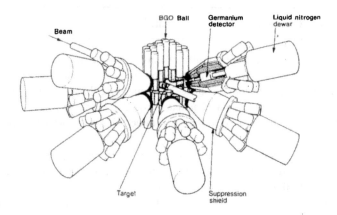

Figure 14b.2. The TESSA-3 detector array. (Taken from Nolan and Twin 1988. Reproduced with permission from the *Annual Review of Nuclear Science* **38** © 1988 by Annual Reviews Inc.)

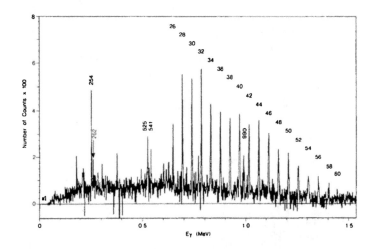

Figure 14b.3. Gamma transitions linking members of the superdeformed band in ^{152}Dy. The spins assigned to these superdeformed band members are given at the top of the gamma peaks. More details on how this band is discriminated from the huge amount of gamma transitions are found in Nolan and Twin (1988). (Reproduced with permission from the *Annual Review of Nuclear Science* **38** © 1988 by Annual Reviews Inc.)

complex gamma spectrum is shown in figure 14b.1: 30% of the reaction is to ^{152}Dy but only 0.3% goes to the superdeformed band. A critical factor therefore is the signal-to-noise ratio, and it is necessary to use high-resolution Ge detectors in order to be able to obtain excellent statistics in experiments that last a few days.

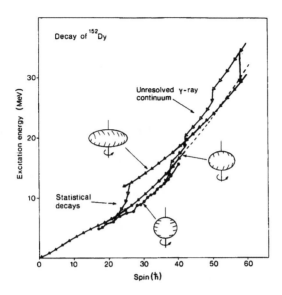

Figure 14b.4. A schematic illustration of the proposed gamma-ray decay paths in ^{152}Dy starting from a high-spin entry point. Only a small ($\sim$10%) branch feeds the superdeformed (SD) band which is assumed to become yrast at a spin of 50–55$\hbar$. The de-excitation of this SD-band shows up near spin 26$\hbar$ when the band is about 3–5 MeV above the yrast band and a statistical gamma decay pattern connects it to the lower-lying oblate states with spins lying between 19$\hbar$–25$\hbar$. (Taken from Nolan and Twin 1988. Reproduced with permission from the *Annual Review of Nuclear Science* **38** © 1988 by Annual Reviews Inc.)

The TESSA-3 spectrometer (total energy suppression shield array) combines 12 or 16 Ge detectors, surrounded by a suppression shield of NaI and/or BGO detectors to form an inner calorimeter or ball. The peak efficiency in the Ge detectors can be increased to 55–65% for Compton suppressed systems, at the same time increasing the coincident γ–γ event rate from 3% to 36%. The central BGO Ball acts as a total energy detector (covering a solid angle of $\simeq$4π) and this sum energy data from the Ball can be used to select particular reaction channels leading to ^{152}Dy. In this particular case (see also figure 14.18) a selection of the channel can be made since essentially all γ-rays decaying from the high-spin states decay through a 60 ns isomeric state with spin 17^{+}. The total ensuing improvement in the signal-to-noise ratio due to the coincidence condition is clearly seen in the spectrum of figure 14b.1. The TESSA-3 set-up is illustrated in figure 14b.2.

The superdeformed γ-spectrum in ^{152}Dy is given in figure 14b.3 and was obtained by setting gates on most members of the band. The spectrum is dominated by a sequence of 19 transitions with a constant spacing of 47 keV.

From their intensities, an average entry spin to the yrast band was determined as $21.8\hbar$ and so the spin of the final state in the superdeformed band should, most probably, be even higher. The energies of the oblate, prolate and superdeformed states are plotted as a function of spin with the assumption that the superdeformed band becomes yrast between $50\hbar$ and $60\hbar$. The precise energy *and* spin values of the superdeformed band are, however, not yet firmly known (figure 14b.4).

Further information was obtained for the quadrupole moment of the band. The data correspond to a value of $Q_0 \simeq 19$ eb, which is equivalent to a B(E2) strength of 2660 sp units and indicates a deformation parameter of $\varepsilon_2 = 0.6$. These values (even taking conservative error bars into account) indicate a collective band structure which does indeed conform with theoretical values and can correctly be called a superdeformed band.

Various other interesting information relating to the moments of inertia, the feeding process into the superdeformed band are discussed by Nolan and Twin (1988).

Chapter 15

Nuclear physics at the extremes of stability: weakly bound quantum systems and exotic nuclei

15.1 Introduction

Nuclear structure has been studied and discussed in chapters 10 to 14 by studying the response of the nucleons to external properties like exciting (heating) to higher excitation energies and rapidly cranking the nucleus. Thereby, we have come across a number of specific degrees of freedom like the existence of an average field that can even be deformed and various reorderings of nucleons in such an average field (elementary vibrational excitations, rotational bands, etc). This has given rise to a very rich spectrum and shows how the nuclear many-body system behaves but is still under the constraint of describing nuclei near to the region of β-stability.

In this chapter, we shall discuss a third major external variable that allows us to map the nuclear many-body system: changing the neutron-to-proton ratio N/Z, or equivalently the relative neutron excess $(N - Z)/A$, thereby progressing outside of the valley of β-stable nuclei. A number of rather general questions have to be posed and, if possible, answered: how do nuclei change when we approach the limits of stability near both the proton and neutron drip lines? In section 15.2 we discuss a number of theoretical concepts like the changing of the mean-field concept and the appearance of totally new phenomena when we reach systems that are very weakly bound (nuclear halos, skins and maybe even more exotic structures) as well as the experimental indications that have given rise to developing totally new and unexpected nuclear structure properties. In section 15.3 we discuss the development of radioactive ion beams (RIBs), the physics of creating new unstable (radioactive) beams with a varying energy span from very low energy (below the Coulomb energy) up to a few GeV per nucleon, as well as the intriguing and challenging aspects related to performing nuclear reactions with unstable beams and reaching into the nuclear astrophysics realm.

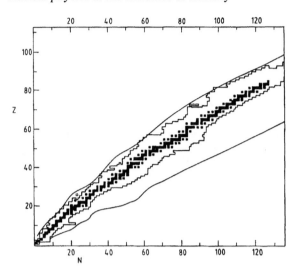

Figure 15.1. Map of the existing atomic nuclei. The black squares in the central zone are stable nuclei, the larger inner zone shows the status of known unstable nuclei as of 1986 and the outer lines denote the theoretical estimate for the proton and neutron drip lines (adapted from Hansen 1991).

Finally, in a short section 15.4, we close this chapter with a personal outlook as to where these new methods and the quickly developing field of nuclear physics in weakly bound quantum systems may bring us.

15.2 Nuclear structure at the extremes of stability

15.2.1 Theoretical concepts and extrapolations: changing mean fields?

In chapter 7 (equations (7.30)), we have defined the edges of stability through the conditions $S_p = 0$ and $S_n = 0$, delineating the proton and neutron drip lines, respectively. In figure 15.1, we draw the system of nuclei in which one can observe the 263 stable isotopes, the approximately 7000 particle-stable nuclei predicted to exist within the drip lines and the change in the upper part because of the rapidly increasing decay rates for fission and α-decay. A nucleus like ^{160}Sn with 50 protons and 110 neutrons is theoretically particle-stable but, at present, no means are available to study the wide span from the doubly closed shell region ^{132}Sn (with 82 neutrons) to the unknown territory towards the neutron drip line.

 Besides a number of binding energy considerations, on the level of the liquid-drop model expression, eventually refined with shell corrections, one of the questions of big importance is to find out about the way in which the nuclear average field concept, and the subsequent description of nuclear excitation modes using standard shell-model methods (see chapters 10, 11 and 12), can be extrapolated when going far out of the region of β-stability. Central to a good working of the shell model is the separation of the nucleons into a closed core (which is associated with the well-known magic numbers at 2, 8, 20, 28, 50, 82, 126, etc) and a number of valence nucleons. It is not unreasonable to imagine that these numbers may change. Experimental studies in the neutron-rich Na, Mg nuclei near $N = 20$ (Caurier *et al* 2002) indicate the appearance of a zone of increased binding energy, relative to the regular 'sd' model space calculations only. Within the context of the shell model, one can study the variation in the single-particle spectrum for protons (with changing neutron number) and for neutrons (with changing proton number). These effective single-particle energies, in which the proton–neutron interaction plays the major role, can be given as

$$\tilde{\epsilon}_{j\rho} = \epsilon_{j\rho} + \sum_{j\rho'} \langle j\rho j\rho' | V | j\rho j\rho' \rangle v_{j\rho'}^2, \tag{15.1}$$

in which ρ and ρ' are the charge labels for the nucleon occupying the orbitals describing the nucleon–nucleon interaction matrix element. Thereby, the extra amount of extra binding energy that comes from residual nucleon–nucleon interactions may well affect the relative proton and neutron single-particle energies.

A number of studies have been carried out for light nuclei using the standard shell-model methods. Otsuka *et al* (Otsuka 2002) have shown that, for the sd shell, moving towards very neutron-rich systems like $^{24}_{8}O_{16}$, a shell gap develops between the $1s_{1/2}$ and $2d_{3/2}$ orbitals giving rise to a new magic number at $N = 16$ (see figure 15.2). The origin lies, in particular, in the strongly attractive proton–neutron interaction between spin–orbit partners (e.g. the proton $1d_{5/2}$ and the neutron $1d_{3/2}$ orbitals). This same mechanism is also at work in making the neutron gap at $N = 20$ vanish for nuclei with a proton number close to $Z = 11, 12$. It can be used to study the relative changes of the $1/2^-$ and $1/2^+$ states going from ^{13}C to ^{11}Be onwards to ^{9}He indicating that, here, $N = 6$ has now become the new magic number (figure 15.2). There is nothing 'magic' about these changes since the shell model itself, through the large extra binding energy effects, takes care of a continuous set of adjustments. The point is that, in order to be able to undertake a serious calculation, one must have the capabilities to handle many major shells at the same time. Until recently, this was totally impossible but the SMMC and the large-scale 'standard' shell-model calculations in light nuclei can, at present attack these imporant issues, as discussed in chapter 11.

Likewise, modifications in the relative single-particle energies have been found and these are also well documented for heavier nuclei too. Here, most attention has gone to the study of variations in the proton single-particle states through a long series of isotopes. The $_{51}Sb$ isotopes form a most interesting example. The most dramatic effect here is the strong lowering of the $1g_{7/2}$ orbital relative to the $2d_{5/2}$ orbital when filling neutrons in the upper part of

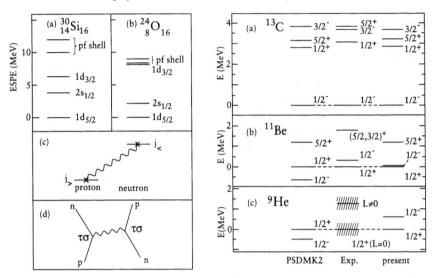

Figure 15.2. Left-hand part: Effective single-particle energies (*a*), (*b*), comparing ^{30}Si with ^{24}O, relative to 1d$_{5/2}$, (*c*) major contributions producing relevant single-particle variations and (*d*) the $\sigma \cdot \sigma\tau \cdot \tau$ nucleon–nucleon force component. Right-hand part: energy levels compared for nuclei with seven neutrons ^{13}C, ^{11}Be, ^{9}He, relative to the state corresponding to the experimental ground state. Results from (Otsuka 2002) are compared with the data. The hatched region denotes resonant states. (Reprinted from Otsuka T *et al* 2002 *Eur. Phys. J.* A **13** 69 © 2002 Springer-Verlag (left-part) and from Otsuka T *et al* 2001 *Phys. Rev. Lett.* **87** 082502 © 2001 by American Physical Society (right-part))

the $N = 50$–82 shell (mainly filling up the neutron 1h$_{11/2}$ orbital). A similar dramatic variation in neutron single-hole excitations at $N = 81$ is found when filling the proton 1g$_{9/2}$ orbital going from $_{40}$Zr$_{81}$ up to $_{50}$Sn$_{81}$. We have to draw attention to a point, already made in the introductory section, namely that the effective nucleon–nucleon force will itself start to change from what we know near stability. So we again enter a self-consistency problem since the two-body effective interaction will most probably be density dependent and change when we consider a long series of isotopes or isotones. This modification will imply extra changes to the single-particle properties (energy, wavefunction, etc) that are generated (see equation (15.1)) when determining the self-energy corrections $\tilde{\epsilon}_{j\rho}$.

The situation at present is such that the exploration of nuclei far from stability can be understood rather well using the spherical shell model and the strong induced proton–neutron energy shifts. Possible strong variations in the shell gaps should become visible in the one- and two-neutron separation energies. In particular, for the S_{2n} values and nuclear binding energies (masses), comparing theoretical numbers and data for the nuclear mass in, for example, the Sn nuclei, goes very well where data exist but large deviations show up amongst various calculations where no data are present. This is not a very comforting situation.

Even an extra complication arises. When moving out towards the very neutron-rich nuclei, the least-bound orbitals move near to the region of zero-binding energy and, as a consequence, the lowest 1p-1h excitations become embedded in the particle continuum. Thus, a clear separation between the region of bound orbitals and unbound scattering states ceases to hold. Recent work in this direction has been carried out (Michel *et al* 2002, Okolowicz *et al* 2003). Even though it is known that the isospin $((N-Z)/A)$ dependence of the average field is determined through the isospin dependence of the basic nucleon–nucleon interaction, the changing single-particle structure as well as the changing properties of nuclear vibrations are largely unknown when approaching those regions where neutron single-particle (or proton-particle) states are no longer bound in the potential well.

We illustrate in figure 15.3 what kind of modifications might be expected when approaching the doubly closed ^{100}Sn nucleus (which becomes the heaviest $N = Z$ nucleus known at present—see Box 15a for a more detailed discussion of the first experimental observation of this nucleus). In studying the single-particle variation of the proton single-particle states when ^{100}Sn is approached, starting from the region of β-stability, one observes a decreasing binding energy of the unoccupied proton orbitals just above $Z = 50$. The best evidence, at present, derived from a number of theoretical studies and making use of extrapolations of experimental information from neutron deficient Zr, Nb, Mo, Tc, Ru, Rh, Pd, Ag, Cd, In nuclei indicates unbound proton $2d_{5/2}$ and $1g_{7/2}$ states (the Coulomb field barely localizes these states) and fully unbound higher-lying proton orbitals $2d_{3/2}$, $3s_{1/2}$ and $1h_{11/2}$. Those proton orbitals appear in the continuum of positive energy states and thus the standard nuclear structure shell-model problem is completely changed: one can no longer separate the unbound and bound energy regions. Bound nuclear structure and reaction channels (proton-decay, proton pair scattering and decay) now form a fully coupled system. This indicates that the separation of the way in which the Hartree–Fock mean field and the pair scattering processes (BCS correlations) could be determined now fails. Calculations have been carried out by the group of Nazarewicz and Dobaczewski (Nazarewicz *et al* 1994, Dobaczewski *et al* 1994, 1995, 1996a, b) and by Meng and Ring (1996).

We present some salient features of those calculations of Hartree–Fock single-particle energies as discussed in detail by Dobaczewski *et al* (1994) (see figure 3 of that paper) for the $A = 120$ isobars. The spectrum corresponding to positive energies (unbound part) remains discrete because the whole system is put in a large, albeit finite, box. One clearly notices that the major part of the well-bound spectrum (the single-particle ordering and the appearance of shell gaps at 50, 82, 126) does not change very much when entering the region of very neutron-rich nuclei. A closer inspection, however, signals a number of specific modifications with respect to energy spectra in the region of stability. At positive energy and for the neutron states, one observes a large set of levels with almost no

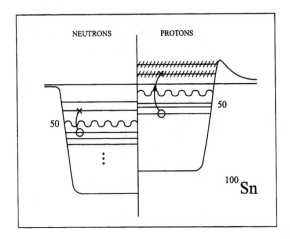

Figure 15.3. Schematic figure showing both the potential and the single-particle states corresponding to the doubly magic nucleus ^{100}Sn. It is shown that for the unoccupied proton particle states, the Coulomb potential causes a partial localization in the vicinity of the atomic nucleus (proton single-particle resonances).

neutron number dependence, for the low angular momentum values. In such states (low angular momentum), no centrifugal barrier can act to localize states and thus there is almost no dependence on the properties of the average potential. For the proton states, a weak Z dependence shows up, because the Coulomb potential (which amounts to 5 MeV in ^{100}Zn and 9 MeV in ^{100}Sn) has a clear tendency to keep single-particle motion within the nuclear interior. This particular effect is even enforced for higher angular momentum states, such that the known shell gaps are not easily modified or destroyed. The latter states can be described as quasi-bound resonances. These results hold also for other A values too.

The pairing force seems to play an important role in these neutron-rich nuclei near the drip line due to scattering of nucleon pairs from the bound into the unbound single-particle orbitals which gives rise to the formation of an unphysical 'particle-gas' surrounding the atomic nucleus (Dobaczewski *et al* 1984). Only a correct treatment of the full Hartree–Fock–Bogoliubov (HFB) problem treating the interplay of the mean-field and the nucleon pair scattering into the continuum can overcome the above difficulties. A detailed treatment of the effect of pairing correlations on a number of observables is discussed by Dobaczewski *et al* (1996b). This generalized treatment has a number of drawbacks such as the fact that the shell structure cannot easily be interpreted as related to a given set of quasi-particle energies E_α as eigenvalues of the HFB Hamiltonian. Without going into technical details (Dobaczewski *et al* 1994, Nazarewicz *et al* 1994), one can calculate the expectation value of the single-particle Hamiltonian, including occupation numbers (pairing correlations

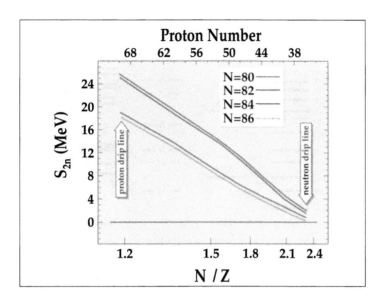

Figure 15.4. Two-neutron separation energies for spherical nuclei with neutron numbers $N = 80, 82, 84$ and 86 and even proton number, determined from self-consistent Hartree–Fock–Bogoliubov calculations. The arrows indicate the approximate positions for the neutron and proton drip lines (courtesy of W Nazarewicz (1998) with kind permission).

included), in the basis which diagonalizes the single-particle density. As the density goes to zero for large distances, these diagonalized (or canonical) states are always localized. These values of $\varepsilon_{\mathrm{HFB}}$ obtained in such a self-consistent calculation have a clear interpretation in the vicinity of the Fermi energy only.

One observes that the shell gap at $N = 82$ decreases in an important way (figure 15.4) on proceeding away from the region of the valley of stability towards the neutron drip line. Calculations have been performed by Dobaczewski *et al* (1996b). This particular quenching at the neutron drip line seems only effective in those nuclei with $N < 82$. This is at variance with nuclei near the proton drip line because the Coulomb potential prevents the low-j unbound states from approaching the bound state spectrum and so does not imply particularly strong modifications.

In conclusion, the essential result of treating the coupling of the bound states to the continuum of the particle spectrum in a self-consistent way, combined with a large diffuseness of the neutron density and subsequently of the central potential, results in a modified single-particle spectrum reminiscent of a harmonic-oscillator well without a centrifugal term ℓ^2 but including the spin-orbit part. It could of course well be, as discussed before, that the opening of reaction channels at very low energy alters the standard description, in terms of a mean field, radically and

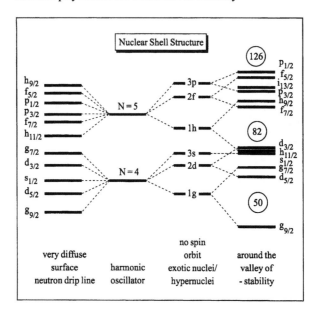

Figure 15.5. Nuclear single-particle ordering in various average fields. At the far left one uses a spin-orbit term only, corresponding to a rather diffuse nuclear surface. Next, the fully degenerate harmonic oscillator spectrum is shown (for $N = 4$ and $N = 5$, only). The following spectrum corresponds to a potential with a vanishing spin-orbit term but including an ℓ^2 term and, at the extreme right, the nuclear single-particle spectrum for nuclei in or close to the region of stable nuclei is given (taken from DOE/NSF Nuclear Science Advisory Committee (1996), with kind permission).

thus would signal the breakdown of putting the nucleon–nucleon interactions in a mean field as an approximate first structure. An illustration of the possible modifications of the nuclear shell structure in exotic nuclei, approaching the neutron drip-line region, is presented in figure 15.5.

Precisely in this region extrapolations will fail and exotica such as a neutron 'stratosphere' around the nuclear core as diluted neutron gas, formation of cluster structures, etc, may well show. Much development work, on a firm theoretical basis, still needs to be carried out.

Another issue in exploring the light $N = Z$ self-conjugate nuclei shows up in the region of nuclei that are very difficult to create (see the search for the doubly-closed shell nucleus ^{100}Sn). These are nuclei in which one can follow how isospin impurities evolve in the 0^+ ground state, starting from the very light $N = Z$ nuclei like ^{4}He, ^{16}O and moving up to the heaviest $N = Z$ systems. For these nuclei, when moving along the line of $N = Z$ nuclei as a function of proton number Z, one can also try to find out about a component of the nucleon–nucleon interaction that is not so well studied in nuclei, i.e. the possible proton–neutron pairing component in both the $T = 0$ and $T = 1$ channels. The $T = 1$ pairing channel is very well studied in many nuclei (chapter 11) with a number of valence protons or neutrons outside closed shells. The signature is the strong binding of these valence nucleons in 0^+ coupled pairs. In $N = Z$ nuclei, other components become dominant that are exemplified by singularities in binding energy relations at $N = Z$. At present, one wants to study how quickly the $T = 1$ pairing nucleon–nucleon interactions take over in the determination of low-lying nuclear structure properties when moving out of $N = Z$ nuclei and, more importantly, to identify clear fingerprints for the appearance of a strong $T = 0$ neutron–proton 'pairing' collective mode of motion. Both theoretical and experimental work is needed to elucidate this important question in nuclear structure. The specific issues that appear in determining the structure of very weakly bound quantum systems and the new phenomena that appear when these conditions are fulfilled will be amply discussed in the next section, indicating some basic quantum-mechanical consequences from the very small binding energy as well as illustrating the compelling and rapidly accumulating body of experimental results. This field is termed 'drip-line' physics.

15.2.2 Drip-line physics: nuclear halos, neutron skins, proton-rich nuclei and beyond

(a) Introduction: neutron drip-line physics

Nuclear stability is determined through the interplay of the attractive nucleon–nucleon strong forces and the repulsive Coulomb force. In chapter 7, these issues were discussed largely from a liquid-drop model point of view in which the stability conditions against protons and neutrons just being bound, fission and α-decay have been outlined. In chapters 8–12, the shell-model methods to explain nuclear binding, stability and excited-state properties have been presented in quite some detail. In the present section, we shall mainly discuss the physics one encounteres in trying to reach the drip line. The essential element, resulting from the basic quantum mechanics involved when studying one-dimensional bound quantum systems, is that wavefunctions behave asymptotically as exponential functions, given by the expression

$$\psi(x) \propto \exp(-\kappa x), \qquad \text{with } \kappa = (2m|E|)^{1/2}/\hbar, \tag{15.2}$$

and $|E|$ the binding energy ($E < 0$). This allows for particles to move far away from the centre of the attractive potential. This idea can be extended easily to more realistic and complex systems but the essentials remain: neutrons can move out into free space and into the classical 'forbidden' region of space. This has given rise to the subsequent observation of neutron 'halo' systems in weakly bound nuclei where a new organization of protons and neutrons takes place which minimizes the energy by maximizing the coordinate space available.

Besides the formation of regions of low nuclear density of neutrons outside of the core part of the nucleus, a number of interesting features can result such as

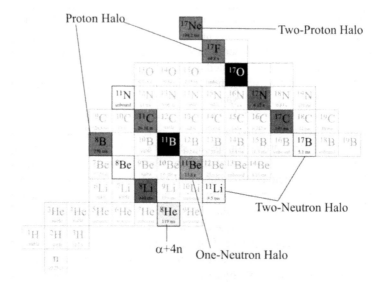

Figure 15.6. Excerpt of the nuclear mass table for very light nuclei. Stable nuclei are marked with the heavy lines. We also explicitly indicate one-neutron and two-neutron nuclei and candidates for proton-halo nuclei. (Reprinted from Jonson B 2004 *Phys. Rep.* **389** 1 © 2004 with permission from Elsevier.)

- because of the large spatial separation between the centre-of-mass and the centre-of-charge, low-energy electric dipole oscillations can result and show up as what are called soft dipole giant resonances (SGDRs),
- very clear cluster effects can show up and thereby all complexities related to three-body (and even higher) components will come into play. This forms the basis of an important deviation from standard mean-field shell-model methods which fail in such a region.

As can be observed in figure 15.1, the neutron drip line is situated very far away from the valley in the mass surface and there is no immediate or short-term hope of reaching the extremes of stability for neutron-rich nuclei (^{160}Sn is still particle-stable with 50 protons and 110 neutrons). So, the present information about the physics at or near the neutron drip line is mainly based on the detailed and extensive experimental studies carried out for neutron-rich very light nuclei. The present mapping of this region is illustrated in detail in figure 15.6 (Jonson 2004) with extensive information about ^{11}Li (two-neutron halo) and ^{11}Be (one-neutron halo). The next heavier nuclei such as ^{12,14}Be, ^{15,17,19}B and ^{18,19,20,22}C have already been reached or will be studied in detail soon. In carefully studying the specific topology of the neutron drip line, a number of interesting physics issues become compelling.

(i) The neutron–neutron 'binding' is dominated by the spin-singlet and isospin-triplet or $^1S_0$ configuration. The di-neutron is not bound but is very close to forming a bound state—only 100 keV away, which is really very small in the light of the binding energy forming the balance between the large and positive kinetic energy and the large but negative potential energy. The nucleus ^{4}He cannot bind another neutron but it can bind a di-neutron; it will also not bind three but can do so for four neutrons forming ^{8}He. This odd–even staggering is nothing but a reflection of the pairing force binding a di-neutron in the presence of a core nucleus.

(ii) Drip-line nuclei are laboratories to test neutron pairing properties in a neutron-rich environment. This can easily be seen from the approximate relation connecting the Fermi-level energy λ, the pairing-gap energy Δ and the particle separation energy S, i.e.

$$S \simeq -\lambda - \Delta. \qquad (15.3)$$

A justification for the above relation can be given as follows. Using equations (7.30) from chapter 7, it follows that the separation energy, in the absence of pairing correlations, would be equal to the absolute value of the Fermi energy-λ of the particle that is removed from the nucleus. Pair correlations will increase this number by an amount of approximately the pairing energy Δ for even-even nuclei (see figure 7.11) and reduce by the amount Δ for an odd-mass nucleus as one should use the quasi-particle separation energy and not the particle separation energy (Heyde 1991). A more detailed discussion is given by Smolańczuk *et al* (1993). At the drip line, the separation energy becomes vanishing and consequently the absolute values of the Fermi energy (characterizing mean-field properties) and the pairing gap (characterizing pair scattering across the Fermi level) are almost equal. So, pairing can no longer be considered a perturbation to the nucleonic mean-field energy and appears on an equal footing with the single-particle energies.

(iii) The central densities in regular nuclei ($\sim$0.17 nucleons fm^{-3}) are almost independent of the given mass number A and the nuclear radius varies with $A^{1/3}$ with a nuclear skin diffusivity roughly independent of mass number too. These properties, almost taken as dogma, in the light of the points (i) and (ii), only apply for stable or nearby nuclei. Studies of neutron-rich and proton-rich nuclei show differences in their properties: neutron halos and neutron skins can form (see figure 15.7 for a schematic view). The former are a consequence of the very low binding energy and thus allow the formation of wavefunctions extending far out into the classical 'forbidden' space region. The latter can result in heavier nuclei for very neutron-rich nuclei and are obtained by minimizing the total energy of such nuclei: the lowest configuration becomes one with a kind of neutron 'stratosphere' around the internal core.

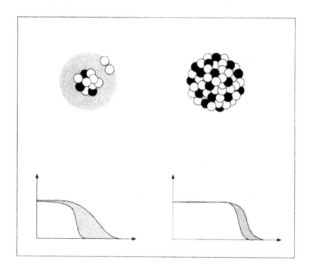

Figure 15.7. A schematic presentation of the spatial nucleon distributions and the corresponding proton and neutron mass density distributions for both a halo nucleus and a nucleus with a neutron skin.

Next, we go on to discuss specific issues for single-neutron halo systems (b), two-neutron halo systems (c) and experimental tests for the existence of halo structures (d), proton-rich nuclei and other exotica (e) and some concluding remarks on drip-line physics (f).

(b) Single-neutron halo nuclei

An elementary description of halo nuclei where the motion of a single nucleon (neutron in this case) constitutes the asymptotic part of the nuclear wavefunction has been given in an early paper by Hansen and Jonson (1987). We use here some of the discussion in presenting the essential physics of those single-nucleon halo systems.

Consider the motion of a neutron, with reduced mass μ_n in a three-dimensional square-well potential with radius R, then the asymptotic radial part of the neutron wavefunction can be described by the following expression (for a relative $\ell = 0$ or s-state orbital angular momentum state)

$$\psi(r) = \sqrt{2\pi\kappa}\,\frac{e^{-\kappa r}}{\kappa r}\,\frac{e^{\kappa R}}{\sqrt{1+\kappa R}}, \qquad (15.4)$$

with $\kappa = (2\mu_n S_n)^{1/2}/\hbar$ and $x = \kappa R$ a small quantity. Starting from this simple wavefunction, the mean-square radius describing the asymptotic radial part gives rise to the value

$$\langle r^2 \rangle = \frac{1}{2\kappa^2}(1+x) = \frac{\hbar^2}{4\mu_n S_n}(1+x), \tag{15.5}$$

with higher-order terms in x neglected. One notices that this mean-square value is inversely proportional to the neutron binding energy. Another interesting quantity is obtained through a Fourier transform of the coordinate wavefunction, giving rise to the momentum wavefunction (neglecting terms in the quantity x) or the momentum probability distribution and results in the expression

$$|F(p)|^2 = \hbar\kappa \frac{1}{\pi^2(\kappa^2\hbar^2 + p^2)^2}. \tag{15.6}$$

This expression reflects Heisenberg's uncertainty principle: the large spatial extension of the neutron halo gives rise to an accurate determination of the neutron momentum distribution, as becomes clear from the p^{-4} dependence for large p values.

The large separation between the external neutron and the core (containing the charged particles) implies large electric dipole polarizability and so gives rise to the possibility of inducing dissociation of the halo nucleus (much like the dissociation in molecular systems) through an external Coulomb field. The Coulomb dissociation cross-section for the collision of a halo nucleus (moving with velocity v, containing charge Z_h and core mass M_h) with a heavy target nucleus with charge Z_t can be derived as

$$\sigma_C = \frac{2\pi Z_h Z_t e^4 \mu_n}{3v^2 M_h^2 S_n} \ln\left(\frac{b_{max}}{b_{min}}\right), \tag{15.7}$$

with maximal and minimal impact parameters b_{max} and b_{min}, respectively. This cross-section becomes large for large Z_h and Z_t values, for slowly moving halo nuclei and for halo nuclei with a very small separation energy.

Extensive discussions of single-neutron halos (and also for more complex halo systems or nuclei with external neutron-rich skins) can be found in some highly readable articles by Hansen 1991, 1993a, Jonson 1995, Hansen *et al* 1995 and Tanihata 1996. In these articles, extensive references are given to the recent but already very large set of results in this field.

Before going on to discuss the extra elements that appear when describing nuclei where the halo is constituted of two neutrons, we give some key data and results for the best example known as yet of a single-neutron halo, i.e. ^{11}Be. In this nucleus, the last neutron is bound by barely 504 ± 6 keV and so the radial decay constant κ^{-1} amounts to about 7 fm, to be compared with typical values of 2.5 fm as the root-mean-square radius for p-shell nuclei. In figure 15.8, the radial wavefunctions for the two bound ($\frac{1}{2}^+$ or s-state and $\frac{1}{2}^-$ or p-state) states in ^{11}Be are given and clearly illustrate the fact that the radial extension is far beyond the typical p-shell nucleus extension (of the order of 2.5 fm). Something

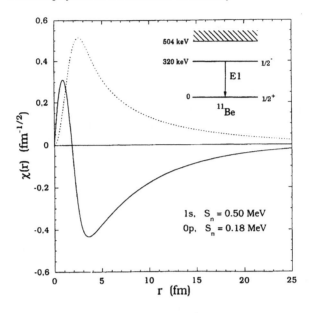

Figure 15.8. The radial wavefunctions for both the $\frac{1}{2}^+$ (1s orbital) and the $\frac{1}{2}^-$ (0p orbital), ground-state and first-excited state in ^{11}Be. In the inset, the level scheme of ^{11}Be is also shown (taken from Hansen *et al* (1995) with permission, from the Annual Review of Nuclear and Particle Science, Volume 45 © 1995 by Annual Reviews).

particularly interesting is happening in this nucleus as the single-particle ordering of the $2s_{1/2}$ and $1p_{1/2}$ states is reversed when compared to the standard single-particle ordering for nuclei closer to the region of β-stability. Calculations going beyond a Hartree–Fock mean-field study, called a variational shell-model (VSM) calculation, carried out by Otsuka and Fukunishi (1996), are indeed able to reproduce the inverted order by taking into account important admixtures of a $1d_{5/2}$ configuration coupled to the 2^+ core state of ^{10}Be. They also showed that this coupling is essential to produce the energy needed to bind the ^{11}Be nucleus.

Another interesting piece of information is the E1 transition probability connecting the $\frac{1}{2}^-$ and $\frac{1}{2}^+$ states, as measured by Millener *et al* 1983, to be the fastest E1 transition known corresponding to 0.36 Weisskopf units. As explained before, from dissociation experiments (see the above references) causing the fragmentation of the halo nucleus, at higher energies, it has been possible to derive the neutron momentum distribution corresponding to neutrons moving in the core and to the halo neutron. The results, here illustrated in figure 15.9, precisely show the complementary characteristics expected from a neutron moving in the highly localized core and non-localized halo parts of coordinate space.

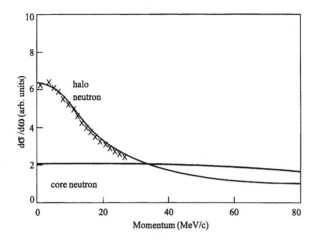

Figure 15.9. Momentum distribution for a core neutron (flat curve) and for a loosely bound neutron in the one-neutron halo nucleus ^{11}Be, measured with respect to the final nucleus after fragmentation reactions (taken from DOE/NSF Nuclear Science Advisory Committee (1996), with kind permission).

(c) Two-neutron halo nuclei

If we consider a neutron pair in a nucleus like ^{11}Li, it was suggested by Hansen and Jonson (1987) that the pair forms a di-neutron which is then coupled to the ^{9}Li internal core. Considering the di-neutron as a single entity, with almost zero binding energy, it is then possible to a first approximation to consider the binding energy of this di-neutron system to the core as the two-neutron separation energy S_{2n}. In the case of ^{11}Li this latter value $S_{2n} = 250 \pm 80$ keV, causes a very extended two-neutron halo system to be formed, as the external part of the di-neutron radial wavefunction will be described by the radial decay constant $\kappa = (2\mu_{2n}S_{2n})^{1/2}/\hbar$, with μ_{2n} the reduced di-neutron mass and S_{2n} the two-neutron separation energy. So, all of the results derived in section (b) above can be used again to a first approximation.

As a way of illustrating the ^{11}Li two-neutron halo nucleus we can draw a diagram like that shown in figure 15.10. In this drawing the halo neutrons are put close together or correlated spatially. In order to decide on such details, one can no longer rely on a two-part wavefunction, consisting only of the core nucleus ^{9}Li on one side and the di-neutron on the other side. Here, the full complexity of the quantum-mechanical three-body system shows up which does not allow for an exact solution. Various approximate studies have been carried out over the years giving a better and more sophisticated view of the internal structure of this loosely bound halo structure, containing a core and two extra neutrons.

Before giving reference to a number of those studies, we point out that three 'particles' interacting via short-range two-body interactions can give rise

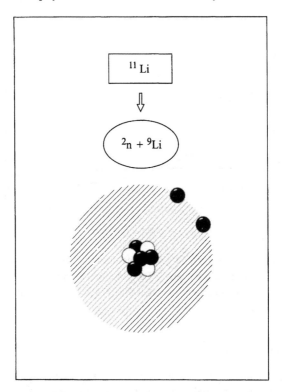

Figure 15.10. Schematic drawing of the halo nucleus ^{11}Li in which the three protons and six neutrons form the ^{9}Li rather inert core system and in which the remaining two loosely bound neutron systems form a halo system radially extending very far.

to a variety of different structures (Fedorov *et al* 1994). Leaving out the details that make the description more cumbersome (like intrinsic spin) one arrives at the following classification as shown in figure 15.11 for a nuclear core system, denoted by A, and two extra neutrons. The interactions between the neutron and core A are denoted by V_{An} and the neutron–neutron interaction by V_{nn}. Using the strengths of these two forces as variables, one can show that the plane separates into regions where the two-body systems (nn) and (An) are bound or unbound. There is, however, a region where the three-body system gets bound but none of its two-body subsystems are bound. This is the so-called Borromean region (Zhukov *et al* 1993). It resembles the heraldic symbol of the Italian princes of Borromeo consisting of three rings interlocked in such a way that if any one ring is removed, the other two separate. Besides ^{11}Li, other Borromean nuclei are, for example, ^{6}He, ^{9}Be and ^{12}C. The extra element in ^{11}Li though, compared to some of these other nuclei, is the extra-low two-neutron binding energy, implying

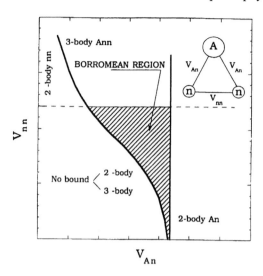

Figure 15.11. Schematic illustration of the three-body system consisting of a core system A and two extra neutrons. The binding of this system is studied with respect to variations of the A–n and the n–n interaction strengths. The thick curve separates the region for bound or unbound three-body systems; the vertical and horizontal lines make a division between bound and unbound two-body systems. The hatched region, also called the Borromean region, forms a three-body bound system in which none of the three two-body systems is bound. (Taken from Hansen *et al* (1995) with permission from the Annual Review of Nuclear and Particle Science, Volume 45 © 1995 by Annual Reviews.)

that most of the ground-state properties of this nucleus will be determined by the asympotic part of the wavefunction.

The theoretical studies, in order of increasing sophistication, after the early study of Hansen and Jonson (1987), involved calculations using shell-model methods, the cluster model, three-body studies, Hartree–Fock techniques as well as calculations concentrating on reaction processes all of which have been performed extensively (see Hansen (1993b) for an extensive reference list). In all these studies, the role of pairing plays an essential part. The pairing property (see chapters 10 and 11 for more detailed discussions) can radically change the properties of a many-body system. In metals, for example, the Coulomb repulsion between electrons is modified into slight attraction through the mediation of lattice vibrations and finally results in superconductivity. In atomic nuclei, pairing condenses nucleons into nucleon pairs giving rise to a superfluid structure at lower energies which is the origin of the stability of many atomic nuclei. Pairing properties in a dilute neutron gas will influence properties in neutron-rich matter and neutron stars. It is also the pairing part that is largely responsible for the existence of the above Borromean structures.

It is interesting to note first that the two extremes—one where the two extra neutrons were treated as a single entity, the di-neutron (Hansen and Jonson (1987)), or the case in which pairing beween the two neutrons is fully ignored and so only the coupling between the neutron and core with no direct nn coupling is considered (Bertsch and Foxwell 1990a, b)—gave a number of results that were quite close (break-up probabilities, size of the halo). This is partly due to the fact that both calculations had to use the experimental binding energy for the two-neutron system as input. Since then, pairing forces have been explicitly included in the theoretical studies and detailed studies of the three-body (or cluster) characteristics have been performed (Bertsch *et al* 1989, 1990, Esbensen 1991, Bertsch and Esbensen 1991). These results showed that when neutrons are far out in the halo they are likely to correlate closely in coordinate space. On the other hand, when moving close to the ^{9}Li core, they are far apart spatially. So, these more realistic studies are able to 'interpolate' between the two extremes of having a single highly correlated di-neutron and two independently bound neutrons. These calculations were highly successful in producing the correct momentum distributions observed in ^{11}Li break-up reactions. Calculations by the Surrey group (Thompson *et al* 1993) and by Zhukov *et al* (1993) then showed the Borromean characteristics of ^{11}Li. Consequently researchers are now confident to study neutron pair correlations in a low-density environment and this may lead to interesting applications in other domains of nuclear physics.

One can now of course ask questions and try to solve problems about the transition region between normal and low-density (halo) structures in atomic nuclei as well as about the issue of possible formation of halo structures containing more than two nucleons. In these latter systems, the precise treatment of pairing characteristics of the nucleon–nucleon force will be very important. Efimov (1970, 1990) pointed out that if the two-body forces in a three-body system are such that binding of the separate systems, two at a time, is almost realized, the full system may exhibit a large, potentially infinite, number of halo states. These are exciting results and may well show up when studying the rich complexity of nuclei when progressing from the region of β-stability towards the drip-line region.

(d) Experimental tests for the existence of halo nuclei

The early research at ISOLDE (CERN), which made it possible to produce elements far from the region of stability, has given access to some of the essential ground-state properties of nuclei.

One interesting idea was to use the experimental methods of measuring the total interaction cross-sections for light nuclei by the transmission through thick targets as a means to determine the matter radii in these light and exotic nuclei. Early experiments, carried out by Tanihata and coworkers at Berkerley (Tanihata *et al* 1985) and later on at GANIL (Mittig *et al* 1987, Saint-Laurent 1989), were able to give access to interaction radii of light nuclei from such cross-

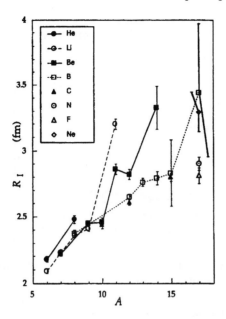

Figure 15.12. Interaction radii for light nuclei determined from interaction cross-sections. A sudden increase of the matter radii is observed for a number of nuclei near the neutron drip line. Data taken from Tanihata *et al* (1988, 1992), Tanihata (1988) and Ozawa *et al* (1994). (Taken from Tanihata © 1996, with kind permission of Institute of Physics Publishing.)

section measurements. It came as a real surprise (see also the results shown in figure 15.12) to observe significantly larger matter radii in nuclei like ^{6,8}He ^{11}Li compared to the matter radii for the more standard p-shell nuclei with a constant mass radius of about 2.5 fm. These results clearly pointed out that nuclear matter must appear much further out than normally forming halo-like structures. It should be possible to find out about the charge distribution in such neutron-rich nuclei by measuring electric quadrupole measurements. Experiments carried out at ISOLDE (Arnold *et al* 1987) unambiguously showed that the charge distribution inside ^{11}Li is almost identical to the charge distribution in ^{9}Li thus bringing in additional and clear evidence that the large matter radii obtained were due to some unexpected behaviour of the last two neutrons, forming a halo structure around the ^{9}Li core.

Another important tool to disentangle the structure of these light neutron-rich nuclei at the edges of stability came from experiments trying to determine (mostly in an indirect way) the neutron momentum distribution with respect to the internal core system. Kobayashi *et al* (1988, 1989) were able to determine the transverse momentum distribution of ^{9}Li recoils from fragmentation of ^{11}Li

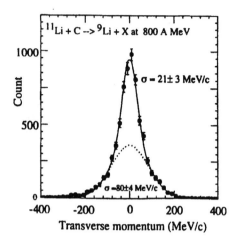

Figure 15.13. Transverse momentum distributions of projectile fragments of neutron halo nuclei. The narrow distributions show that these distributions correspond to spatially extended distributions for the loosely bound neutrons (taken from Tanihata © 1996, with kind permission of Institute of Physics Publishing).

on a carbon target. In figure 15.13, these results are shown for the reaction ^{11}Li + C. It is very clear that, besides a broad bump, quite narrow momentum distributions are obtained reflecting the sharply momentum-peaked outer or halo-neutron momentum distributions. These results are a good test for Heisenberg's uncertainty relation connecting momentum and coordinate for a given quantum-mechanical system: the neutrons in ^{11}Li or ^{11}Be in the present cases. The very weak spatial localization of the outer halo neutrons in these nuclei correspond to a rather precise momentum localization as verified in the above experiments and, even more so, in more recent ingenious sets of experiments elucidating the momentum distribution of nucleons inside these weakly bound nuclei. Here, simple shell-model-type wavefunctions totally fail to produce the observables.

More recent experiments, carried out by Orr *et al* (1992), determined the longitudinal momentum distribution of the ^{9}Li fragment from break-up reactions of ^{11}Li on a number of targets. In these experiments, a Lorentzian curve characterizes the central part of the momentum distribution with a width of only 37 MeV/c and this stringently and unambiguously determines the small spread in the outer neutron momentum distribution in ^{11}Li which implies that there is a large spatial extension for the weakly bound nucleus ^{11}Li. In an experiment to disentangle the various nuclear properties in such a dilute nuclear system, Ieki *et al* (1993) carried out a kinematically complete experiment: they determined the directions and energies of both the outgoing neutrons and the recoil nucleus in electric break-up of ^{11}Li on a Pb target.

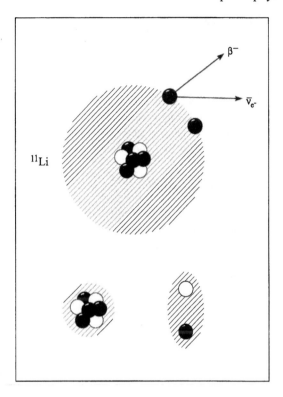

Figure 15.14. Schematic illustration of the β-decay of a neutron out of the halo region. This process probably results in the formation of the core nucleus ^{9}Li and a deuteron. This decay rate should approach the decay rate of an almost free neutron.

These two important experiments also gave answers to the issue of how the extra neutrons are spatially correlated: moving independently or moving in a highly correlated spatial mode. Now, with quite high precision, experiments show that a correlation between the momenta of the neutrons emitted in the break-up reactions does not seem to exist. It seems that the reconstructed decay energy spectrum is consistent with the idea that energy is shared between ^{9}Li and the two remaining neutrons according to three-body phase space (statistical) only and, as a consequence, that no particular neutron–neutron correlations in the outer spatial region are needed.

A large number of complementary and more recent experiments are amply treated in the following review papers: Hansen (1993b), Jonson (1995), Hansen *et al* (1995b) and Tanihata (1996).

Interesting experiments in those halo nuclei in which a rather dilute neutron halo structure appears involve the study of beta-decay of neutrons in this halo structure (see chapter 5 for details on beta-decay). In figure 15.14, we illustrate in a schematic way the decay of a quasi-free neutron. We expect to find the decay rate for 'free' neutrons. A possibility, as expressed by figure 15.14, takes into account the fact that a deuteron structure may be formed and subsequent deuteron emission results. Such a process has been observed in the decay of ^{6}He by Riisager *et al* (1990). This domain is highly important as it may give access to the exotic mode of decay of neutrons in a quasi-free environment and may reflect properties that make the systems different from an independent gas of free neutrons. Access to corrections from pairing correlations or other spatial correlations between the neutrons may thus become possible.

(e) Proton-rich nuclei and other exotica

By now, interesting experimental data for nuclei far from stability have originated from the light neutron-rich nuclei, mapping the neutron drip-line structure. Moreover, there exists a vast amount of fission product data for nculei in the $80 < A < 150$ mass region. Except for those regions, very little is known about neutron-rich nuclei at medium and heavy masses.

A first, very natural question is: if a very low separation energy of the last bound particle implies wavefunctions that fall off with a very small factor (related to the radial decay constant κ, see also equation (15.4)), then can one also find nuclei with a proton-halo structure? There is, however, a very important and essential difference in the case of uncharged particles in the appearance of the Coulomb potential. When a proton tries to extend away from the inner core part of the loosely bound nucleus it encounters the Coulomb potential at the nuclear surface region (which is about 2.5 fm for typical p-shell nuclei). Although the wavefunction will still extend far out of the nucleus, due to the basic theorem on loosely bound quantum systems and its radial structure, the amplitude will be largely quenched because of the Coulomb barrier. So, the problem to be solved, even in the simplest formulation, now becomes one of a three-dimensional spherical potential well with a Coulomb potential added from the nuclear radius outwards to infinity.

At the limit of the bound proton-rich nuclei, many nuclei have proton separation energies as low as 1 MeV. In taking the spatial extension measured from break-up and dissociation reactions, no clear-cut evidence for significant increases in mean-square charge radii has been observed as yet (Tanihata 1996 and references therein).

So, the observation of proton-halo structures as yet is rather unclear and more systematic data are needed. The aforementioned issue of the presence of a Coulomb barrier, however, may provide interesting spectroscopic information at the proton-rich region of the nuclear mass table approaching the proton drip line. The Coulomb barrier here prevents otherwise unbound proton states from decaying and so retains a quasi-bound character for these states with respect to the proton decay channel (Hofmann 1989) (see also figure 15.15 for a schematic view). The corresponding lifetimes, which can range from a millisecond to a

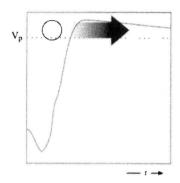

Figure 15.15. Illustration of the quasi-bound characteristics for very weakly bound proton states near the proton drip line. These states remain quasi-bound because of the Coulomb field and the properties will be determined by the appropriate tunnelling amplitude for the proton decay channel.

few seconds, are long compared to the time scale set by the strong nuclear force and may provide very useful tools for extracting spectroscopic information about the very proton-rich nuclei. Proton radioactivity will most probably become an interesting probe in that respect and will be, both experimentally and theoretically, a tool for future exploration of nuclear structure.

Due to the excess of neutrons, new types of collective modes might be conceived: the coordinate connecting the centre-of-mass and the centre-of-charge is of macroscopic measure and thus oscillatory motions can give rise to a low-lying electric dipole mode (also often called the soft dipole mode). The regular E1 mode corresponds to the oscillatory motion of the core protons versus the core neutrons whereas the soft, second mode, at much lower energy, would then correspond to an oscillatory mode of the outer neutrons versus the core protons (Ikeda 1992). These processes are illustrated in a schematic way in figure 15.16.

Experimental searches for such states with large E1 strength have been carried out in ^{11}Li and possible candidates were observed at an energy of 1.2 ± 0.1 MeV (Kobayashi 1992) but no clear-cut or unique conclusions could be drawn. Similar experiments have been performed in the nucleus ^{11}Be, a single-neutron halo nucleus, with both the ground and first excited states still bound. In this nucleus, the E1 excitation distribution was measured by Nakamura *et al* (1994) at RIKEN in Japan. The first excited state in this nucleus appears at 0.32 MeV (spin $\frac{1}{2}^-$) and thus, it is expected that the E1 strength of the low-energy E1 resonance is mainly located in the bound state. The experimental result, observed in electromagnetic dissociation and which gives information on possible collective motion in the halo nucleus ^{11}Be, is shown in figure 15.17. The data are compared with the results of a direct break-up model calculation with a

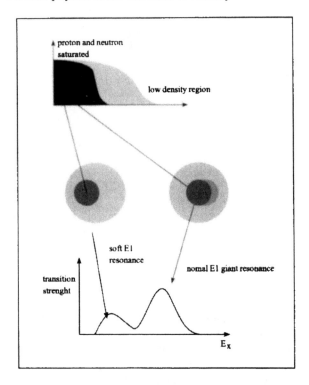

Figure 15.16. Besides the regular giant dipole resonance (oscillatory motion of core protons versus core protons) in the more tightly bound core part, the possibility exists of forming a soft E1 dipole mode in which the oscillatory motion acts between the inner core and the outer neutron halo. Both the density profiles, the spatial oscillatory possibilities and the resulting E1 strength distribution are illustrated (adapted from Tanihata © 1996, with kind permission of Institute of Physics Publishing).

corresponding E1 strength function given by the expression

$$dB(E1)/dE_x = S \frac{e^{2\kappa r_0}}{1 + \kappa r_0} \frac{3\hbar^2}{\pi^2 \mu} e^2 \left(\frac{Z}{A}\right)^2 \frac{\sqrt{E_n}(E_x - E_n)^{3/2}}{E_x^4}. \tag{15.8}$$

In this expression, S denotes a normalization given by the spectroscopic factor in the $\frac{1}{2}^+$ state, $\kappa = (2\mu E_n)^{1/2}/\hbar$ gives the radial exponential wavefunction decay rate, μ describes the neutron reduced mass, r_0 the radius of the square-well potential used to describe the halo wavefunction, E_n the neutron separation energy and E_x is the excitation energy. There is good agreement between the data and the model calculation of figure 15.17, showing that the transition is dominated by the non-resonant E1 of the halo neutron.

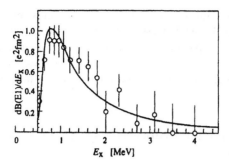

Figure 15.17. The E1 strength distribution deduced from electromagnetic dissociation cross-section measurements of ^{11}Be. The calculated break-up model calculation for a soft E1 mode is also shown (the solid curve) (taken from Tanihata © 1996, with kind permission of Institute of Physics Publishing).

(f) Some concluding remarks on drip-line physics

The present field of exotica near drip lines, in particular the extensive experimental set of studies carried out on light neutron-rich nuclei, is merely 10 years old and is still rapidly developing. The first nucleus at the neutron drip line, ^{6}He, was discovered more than 60 years ago but the last 10 years have seen an explosive increase in information about these loosely bound quantum systems.

This section has tried to give a flavour of this type of research and physics. The reader is referred to the references given in the present chapter for a more detailed study. It appears quite clear that, in the light of the new forms of nucleon distributions found in the light neutron-rich nuclei, a number of surprises are almost sure to be encountered when exploring the nuclear mass table towards the edges of stability, and even beyond for the proton-rich side. We present, in figure 15.18, the territory of exotica with, be it very schematic because of a lack of detail and also since part is still speculative, a number of insets showing what kind of phenomena may be observed.

An important point to make here is the fact that what happens in the element synthesis and the energy production inside stars is strongly related to the physics of unstable nuclei. We shall devote a short section (15.3.3) to these nuclear astrophysics applications.

We would also like to mention that a number of opportunities for the use of nuclei far from stability are situated somewhat outside of the major themes of nuclear physics research but are of great importance. A number of tests on the Standard Model (strength of the weak interaction, precision measurements of parity and time-reversed violations, ...) fall into this category. These are discussed in a more detailed way in the OHIO Report (1997) and by Heyde (1997).

We end this section by honestly stating that most probably state-of-the-art experimental work, making use of the most recent technical developments on the production, separation and acceleration of radioactive ion beams, will be rewarding and lead to a number of real discoveries, way beyond present theoretical (biased) extrapolations that normally fail when reaching *terra incognito*.

15.3   Radioactive ion beams (RIBs) as a new experimental technique

15.3.1   Physics interests

In figure 15.18, a pictorial overview of the various types of physics phenomena that can be expected or have barely been noticed experimentally is given (Nazarewicz *et al* 1996). The proton drip line in this mass landscape, for the proton-rich nuclei, is rather well delineated, up to the $Z = 82$ number. As was amply discussed in section 15.2, except for the light neutron-rich nuclei, one does not know very precisely where the corresponding neutron drip line is situated. Various extrapolations using mass formulae and fits to the known nuclei in the region of stability, diverge rather wildly when approaching the drip-line region. So, it is clear that a major part of the physics interest is found in studying and trying to come to grasp with new ways in which protons and neutrons are organized in forming nuclei that are either very neutron or proton rich.

In order to reach those regions, new production and separation techniques are under development (or have to be developed) which will lead to intense and energetic radioactive ion beams of short-lived nuclei. Eventually, these beams can be used to go even further out of stability. This will be discussed in more detail in section 15.3.2.

A very important driving idea in the production of these radioactive beams is the access one gets to a number of important nuclear reactions that have played a role in the element formation inside stars. The domain at the borderline between nuclear physics and astrophysics, called 'nuclear astrophysics', can thus be subjected to laboratory tests and one can of course select the key reactions that have led to the formation of the lightest elements first: light-element synthesis, understanding the CNO catalytic cycle, etc. In stars, at the high internal temperatures present, corresponding energies are of the order of the Coulomb energy so that the appropriate fusion for charged particle reactions can take place. Cross-sections are typically of the order of μb to mb and nuclear half-lives are typically of the order of a few minutes, even down to some seconds. So, the way ahead becomes well defined: selective radioactive ion beams of high purity with the correct energies (in the interval 0.2 to 1.5 MeV/nucleon) must be produced. In the domain of nuclear astrophysics, the neutron-rich elements to be explored form an essential part: it is in this region that the formation of the heavier elements has occurred during the r-process (the rapid capture of a number of neutrons in

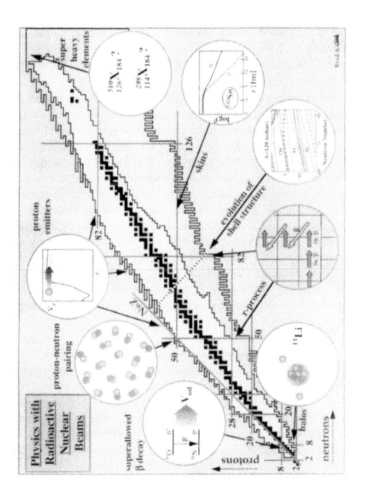

Figure 15.18. The nuclear mass region in which the many physics issues, appearing near or at the proton and neutron drip lines, are indicated schematically (taken from Nazarewicz *et al* 1996). A large number of these issues are discussed in the present chapter and we refer to the many references given here for more details. (Reprinted from Nazarewicz *et al* © 1996, with permission from Gordon and Breach.)

capture reactions (n, γ) before β-decay back to the stable region could proceed (see Clayton 1968, Rolfs and Rodney 1988)). The present status and important results will be discussed and illustrated in more detail in section 15.3.3.

15.3.2   Isotope separation on-line (ISOL) and in-flight fragment separation (IFS) experimental methods

For the purpose of creating beams consisting of radioactive elements, two complementary techniques have been developed over the last decades: the isotope separation on-line (ISOL) technique, eventually followed by post-acceleration and the in-flight separation (IFS) technique.

The idea of producing radioactive elements in a first reaction process at a high enough intensity, separating the radioactive nuclei and ionizing into charged ions with a subsequent second acceleration to the energy needed, was put forward. The method relies on having the ions produced at thermal velocities in a solid, liquid or gas. The method is in fact the same as standard accelerator technology albeit with a much more complex 'ion' source producing radioactive ions. The big difference is that in order to produce the necessary ions, one needs an initial accelerator (also called a 'driver' accelerator) that provides a primary beam producing exactly the required secondary ions through a given nuclear reaction process. Observing the various steps in the whole process, from the primary ions of the first accelerator up to the accelerated secondary ion beam to be used to study a given reaction, it is clear that each element of this chain needs as high an efficiency as possible and as fast as possible in order to make this whole concept technically feasible. The ISOL type of production is illustrated schematically in the upper part of figure 15.19.

The complementary idea was put forward of trying to form unstable elements by accelerating a first, much heavier element (projectile) and colliding it with a stationary target nucleus, thereby fragmenting the projectile, and later on separating the right fragments (fragment mass separator) via electric and magnetic fields (and also involving sometimes atomic processes). This so-called projectile fragmentation (PF) is the most common of IFS methods. It is also illustrated schematically in the lower part of figure 15.19. Under this same heading of in-flight separation, light-ion transfer reactions and heavy-ion fusion-evaporation reactions have also been put forward as a good idea to produce appropriate radioactive ion beams. Here too, the ions need to be selected by some separator set-up in order to be selective.

It has been through the enormous advances in accelerator technology at existing facilities combined with the construction of efficient and selective ionizing sources and high-quality mass separation over the last 20 years that this field of producing radioactive ion beams (RIBs) is now in a flourishing and expanding period. Recent reviews by Boyd (1994) and Geissl *et al* (1995) on this issue of RIBs give more detail.

HIGH INTENSITY RADIOACTIVE BEAM PRODUCTION METHODS

ISOL

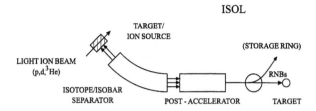

PROJECTILE FRAGMENTATION

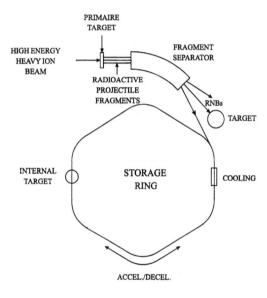

Figure 15.19. Schematic illustration comparing the typical isotope separation on-line (ISOL) and the projectile fragmentation in-flight separation (IFS) methods for producing high-intensity radioactive ion beams.   (Taken from Garrett and Olson (ed) © 1991 *Proposal for Physics with Exotic Beams at the Holifield Heavy-Ion Research Facility.*)

The above two methods, called the isotope separation on-line (or ISOL) method and the in-flight separation (or IFS) method are very complementary methods and have been developed in much technical detail. This complementarity is best illustrated in figure 15.20 in which the typical energy ranges obtained with the ISOL and IFS methods are shown and compared. What is clear is that for the ISOL method one needs post-acceleration of the radioactive ions up to the desired

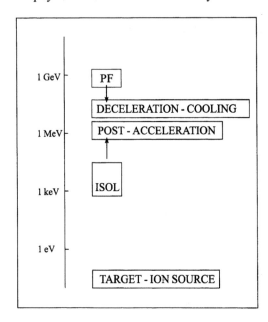

Figure 15.20. Schematic presentation of the energy domains covered by the isotope separation on-line (ISOL) and subsequent post-acceleration on one side and on the other by the fragmentation technique (PF), which can be followed by deceleration-cooling in storage rings.

energies with good beam qualities. From the higher-energy ions, produced by projectile fragmentation and IFS, in general, energies are quite often too high and one needs deceleration and cooling (by putting the ions in storage rings) to come up with high enough intensities.

We now discuss some typical characteristics for each of the two methods of producing radioactive ion beams: the ISOL and IFS methods. In the ISOL line of working, almost all projects (see also Box 15b) are based on existing accelerator facilities which can be the driver machine or, in constructing a new one, even be rebuilt into a post-accelerator machine. The very extensive set of ISOL projects are complementary because of the specific aims at reaching a given number of radioactive beams at specified energies. Moreover, the complementarity is present in both the production, ionizing and post-acceleration phases of the actual realization (Ravn *et al* 1994). A range of choices is illustrated in the line-drawing of figure 15.21. One of the major technical but essential points in reaching highly efficient and cost-effective accelerated radioactive ion beams is *the production of low-energy secondary ions in a high charge state*. A long-time experience has been gained at existing isotope separators in the production of singly charged ions. The knowledge gained at ISOLDE (CERN) over the last 25 years has been

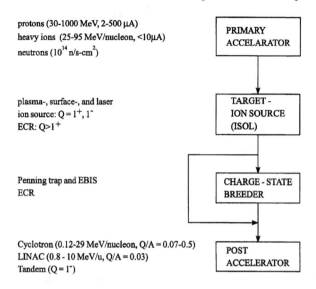

protons (30-1000 MeV, 2-500 μA)

heavy ions (25-95 MeV/nucleon, <10μA)

neutrons (10^{14} n/s-cm^2)

	PRIMARY ACCELARATOR

plasma-, surface-, and laser

ion source: Q = 1$^+$, 1$^-$

ECR: Q>1$^+$

	TARGET - ION SOURCE (ISOL)

Penning trap and EBIS

ECR

	CHARGE - STATE BREEDER

Cyclotron (0.12-29 MeV/nucleon, Q/A = 0.07-0.5)

LINAC (0.8 - 10 MeV/u, Q/A = 0.03)

Tandem (Q = 1$^-$)

	POST ACCELERATOR

Figure 15.21. The different steps, shown in a schematic way, needed to produce a post-accelerated radioactive beam (reprinted from Nazarewicz *et al* © 1996, with permission from Gordon and Breach).

instrumental in further developments: it has resulted in a wide range of ions with intensities going up to 10^{11} ions per second for specific ions. These developments are rigorously continued to further increase the intensity and the purity of the low-energy RIBs. For instance, resonant photo-ionization using lasers has recently been implemented at ISOLDE (CERN) and LISOL (Louvain-la-Neuve). Now, in order for the acceleration in the second stage of the radioactive ions to be highly efficient and effective, one needs to have the initial ion in a higher charged state with the ratio Q/A (charge/mass) from $\frac{1}{9}$ to $\frac{1}{4}$, typically. At this stage, one enters the domain of producing multiply charged ions. Here, a large body of knowledge exists with different types of ion sources: electron cyclotron resonance (ECR) sources, electron beam ion sources (EBIS) and laser ion sources (NUPECC Report 1993, Ravn 1979, Ravn and Allardyce 1989). The former two are currently used in a standard way to produce and inject ions corresponding to the known stable elements into accelerators.

There are, at present, worldwide efforts to set up and plan RIB facilities (see Box 15b for a more detailed overview on the world-scale). All or most are still first-generation facilities that will bring in a lot of information on both the various technical steps as well as on physics issues. The results obtained at these facilities will guide us to set up a large-scale next-generation RIB facility, producing high-intensity and high-quality selective beams in a large energy interval.

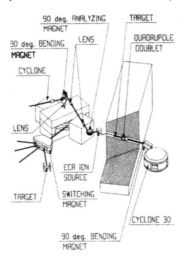

Figure 15.22. Schematic drawing of the two-cyclotron set-up, used and operational at the radioactive ion beam facility at Louvain-la-Neuve (taken from Darquennes *et al* © 1990 by the American Physical Society, with permission).

We discuss in some more detail the ISOL facility at Louvain-la-Neuve. This facility is based on a two-cyclotron concept: the CYCLONE30 is the driver accelerator, which is coupled to the $K = 110$ CYCLONE cyclotron[1]. The outline of this set-up is given in figure 15.22.

The first cyclotron produces a large number of radioactive nuclei that are extracted subsequently from the first target, transformed in atomic or molecular form (depending on the Z value of the produced element) before being ionized in an ECR ion source. These secondary ions are transported and injected into the second cyclotron which accelerates the radioactive ions up to the desired energy. One of the advantages of this two-cyclotron set-up is the possibility of generating a broad spectrum of radioactive ion beams over an energy interval between 0.2 and 2 MeV/A. The first radioactive beam consisted of ^{13}N ($T_{1/2} \simeq 10$ minutes) in a 1^+ charge state, produced with an energy of 0.65 MeV/A and an intensity of 1.5×10^8 ions per second. This was made feasible by the use of the driver 30 MeV, 100 μA proton beam (CYCLONE30). This beam was used to create the necessary ^{13}N 1^+ isotopes of atoms from a target, positioned in a concrete wall that was used to separate the vaults of the two cyclotrons. The reaction used was ^{13}C(p, n) ^{13}N. The ^{13}N could be extracted from the target as a N$_2$ molecule (^{13}N ^{14}N). These molecules were transferred to the ECR ion source specially designed to have optimal production of nitrogen in the 1^+ charge state.

---

[1] The K value of a cyclotron is connected to the maximum energy per nucleon delivered W_{max} through the relation $W_{max} = K(Q/A)^2$ with Q the charge of the ion and A the atomic mass number of the ion. For a proton, one obtains $W_{max} = K$.

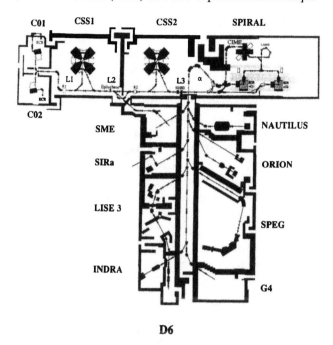

Figure 15.23. Layout of the future installations SPIRAL at GANIL. The primary beam coming out of the second cyclotron CSS2 can be directed either to the various experimental halls or to the SPIRAL part. After production, ionization and magnetic separation, the secondary beam is injected into the new CIME cyclotron. The accelerated RIBs are then sent into the experimental halls (reprinted from Villari *et al* © 1995, with permission from Elsevier Science).

Extraction from the ion source, followed by a mass analysis were performed before transporting the ions for injection into the second cyclotron, the standard $K = 110$ cyclotron of Louvain-la-Neuve. This latter cyclotron has, however, been modified in order to allow acceleration of the RIB in the energy region near 0.65 MeV/A. The production of light radioactive ions in this particular energy region is very well adapted to study reactions with astrophysics interests. Thus, the first beam was a ^{13}N 1^+ beam used to study the ^{13}N(p, γ) ^{14}O reaction, crucial for a detailed and precise understanding of the CNO stellar cycle. This point will be discussed in some more detail in section 15.3.3. A number of dedicated RIB beams have been developed during recent years ranging from ^{6}He (half-life of 0.8 s) up to ^{35}Ar (half-life of 1.7 s) with energies ranging in the interval 0.6 to 4.9 MeV/A. Plans to have a different cyclotron set up with the construction of a new cyclotron CYCLONE44 will be discussed in Box 15b.

We shall next go into some more detail in a presentation of the GANIL accelerator facility at Caen in France. This facility has already had a long-standing tradition in the field of accelerating a wide spectrum of heavy ions. Over the last 20 years much has been learned in the field of projectile fragmentation as a means to produce accelerated radioactive beams for the study of exotic nuclei through a multitude of nuclear reactions. At present, GANIL produces high-intensity beams of ions ranging from ^{12}C up to ^{238}U at energies varying between 24–96 MeV/A. The facility of three cyclotrons to reach this broad spectrum (C01, CSS1 and CSS2) as well as the very versatile and large number of experimental set-ups is presented in figure 15.23. Considerable efforts have been made to improve the present beam intensities with the aim of producing radioactive ion beams in an ISOL environment (SPIRAL to be discussed more in Box 15b). Thus, at GANIL the combination of producing, at the same place, a number of ISOL radioactive beams through the SPIRAL facility as well as the possibility of obtaining in-flight projectile fragmentation and the subsequent mass-separated beams, exists.

We would just like to mention that IFS is not limited to projectile fragmentation only: the Notre Dame facility is using light-ion transfer reactions in order to produce the secondary ion beams and makes use of two superconducting solenoids in order to separate the ions. The tandem accelerator has been used to carry out, very early on, some pioneering studies using very specific ions. At ANL (Argonne National Laboratory), there is an in-flight separator which can separate beams from heavy-ion fusion-evaporation reactions. An extensive overview of in-flight separation methods is discussed by Geissl *et al* (1995).

15.3.3 Nuclear astrophysics applications

(a) Introduction

Recently, a number of interconnections between the fields of nuclear physics research and developments in astrophysics have brought new areas of research into being. One of those fields, nuclear astrophysics, has benefited explosively from the technical possibilities of developing radioactive ion beams and accelerating (or decelerating) them into the appropriate energy region for the study of reactions that were and still are essential in the synthesis of elements (nucleosynthesis). Therefore, in this chapter on the physics of nuclei far from stability, weakly bound quantum systems and exotic nuclei, we devote a section to an introduction to this very rapidly developing field and just bring in some of the flavour that could ignite a further study of the field.

For extra reading and study into nucleosynthesis we refer to the work of Clayton (1968) and more recent books (an introductory text by Phillips (1994) and a book largely concentrating on experimental methods by Rolfs and Rodney (1988)). In a monumental article on nucleosynthesis (Burbridge *et al* 1957) and a more recent one by Fowler (1984), one can find good study work to get to know the physics of element synthesis. A more extensive list of references can be found in those books and in Heyde (1996).

(b) Nuclear astrophysics: a glimpse inside stars

When studying the constitution of some of the oldest stars, we learn that only the lightest nuclei like deuterium, ^{3}He, ^{4}He and ^{7}Li were produced in appreciable amounts out of the early substance of the universe which consisted of quarks, leptons and the fundamental force carriers. The remainder of the elements have been synthesized in stars which behave like fusion reactors, fusing hydrogen into helium. Many of these reactions are rather well known but there remain a number of key processes that are not so well understood, in particular the reaction rates and temperature (energy) conditions as well as the corresponding probabilities to form heavier elements (cross-sections). Some of the most uncertain of those reactions are, for example, the reaction to fuse hydrogen with ^{7}Be to form ^{8}B (the ^{7}Be(p, γ) ^{8}B reaction) and the fusion of helium with carbon to form oxygen (the ^{12}C(α, γ) ^{16}O reaction). They determine in a decisive way the relative amount of carbon to oxygen in massive stars, which, in retrospect, has important consequences on heavy-nuclei formation in explosive phases (supernovae processes). Also, the ^{13}N(p, γ) ^{14}O reaction, which is a key reaction in the CNO catalytic cycle in which four protons are transformed into a ^{4}He nucleus, is important. This reaction has recently been studied using RIB techniques and will be discussed in a later paragraph.

The various processes that form the heavier elements, the s-process (slow neutron capture reactions), the r-process (rapid neutron capture reactions) and other processes that are not as dominant but specific to the formation of certain elements (the p-process and rp-process: proton capture and rapid proton capture reactions) determine paths in the nuclear mass (N, Z) plane (see the former astrophysics references for detailed discussions). In figure 15.24, the most important s and r neutron capture paths are indicated. Current studies, using increasingly neutron-rich stable nuclei, do not allow us to explore this region. Making use of the increased technical possibilities of forming radioactive ion beams and inducing reactions with neutron-rich targets, the hope is to enter this field of very neutron-rich systems and come to unravel bit by bit the territory of neutron-rich nuclei proceeding towards the drip line.

(c) Stellar reactions in the laboratory

Experiments at Louvain-la-Neuve, GANIL, ANL and RIKEN have used selective radioactive ion beams to study a number of reactions that form critical steps in light-element synthesis.

The problem is producing the appropriate beams of high-enough intensity and at the correct energy. Two reactions: ^{13}N(p, γ) ^{14}O and ^{11}C(p, γ) ^{12}N have been studied at GANIL in an inverse process. The nuclei ^{14}O and ^{12}N have been produced by projectile fragmentation and have been sent through the strong Coulomb field of a heavy nucleus (^{208}Pb) (see figure 15.25). The subsequent exchange of quanta in the electromagnetic field makes the incoming nucleus

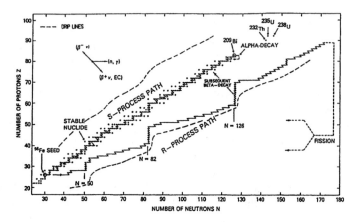

Figure 15.24. Neutron-capture paths for the s-process and the r-process projected onto the (N, Z) plane (taken from NUPECC report, May 1993, with kind permission).

dissociate (Coulomb dissociation). This experiment was also carried out in an inverse direction by Motobayashi *et al* (1991) at RIKEN. The direct reaction, using a RIB of ^{13}N 1^+ radioactive ions (half-life of 600 s) accelerated to an energy of 0.63 MeV/A impinging on a proton target, was performed at the Louvain-la-Neuve two-cyclotron facility (see also section 15.3.2). The resonant capture gamma-rays could subsequently be detected using Ge detectors (results for these reactions are illustrated in figure 15.26). One knows at present that various cycles are responsible for the element formation up to ^{56}Fe. The precise way in which these cycles are interconnected (escape from cycles) is still not very well understood. A key nucleus that allows escape out of the CNO cycle is ^{19}Ne. The reaction ^{19}Ne(p, γ) ^{20}Na(β^+) ^{20}Ne allows escape out of the CNO cycle. There is, however, a competition with the β^+ decay of ^{19}Ne leading to ^{19}F and keeping within the CNO cycle. Measurements on the proton capture have been studied in detail at GANIL. This same proton capture reaction, and ^{18}F(p, α) and ^{8}Li(α, n) reactions have been studied at Louvain-la-Neuve, ANL and RIKEN.

   As the intensity of the currently available RIB is still rather low and nuclear reactions proceed at very small cross-sections, a big effort needs to be invested in order to detect the very minute signals in a large background regime. Therefore, new detection systems are being developed in order to allow the detection of gamma-rays (using Ge mini-ball set-ups) and particle detector systems. Much work is in progress in this direction.

15.4 Outlook

The study of quantum systems at the limit of stability is an intriguing and most interesting domain of physics. This holds for the study of atomic nuclei too

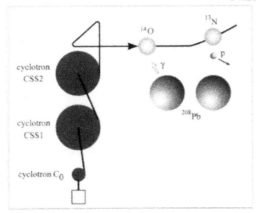

Figure 15.25. Dissociation of the exotic nucleus ^{14}O in the inverse ^{13}N(p, γ) ^{14}O reaction. The radioactive nuclei ^{14}O are produced in the two-cyclotron set up at GANIL and are dissociated in the strong Coulomb field of a heavy nucleus (^{208}Pb) (adapted from Saint-Laurent 1997 © GANIL/IN2P3/CNRS-DSM/CEA).

when progressing and exploring physics towards the drip lines. The difficulty in obtaining a good description of those more exotic regions of the nuclear mass table using extrapolations from the known region of β-stable nuclei will fail. It is not very clear as yet how the nuclear many-body system will rearrange itself if a very large neutron or proton excess develops. In the region where experimental access has now opened up possibilities to study drip-line physics, i.e. the very light nuclei, totally unexpected results such as the appearance of halo systems in the neutron distribution have been obtained. Theoretical efforts will most probably have to start from 'scratch' or, stated slightly differently, using general theoretical methods to study the binding conditions for nuclear matter where the neutron-to-proton ratio is extreme, will have to be used. The studies that have been made so far on the pairing properties of neutron matter may well be helpful to understand finite nuclei too. Much work and effort will be needed in the coming years.

On the experimental side, it has been illustrated that the great effort put into developing radioactive beams of various ions and accelerating them into a diversified energy interval has passed an initial exploratory phase. Various experimental efforts and projects to gain a deeper understanding of the various technical aspects related to both the ISOL and IFS methods are needed in order to construct a firm basis for the development of a full-fledged next-generation RIB facility. At the same time one needs to develop 'dedicated' detection systems like the Ge-miniball at REX-ISOLDE, EXOGAM at GANIL, ..., study secondary-beam detection systems (primary beam suppression), engineer new spectrometers, ... Such a facility will certainly open new avenues in nuclear physics research as another illustration of the theme that, whenever new technical possibilities

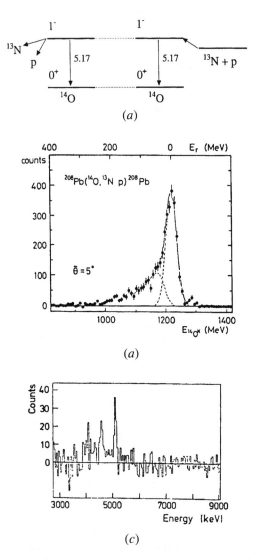

Figure 15.26. (*a*) A schematic illustration of the proton capture on ^{13}N (right-hand side) and of the Coulomb dissociation of ^{14}O radioactive beams on a ^{208}Pb heavy nucleus. In parts (*b*) and (*c*) the essential data resulting from the proton capture reaction, carried out at Louvain-la-Neuve ((*c*) Decrock *et al* © 1991, by the American Physical Society), as well as from the Coulomb dissociation reaction ((*b*) reprinted from Motobayashi *et al* © 1991, with permission from Elsevier Science) are given (adapted from Nazarewicz *et al* © 1996, with permission from Gordon and Breach).

to study the atomic nucleus have become available, the nucleus has always responded with unexpected physics answers. The nuclear physics community is ready to make this next step successful.

Box 15a. The heaviest $N = Z$ nucleus ^{100}Sn and its discovery

---

The study and production of doubly-magic nuclei has formed a very specific goal in experimental nuclear physics in particular. Besides the well-known lighter examples of ^{4}He, ^{16}O, ^{40}Ca, ^{56}Ni the search for the doubly-magic nucleus ^{100}Sn has been particularly difficult because this nucleus is situated quite far from the region of β-stable Sn nuclei. Also, from a theoretical point of view, a number of interesting questions are connected with the study of this doubly-closed ^{100}Sn nucleus, just near to the proton drip line. Questions about the stability (proton binding in a potential well is helped because of the Coulomb barrier) of this nucleus, the isospin or charge-independent characteristics of the nuclear ground state as well as the study of Gamow–Teller β-decay properties are all highly interesting.

Nuclei in this $A = 100$ region were mainly produced in heavy-ion fusion-evaporation reactions and investigated using on-line mass separation and in-beam techniques. The search for ^{100}Sn has been going on for almost two decades at the GSI and GANIL laboratories without success (Ryckaczewski 1993, Lewitowicz *et al* 1993). It was only with the advent of in-flight projectile fragmentation and subsequent separation methods with high resolving power, first into given Q/A selection combined with energy-loss (degrading) systems, in order to select the specific charge isotopic value (Z value) that ^{100}Sn was discovered.

The first unambiguous production and identification was almost simultaneously made by groups at GSI and GANIL during experiments carried out in the period March-April 1994. The group at GSI, a TU München-GSI collaboration, was using a ^{124}Xe beam accelerated at the heavy-ion synchrotron SIS up to an energy of 1095 MeV/A. Subsequent collisions on a Be target fragment, the projectile, in-flight, and the projectile-fragment separator FRS then allowed for the identification of the nucleus ^{100}Sn. The GSI team needed 277 hours, about 1.7×10^{13} ^{124}Xe ions to produce just about seven ^{100}Sn nuclei (see figure 15a.1). The experiment was carried out between 10 March and 11 April 1994. The technical details of the projectile fragment separation and identification are discussed in the paper identifying the first production of the doubly-magic nucleus (Schneider *et al* 1994) and, in a more popular way, by Friese and Sümmerer (1994). The paper announcing the results was received by Zeitschrift Für Physik on 27 April 1994.

At GANIL, in November 1993, experiments were carried out using a ^{112}Sn beam with an energy of 58 MeV/A leading to the identification of elements down to ^{101}Sn. In experiments, carried out a few months later, in April 1994, but now at an energy of 63 MeV/A, production rates could be increased by an order of magnitude using a thicker target of natural Ni placed inside two high-acceptance superconducting solenoids (SISSI). The experiments were able to identify eleven events of ^{100}Sn. These events were observed, within 44 hours, in a charge state $Q = +48$ and these most dramatic results are shown in figure 15a.2. The paper,

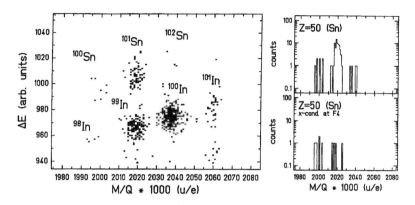

Figure 15a.1. Left-side: the energy deposition versus nuclear-mass-to-charge ratio obtained at the focal plane of the fragment mass separator (FRS) at GSI, Darmstadt. Right-side: precise distribution of the Sn isotopes (upper part). The lower part is obtained from the upper part by putting an independent selection at the focal position of the FRS on the horizontal position within a window of +2.5 cm. It is seen that this constraint removes most of the counts with mass $A = 100$. More details can be found in Schneider *et al* (1994). (Adapted and reprinted from Schneider *et al* © 1994, with permission from Gordon and Breach.)

announcing these results, was received by *Physics Letters* B on 7 June 1994. These conclusive experiments at GANIL were run almost exactly on the date the GSI group announced its experimental finding on the detection of ^{100}Sn nuclei, thus ending a most competitive search for this heaviest doubly magic nucleus. More technical details can be found in the paper discussing this observation (Lewitowicz *et al* 1994) and in a short announcement article by Ryckazewski (1994).

The results obtained at both GSI and GANIL confirm that medium-energy projectile fragmentation reactions combined with projectile-fragment separator techniques offer the most efficient method to produce and study the nuclei up to mass number $A = 100$ at the proton drip-line region. A number of papers discussing both the observation of ^{100}Sn and some nearby nuclei are Lewitowicz *et al* (1995a, b), Schneider *et al* (1995) and Ryckazewski *et al* (1995).

This first unambiguous signal was the start of an intensive search for extra information about the ^{100}Sn nucleus including the determination of its mass and life-time. A standard method uses time-of-flight (TOF) measurements for the fragments over a given linear flight path of up to 100 m, using high-precision magnetic spectrometers (SPEG at GANIL, TOFI at Los Alamos). The resolving power obtained in such methods is clearly not high enough to carry out a high-precision mass determination for ^{100}Sn. Instead a much increased path length was used and the ions were put in a spiral trajectory using a cyclotron (in the

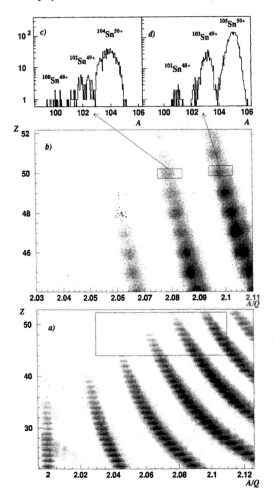

Figure 15a.2. Identification of the fragments produced in the ^{112}Sn fragmentation reaction at GANIL. In part (a), the identification of the atomic number Z versus the mass-to-charge ratio A/Q is given. In part (b) (blow-up of the excerpt shown (a)), two groups of Sn isotopes are indicated (the small boxes) for which the more detailed mass distributions have been determined and are shown in parts (c) and (d). More details can be found in Lewitowicz *et al* (1994). (Reprinted from Lewitowicz *et al* © 1994, with permission from Elsevier Science.)

case of GANIL, using the second cyclotron CSS2), and thereby a method was developed transforming such a cyclotron into a very high-precision spectrometer. The mass resolution at the CSS2 in GANIL was shown to go down to 10^{-6} for light ions (Auger *et al* 1994). Using the fusion-evaporation reaction ^{50}Cr + ^{58}Ni at 255 MeV, and thus optimizing production of elements near mass $A = 100$, the various elements formed are injected into the second cyclotron and are accelerated at the same time. Starting from the known mass of ^{100}Ag, Chartier *et al* (1996) were able to determine for the first time the mass of ^{100}Sn with a precision in $\Delta m/m$ of 10^{-5} resulting in the value of the mass (mass excess to be more precise) of -57.770 ± 0.300 (syst.) ± 0.900 (stat.) MeV.

For the seven ^{100}Sn events in the GSI experiment that were implanted into the final detector system, a first analysis of subsequent decay events was performed and a value for the half-life deduced as $T_{1/2}(^{100}\text{Sn}) = 0.66^{+0.59}_{-0.22}$ s (Schneider *et al* 1995).

Box 15b. Radioactive ion beam (RIB) facilities and projects

At present, seen on the world scale (see figure 15b.1), a large number of the facilities where tandem accelerators, cyclotrons, synchrotrons, etc, are operational are modifying and/or upgrading their present capabilities so as to be able to produce radioactive beams (RB, RIB or RNB are commonly found acronyms standing for radioactive beams, radioactive ion beams or radioactive nuclear ion beams but they all mean the same). A number of new projects are still under study in the scientific community, some projects are through the scientific study phase but are awaiting possible agreement on funding and a number are in the construction phase. These facilities are still very much complementary in order to learn how to control the various 'parameters' that enter the many phases when producing the accelerated radioactive beams (production, ionization, fragmentation, separation, post-acceleration, etc). They all aim at bringing together the necessary technical expertise to set up a full second-generation large-scale versatile RIB laboratory or facility.

In the light of the specific differences between isotope separation on-line (ISOL) and in-flight separation (IFS) methods, we shall discuss the various present-day operational and planned projects in these two classes, as much as possible, although some of the future plans aim at utilizing both ISOL and IFS

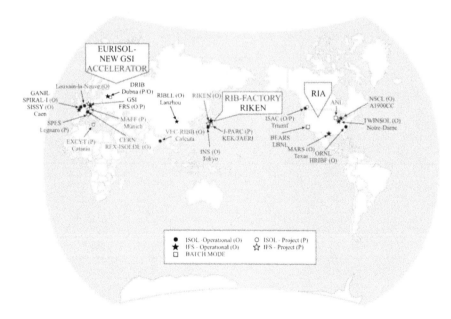

Figure 15b.1. World map indicating the many worldwide efforts in radioactive ion beams (RIB). (Updated Heyde K 2004.)

production mechanisms. In each division, we discuss both European, North American and Japanese as well as other efforts.

In the present discussion, the 'closing' date has been taken as january 2004. We further use the notation (O) for an operational facility, (P) for a project on a middle-term scale. Moreover, at present, there are upgrades and new developments at a number of existing facilities which we denote by (O/P). At the end, we discuss, succinctly, a number of long-term projects that aim at setting up second-generation RIB facilities, both in Europe, North America and Japan. For most of these facilities and projects, the home page is given since there are, at present, such rapid developments that the reader is advised to look at these home pages and, thus, be able to follow the progress made in this field. Moreover, a number of reports have been written or are in progress and are given in the present reference list at the end.

ISOL facilities and projects

There exist, at present, a large number of first-generation ISOL-type facilities. This then forms a good basis for studying the construction of the next-generation facilities. One can make a distinction in the type of ISOL facilities according to the type of driver and post-accelerator devices that are used. In general, the energy is rather low and can only go up to ≈25 MeV.A.

Combination of driver cyclotron and post-acceleration cyclotron

Louvain-la-Neuve (O)

In the text of this chapter (15), we have discussed this working ISOL facility in some detail, outlining both the technical aspects and the physics research program. The facility has recently been complemented by a new, post-acceleration cyclotron (CYCLONE44) which is used for accelerating radioactive ions in the energy range 0.2–0.8 MeV.A and is particularly suited for research into nuclear astrophysics.This then results in an increased intensity and mass resolving power for the accelerated beams. In addition to the CYCLONE30 proton accelerator, the option of using the present K110 cyclotron (now used as the post-accelerator) as the driver accelerator, is retained. See http://www.cyc.ucl.ac.be.

GANIL-SPIRAL at Caen (O)

SPIRAL at GANIL has been producing radioactive beams for experiments since September 2001 and complements the IFS facility (see next section). The primary accelerated light- or heavy-ion beams, bombard a production target (at high temperatures of 2300 K), then pass into an ECRIS source. After extraction, the low-energy RIB will be selected by a mass separator with a relatively low analyzing power ($\Delta m/m \cong 4 \times 10^{-3}$) before injection into the compact CIME cyclotron and accelerated up to energies of 1.7A $\sim$ 25A MeV. A special magnetic selection is made before sending the RIB into the experimental halls. GANIL publishes an interesting journal *Nouvelles du GANIL* and more information can be found at the web pages http://www.ganil.fr/research/developments/spiral and also at http://www.ganil.fr/spiral/overview.html.

DRIB project FLEROV Laboratory FLNR at Dubna (P)

This project, already outlined in the former edition of this book has remained essentially the same. It aims at using the U400M cyclotron accelerator to produce radioactive nuclei and the U400 cyclotron in order to accelerate the low-energy beam(s) up to an energy of 5–20 MeV.A. There are also plans to have RIB produced from the photofission of uranium. Here, a 25 MeV microtron MT-25 would be used to produce the fission products and the U400 cyclotron would be used as the post-accelerator. More details can be seen at the web page: http://159.93.28.88/flnr/index.html.

Combination of a driver proton synchrotron and a post-acceleration linac

REX-ISOLDE at CERN (O)

This project was agreed by the CERN Research Board in the Spring of 1995 and has been set up as a pilot project for post-accelerating the already existing radioactive ions produced at the PS Booster by ISOLDE. The singly-ionized ions, coming from ISOLDE, are stopped in a buffer gas-filled Penning trap and are cooled and ejected as ion bunches in the EBIS source which acts as a charge breeder before post-acceleration up to 3.1 MeV.A using a linear accelerator. The facility has been operational since the end of 2002/early 2003. See the web page at http://isolde.web.cern.ch/isolde.

ISAC-TRIUMF at Vancouver (O)

After the agreement to build the Isotope Separator and Acceleration (ISAC) facility in 1995, the first radioactive beams were produced in the fall of 1998. The facility has become operational and is a high-intensity ISOL facility. It uses one of the 500 MeV proton synchrotron beam lines and uses up to 100 μA of a continuous-wave primary proton beam current. Whereas the Triumf Isotope Separator on Line (TISOL) continues to operate, it is within the ISAC facility and the upgrade project of ISAC-II that the production of accelerated RIB's is directed towards. The post-acceleration is performed using a linear accelerator and this can bring the RIB energy up to 1.5 MeV.A. A recoil-mass separator (DRAGON) has been installed since. The recently aproved ugprade for ISAC-II will largely extend the mass range of the ions used (up to $A = 150$) and the energy will move up to almost 6.5 MeV.A. For recent information on both ISAC and the recent

upgrade ISAC-II, the reader can consult the web pages http://www.triumf.ca/ and
http://www.triumf.ca/people/baartman/ISAC.

Combination of a driver linac and a post-acceleration linac

SPES project at Legnaro (P)

The SPES (Study and Production of Exotic Species) project is a middle-term
project and consists of a two-accelerator ISOL-type of facility in order to produce
intense beams of neutron-rich nuclei in the mass range of $A = 80$–160. The
conceptual design, based on a high-intensity 50 MeV proton linac as the driver
(this could be later changed into a deuteron primary beam), will produce an
intense beam of neutrons in a Be converter and induce fission. The neutron-
rich fragments will be extracted from the target and ionized. The existing heavy-
ion linear accelerator (ALPI) will then act as the post-accelerator and produce
energies up to 7 MeV.A and deliver the beams to the experimental area. More
details can be found at http://www.lnl.infn.it/~spes/.

Combination of a driver cyclotron and a post-acceleration linac

KEK-Tanashi RNB facility (O)

The facility consists of a target and ion-source system where radioactive nuclei
are produced using the light-ion beam from the SF cyclotron. The construction of
this ISOL-based facility at INS started in 1992 and it was completed in 1996. The
mass-analyzed beam is transported to a heavy-ion linac complex and accelerates
the radioactive ions up to 1 MeV.A. This facility acts as the prototype for an exotic
nuclei project that is planned as a second-generation RIB facility (see later). See
also http://accelconf.web.cern.ch/AccelConf/e98/PAPERS/MOP07B.pdf.

VEC-RIB at Calcutta (P)

At the Variable Energy Cyclotron Centre, plans have been made to start
accelerating protons and α particles from the K130 cyclotron, in order to produce
radioactive elements in a thick target. Using an ECR source and charge breeding,
the aim is to use an RFQ and also a post-accelerating linac in order to be able
to produce beams with an energy up to 5 MeV.A. The status can be seen at the
webpage http://www.veccal.ernet.in/~vecpage/rib.htm.

Combination of a driver cyclotron and a post-acceleration tandem

HRIBF at Oak-Ridge (O)

The Holifield Radioactive Ion Beam Facility (HRIBF) at Oak-Ridge uses the
classical ISOL approach starting with ORIC, the cyclotron, as the driver and
the 25 MeV Tandem as the post-accelerator. After the driver, there is an ISOL

target and ion-source installation. Note here the specific need for negative ions to be accelerated in the tandem afterwards. In 1996, a first successful accelerated beam of ^{70}As (produced in a (p, n) reaction using 42 MeV protons from the cyclotron) was produced with an energy of 140 MeV. A proposal for a second-generation ISOL facility that may use a National Spallation Neutron Source as a driver, named NISOL, has been proposed. See also the webpage http://www.phy.ornl.gov/hribf/hribf.html.

EXCYT at Catania (O/P)

The name stands for 'Exotics at the Cyclotron Tandem Facility'. At present, the facility—using the K800 superconducting cyclotron—delivers intermediate-energy heavy-ion beams that can be used with the fragment mass separator ETNA. The new project (using an ISOL system),will use the tandem as a post-accelerator up to energies of 8 MeV.A. The project is now entering its final stages and the first RIB transport test and RIB experiment is expected before the end of 2004. For up to date information about the project, consult the web page at http://www.lns.infn.it/excyt/index.html.

Combination of a driver reactor and a post-acceleration linac

MAFF at München (P)

There exist, at present, facilities where low-energy fission fragment beams are operational such as the OSIRIS facility in Studsvik (see http://www.studsvik.uu.se/ Facilities/Osiris/osiris.htm) but post-acceleration of fission products was first planned at the high-flux reactor at the Institut Laue-Langevin (ILL) within the project PIAFE. This concept, using a combination of a reactor as the driver and a linac as post-accelerator, will use the MAFF fragment accelerator. More information can be found at the web page http://www.ha.physik.uni-muenchen.de/maff.

'Batch-mode' facilities

ATLAS at ANL (O)

At the Argonne Tandem/Linac Accelerator System (ATLAS),vigorous plans exist for setting up schemes based on the present ATLAS complex as the post-accelerator. At present, heavy-ion beams ranging over all possible elements, from hydrogen to uranium, can be accelerated to energies as high as 17 MeV.A and be delivered to three target stations. Since 1994 the facility has been used to study some very specific RIB experiments in a 'batch' mode. An in-flight production system has also been developed at ATLAS to supplement the 'batch' mode and consists of gas cells in which inverse reactions are induced by the primary heavy-ion beams. The reaction products are then collected and focused with a superconducting solenoid. For more detailed information, see the webpage http://www.phy.anl.gov/atlas/fac-acc.html and also http://www.phy.anl,gov/atlas/index.html

BEARS at LBNL (O/P)

A project called Berkeley Experiments with Accelerated Radioactive Species (BEARS) has been implemented at the 88″ cyclotron. Here, specific radioactive elements are made at the 11 MeV medical cyclotron and transported using a gas carrier into the ECR source for subsequent acceleration using the 88″ cyclotron. Energies up to 5–10 MeV per nucleon could be reached. Some first experiments have shown the feasibility of this particular mode of operation. See also http://www-nsd.lbl.gov/nsd/programs/programs.html.

IFS facilities and projects

GANIL at Caen (O)

Here, two K380 cyclotrons provide intermediate-energy heavy ions up to an energy of 95 MeV.A. The subsequent fragment separation, in flight, is made in a high-power thin target, preceded and followed by magnetic lenses (the SISSI device) in order to increase the intensity of the secondary beams. The fragments are collected and identified by the LISE device. See the web page http://www.ganil.fr/operation/sissi/sissi.html and also http://www.ganil.fr/lise.

GSI at Darmstadt (O)

At the GSI (Darmstadt), the whole mass range of heavy ions, up to uranium, is available, using the UNILAC and SIS synchroton combination with energies up to 1 GeV.A. These ions are then fed into the high-transmission fragment mass separator (FRS). The storing and cooling of secondary beams in the ESR (Experimental Storage Ring) provides a number of unique possibilities. The upgrade of the facility will be discussed later. See also the web page http://www-w2k.gsi.de/frs/index.asp and http://www-aix.gsi.de/~msep/isol.html.

DRIB project FLEROV Laboratory FLNR at Dubna (O)

The Flerov Laboratory operates two cyclotrons, i.e. the U400 and U400M. The delivered beams can be used for fragmentation and the resulting nuclei are studied in the two separators ACCULINNA and COMBAS. More details can be seen on the web page http://159.93.28.88/flnr/index.html.

NSCL at Michigan State University (O)

Michigan State University has been using radioactive beams since 1990. Secondary beams have been produced from reactions using the 30–200 MeV.A primary beams produced by the K1200 cyclotron. These secondary beams have been collected and separated in-flight by the A1200 mass separator. In 1996, the NSF agreed upon an important upgrade to have the refurbished K500 cyclotron coupled with the K1200 cyclotron. The first one accelerates low-charge but intensive beams that are stripped in the K1200 cyclotron and accelerated up to 200 MeV.A. This upgrade has been carried out and has been accompanied by a replacement of the existing A1200 beam analyzing system with the more performant A1900 fragment mass separator. In this transformation, a very large gain of two to three orders of magnitude (depending on the precise needs) has been achieved. This program then allows the proton drip-line for nuclei as heavy as $A = 100$ to be approached and for the neutron drip-line for nuclei up to the Ca nuclei to be covered. At this facility, essentially all elements can be accelerated up to high energies. More detailed information can be obtained from the webpage http://www.nscl.msu.edu/tech/devices.

TWINSOL at Notre-Dame (O)

At the FN Notre-Dame tandem Van de Graaff accelerator (10–11 MeV), secondary beams have been produced since 1987 using in-flight light-ion transfer reactions. A 3.5 T superconducting solenoid is used to separate the ions. At present, a pair of such solenoids, called TWINSOL, operating at 6 T, are used to collect and focus the low-energy RIB produced in the direct reaction, using the primary beam of the tandem (see also at http://www.physics.lsa.umich.edu/twinsol and http://www.nd.edu/~nsl/Research_Facilities/RNB_Facility/rnb.html).

MARS at Texas A&M (O)

Groups at the Texas A&M Cyclotron Institute have been using primary heavy-ion beams from the K500 superconducting cyclotron to perform reactions in a gas target in the beam line.The experiments then use the recoil mass separator called MARS to achieve very pure beams at the focal plane where the secondary reaction targets are positioned (see also at http://cycnt.tamu.edu).

RIBLL at Lanzhou (O)

The facility in Lanzhou makes use of a K540 separated sector cyclotron in order to produce radioactive beams of light elements up to 80 MeV.A by in-flight (IF) fragmentation using the separator RIBLL. Plans for installing of a new storage-ring facility are present.

Second-generation projects

RIKEN in Japan (O/P)

At present, RIKEN has an operational IFS system (RIPS) at the K540 ring cyclotron. One of the most advanced second-generation proposals for a RIB factory has been put forward at the RIKEN Accelerator Research Facility (RARF) in order to provide radioactive beams over the full range of atomic masses. The new accelerator system will consist of three ring cyclotrons: a K510 (fixed frequency) (fRC), a K980 one as the intermediate stage (IRC)and a K2500 superconducting cyclotron (SRC). Commissioning is foreseen for 2006. This system will act as a post-accelerator and boost the energy of the heavy-ion beams of the operational K540 cyclotron up to 350 MeV.A and 400 MeV.A for light ions. Intense radioactive beams are then produced via the fragmentation of the heavy-ion projectiles or using in-flight fission of the uranium ions. Various experimental setups for use of these mass-separated beams will then be built, e.g. an accumulator cooler ring (ACR) and an electron-RIB collider and MUSES: a set of multi-use experimental storage rings. This Radioactive Ion Beam Facility (RIBF), the result of upgrading the present facility, is a very ambitious project. See http://ribfweb1.riken.go.jp.

JAERI/KEK Joint RNB facility in Japan (P)

At the JAERI/KEK (Japan Atomic Energy Research Institute/High Energy Accelerator Research Organization) Joint facility for high-intensity proton accelerators (called the J-PARC project), which is new and which should produce high-power proton beams at both 3 and 50 GeV, a KEK-JAERI Joint RNB Facility is planned. This facility is based on the ISOL and post-acceleration scheme in order to produce heavy, neutron-rich beams of up to 9 MeV.A. It will make use of the 3 GeV proton beam-line. At present, the ISOL and SCRFQ as well as the IH1 linacs have already been constructed and are working as discussed under the KEK-Tanashi facility.These will then be moved to the proposed facility, when completed, together with the SC linac presently working at the Tandem facility of JAERI. The project would then allow the production of both low-energy, medium-energy and higher-energy experiments in various experimental halls. Information on this ISOL-type of large-scale facility can be obtained from the web page http://jkj.tokai.jaeri.go.jp.

GANIL—SPIRAL-II at Caen (P)

Here, the aim is to set up a facility for the production and acceleration of neutron-rich fission fragments. Two possible production methods have been in competition. The first one is based on the use of a primary deuteron beam (80 MeV, 500 μA current) followed by a Be converter to produce the very high neutron flux that will induce fission in a thick uranium target. This possibility has been tested over the last years at Orsay with the PARRNE system and promising results have been obtained. Another possibility is the use of a linear electron accelerator (50 MeV, 500 μA current). In this case, one uses photon-induced fission on uranium starting from a bremsstrahlung beam generated by the stopped electrons. The main difference in the two methods results from the induced fission fragment mass distributions obtained. Finally, the option for a superconducting driver linac—LINAG1—has been made.

The post-acceleration will then proceed by the Ganil cyclotrons CSS1 and CSS2. Combining both these possibilities for the original SPIRAL facility with this SPIRAL-II project would lead to very extensive opportunities in the research programs. See the Spiral web page http://www.ganil.fr/research/developments/spiral.

GSI at Darmstadt (P*)

The GSI, at Darmdstadt, in February 2003, was given the green light by the German Government (which will provide 75% of the construction costs and with the other 25% being provided by European and other international collaborations) to start building an International Accelerator Facility for Research with Ions and Antiprotons. The choice has been made for the driver to be a new synchrotron (SIS 100/200). This new high-intensity accelerator will consist of fast-cycling superconducting magnets and will provide a beam of 10^{12} $^{238}U^{7+}$ per second at 1 GeV. The fragmentation of this beam will produce intense RIB beams over a broad mass spread. For a detailed outline of the accelerator complex and the possible experimentation possibilities, one can obtain the most recent information from the homepage of GSI http://www.gsi.de.

EURISOL (P)

Next to the IFS production facility that will be built at GSI, an extensive research program, supported by the European Commission, has been launched in order to carry out a preliminary design study for a second-generation ISOL-type facility. The best option for the driver seems to be to construct a 1 GeV proton accelerator, most probably a superconducting linac with high intensity. Decisions on the target–ion source assembly that will be used, the mass selection techniques to be implemented, the post-acceleration and the necessary instrumentation have yet to be made. At present, a detailed report is ready. The proposed design study and the proposal have been discussed at CERN on 16 February 2004 for submission within the Sixth framework of the European Commission. A detailed discussion of this project as well as the present version of the proposal can be consulted at the homepage http://www.ganil.fr/eurisol.

RIA in the USA (P)

To study the possibility for a second-generation high-intensity RIB facility, a task force was set up in 1998. As a result of extensive study, a proposal has been made for a highly flexible Rare-Isotope Accelerator (RIA) Facility. The RIA would offer a selectable driver beam, going all the way from protons through uranium, with the possibility of reaching 400 MeV.A uranium. It would then also offer varied target styles and methods encompassing both classical ISOL, two-step ISOL, IFS and an isobar-separated gas catcher/ion guide. The fundamental technology needed for RIA has been demonstrated though further R&D is needed. It is expected that the RIA will be able to deliver a broad variety of intense exotic beams. A number of detailed scenarios for implementing the RIA at either the ANL or the MSU-NSCL site have been carried out and these can be consulted on the web pages http://www.nscl.msu.edu/ria/index.php and http://www.phy.anl.gov/ria.

At present, at a large number of the laboratories strongly engaged in research with and the development of RIB projects as discussed in the previous outline, there is much interest in studying these radioactive ions using various traps. In recent years, activity in this field has grown rapidly, since it will not only allow these radioactive ions to be stored. It will also be possible to manipulate them in various ways, thus enabling the basic properties of nuclear physics and physics, in general to be studied, within the short life span of these unstable ions or atoms.

This presentation should have covered the existing facilities that produce accelerated radioactive beams. Standard isotope separators have not been discussed here. We have tried to be as complete as possible in covering both operational facilities and upgrades to existing and working facilities that have been agreed or are still in a proposal phase. We have also presented a number of new projects at existing accelerator facilities at the middle-term stage. Finally we presented major projects that aim to implement the present state of the art in second-generation facilities. A good number of these are still heavily discussed but it is clear that many efforts are progressing in both Europe, North America and Japan.

The discussion has always been presented in a condensed way. Therefore we give, for the reader in essentially all cases, the appropriate web pages in order to find out more about these facilities, both operational and planned. A number of references can be found in the NuPECC report (2000) and 'Radioactive nuclear beams' (2002) which can act as references for further learning.

Chapter 16

Deep inside the nucleus: subnuclear degrees of freedom and beyond

16.1 Introduction

In nuclear physics research a number of new directions has been developed during the last few years. An important domain is intermediate energy nuclear physics, where nucleons and eventually nucleon substructure is studied using high-energy electrons, as ideal probes, to interact in the nucleus via the electromagnetic interaction (Box 16a). It has become clear that the picture of the nucleus as a collection of interacting nucleons is, at best, incomplete. The presence of non-nuclear degrees of freedom, involving meson fields, has become quite apparent. Also, the question whether nucleons inside the nucleus behave in exactly the same way as in the 'free' nucleon states has been addressed, in particular by the European Muon Collaboration (EMC) and the New Muon Collaboration (NMC), with interesting results being obtained, in particular, at CERN. These various domains, once thought of as being outside the field of nuclear physics have now become an integral part of recent attempt to the nucleus and the interactions of its constituents in a broader context.

Since the last edition of this book, in 1998, important new results have been obtained, in particular making use of the continuous electron beam facilities. Two interesting text books have addressed this particular field of intermediate-energy physics which spans the region in between nuclear and nucleon structure (Povh *et al* 2002, Mosel 1999) and, in the present chapter, no attempt will be made towards completeness.

We shall discuss and illustrate the early studies and observations that lead to unambiguous evidence for the presence of mesons in atomic nuclei (section 16.2). In section 16.3, we concentrate on the dedicated project to probe inside nucleons at the 4–6 GeV energy scale at CEBAF (or JLab standing for Thomas Jefferson National Accelerator Facility—also called the Jefferson Lab for short)) and discuss some of the recent results as well as future plans. In section 16.4, we discuss a number of facets related to nucleon and hadron structure and highlight this with the recent discovery of pentaquark systems (Boxes 16b and 16c). In section 16.5, at much higher energies still, we concentrate on the possible formation of a new state of matter, i.e. the quark–gluon phase and its implications for understanding early phases of matter formation in the universe at large. Here, too, we shall discuss some of the very recent results obtained at the dedictated RHIC facility at Brookhaven and point out future plans at the LHC facility at CERN with the ALICE experiment.

16.2   Mesons in nuclei

The initial strong evidence for the presence of meson degrees of freedom inside light nuclei (deuterium) came from experimental studies at the Linear Accelerator of Saclay (ALS). In theoretical studies, these systems have the advantage of containing just a few nucleons. It is possible, starting from a two-body interaction, to calculate various observable quantities of these very light systems, e.g. charge and magnetism inside the nucleus. The lightest mirror nuclei ^{3}He–^{3}H show a number of interesting facets on the detailed nuclear interaction processes. Experiments on ^{3}H, a radioactive nucleus, have been quite difficult to carry out and it took the presence of the Saclay acclerator ALS and the building of a very special ^{3}H target with an activity of $\approx 10\,000$ Curie in order to carry out experiments in a successful way. Experiments concerning nuclear currents turned out to be radically different from a standard pure-nucleon picture and the measured magnetic form factor is shown in figure 16.1.

In general, it is assumed that the nucleon–nucleon interaction originates from meson exchange. For charged meson exchange, the appropriate exchange currents are created. It is this current, superimposed on the regular nucleon currents that have been observed in an unambiguous way via electron scattering. It has been proven that a class of exchange currents arise because of underlying symmetries, which, in retrospect influence processes like pion production, π meson scattering off nucleons and muon capture in nuclei. All these experiments have given firm evidence for the presence of meson degrees of freedom that describe a number of nuclear physics observables.

Probing the atomic nucleus with electrons and so using the electromagnetic interaction, is an ideal perturbation with which to study the nucleus and its internal structure (see chapters 1 and 10). With photons with an energy of 10–30 MeV, the nucleus reacts like a dipole and absorbs energy in a giant resonant state (chapter 13). The absorption cross-sections, using the dipole sum rule, clearly indicated the effect of meson exchange in an implicit way.

At higher energies, the nucleus behaves rather well like a system of nucleons moving independently from each other. Nucleon emission can originate in a 'quasi-free' way and indicates motion of nucleons in single-particle orbitals (chapter 10). An important result has been the study of the nuclear Coulomb

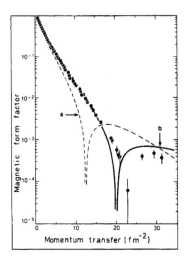

Figure 16.1. Magnetic form factor for ^{3}H as measured from electron scattering. The data points conform with (*a*) an interpretation where nuclear currents are carried by the three nucleons only and, (*b*) an interpretation where the meson degrees of freedom, responsible for the nuclear force, are also considered (taken from Gerard 1990).

response, i.e. the reaction of the nucleus to a perturbation of its charge distribution. In contrast to the expected result that this response would vary with the number of protons present in a given nucleus, the variation is much less, indicating about a 40% 'missing' charge in Fe and Ca nuclei. This feature is still not fully understood and is presented in figure 16.2 for ^{40}Ca.

A number of theoretical ideas have been put forward relating to a modification of the proton charge by dressing it with a meson cloud or modifying the confinement forces of quarks inside a nucleon. The results of a comparison of electron scattering cross-sections on ^{40}Ca with similar scattering on isolated protons indicates that even a very small modification of the proton characteristic is highly improbable and thereby invalidates many theoretical models.

Whenever the energy transferred by a photon to the nucleus becomes of the order of the π meson rest mass ($\simeq 140$ MeV/c^2), these mesons can materialize inside the nucleus. At 300 MeV incident energy, the nucleon itself becomes excited into a Δ state. This is a most important configuration which modifies the nuclear forces inside the nucleus. They might be at the origin of understanding 'three-body' forces. Using photo-absorption at the correct energy, the Δ resonance can be created inside the nucleus and its propagation studied. Experiments at the ALS, on light nuclei, have indicated that the Δ nucleon excitation gives rise to a sequence of π meson exchanges between nucleons thereby modifying the original n–n interaction.

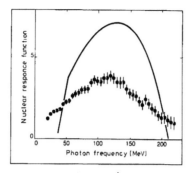

Figure 16.2. Nuclear response for ^{40}Ca after Coulomb exciting this nucleus. The amplitude is given as a function of the exciting photon frequency absorbed in this process. The full line corresponds to absorption by A independent nucleons described by a Fermi-gas model (taken from Gerard 1990).

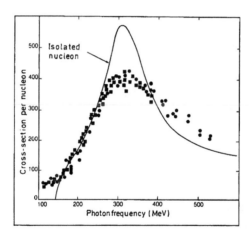

Figure 16.3. The cross-section per nucleon as a function of the frequency $\hbar\omega$ that excites the Δ resonance in a range of nuclei lying between ^{9}Be and ^{208}Pb. A comparison with exciting a single nucleon (full curve) is also given (taken from Gerard 1990).

The absorption cross-section remains the same, independent of the nuclear mass (see figure 16.3) and its magnitude increases according to the number of nucleons present, indicating an 'independent' nucleon picture, even though the spectrum deviates strongly from that of a free nucleon.

In all of the above, a pure nucleonic picture fails to describe correctly many of the observations. The question on how many nucleons the photon is actually absorbing is not an easy one to answer. Light nuclear photodisintegration has been studied in that respect. It has thereby been shown that ejection of a fast-moving nucleon from the nucleus results in a three-body decay. The resulting distributions, however, are comparable to the mechanism of photon absorption on a pair of correlated nucleons. This is a most important result.

Even though the basic nuclear structure remains that of protons and neutrons, the substructure of nucleons is a meson cloud, in which Δ resonances play an important rôle too and this importance allows a new and more fundamental picture of the nucleus to emerge.

In order to elucidate a number of questions relating to the precise role of mesons inside the atomic nucleus and even, at a deeper level, to probe the quark substructure, new electron accelerators had to be built. A project to build a multi-GeV continuous electron facility was set up and indeed built at Newport News with an initial starting energy of 4 GeV, which was reached in 1995. It was called CEBAF and renamed as the Thomas Jefferson National Acclerator Facility (TJNAF or JLab for short).

16.3   CEBAF: Probing quark effects inside the nucleus

Electron energies up to $\simeq 4$ GeV are needed to probe deep inside the nucleus with energies and momenta high enough to 'see' effects at the quark level. The machine requires a 100% duty cycle with a high current ($\simeq 200$ μA) in order to allow coincidence experiments that detect the scattered electron and one or more outgoing particles, so that the kinematics of the scattering processes may be charactarized in a unique way.

A consortium of South-eastern Universities (USA) in a Research Association (SURA) have created the possibility for a Continuous Electron Beam Accelerator Facility (CEBAF) at Newport News, Virginia.

JLab's physics interests concentrate on the investigation of the boundary between nuclear and particle physics, with the aim of understanding the forces between quarks and also exploring how hadrons are constructed out of interacting quarks and gluons. So, the study of the proton, the neutron, proton–neutron correlations (six-quark correlations), nucleon resonances (excited three-quark systems and potentially more complex quark systems) and hypernuclei form some of the major research topics. Besides this, interests in the study of transitions between the nucleon-meson and quark–gluon description of nuclear matter and of finite light nuclei form a 'red'-line in the research program at JLab.

Coincidence experiments detecting both the emerging, inelastically scattered electrons and the nucleon fragments (an emerging nucleon, an emerging nucleon pair, emerging nucleon(s) accompanied by meson(s), etc) have given access to new data that will allow the properties of quark clusters and their dynamics inside the nucleus to be resolved. In 2000, the Nuclear Science Advisory Committee (NSAC) endorsed the study for an upgrade of the facility running at 6 GeV (since 2000) towards reaching 12 GeV. This upgrade is now one of the 12 near-term

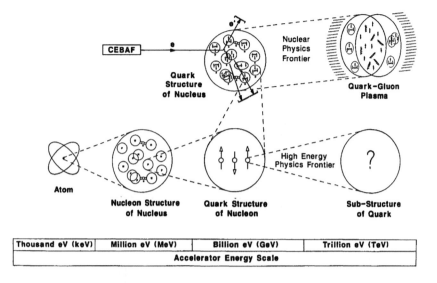

Thousand eV (keV)	Million eV (MeV)	Billion eV (GeV)	Trillion eV (TeV)
		Accelerator Energy Scale	

Figure 16.4. Diagrammatic illustration of the various energy scales in studying the atomic nucleus. The physics, that might be reached with the 4 GeV electron facility at CEBAF, is presented in the upper figures (taken from CEBAF report 1986).

top priorities in DOE's future plans. This then would allow the fundamental characteristics of the photon–quark coupling process to be studied. A diagram, presenting a number of energy scales in the study of the physics of the atomic nucleus and its nucleon substructure, is shown in figure 16.4.

After more than a decade of planning and preparation, CEBAF began operating in November 1995 with experiments using the 4 GeV continous-wave (cw), 100% duty-cycle electron accelerator facility. The facility has now been renamed the Thomas Jefferson National Accelerator Facility (TJNAF) or JLab for short. The accelerator has been designed to deliver independent cw beams to the three experimental areas, called Halls A, B and C. These extensive and complementary Halls allow researchers to probe both the internal nucleon and nuclear structure characteristics. An interesting, more technical but very readable description of the recent status of JLab is given by Cardman (1996) and readers are suggested to look at JLab's informative web page http://www.jlab.org, in order to get information on the highlights of current research.

Over the next decade, it is expected that the maximum energy of the accelerator will be upgraded to 8–10 GeV. With the present radii in the accelerator set-up the large recirculation arcs even permit upgrading to 12 GeV before synchrotron radiation becomes important.

The present CEBAF or TJNAF accelerator facility uses a recirculating concept that requires superconducting accelerator cavities and therefore a huge

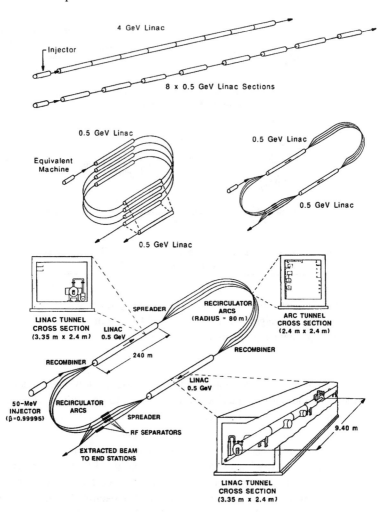

Figure 16.5. Comparison of the recirculating principle to obtain 4 GeV electrons with the linear concept of a 4 GeV accelerator. The 8 × 0.5 GeV sections can be reduced by an ingenious recirculating principle to only 2 × 0.5 GeV sections. The recirculating structure is illustrated in the lower part together with some linac tunnel cuts (taken from CEBAF report 1986).

liquid helium refrigeration plant had to be constructed. The accelerating and recirculating mechanism is illustrated in figure 16.5. This ingenious mechanism has been working in a 'mini' version at the DALINAC in Darmstadt since 1991.

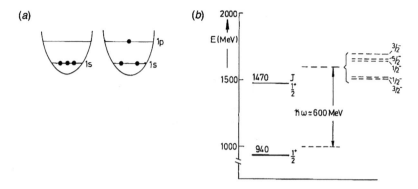

Figure 16.6. The excitation spectrum of the nucleon starting from a description of three quarks moving in oscillator potential with oscillator energy $\hbar\omega = 600$ MeV (*b*). In (*a*) the odd-parity states are shown at the right and result from promoting a quark by one oscillator quantum (from a 1s into a 1p state).

16.4   The structure of the nucleon

In the present section we will discuss some of the implications of our understanding of the nucleon and of its motion inside the nucleus. The 'bag' model allows us to describe quite a number of nuclear observables and is related to the question of how 'large' a nucleon is.

The nucleon appears to be made up of two regions: a central part, the part of freely moving quarks in the asymptotic free regime, and the outer region of the meson cloud where pions and other heavy mesons can exist. Recent theoretical studies build on the concept of 'chiral symmetry' and play an important role in understanding this division into two regions (Bhadhuri 1988).

A number of early results on the internal proton structure became accessible through highly inelastic electron scattering carried out at the Stanford Linear Accelerator Center (SLAC). A MIT-SLAC collaboration showed clearly a structure of point-like constituents but this was puzzling because such a structure seemed to contradict a number of well-established descriptions. Later work helped to identify these structures with quarks inside the proton (Kendall *et al* 1991) (see also chapter 1).

A good approach to study the nucleon and the internal structure stems from measuring the excited states of a nucleon made out of three quarks. Since quarks are confined, a single quark cannot move very far out and so, the excitation spectrum of the nucleon remains very simple as indicated in figure 16.6(*a*), which shows the lowest-lying excited states. Other excitations begin above $\simeq 1700$ MeV.

The negative parity excited states can be well described by considering quarks to move in a harmonic oscillator potential (figure 16.6(*b*)). There are five odd-parity excited states since the spins of the three quarks can be coupled to spin

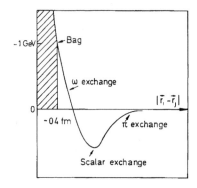

Figure 16.7. Nucleon–nucleon interaction versus the separation between the two interacting nucleons. The coupling (meson exchange) at the various separations is expressed by the type of meson exchange: π-exchange (OPEP), scalar exchange, ω-exchange eventually reaching the nucleon bag radius (taken from Brown and Rho 1983).

$\frac{1}{2}$ or $\frac{3}{2}$; the total is coupled to the orbital angular momentum of spin 1 yielding $J^{\pi} = \frac{1}{2}^{-}, \frac{3}{2}^{-}$ and $\frac{1}{2}^{-}, \frac{3}{2}^{-}, \frac{5}{2}^{-}$. These states would all be degenerate were it not for the presence of spin–spin residual interactions

$$\hat{H}_{\text{int}} = \sum_{i \neq j} g(|\vec{r}_i - \vec{r}_j|) \vec{\sigma}_i \cdot \vec{\sigma}_j. \tag{16.1}$$

This gives rise to a repulsive energy correction in states with all spins parallel and corresponds to the observed triplet and doublet states, as indicated in figure 16.6. The centroid energy is at $\sim$1600 MeV with the nucleon ground state at 940 MeV. Through the residual interactions as described by equation (16.1), the spin $S = \frac{1}{2}$ ground state will be pushed down from its unperturbed energy close to 1000 MeV such that we can approximate the oscillator energy quantum $\simeq$600 MeV from the energy difference between the lowest s state and the energy corresponding to the centroid of p states (see also figure 16.6). In the harmonic oscillator approximation we can equate

$$\frac{\hbar^2}{M_q r_q^2} = \hbar\omega, \tag{16.2}$$

with M_q the quark mass and r_q the radius parameter. The mass problem is a difficult one but, using arguments relating to magnetic moments, a mass of $M_q \simeq \frac{1}{3} M_N$ can be deduced. Such a model with massive quarks is called the 'constituent quark' model. With such a massive quark, a radius $r_q \simeq 0.45$ fm results with a mean-square value of $\langle r^2 \rangle_N = \frac{3}{2} a^2$. Correcting for spurious centre-of-mass motion, the value is reduced to $\langle r^2 \rangle_N = a^2$, or a root-mean-square radius of $\simeq 0.45$ fm.

The effect on the nucleon structure of gluon exchange between quarks has been discussed by Rujula *et al* (1975). A simple introduction to the internal structure of nucleons using the terminology of quarks, gluons and the like can be found in Bhadhuri (1988). In lowest order, the coupling is depicted by a Lagrangian very similar to the one describing photon–electron coupling with g the quark–gluon coupling strength. The potential becomes identical to that describing the electromagnetic interaction between two relativistic particles. The coupling strength, here, becomes a function of the momentum. The spin–spin interaction, in addition to other spin-orbit and tensor components, has the Fermi–Breit expression (Bhadhuri 1988). The nucleon is made up of three quarks in the s-state coupled to $S = \frac{1}{2}$, while the $S = \frac{3}{2}$ configuration corresponds to a three s-quark configuration at Δ (1230 MeV). To lowest order, the detailed structure of the Fermi–Breit interaction does not intervene and for the nucleon (or isobar) one has

$$\hat{H}_{\text{int}} = C \sum_{i \neq j} \vec{\sigma}_i \cdot \vec{\sigma}_j, \tag{16.3}$$

with C containing constants and radial strength integrals. A value of $C = 25$ MeV is deduced from the N–Δ energy separation if we contribute the full energy splitting to the effect of (16.3). These ideas can also be used to deduce a value of $\simeq 150$ MeV for the splitting between the triplet and doublet negative parity states as indicated in figure 16.6(*b*). This is in rather good agreement with the data. A puzzle, though, is the fact that no, or at most a weak, spin-orbit force is needed to describe the nucleon and its low-lying excited states. The spectroscopy of various other mesons and baryons can be understood rather well along similar lines and so gives a simple, unifying lowest-order picture. Some exciting new results are described in Box 16b

Using the simple quark model, a ratio of the proton to neutron magnetic dipole moment of $-\frac{3}{2}$ (compared to -1.46 experimentally) is derived in an $SU(4)$ model. Here, the proton has two up quarks (u) and one down quark (d) and the neutron one up and two down quarks. Calling μ_a the magnetic moment of the two like quarks and μ_b the magnetic moment of the remaining quark one derives (Perkins 1987)

$$\mu_N = \tfrac{4}{3}\mu_a - \tfrac{1}{3}\mu_b. \tag{16.4}$$

In this quark $SU(4)$ model, the u(d) quark moments are

$$\mu_{\text{u(d)}} = \frac{1}{2}\left(\tau_3 + \frac{1}{3}\right)\frac{M_N}{M_q}(\mu_B) \tag{16.5}$$

with $\tau_3 = +1(-1)$ for u(d) quarks. Here, τ_3 denotes the third component of isospin. The isospin variable is, just like the ordinary spin variable for a two valued quantity which can be used to classify, in a most elegant way, a large collection of protons and neutrons constituting a nucleus or other composite systems (Bhadhuri 1988). One can then derive

$$\mu_n = M_N/M_q$$

$$\mu_p = -\tfrac{2}{3} M_N / M_q, \qquad (16.6)$$

where we have taken $M_u = M_d = M_q$. For $M_q = \tfrac{1}{3} M_N$; one gets $\mu_p = 3$, $\mu_n = -2$, quite close to the data.

This picture has been extended to other particles (Brown and Rho 1983). Recent studies on the precise origin of the spin structure of the nucleon have been carried out in muon scattering experiments at CERN and lead to some slight problems relating to the simple three-quark picture of nucleon structure.

Early models using the construction of a nucleon with three 'constituent' quarks, were able to get a rather good description of the nucleon gyromagnetic ratios (Perkins 1987). More detailed studies have tried to find out where the nucleon spin resides: in the quark structure, what would be the gluon spin content and can the quark-sea become polarized? Experiments carried out at CERN, within the EMC (the European Muon Collaboration) and at SLAC have shown that the spin contributions of all quarks and antiquarks present in the nucleon almost cancel to zero. A more detailed outline of this search for a better understanding of the original spin content of protons and neutrons is given in Box 16c.

As pointed out earlier, the deep-inelastic electron scattering experiments at SLAC were the decisive in demonstrating the nuclear quark substructure, with quarks moving almost as free particles. The MIT-bag model (Chodos *et al* 1974) considers quarks confined by a boundary condition at radius R. No particles can escape the bag which translates into a condition on the normal component of the vector current at $r = R$. From this model, one can construct quark wavefunctions which move inside the bag as massless Dirac fermion particles. The amount of energy inside the bag (volume term) is $\tfrac{4}{3}\pi R^3 B$ (with B a 'bag' constant). Quark energies can then be obtained using stationary states inside a bag with the necessary boundary constraints. So one gets e.g. for the $1s_{1/2}$ ground-state quark the value $E_q = 2.04\hbar c / R$ and a corresponding bag energy

$$E_{\text{Bag}} = \tfrac{4}{3}\pi R^3 B + 3(2.04\hbar c / R). \qquad (16.7)$$

The parameter B can be determined by requiring that the energy (16.7), is minimized with respect to R, and corresponds to the experimental nucleon mass. Of course, a sharp bag surface remains a crude approximation but this picture greatly simplifies meson coupling to the bag.

A great advantage of the MIT bag model is that it gives a specific way to make allowing a number of detailed calculations that give rise to quark wavefunctions. Gluon exchange can be handled using the quark wavefunctions that replace the function of the constituent quarks. In the bag model, tensor and spin-orbit forces are then the result of gluon exchange. The spin-orbit terms can be evaluated and it turns out that the confinement contribution almost fully cancels the gluon exchange part. This MIT bag model, in which one starts with massless up and down quarks, has great advantages even if on a phenomenological level it does no better than the constituent quark model.

It is interesting to apply the idea of the nucleon as a small quark core surrounded by a meson cloud to interacting nucleons and which leads to the Yukawa idea of a meson-exchange description. By coupling the pion to the nucleon bag, one is able to construct the boson exchange model. The intermediate range interaction comes from the exchange of two-pion systems in a relative $l = 0$ state; the ρ meson consists of two pions in a relative $l = 1$ (p-state) state. Both the ρ and ω mesons exhibit an underlying q$\bar{\text{q}}$ structure at short distances. The repulsive part is related to ω exchange (figure 16.7) which gives rise to a potential of the form

$$V_\omega(|\vec{r}_1 - \vec{r}_2|) = \frac{g_{\omega\text{NN}}^2}{4\pi} M_\omega c^2 \frac{\exp(-M_\omega|\vec{r}_1 - \vec{r}_2|)}{M_\omega|\vec{r}_1 - \vec{r}_2|}, \tag{16.8}$$

where $g_{\omega\text{NN}}^2/4\pi$ is $\simeq 10\text{--}12$, and $M_\omega c^2 = 783$ MeV. This interaction quickly decreases with increasing distance $|\vec{r}_1 - \vec{r}_2|$ and has a value of $\simeq 1$ GeV at 0.4 fm; it is still appreciable at a distance of $|\vec{r}_1 - \vec{r}_2| \simeq 1$ fm. In calculations with the bag model, the ω-exchange potential will begin to cut off when the nucleon bags start merging but a repulsive part will remain even for very short distances. So, bags in nuclei do not move freely in an uncorrelated way because of the highly repulsive interaction which keeps the bags (nucleons) separated. This repulsion between bags may be the major reason why the quark substructure of the nucleon seems to show so little effect on genuine nuclear physics and nuclear structure phenomena. Proceeding to much higher energies and densities (see section 16.5) can create conditions to overcome the repulsive part.

The precise structure of the repulsion that remains after two bags have merged still needs to be determined. In general, effects from quark substructure are very difficult to be observed in low-energy nuclear physics processes. Not only does the quark core of the nucleon have a small radius compared with the average nucleon–nucleon separation inside the nucleus, but the repulsion coming from the vector–meson exchange potentials works in the direction of keeping the nucleons apart. It will take some time longer before an understanding of the nucleon–nucleon interaction can be put into the language of non-perturbative properties of QCD.

At present, there are essentially three lines of research that are rather promising in order to reach this goal. At first, there is the method of trying to solve the equations of the strong interaction (QCD) numerically by what has been known as 'lattice QCD' (Creuz 2001). Another approach focuses on the low-energy regime of QCD and aims at studying the implications of the spontaneous breaking of chiral symmetry. At present, interesting results have been obtained for few-nucleon systems (Bedaque and van Kolck 2002). A third method tries to use perturbative methods applied to QCD in order to extract the structure of the nucleon starting from high-energy scattering processes (Altarelli 1994, Goeke *et al* 2001). A more popular review of the status of our understanding of the nucleon and nucleon–nucleon forces is to be found in (Davies and Collins 2000).

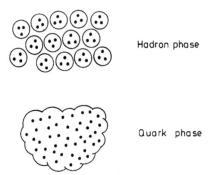

Hadron phase

Quark phase

Figure 16.8. Comparison of a collection of nucleons in (*a*) a hadronic or nuclear matter phase, and (*b*) within a quark–plasma description.

16.5   The quark–gluon phase of matter

In the present chapter, it has become clear that protons and neutrons are no longer considered as elementary but are composed of quarks in a bound state. The binding forces are quite distinct from forces normally encountered in physics problems: at very short distance, the quarks appear to move freely but, with increasing separation, the binding forces increase in strength too. So, it is not possible to separate the nucleon into its constituent quarks. Quarks seem to be able to exist only in combination with other quarks (baryons) or with anti-quarks (mesons).

This picture has also modified our ultimate view of a system of densely packed nucleons. For composite nucleons, interpenetration will occur if the density is increased high enough and each quark will find many other quarks in its immediate vicinity (see figure 16.8). The concepts of a nucleon and of nuclear matter become ill-defined at this high-energy limit and a new state of matter might eventually be formed: a quark plasma whose basic constituents are unbound quarks (Satz 1985).

One might ask the question why would the quarks become an 'unconfined' state at high energy and high density. The reason is to be found in charge screening which is also known in atomic physics. The force between the charged partners appearing in a bound-state configuration is influenced in a decisive way if *many* such objects are put closely together. The Coulomb force has, in vacuum, the form e^2/r and in the presence of many other charges it becomes subject to Debye screening and we obtain the form $e^2 \exp(-r/r_D)/r$ with r_D, the screening radius inversely proportional to the overall charge density of the system. If, in atomic systems, the Debye radius r_D becomes less than the typical atomic radius r_{Bohr}; the binding force between the electron and the nucleus is effectively screened and the electrons become 'free'. The increase in density then results in an insulator to conductor transition (Mott 1968). Screening is thus a short-range mechanism that can dissolve the formation of bound states and one expects this to apply to quark systems too.

The force acting between quarks acts on a 'colour' charge which plays the rôle of the electric charge in electromagnetic interactions. Nucleons are 'colour'-neutral bound states of 'colour'-charged quarks and so, nuclear matter corresponds in the above terminology to a 'colour' insulator. With increasing density though, 'colour' screening sets in and leads to a transition towards a 'colour' conductor: the quark plasma. So the new state is conductive in relation to the basic charge of the strong interaction (and thus of QCD). The experimental investigation of these phenomena is a major study topic for ultra-relativistic heavy ion colliding projects (RHIC at Brookhaven National Laboratory, LHC at CERN, etc) (Gutbrod and Stöcker 1991).

Quantum chromodynamics (QCD) describes the interaction of quarks in much the same way as QED for electrons. Matter fields interact through the exchange of massless vector gauge fields (here gluons and quarks correspond to photons and electrons). The interaction is determined by an intrinsic charge, the 'colour', and each quark can be in one of the three possible quark colour states. In contrast to QED, however, the gauge fields also carry charge and as a result the gluons interact with the quarks but also amongst themselves which is *the* crucial feature of QCD. It results in a confining potential $V(r) \sim |\vec{r}|$. QCD can now predict e.g. the transition temperature from nuclear matter to a quark plasma in units of the nucleon mass and this mass itself is calculated as the lowest bound state of three massless quarks. Starting from the QCD Lagrangian, one has to determine the Hamiltonian $\hat{H}$ and then construct the partition function in the usual way for a thermodynamic system at temperature T and within a given volume V as,

$$Z(T, V) = \text{Tr}(e^{-(\hat{H}/T)}). \tag{16.9}$$

Once the function Z is known, the desired thermodynamic quantities can subsequently be determined, such as

$$\varepsilon = T^2/V (\partial \ln Z/\partial T)_V$$
$$P = (\partial \ln Z/\partial V)_T, \tag{16.10}$$

with ε the energy density, P the pressure. At high density (the quark-free regime) calculations are rather straightforward and known from QED. Since we are interested in the transition between strongly coupled nuclear matter and the weakly coupled plasma, a much more difficult problem is posed. The only approach possible is that of using a lattice formulation (Wilson 1974). Replacing $(\vec{r}, t)$ with a discrete *and* finite set of lattice points, the partition function becomes very similar to the corresponding partition function for spin systems in statistical mechanics. So, computer simulations can be carried out in the evaluation of lattice QCD studies (Creutz 1983). Renormalization methods then allow us to take the

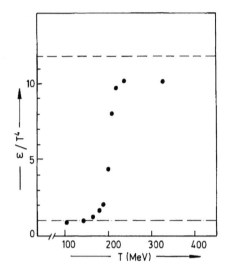

Figure 16.9. The energy density (ε/T^4) of strongly interacting matter as a function of the temperature. We also compare the horizontal lines for the ideal quark–gluon plasma (dashed line) and for the ideal pion gas (dot-dashed line) (taken from Celik *et al* 1985).

limit and recover the continuum limit. These lattice QCD states have been used extensively in the study of the phase transitions in strongly interacting matter.

We shall not discuss computer simulations of statistical systems (Binder 1979) but concentrate on the major results obtained. Most results are obtained for 'hot mesonic matter' of vanishing baryon number density. The other extreme, the compression of 'cold nuclear matter' is the subject of much research.

Starting with matter of vanishing baryon density, the energy density of a non-interacting gas of massless quarks and gluons is

$$\varepsilon = (37\pi^2/30)T^4 \simeq 12T^4. \tag{16.11}$$

Just like in the Stefan–Boltzmann form for a photon gas, the numerical factor in (16.11) is determined by the number of degrees of freedom of the constituent particles: their spins, colours and flavours. The energy density obtained via computer simulations is shown in figure 16.9; there is a rapid change from the mesonic regime into the plasma regime. The transition temperature is around 200 MeV which means an energy density of at least 2.5 GeV fm^{-3} in order to create a quark–gluon plasma phase.

The thermodynamics of strongly interacting matter is of importance to cosmology. Very soon after the big bang, the universe most probably looked like a hot, dense quark–gluon plasma; its rapid expansion then reduced the density such that after $\simeq 10^{-3}$ s, the transition to the hadronic phase took place. The collision

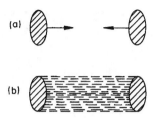

Figure 16.10. High-energy collision of two nucleons in a schematic way (*a*) before and (*b*) after the collision (taken from Satz 1986).

of two heavy nuclei at high enough energies should then be able to recreate the conditions for a 'little bang'. The important questions here are: (i) can a high enough energy density be obtained for the transition to take place; (ii) whether the system is extended and long-lived enough to reach the thermodynamical equilibrium required for a statistical QCD description; and (iii) how can we determine that a quark–gluon plasma has indeed been produced and how to measure its properties.

When two nucleons collide at high energy they almost always pass just through each other. In the process, the nucleons are excited but in addition they leave a 'zone' of deposited energy in the spatial region they passed through (figure 16.10). This 'vacuum' energy is rapidly hadronized; the secondary particles usually produced in such collisions. Analysing p–p collision data, a deposition of energy at high enough energies is about $\frac{1}{3}$ GeV fm^{-3}. If instead, heavy ions are used, a much larger energy density can be reached, and sample geometric arguments lead to densities of at least $\varepsilon \propto A^{1/3}$ GeV fm^{-3}. For two ^{238}U nuclei, this means a value of about ~ 6 GeV fm^{-3}. Computer codes simulating such collisions have been studied.

Very important is the signature for the formation of quark matter. Various signals have been suggested and are sought for in the current experiments. The study of photons and leptons, particles not affected by the strong interaction, carry information about *conditions* at the point of formation. Thermal lepton particles which can be created at any stage of the collision have production rates strongly dependent on the temperature in the interior of the fireball. The production of heavy vector mesons and the J/ψ particle is a much suggested signal. If the plasma formed is not only in thermal equilibrium but also in equilibrium with respect to the production of different flavours of quarks, a large number of strange $q\bar{q}$ pairs will be created: hyperons (Λ-particle, etc); and mesons ($K^+, K^-, \ldots$) may survive the expansion of the interaction volume and give an indicator of the quark–gluon plasma formation. A variety of other signals have been suggested and so this very important set of questions is still in a stage of evolution.

Finally we discuss the experiments performed up to now and the planned projects. The first experiments were carried out using beams of light ions

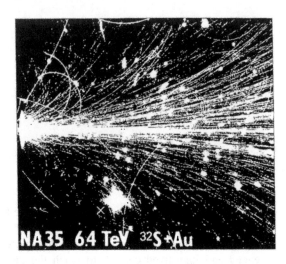

Figure 16.11. The illustration of a streamer–chamber photograph showing a central collision between a ^{32}S nucleus and an Au target nucleus from the NA35 CERN collaboration (taken from Karsch and Satz 1991).

(^{16}O, ^{28}Si, ^{32}S) at both the AGS (Brookhaven) and at CERN (the SPS). In the experiments using 200 GeV A^{-1} ^{16}O and ^{32}S ions, an energy density of $\varepsilon \sim 2$ GeV fm^{-3} has been reached. These experiments indicate very large hadron multiplicities (figure 16.11). As a result of experiments started in 1994 at SPS CERN and eventually continuing to study the collision of a beam of Pb ions (accelerated up to 33 TeV) with Pb and Au targets, data have been taken at seven large experiments. Some of these detectors were multi-purpose detectors, others have been set up to detect rare signals up to very high statistics. A special seminar was held at CERN on 10 February 2001 in which the results of the seven experiments were combined, leading to statements by the spokespersons of 'circumstantial' evidence for the creation of a quark–gluon plasma state of matter. The point was that at an energy density in the initial phase of $\epsilon \approx 2.5$ GeV fm^{-3}, well above the critical point, hints for a strong suppresion of the J/ψ particle were observed. The fact that many strange quarks had also been produced and that indications for the presence of hadronic resonances were absent (pointing towards a 'melting' of the regular hadrons) lead to this special event at CERN. The results of these various presentations can be found back at the CERN webpage.

In the process of reaching the very high energy needed, Brookhaven has built a Relativistic Heavy Ion Collider (RHIC), which is a dedicated accelerator facility for probing the formation of a quark–gluon state of matter under the optimal conditions. Making use of the existing Alternating Gradient Synchrotron (AGS) as an injector, heavy atoms (Au nuclei) can be accelerated in the 4 km-long rings using superconducting magnets. In these rings, counter-rotating beams are

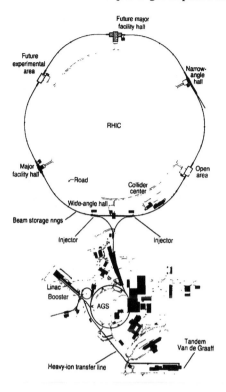

Figure 16.12. The RHIC collider at Brookhaven National Laboratory (BNL) under construction. The rounded, hexagonal curve indicates the RHIC storage rings, carrying two counter-circulating beams of high-energy heavy ions (HI) around the 4 km circumference. These beams can be made to collide at six interaction points. The acceleration and stripping of ions will begin at the Tandem Van de Graaff machine and continue in the new booster and the AGS (Alternating Gradient Synchrotron) before injection into the RHIC rings (taken from Schwarzschild 1991).

accelerated up to 100 GeV per nucleon (200 GeV in the COM (centre-of-mass) system). For protons, the maximal energy is 250 GeV (see also figure 16.12 and (Harrison *et al* 2002) for an outline of the RHIC facility). There are four interaction points at which four detectors take data for these collisions: STAR and PHENIX are two large multipurpose detectors whereas the other two, PHOBOS and BRAHMS, are smaller and have been designed for the detection of specific particles. The energy in these beams is such that for the collision of Au on Au beams at the highest energy possible, an intial energy density of $\epsilon \approx 10 \, \mathrm{GeV \, fm^{-3}}$ can be reached, a value that is thought to be more than adequate for the creation of an initial quark–gluon plasma state of matter.

Since the inaugural run of Au on Au at 70 GeV per nucleon (12 June 2000 to September 2000), the runs starting in June 2001 with the design energy of 100 GeV per nucleon and the most recent experimental runs (spring of 2003 with Au on deuterium experiments), impressive new results have been obtained (see also Box 17d). A number of more popular articles on the RHIC facility and the very recent results can be found in (Wolfs 01, Schaefer 2003, Ludlam and McLerran 2003) and also by consulting the RHIC web page at http://www.bnl.gov/rhic.

Still higher energies are expected to become available when the LHC facility becomes operational which is expected to be by 2007. One of the detectors, ALICE, will take data from colliding Pb on Pb nuclei at CERN. This highly complex detector is currently in the building stage. More details for the interested reader can be found at the ALICE webpage http://www.cern.ch/alice.

Box 16a. How electrons and photons 'see' the atomic nucleus

Photons are the force carriers of the electromagnetic interaction (see chapter 6) and allow us to probe the nucleus in two different ways: by photoabsorption of real photons or by virtual photon exchange in electron scattering. In the latter process, the electron entering the electromagnetic field of the nucleus emits a virtual photon ($E_\gamma \neq p_\gamma c$) which is absorbed by a charged particle in the nucleus. The waves associated with the photon, characterized by λ, ν determine the characteristic length and time scale for probing the atomic nucleus. In contrast to the real photon, λ and ν can be freely varied. The resolution, both in length and time, becomes better with increasing photon and/or electron energy and both λ and ν can vary independently, depending on the kinematical conditions characterizing the reaction. Because of the point characteristics of the electron, the outcome of these 'electromagnetic' interactions are very clear results that are amenable to an unambiguous analytical processes.

The interaction in electron scattering off a nucleon in the atomic nucleus is illustrated in figure 16a.1.

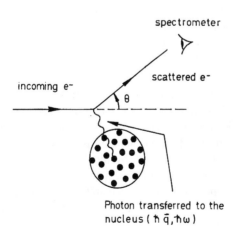

Figure 16a.1. Illustration of the electromagnetic scattering process where an incoming and outgoing electron (e^-) are present. In the electromagnetic interaction, energy $\hbar\omega$ and momentum $\hbar\vec{q}$ are transferred via the exchange of a virtual photon to the atomic nucleus, here depicted as a collection of A nucleons.

Box 16b. Exotic baryons with five quarks discovered: pentaquark systems

---

For a very long time, all known mesons and baryons could be attributed to a bound state of either a quark and an antiquark or of a three-quark system. An exception to this rule would then be called 'exotic'. Physicists had been searching for five-quark systems but recent theoretical work Diakonov *et al* (Diakonov *et al* 1997), predicting the existence of an exotic, isoscalar baryon with spin and parity assignment $1/2^+$ and strangeness $S = +1$ gave it a new thrust. The lowest mass of the 5-quark system turned out to be 1530 MeV, a mass only 60% more than the nucleon mass, moreover having a width of only 15 MeV. The suggestion was that the 'particle' would appear as a sharp peak in the nK^+ or pK^0 mass spectrum with a structure of $uudd\bar{s}$.

Now, it seems that just very recently (2003), experimental evidence in at least four different laboratories has emerged for the existence of such a pentaquark, called the Θ^+ particle. The first result came from a team at the Laser Electron Photon Facility at SPring-8 (LEPS) collaboration in Osaka (Nakano *et al* 2003). This team studied the reaction $\gamma n \to K^+ K^- n$ on a ^{12}C target by measuring the K^+ and K^- at forward angles. A positive signal had already been obtained in 2002 but thorough analyses of the data implied a delay with the announcement at the International Conference on Particles and Nuclei in October 2002 of a baryon with mass 1540 MeV and a width less than 25 MeV.

At the Insitute of Theoretical and Experimental Physics (ITEP) in Moscow, the DIANA collaboration was examining older 1986 data from low-energy K^+Xe collisions in a xenon bubble chamber. They studied the effective mass of the pK^0 system in charge-exchange reactions in the exit channel, giving rise to a baryon resonance with a mass of 1539 MeV and a width of less than 9 MeV (Barmin *et al* 2003).

Experiments at JLab produced results with the best statistical confidence for the presence of the announced baryon. The experiment, carried out with the CLAS magnetic spectrometer in Hall B at JLab, used a multi-GeV beam of photons impigning on a ^{2}H target inducing the reaction $\gamma d \to K^+ K^- pn$. The particles thus created were detected by the CLAS detector that also detected, besides the K^- meson and the proton, a five-quark intermediate baryon, that appeared as a resonance in the K^+n final state. Here, an invariant mass for the baryon of 1542 MeV carrying a width of 21 MeV (Stepanyan *et al* 2003) (see figure 16b.1) was found.

Still another group—the SAPHIR collaboration at the Electron Stretcher Accelerator (ELSA) at Bonn—using data taken in the period 1997-98 found a peak in the K^+n spectrum from the $\gamma p \to nK^+ K_s^0$ reaction with the subsequent decay of the K_s^0 meson into a $\pi^+ \pi^-$ pair. These data were also in line with the former results (Barth *et al* 2003).

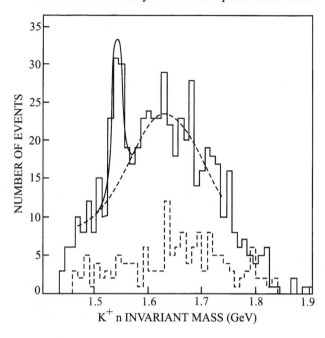

Figure 16b.1. Experimental evidence for the exotic Θ^+ baryon is given by the large peak (full line fit) centered at 1542 ± 5 MeV in the K^+n invariant-mass distribution (Stepanyan *et al* 2003).The bold broken line describes a fit to the background and the broken histogram events excluded because a Λ^0 baryon had been produced. (Reprinted with permission from Search and Discovery September 2003 *Physics Today* p 20 (figure 2) © 2003 American Institute of Physics and from Stepanyan S *et al* 2003 *Phys. Rev. Lett.* **91** 252001 © 2003 by the American Physical Society. Credit is also given to the CLAS Coll. at JLab.)

The precise correlations that make up for such a narrow 5-quark baryon is not fully understood as yet but it may hint at interesting structures in which the bound state results from two highly correlated *ud* di-quark pairs plus a single s antiquark (Jaffe and Wilzeck 2003, Bijker *et al* 2003).

In view of a very recent analysis of data taken by the NA49 experiment at the CERN SPS (Alt *et al* 2003), giving strong indications for a new 5-quark baryon at 1862 MeV, it seems like a new sector of strongly interacting matter is being opened up. These experiments may then shed new light on the way in which multi-quark configurations result in baryon structures.

Box 16c. What is the nucleon spin made of?

Protons and neutrons are fermions with an intrinsic spin of $\frac{1}{2}$ (in units of $\hbar$). There does not seem to be much doubt about that. In the fundamental theory of quantum chromodynamics (QCD), protons and neutrons involve more fundamental objects such as quarks, gluons and a sea of quark–antiquark pairs. Looking from this fundamental side, a nucleon shows up as a very complex interacting many-body system with all the various constituents contributing some part to the full nucleon angular momentum of spin $\frac{1}{2}$ (Stiegler 1996).

Early ideas, in order to understand nucleon magnetism expressed through the particular gyromagnetic spin ratios, pictured the nucleon as the addition of just three 'constituent' quarks. These early arguments have been discussed, for example, by Perkins (1987) and Bhadhuri (1988).

Physicists have tried to gain a better insight into the precise balance between the various elementary constituents in order to obtain the total nucleon spin value. It came as a surprise that the addition of the spin carried by the quark degree of freedom almost added up to a vanishing spin value. Intensive and detailed experiments have been carried out at the Stanford Linear Accelerator (SLAC), CERN and DESY. The results show that the spin part carried by the quark degree of freedom adds up to only $\frac{1}{3}$ of the nucleon spin, accentuating the very complicated internal structure of a single nucleon with respect to generating angular momentum.

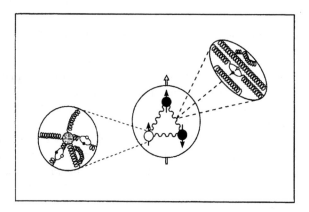

Figure 16c.1. Schematic view of the spin structure of the proton. The three valence quarks interact via the exchange of a gluon. The 'microscopic' loop into the quark indicates the existence of a sea of virtual particles that can be created through vacuum fluctuations (left part). The gluon then can be modified into a quark–antiquark fluctuation (right part). The arrows indicate the spin direction of the contributing particles.

This so-called 'nuclear spin crisis' is being addressed at present in a number of experiments in order to find out the contribution of (i) the 'valence' quarks discriminating between both the intrinsic and the orbital part, (ii) the gluons and (iii) the part coming out of the polarization of the quark–antiquark sea. Also an experiment called HERMES at Desy (Hamburg), SLAC and other planned experiments at MIT, CEBAF and RHIC are trying to investigate not only the summed spin parts but also the differential parts. All of these experiments, aimed at a deeper understanding of nucleon properties, clearly indicate that supposedly rather simple systems are turning out to be complicated many-body systems.

Very recently, experiments at both Hermes and at JLab have given evidence that the valence u-quarks (by an appropriate choice of the energy for the incoming electrons) are polarized in the direction of the proton spin whereas the d-quarks have their spin polarized in the opposite direction. It seems that the sea-quarks (essentially the $\bar{d}$, the $\bar{u}$ and s quarks) show no particular polarization in these experiments (Zheng *et al* 2003, FOM 2003). These experiments have given unambiguous results for the fact that—contrary to what was expected some years ago—the strange quarks do not contribute substantially to the proton spin. So, the expectation now is that a good fraction of the proton spin comes from the fact that gluons are, to a large extent, polarized and, moreover, that the orbital part of the valence quarks adds another important part to the total spin of the proton.

Experiments now planned at CERN, SLAC and RHIC aim to determine precisely the gluon contribution to the spin of the proton. At CERN, the COMPASS experiment will make use of beams of longitudinally polarized muons interacting with longitudinally polarized proton targets. SLAC will concentrate on the study of photon–gluon fusion processes, using real photons while at RHIC, quark–gluon Compton interactions will be used, starting from polarized proton beams to probe the spin polarization. When completed, these experiments should bring in new information on the gluon content in polarization within the proton.

In figure 16c.1, a schematic picture is shown that tries to convey this 'elementary' idea of a nucleon being built out of 'three' constitutent quarks which, looking at a more fundamental level, themselves look like mere 'quasiparticle' quarks, indicating important dressing with quark–antiquark sea polarization excitations as well as with the gluon degrees of freedom. It is important that experiments are carried out, but detailed theoretical studies are also being planned and it is hoped that the various results will be collated in the coming years. At the same time there is a most interesting element in that one has the constraint that all spin parts have to add up to $\frac{1}{2}$ which is incredible in the light of the present day experiments. This may indicate that some underlying symmetries, unknown at present, are playing a role in forming the various contributing parts such that the final sum rule gives the fermion $\frac{1}{2}$ value.

Box 16d. The quark–gluon plasma: first hints seen?

There is now quite clear evidence that there exist various phases in which to consider the collection of protons and neutrons, bound by the strong nuclear force, inside the atomic nucleus. At a first approximation, one can picture the nucleus as a collection of nucleons moving as independent particles in a spherical (or deformed) average field. This field can be understood better using self-consistent Hartree–Fock methods (see chapter 10) as a microscopic underpinning of this average field. Residual forces of course modify this independent particle motion: the short-range pair correlations modify this independent particle motion such that in many even–even nuclei a pair-correlated kind of superfluid picture of nucleons shows up as a better approximation to the low-energy structure of many atomic nuclei. This superfluid structure is rather rigid against external perturbations but at the high rotational frequency with which the nucleus gets cranked (Coriolis forces) and/or high internal excitation energy (heating the nucleus) this paired system eventually breaks up.

A long search has been made in order to see if a phase transition can be realized by heating nuclei such that only a gaseous phase of interacting protons and neutrons survives. Colliding heavy ions has been the way to probe this phase transition from a liquid (superfluid) into a gaseous phase of nuclear matter. By precisely identifying the number and species of fragments made in these heavy-ion collisions, researchers were able to derive the excitation energy and corresponding 'temperature' reached in this heavy-ion collison at the time of break-up. Most interesting results are shown in figure 16d.1 in which the nuclear temperature versus excitation energy per nucleon has been plotted and compared with the corresponding temperature versus excitation energy per molecule of water. In recent experiments, using 8 GeV pions, colliding with a Au beam at the AGS accelerator of BNL, the multifragmentation producing fragments of various sizes has been studied in great detail (Elliott *et al* 2002) and has allowed the second order continuous nuclear matter phase transition in finite nuclear systems to be traced to a high degree of accuracy (see also figure 16d.2). The similarity is remarkable and gives a clear indication of a phase transition appearing in nuclear matter changing from a fluid into a gas-like phase. It has been suggested that at even higher temperatures and simultaneously higher internal pressures still another phase transition might occur in which the nuclear building blocks (protons and neutrons) are 'melted' into a plasma phase and a totally new state of matter, a quark–gluon phase emerges (Schukraft 1992).

As has been discussed in the text (section 16.5) this process is of immense importance because the observation and detection bears on many other fields of physics too (cosmology, particle physics, etc). Theoretical signals for the occurrence of such a phase, which will only last for very short time spans, have been amply discussed in the present literature and also a large

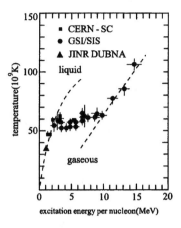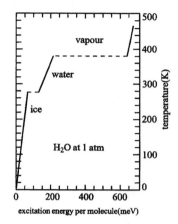

Figure 16d.1. Comparison of the phase diagrams for nuclear matter and water. The data were taken at JINR (Dubna), CERN (SC) and at the GSI (Darmstadt) (taken from GSI Nachrichten 1996b, with permission).

number of experiments, using asymmetric heavy-ion combinations at the AGS (Brookhaven), the SPS (CERN), and RHIC, have been carried out.

Experiments have been carried out at the CERN SPS accelerator, have used Pb + Pb collisions with the aim of searching for a particular signal: the J/ψ suppression. It is expected that charmonium production ($c\bar{c}$ structures) will form a suitable probe for the plasma phase of deconfinement. At the early phase of formation, a suppression of the J/ψ and ψ' resonances is expected from Debye screening in the quark–gluon plasma. Measurement of charmonium production cross-sections are therefore used as a very important signal with respect to the formation of a quark–gluon plasma state. The CERN NA38 experiments were the first to study in a systematic way the charmonium production cross-sections in p + B$_{target}$, O + B$_{target}$ and S + U$_{target}$ collisions. These data can then be used as a reference when comparing the Pb + Pb data that were taken at an energy of 158A GeV/c recently. Recent publications (Abreu *et al* 1997a, b, Gonin *et al* 1996, Sorge *et al* 1997) discuss the implication of these results in which for the Pb + Pb collisions (see figure 16d.3) the J/ψ production cross-section is plotted. In this figure, results from the NA38 collaboration are also given. The heaviest collision data are significantly below the 'expected' trend and this gives a first clear hint that the J/ψ production is anomalously suppressed in the Pb + Pb collisions. So, at present there are clear indications that for most central interactions of Pb on Pb at 158A GeV/c, the onset of a very dense state of matter is signalled rather than indications of possible 'new' physics.

With the start of the RHIC research program (see also section 16.5) and using a dedicated facility with Au beams in colliding experiments (during the

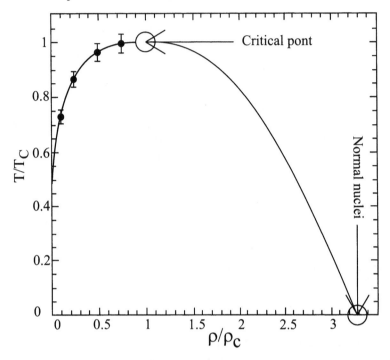

Figure 16d.2. A figure analyzing the 8 GeV pion-induced multifragmentation of a Au beam at the AGS. The analyses in (Elliott *et al* 2002) allows a phase diagram using reduced density and the temperature as axes to be constructed. The bold line describes the calculated low-density branch of the coexistence curve (the vapour phase) coming up to the critical point (the points denote selected calculated errors). The thin line then describes a fit to the high-density (liquid) phase. (Reprinted from Elliot *et al* 2002 *Phys. Rev. Lett.* **88** 042701 © 2002 by the American Physical Society.)

first runs in 2000, in runs with the designed energy of 100 GeV/nucleon in each of the colliding beams and, very recently, during a Au on deuterium run in the spring of 2003), very convincing results have been obtained that point towards the formation of a quark–gluon plasma state of matter in the initial stage of the collisions.

The Au on Au experiments, creating nuclear matter densities about a 100 times higher than normal density, at an energy of 130 GeV per nucleon pair, detected fewer particles with high p_T (particles moving at right angles with respect to the collision direction) as compared to standard theory. In the standard approach, an $A + A$ collision is seen as a sum of independent nucleon–nucleon collisions and this scenario is called 'binary scaling' (Adcox *et al* 2002). It seems that, for the peripheral high p_T events, there is no substantial deviation from the expected results but for central collisions, there is a net reduction for both

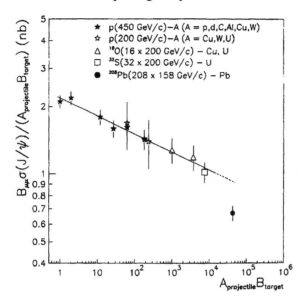

Figure 16d.3. The J/ψ 'cross-section per nucleon–nucleon collision', as a function of the product AB of the projectile and target atomic mass numbers. The results obtained at 450 GeV/c and the Pb–Pb cross-sections are rescaled (reprinted from Abreu *et al* © 1997, with permission from Elsevier Science).

charged and neutral particles. By appropriate normalization, per event, a 'nuclear modification' factor $R_{AA}(p_T)$ can be defined that should approach that for 'hard' scattering. This reduction is hinting at a modification of the medium at the time these particles were formed.

In order to reach conclusive arguments on the medium modification, experiments using protons colliding with Au nuclei, at the same energy regime, should be performed. During the spring of 2003, experiments have been carried out using Au nuclei colliding with deuterons (Adler *et al* 2003, Adams *et al* 2003, Arsene *et al* 2003). By studying both the azimuthal angular dependence of the detected particles as well as the transverse momentum p_T distributions of the particles and comparing the d + Au, Au + Au central collisions with p + p hard scattering data (see figure 16d.4), the formerly observed quenching of high p_T events in the Au + Au case is absent in the d + Au processes. Moreover, in comparing the emission of particles in opposite directions, in which a single pair of partons from the incoming nuclei strike each other directly and thus scatter at large angles with high momentum transfer, the results are totally different if we compare the two experiments: one speaks of jet-quenching in the Au + Au case. This kind of hard parton–parton scattering has been seen for the first time in nuclear collisions and indicates the clear appearance of high-energy quarks or

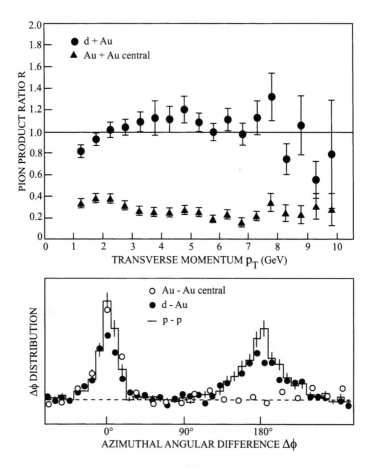

Figure 16d.4. Upper part: the ratio R of π^0 production in d + Au and central events in Au + Au collisions compared to that seen in p + p collisions (however scaled to account for the number of particles participating in the collision), plotted against the transverse momentum p_T of the pion. Whereas the ration in d + Au is close to 1, the ratio in the Au + Au collision is largely quenched with respect to $R = 1$ (see the text and (Adler *et al* 2003, Adams *et al* 2003, Arsene *et al* 2003)). Bottom part: Correlation of the azimuthal angles ϕ between high-p_T particles produced in the same event. Whereas the p + p and d + Au present very similar behaviour (back-to-back jets), the recoil peak at 180° is absent in the Au + Au collisions (see text and (Adler *et al* 2003, Adams *et al* 2003, Arsene *et al* 2003)). (Reprinted from 'What have we learned from the Relativistic Heavy Ion Collider (Ludlam T and McLerran L) *Physics Today* October 2003, p 48 © 2003 American Institute of Physics.)

gluons emerging from the initial collision stage. In p + p scattering, in vacuum, the struck quark flies off creating a jet of hadrons. In the Au + Au collision, however, most probably the scattered parton is embedded in a large volume of newly formed, hot and dense matter giving rise to a slowing down and quenching of the most energetic quarks as they propagate through the medium.

The emerging picture is converging into a consistent one. Earlier in the collision, the energy density far exceeds the theoretical threshold to create the new state of matter. The present experiments may be giving us a new look at matter as predicted by QCD as it might have appeared in the early stages of the universe.

Chapter 17

Outlook: the atomic nucleus as part of a larger structure

In the present discussion we take a rather detailed look at the various facets of the atomic nucleus and the many ways it interacts with various external probes (via electromagnetic, weak and strong interactions). This is the way of learning how the nucleons move in the nucleus and at the same time determine its characteristics. These multi-faceted 'faces' show up in various layers. At the first level, the individual nucleon excitations characterize the average nucleon motion in a mean field. There is evidence for the presence of single-particle orbits near the Fermi level and fragmentation of deep-lying single-particle states. The nuclear shell-model techniques have allowed a detailed description of many nuclear properties, in particular whenever the number of valence protons and neutrons outside closed shells remains small. The appearance of coherent effects in nuclear motion are largely discussed in terms of nuclear vibrational and rotational excitations. The energy scale at which one may to observe collective, pair-breaking and, eventually, statistical nuclear properties is illustrated in figure 17.1.

Here, once again, we stress the importance of using a multitude of probes to test the various 'layers' of nuclear structure. This too is schematically illustrated in figure 17.2.

Also, we point out the major approaches used to describe nuclear properties: starting from nuclear independent-particle motion and building in the various residual interactions, a detailed and rather complete description of nuclear phenomena can be obtained. This is a rather fundamental point of view whereby nuclear motion is organized starting from a microscopic, single-particle level illustrated in figure 17.3(a). A different approach uses a deformed mean field with its time-dependent properties like vibrations, rotations, etc. Incorporating the consequences of nuclear shape, shape variables and the total energy variations related to these different shape changes, insight can be obtained into nuclear dynamics and nuclear collective motion (figure 17.3(b)). A third approach starts from the underlying symmetries which are realized in the nuclear many-body system. Through various classification schemes and related dynamical

584

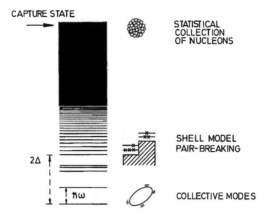

Figure 17.1. Various energy regions with their most typical modes of motion characterizing nuclear structure up to the nucleon binding energy limit of $\simeq 8$ MeV. Starting from regular collective modes and over the pair-breaking region (near $2\Delta \simeq 2$–2.5 MeV), one gradually enters the very-high density region where statistical models are applicable.

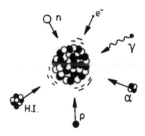

Figure 17.2. Illustration of the many ways of probing the internal nuclear structure: the many different excitation processes are mostly selective *and* complementary ways to explore the atomic nucleus.

symmetries, deep insight can be obtained into the organization of nucleons in the nucleus. The approach makes extensive use of the language of group theory and algebraic methods has been developed during the last decade to a high level (see figure 17.3(*c*)).

Proceeding to higher and higher energies to disentangle the finer details of nuclear motion, one eventually reaches the zone where sub-nucleonic characteristics start playing a major role. In light systems (deuteron, ^{3}He, ^{4}He, etc), mesonic effects have unambiguously been identified in electromagnetic interactions of electrons and protons with such nuclei. It is clear that the internal nucleon structure may even start playing a role when the available energy, e.g. in heavy-ion collisions, is very high. The hope is to break up the nucleon structure

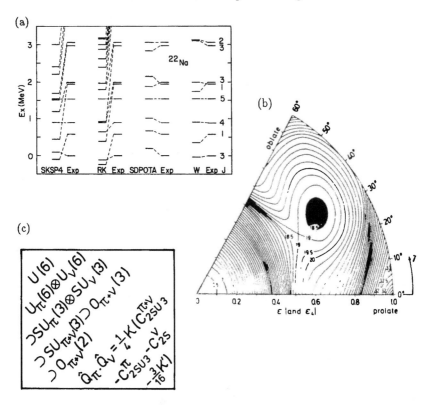

Figure 17.3. Three large categories that illustrate ways of describing the atomic nucleus. (*a*) The nuclear shell model treats the various nucleonic orbitals and their residual interaction. (*b*) The collective model approach depicts the nucleus as an object able to perform collective vibrational, rotational and intermediate excitations. (*c*) An algebraic approach where symmetries governing the nuclear A-body modes of motion are accentuated.

and reach a deconfined quark–gluon phase, a phase that might resemble some early phases after our Universe was formed. So, increasing the energy scale and at the same time decreasing the distance scale we may reveal the ultimate aspects of nuclear structure; its binding and basic properties within the first few moments of the constitution of our Universe. At this level, relativity and quantum mechanics can no longer be separated and a more rich and complete level of understanding the atomic nucleus, its constituents and interactions, is developed.

Many of the above features are displayed in the atomic nucleus in which protons and neutrons can give rise to a most interreresting interplay, building up more complex structures relative to the single-particle concept (vibrational modes, rotational degrees of freedom, large non-spherical shape excitations, etc).

We would like to put forward the idea that with the advent of new experimental techniques over the years, the nucleus has time and time again revealed new, often unexpected features. In the 50–60 years of experimental nuclear physics, many observations were not put forward through theoretical concepts. So, before ending the present discussion, it would be good to have a look through some major technical developments that have been at the origin of new aspects of nuclear structure (Bromley 1979, Weisskopf 1994, van der Woude 1995).

The development of large gamma-detection arrays in the last 10 years has made it possible to observe the full 4π solid angle of detection. It has become clear to physicists, both in particle physics and nuclear physics, that taking minute slices that are open to a given nuclear process gives a very distorted view of the actual phenomena being examined. The combination in these large arrays (Gammasphere, Nordball, Eurogam, Gasp, Euroball, and GRETA, AGATA to come) of Ge detectors with Compton suppression techniques, with the simultaneous detection of particles emitted in nuclear reactions and with the electronic advances made in performing multiple-coincidence experiments, has led to the observation of rotational bands at high spin and high excitation energy in atomic nuclei (Beausang and Simpson 1996). Superdeformed, and possibly also hyperdeformed, shapes have been detected and mapped out systematically in the rare-earth region, near the Pb closed shell and also in the mass $A = 100$ region (Firestone and Singh 1994).

The technical development of continuous wave (cw) 100% duty-cycle electron accelerators has opened new channels to view the atomic nucleus. In particular, 180° electron scattering, intensively studied at the Dalinac accelerator at Darmstadt, has allowed researchers to excite the atomic nucleus in a very selective way which maps the nuclear magnetic properties. These experiments have established the existence of a proton–neutron non-symmetric but collective mode of motion associated with orbital dipole magnetism. These so-called 'scissors' excitations had been discussed in the literature on nuclear collective motion only as a possibility but electron scattering experiments have now proved their unambiguous existence. Over the years, these magnetic dipole excitation modes have been extensively studied and mapped all over the nuclear mass region.

A very interesting example has been brought to light by the coincidence detection of scattered electrons and emitted nucleons (protons, neutrons) or clusters of nucleons in the newest generation of cw 100% duty-cycle electron accelerator facilities. Work at NIKHEF Amsterdam, MAMI at Mainz, BATES at MIT and JLab has allowed researchers to probe the motion of nucleons inside the atomic nucleus. The results of (e,e′p) reactions have given detailed information about the momentum distribution of nucleons as they move inside the atomic nucleus and have made it possible to test the concept of an average field that contains most of the interactions amongst the nucleons in the A-body system. More recently, technological advancement has allowed the study of reactions like (e, e′pp) where besides the scattered electron two nucleons are detected separately. These methods even allow us to extract the precise way in which nucleon short-range correlations modify the original independent motion of the nucleons (Pandharipande *et al* 1997).

It is known that the lifetime of a given nuclear level carries a lot of information about the internal organization of nucleons in such states. Traditional methods could not access given windows of lifetime. A new technique called GRID, which stands for gamma-ray induced Doppler broadening, opens up possibilities to measure lifetimes beyond the nanosecond region (Börner *et al* 1988). Developed at the ILL (Grenoble), this method detects gamma-rays emitted after the neutron capture process. If detected with a high enough resolving power, the line shapes of the detected gamma-rays contain a large 'history' of information about the decay process of the nucleus. Every time a photon is emitted, the rest nucleus recoils so gamma-rays are emitted from moving nuclei and thus contain Doppler-broadening information which is dependent on the lifetime of the level which has been emitting a given gamma-ray. In working backwards, one can eventually unravel the whole de-excitation path and the subsequent nuclear lifetimes. This method has been highly successful and has opened up new possibilities to test nuclear model concepts in regions hitherto 'closed' to detailed observations.

Experimental work and the rapid increase in possibilities to select nuclei from a production process that creates a multitude of nuclei in isotope separation methods (ISOL technique) or starts by fragmenting a rapidly moving projectile on a stationary target nucleus (IFS technique) has opened up windows to very short-lived nuclei. Combined with laser spectroscopic methods, one is at present able to measure gamma-rays emitted by an exotic atomic nucleus and to detect its nuclear moments (dipole and quadrupole moment information) on the time scale of milliseconds and even shorter. Storage and cooling techniques have been developed largely at GSI (Radon *et al* 1997, Geissl *et al* 1992) and mass measurements using traps (Thompson 1987, Stolzenberg *et al* 1990) and at ISOLTRAP (ISOLDE-CERN) have been realized recently. In this rapidly moving domain of research with new technical possibilities, new forms of organizational structures for protons and neutrons have been discovered. Light nuclei with radii much like the heaviest Pb nuclei have been detected in the Li region. The concept of weakly and loosely bound quantum systems with neutron halo structures has been proposed. Here, once again, technical developments have led to unexpected forms in which protons and neutrons become bound while approaching the line of stability of atomic nuclei. Much of this very recent physics has already been discussed in chapter 15.

Related to this issue and in order to reach both the proton and neutron drip lines, a multitude of accelerated radioactive ions have been produced. According to the energy of these secondary beams, both nuclear structure at the extremes of stability as well as nuclear reactions that are going on in the interior of stars in nucleosynthesis can be probed. This field is still rapidly developing and so we refer the readers to the literature in order to follow the many projects that are in a construction phase or are still on the 'drawing board' (see chapter 15,

Box 15b). New physics will clearly evolve as was illustrated in this section. Recently, in various countries, long-range plans relating to setting up larger experimental facilities and bringing together the nuclear physics community in a number of ground-breaking endeavours have been discussed and formulated. We give a number of references to such reports (DOE/NSF 1996, 2002, NUPECC 1991, 1993, 1997, 2000, 2004, ISL Report 1991, OHIO Report 1997, NAP 1999, Mackintosh *et al* 2001, GSI 2001, Bromley 2002, RNB 2002, EURISOL 2003). A number of interesting conference concluding remarks discuss some of the above issues, albeit from a personal viewpoint (Richter 1993, Detraz 1995, Fesbach 1995, Siemssen 1995, Koonin 1994).

I have not been exhaustive in this discussion but one could put all nuclear physics developments within a scheme that strongly correlates them with breakthrough points in technical developments. Every time new techniques and methods have been put into the laboratory, the nucleus has responded with unexpected reactions and answers. Thus the continuous investment in trying to set up new technical methods to reach the extremes will surely pay off and bring in new ways of thinking about and observing the atomic nucleus.

Appendix A

Units and conversion between various unit systems

When discussing equations, in particular involving electromagnetism, we have to decide which system of units to use. The SI (or MKS) system is the obvious choice in the light of present-day physics courses as this is the most familiar system. However, many of the advanced treatments that refer to subatomic physics still mainly use CGS units. We therefore include conversion tables containing the most important equations and relations on electromagnetic properties for SI (MKS, rationalized), Gaussian (CGS), Heaviside–Lorentz (CGS) and natural (using $c = \hbar = 1$, rationalized) units. It is quite possible to write all equations in a form independent of the particular system of units by using the fine structure constant α and by measuring the charge in e units, the absolute value of the charge carried by an electron, and magnetic dipole moments in units of μ_N, the nuclear magneton.

At the same time, we give a list of very useful constants that most often occur when solving numerical problems.

Table A.1. Conversion between SI and CGS unit systems. *Examples*: One metre equals 100 centimetres. One volt equals 10^8 electromagnetic units of potential. (From *Electricity: Principles and Applications* by Lorrain and Corson, © 1978 Paul Lorrain and Dale R Corson. Reprinted with permission of WH Freeman and Co.)

Quantity	SI	CGS Systems esu	emu
Length	metre	10^2 centimetres	10^2 centimetres
Mass	kilogram	10^3 grams	10^3 grams
Time	second	1 second	1 second
Force	newton	10^5 dynes	10^5 dynes
Pressure	pascal	10 dynes/centimetre^2	10 dynes/centimetre^2
Energy	joule	10^7 ergs	10^7 ergs
Power	watt	10^7 ergs/second	10^7 ergs/second
Charge	coulomb	3×10^9	10^{-1}
Electric potential	volt	1/300	10^8
Electric field intensity	volt/metre	$1/(3 \times 10^4)$	10^6
Electric displacement	coulomb/metre^2	$12\pi \times 10^5$	$4\pi \times 10^{-5}$
Displacement flux	coulomb	$12\pi \times 10^9$	$4\pi \times 10^{-1}$
Electric polarization	coulomb/metre^2	3×10^5	10^{-5}
Electric current	ampere	3×10^9	10^{-1}
Conductivity	siemens/metre	9×10^9	10^{-11}
Resistance	ohm	$1/(9 \times 10^{11})$	10^9
Conductance	siemens	9×10^{11}	10^{-9}
Capacitance	farad	9×10^{11}	10^{-9}
Magnetic flux	weber	1/300	10^8 maxwells
Magnetic induction	tesla	$1/(3 \times 10^6)$	10^4 gausses
Magnetic field intensity	ampere/metre	$12\pi \times 10^7$	$4\pi \times 10^{-3}$ oersted
Magnetomotance	ampere	$12\pi \times 10^9$	$(4\pi/10)$ gilberts
Magnetic polarization	ampere/metre	$1/(3 \times 10^{13})$	10^{-3}
Inductance	henry	$1/(9 \times 10^{11})$	10^9
Reluctance	ampere/weber	$36\pi \times 10^{11}$	$4\pi \times 10^{-9}$

Note: We have set $c = 3 \times 10^8$ metres/second.

Table A.2. SI units and their symbols. (From *Electricity: Principles and Applications* by Lorrain and Corson, © 1978 Paul Lorrain and Dale R Corson. Reprinted with permission of WH Freeman and Co.)

Quantity	Unit	Symbol	Dimensions
Length	metre	m	
Mass	kilogram	kg	
Time	second	s	
Temperature	kelvin	K	
Current	ampere	A	
Frequency	hertz	Hz	$1/s$
Force	newton	N	$kg\,m\,s^{-2}$
Pressure	pascal	Pa	$N\,m^{-2}$
Energy	joule	J	$N\,m$
Power	watt	W	$J\,s^{-1}$
Electric charge	coulomb	C	$A\,s$
Potential	volt	V	J/C
Conductance	siemens	S	A/V
Resistance	ohm	Ω	V/A
Capacitance	farad	F	C/V
Magnetic flux	weber	Wb	$V\,s$
Magnetic induction	tesla	T	$Wb\,m^{-2}$
Inductance	henry	H	$Wb\,A^{-1}$

Table A.3. SI prefixes and their symbols. (From *Electricity: Principles and Applications* by Lorrain and Corson, © 1978 Paul Lorrain and Dale R Corson. Reprinted with permission of WH Freeman and Co.)

Multiple	Prefix	Symbol
10^{18}	exa	E
10^{15}	peta	P
10^{12}	tera	T
10^{9}	giga	G
10^{6}	mega	M
10^{3}	kilo	k
10^{2}	hecto	h
10	deka[†]	da
10^{-1}	deci	d
10^{-2}	centi	c
10^{-3}	milli	m
10^{-6}	micro	μ
10^{-9}	nano	n
10^{-12}	pico	p
10^{-15}	femto	f
10^{-18}	atto	a

[†] This prefix is spelled 'déca' in French. Caution: The symbol for the prefix is written next to that for the unit *without* a dot. For example, mN stands for millinewton, and m · N is a metre newton, or a joule.

Table A.4. Fundamental electromagnetic relations valid 'in vacuo' as they appear in the various systems of units. (Wolfgang Panofsky and Melba Philips, *Classical Electricity and Magnetism*, 2nd edn © 1962 by Addison-Wesley Publishing Company Inc. Reprinted with permission.)

MKS (rationalized)	Gaussian* (CGS)	Heaviside–Lorentz (CGS)	'Natural' units $c = \hbar = 1$ (rationalized)
$\nabla \cdot E = \rho/\epsilon_0$	$\nabla \cdot E = 4\pi\rho$	$\nabla \cdot E = \rho$	$\nabla \cdot E = \rho$
$E = \dfrac{1}{4\pi\epsilon_0}\displaystyle\int \dfrac{\rho r}{r^3}\,dv$	$E = \displaystyle\int \dfrac{\rho r}{r^3}\,dv$	$E = \dfrac{1}{4\pi}\displaystyle\int \dfrac{\rho r}{r^3}\,dv$	$E = \dfrac{1}{4\pi}\displaystyle\int \dfrac{\rho r}{r^3}\,dv$
$\nabla \times B = \mu_0\left(j + \epsilon_0 \dfrac{\partial E}{\partial t}\right)$	$\nabla \times B = 4\pi j + \dfrac{1}{c}\dfrac{\partial E}{\partial t}$	$\nabla \times B = j + \dfrac{1}{c}\dfrac{\partial E}{\partial t}$	$\nabla \times B = j + \dfrac{\partial E}{\partial t}$
$B = \dfrac{\mu_0}{4\pi}\displaystyle\int \dfrac{\left(j + \epsilon_0\dfrac{\partial E}{\partial t}\right) \times r}{r^3}\,dv$	$B = \displaystyle\int \dfrac{\left(j + \dfrac{1}{4\pi c}\dfrac{\partial E}{\partial t}\right) \times r}{r^3}\,dv$	$B = \dfrac{1}{4\pi}\displaystyle\int \dfrac{\left(j + \dfrac{1}{c}\dfrac{\partial E}{\partial t}\right) \times r}{r^3}\,dv$	$B = \dfrac{1}{4\pi}\displaystyle\int \dfrac{\left(j + \dfrac{\partial E}{\partial t}\right) \times r}{r^3}\,dv$
$\nabla \cdot B = 0$	$\nabla \cdot B = 0$	$\nabla \cdot B = 0$	$\nabla \cdot B = 0$
$\nabla \times E = -\dfrac{\partial B}{\partial t}$	$c\nabla \times E = -\dfrac{\partial B}{\partial t}$	$c\nabla \times E = -\dfrac{\partial B}{\partial t}$	$\nabla \times E = -\dfrac{\partial B}{\partial t}$
$F = e(E + u \times B)$	$F = e\left(E + \dfrac{u}{-c} \times B\right)$	$F = e\left(E + \dfrac{u}{-c} \times B\right)$	$F = e(E + u \times B)$
$B = \nabla \times A$	$B = \nabla \times A$	$B = \nabla \times A$	$B = \nabla \times A$
$E = -\nabla\phi - \dfrac{\partial A}{\partial t}$	$E = -\nabla\phi - \dfrac{1}{c}\dfrac{\partial A}{\partial t}$	$E = -\nabla\phi - \dfrac{1}{c}\dfrac{\partial A}{\partial t}$	$E = -\nabla\phi - \dfrac{\partial A}{\partial t}$
$\nabla \cdot j + \dfrac{\partial \rho}{\partial t} = 0$	$\nabla \cdot j + \dfrac{1}{c}\dfrac{\partial \rho}{\partial t} = 0$	$\nabla \cdot j + \dfrac{1}{c}\dfrac{\partial \rho}{\partial t} = 0$	$\nabla \cdot j + \dfrac{\partial \rho}{\partial t} = 0$
$\nabla \cdot A + \dfrac{1}{c^2}\dfrac{\partial \phi}{\partial t} = 0$	$\nabla \cdot A + \dfrac{1}{c}\dfrac{\partial \phi}{\partial t} = 0$	$\nabla \cdot A + \dfrac{1}{c}\dfrac{\partial \phi}{\partial t} = 0$	$\nabla \cdot A + \dfrac{\partial \phi}{\partial t} = 0$
$F^{ij} = \begin{pmatrix} 0 & -cB_z & cB_y & +E_x \\ cB_z & 0 & -cB_x & +E_y \\ -cB_y & cB_x & 0 & +E_z \\ -E_x & -E_y & -E_z & 0 \end{pmatrix}$	$F^{ij} = \begin{pmatrix} 0 & -B_z & B_y & +E_x \\ B_z & 0 & -B_x & +E_y \\ -B_y & B_x & 0 & +E_z \\ -E_x & -E_y & -E_z & 0 \end{pmatrix}$	F^{ij} has same form as in Gaussian units	F^{ij} has same form as in Gaussian units
$\partial F^{ij}/\partial x^i = j^j/\epsilon_0$	$\partial F^{ij}/\partial x^i = 4\pi j^j$	$\partial F^{ij}/\partial x^i = j^j$	$\partial F^{ij}/\partial x^i = j^j$

* In some textbooks employing Gaussian units, j is measured in esu. In that case the equations are those given here except that j appears with a factor $1/c$.

Table A.5. Definition of fields from sources (MKS system). (Wolfgang Panofsky and Melba Philips, *Classical Electricity and Mangnetism*, 2nd edn © 1962 by Addison-Wesley Publishing Company Inc. Reprinted with permission.)

	Electric	Magnetic	Equivalent covariant description
Vacuum (all sources) 'accessible'	$\nabla \cdot E = \rho/\epsilon_0$	$\nabla \times B = \mu_0 \left(j + \epsilon_0 \dfrac{\partial E}{\partial t} \right)$	$\dfrac{\partial F^{ij}}{\partial x^i} = \dfrac{j^j}{\epsilon_0}$
Material media, sources separated	$\nabla \cdot E = \dfrac{\rho + \rho_P}{\epsilon_0}$	$\nabla \times B = \mu_0 \left(j + j_M + j_P + \epsilon_0 \dfrac{\partial E}{\partial t} \right)$	$\dfrac{\partial F^{ij}}{\partial x^i} = \dfrac{1}{\epsilon_0}(j^j + j_M^j)$
Inaccessible sources defined from auxiliary function	$\nabla \cdot E = \dfrac{\rho}{\epsilon_0} + \dfrac{(-\nabla \cdot P)}{\epsilon_0}$	$\nabla \times B = \mu_0 \left(j + \nabla \times M + \dfrac{\partial P}{\partial t} + \epsilon_0 \dfrac{\partial E}{\partial t} \right)$	$\dfrac{\partial F^{ij}}{\partial x^i} = \dfrac{1}{\epsilon_0}\left(j^j + \dfrac{\partial M^{ij}}{\partial x^i} \right)$
Definition of partial field	$D = \epsilon_0 E - (-P)$	$H = \dfrac{B}{\mu_0} - M$	$H^{ij} = \epsilon_0 F^{ij} - M^{ij}$
Field equations in media	$\nabla \cdot D = \rho$	$\nabla \times H = j + \dfrac{\partial D}{\partial t}$	$\dfrac{\partial H^{ij}}{\partial x^i} = j^j$

Table A.6. Universal constants. (Samuel S M Wong, *Introductory Nuclear Physics* ©
1990. Reprinted by permission of Prentice-Hall, Englewood Cliffs, NJ.)

Quantity	Symbol	Value	
Speed of light	c	$299\,792\,458$ m s^{-1}	
Unit of charge	e	$1.602\,177\,33(49) \times 10^{-19}$C	$4.803\,206\,8 \times 10^{-10}$ esu
Planck's			
constant	h	$6.626\,075\,5(40) \times 10^{-34}$ J s	$4.135\,669 \times 10^{-21}$ MeV s
(reduced)	$\hbar = h/2\pi$	$1.054\,572\,66(63) \times 10^{-34}$ J s	$6.582\,122\,0 \times 10^{-22}$ MeV s
Fine structure constant	$\alpha = \dfrac{e^2}{4\pi\epsilon_0\hbar c}$	$1/137.035\,989\,5(61)$	
	$\hbar c$		$197.327\,053(59)$ MeV fm

Table A.7. Conversion of units. (Samuel S M Wong, *Introductory Nuclear Physics* ©
1990. Reprinted by permission of Prentice-Hall, Englewood Cliffs, NJ.)

Quantity	Symbol	Value		
Length	fm	10^{-15}	m	
Area	barn	10^{-28}	m^2	
Energy	eV	$1.602\,177\,33(49) \times 10^{-19}$	J	
Mass	eV c^{-2}	$1.782\,662\,70(54) \times 10^{-36}$	kg	
	amu	$1.660\,540\,2(10) \times 10^{-27}$	kg	$931.494\,32(28)$ MeV c^{-2}
Charge	C	$2.997\,924\,58 \times 10^9$	esu	

Table A.8. Lengths. (Samuel S M Wong, *Introductory Nuclear Physics* © 1990.
Reprinted by permission of Prentice-Hall, Englewood Cliffs, NJ.)

Quantity	Symbol	Value
Classical electron radius	$r_e = \dfrac{\alpha\hbar}{m_e c}$	$2.817\,940\,92(35) \times 10^{-15}$ m
Bohr radius	$b_0 = \dfrac{r_e}{\alpha^2}$	$5.291\,772\,49(24) \times 10^{-11}$ m
Compton wavelength	$\lambda_{C_e} = \dfrac{h}{m_e c}$	$2.426\,310\,585(22) \times 10^{-12}$ m
	$\lambda_{C_p} = \dfrac{h}{M_p c}$	$1.321\,41 \times 10^{-17}$ m

Table A.9. Masses. (Samuel S M Wong, *Introductory Nuclear Physics* © 1990. Reprinted by permission of Prentice-Hall, Englewood Cliffs, NJ.)

Quantity	Symbol	Value	
Electron	m_e	$9.109\,389\,74 \times 10^{-31}$ kg	$0.511\,099\,906(15)$ MeV c^{-2}
Muon	m_μ	$1.883\,532\,7 \times 10^{-28}$ kg	$105.658\,39(6)$ MeV c^{-2}
Pions	$\pi^\pm$	$2.488\,018\,7 \times 10^{-28}$ kg	$139.567\,55(33)$ MeV c^{-2}
	π^0	$2.406\,120 \times 10^{-28}$ kg	$134.973\,4(25)$ MeV c^{-2}
Proton	M_p	$1.672\,623\,1 \times 10^{-27}$ kg	$938.272\,31(28)$ MeV c^{-2}
			$1.007\,276\,470(12)$ amu
Neutron	M_n	$1.649\,286\,0 \times 10^{-27}$ kg	$939.565\,63(28)$ MeV c^{-2}
			$1.008\,664\,904(14)$ amu

Table A.10. Other measures. (Samuel S M Wong, *Introductory Nuclear Physics*, © 1990. Reprinted by permission of Prentice-Hall, Englewood Cliffs, NJ.)

Quantity	Symbol	Value
Rydberg energy	$R_\gamma = \frac{1}{2} m_e c^2 \alpha^2$	$13.605\,698\,1(40)$ eV
Bohr magneton	$\mu_B = \dfrac{e\hbar}{2m_e}$	$5.788\,382\,63(52) \times 10^{-11}$ MeV T^{-1}
Nuclear magneton	$\mu_N = \dfrac{e\hbar}{2M_p}$	$3.152\,451\,66(28) \times 10^{-14}$ MeV T^{-1}
Dipole moment (magnetic)		
Electron	μ_e	$1.001\,159\,652\,193(10)\ \mu_B$
Proton	μ_p	$2.792\,847\,386(63)\ \mu_N$
Neutron	μ_n	$1.913\,042\,75(45)\ \mu_N$
Fermi coupling constant	$\dfrac{G_F}{(\hbar c)^3}$	$1.166\,37(2) \times 10^{-5}$ GeV^{-2}
		$1.435\,84(3) \times 10^{-62}$ J m^3
Gamow–Teller/	$\dfrac{G_{GT}}{G_F}$	$1.259(4)$
Fermi coupling constant		
Avogadro number	N_A	$6.022\,136\,7(36) \times 10^{23}$ mol^{-1}
Boltzmann constant	k	$1.380\,658 \times 10^{-23}$ J K^{-1}
		$8.617\,385\,(73) \times 10^{-11}$ MeV K^{-1}
Permittivity (free space)	ϵ_0	$8.854\,187\,817 \times 10^{-12}$ C^{2}N^{-1}m^{-2}
Permeability (free space)	μ_0	$4\pi \times 10^{-7}$ N.A^{-2}
		$\epsilon_0 \mu_0 = c^2$

Appendix B

Spherical tensor properties

In this book, use is made quite often of spherical harmonics, angular momentum coupling and Wigner $\mathcal{D}$ matrices for describing rotational wavefunctions the Wigner–Eckart theorem. Without going into a lengthy discussion of angular momentum algebra, we briefly discuss the major properties and results needed here. For derivations and a consistent discussion we refer to text books on angular momentum algebra and tensor properties.

B.1 Spherical harmonics

The Hamiltonian describing the motion of a particle in a central field $U(|\vec{r}|)$ is

$$\hat{H} = \frac{\hat{p}^2}{2m} + U(|\vec{r}|). \tag{B.1}$$

Using spherical coordinates, the Laplacian is obtained as

$$\Delta \equiv \frac{1}{r^2} \frac{\partial}{\partial r} r^2 \frac{\partial}{\partial r} + \frac{1}{r^2} \left[\frac{1}{\sin\theta} \frac{\partial}{\partial \theta} \left(\sin\theta \frac{\partial}{\partial \theta} \right) + \frac{1}{\sin^2\theta} \frac{\partial^2}{\partial \varphi^2} \right]$$

$$= \frac{1}{r^2} \frac{\partial}{\partial r} r^2 \frac{\partial}{\partial r} - \frac{\hat{l}^2}{\hbar^2 r^2}, \tag{B.2}$$

with

$$\hat{l} = \hat{r} \times \hat{p}, \tag{B.3}$$

the angular momentum operator, for which the z component is

$$\hat{l}_z = -i\hbar \frac{\partial}{\partial \varphi}. \tag{B.4}$$

It is now clear that

$$[\hat{H}, \hat{l}^2] = 0, \qquad [\hat{H}, \hat{l}_z] = 0, \tag{B.5}$$

and that the eigenfunctions $\varphi(\vec{r})$ of $\hat{H}$ will also be eigenfunctions of $\hat{l}^2$ and $\hat{l}_z$. For a central potential $U(r)$, the separation in the radial and angular variables

$$\varphi(\vec{r}) = R(r)Y(\theta, \varphi), \tag{B.6}$$

leads to the angular momentum eigenvalue equations

$$\hat{l}^2 Y(\theta, \varphi) = \lambda\hbar^2 Y(\theta, \varphi)$$
$$\hat{l}_z Y(\theta, \varphi) = m\hbar Y(\theta, \varphi). \tag{B.7}$$

Functions that are solutions of equations (B.7) are proportional to the spherical harmonics $Y_l^m(\theta, \varphi)$ with the explicit form (with $m > 0$)

$$Y_l^m(\theta, \varphi) = \frac{(-1)^m}{2^l l!} \sqrt{\frac{(2l+1)}{4\pi} \frac{(l-m)!}{(l+m)!}}$$
$$\times e^{im\varphi}(1 - \cos^2\theta)^{m/2} \frac{d^{l+m}}{d\cos\theta^{l+m}}(\cos^2\theta - 1)^l, \tag{B.8}$$

leading to the result $\lambda = l(l+1)$. For negative m values one has the relation

$$Y_l^{-m}(\theta, \varphi) = (-1)^m Y_l^{m*}(\theta, \varphi). \tag{B.9}$$

The spherical harmonics can also be given for a complete, orthonormal set of functions on the unit sphere, i.e.

$$\int_0^{2\pi} d\varphi \int_0^{\pi} \sin\theta \, d\theta \, Y_l^{m*}(\theta, \varphi) Y_{l'}^{m'}(\theta, \varphi) = \delta_{ll'}\delta_{mm'}. \tag{B.10}$$

The symmetry of these spherical harmonics under the parity transformation $\vec{r} \to -\vec{r}$ (or $\theta \to \pi - \theta$; $\varphi \to \varphi + \pi$; $r \to r$) gives

$$Y_l^m(\pi - \theta, \pi + \varphi) = (-1)^l Y_l^m(\theta, \varphi). \tag{B.11}$$

A number of other, useful relations are

$$Y_l^m(\theta, \varphi) = \sqrt{\frac{2l+1}{4\pi}} \delta_{m,0}$$
$$Y_l^0(\theta, \varphi) = \sqrt{\frac{2l+1}{4\pi}} P_l(\cos\theta), \tag{B.12}$$

with

$$P_l(\cos\theta) = \frac{1}{2^l l!} \frac{d^l}{(d\cos\theta)^l}(\cos^2\theta - 1)^l. \tag{B.13}$$

Finally, we give for $l = 0, 1, 2$ (most often used) the explicit expressions for $Y_l^m(\theta, \varphi)$

$$Y_0^0(\theta, \varphi) = \sqrt{\frac{1}{4\pi}}$$

$$Y_1^0(\theta, \varphi) = \sqrt{\frac{3}{4\pi}} \cos\theta$$

$$Y_1^{\pm 1}(\theta, \varphi) = \mp\sqrt{\frac{3}{8\pi}} \sin\theta e^{\pm i\varphi}$$

$$Y_2^0(\theta, \varphi) = \sqrt{\frac{5}{16\pi}} (3\cos^2\theta - 1)$$

$$Y_2^{\pm 1}(\theta, \varphi) = \mp\sqrt{\frac{15}{8\pi}} \cos\theta \sin\theta e^{\pm i\varphi}$$

$$Y_2^{\pm 2}(\theta, \varphi) = \sqrt{\frac{15}{32\pi}} \sin^2\theta e^{\pm 2i\varphi}. \tag{B.14}$$

B.2   Angular momentum coupling: Clebsch–Gordan coefficients

Here, we concentrate on the coupling of two distinct angular momenta $\hat{j}_1$, $\hat{j}_2$ (with z-projection components $\hat{j}_{1,z}$, $\hat{j}_{2,z}$). A set of four commuting operators is

$$\{\hat{j}_1^2, \hat{j}_{1,z}, \hat{j}_2^2, \hat{j}_{2,z}\}, \tag{B.15}$$

characterized by the common eigenvectors

$$|j_1 m_1, j_2 m_2\rangle \equiv |j_1 m_1\rangle |j_2 m_2\rangle. \tag{B.16}$$

A different set of commuting operators becomes (using the sum $\hat{j} = \hat{j}_1 + \hat{j}_2$)

$$\{\hat{j}^2, \hat{j}_z, \hat{j}_1^2, \hat{j}_2^2\}, \tag{B.17}$$

with eigenvectors

$$|j_1 j_2, jm\rangle. \tag{B.18}$$

The latter eigenvectors can be expressed as linear combinations of the eigenvectors (B.16) such that the total angular momentum operator $\hat{j}^2$ is diagonalized. The transformation coefficients performing this process are called the Clebsch–Gordan coefficients

$$|j_1 j_2, jm\rangle = \sum_{m_1, m_2} \langle j_1 m_2, j_2 m_2 | j_1 j_2, jm\rangle |j_1 m_1\rangle |j_2 m_2\rangle, \tag{B.19}$$

or, in a simplified notation,

$$\langle j_1 m_1, j_2 m_2 | j_1 j_2, jm\rangle \rightarrow \langle j_1 m_1, j_2 m_2 | jm\rangle. \tag{B.20}$$

Table B.1.

j	$m_s = +\frac{1}{2}$	$m_s = -\frac{1}{2}$
$l + \frac{1}{2}$	$\sqrt{\dfrac{l + \frac{1}{2} + m}{2l + 1}}$	$\sqrt{\dfrac{l + \frac{1}{2} - m}{2l + 1}}$
$l - \frac{1}{2}$	$\sqrt{\dfrac{l + \frac{1}{2} - m}{2l + 1}}$	$-\sqrt{\dfrac{l + \frac{1}{2} + m}{2l + 1}}$

In table B.1, we list the Clebsch–Gordan coefficients for coupling an orbital angular momentum l with an intrinsic spin $s = \frac{1}{2}$ to a total angular momentum $j = l \pm \frac{1}{2}$.

The algebra for constructing the various Clebsch–Gordan coefficients according to the Condon–Shortley phase convention is discussed in Heyde (1991) and is not elaborated on at present.

A number of interesting orthogonality properties can be derived such as

$$\sum_{m_1, m_2} \langle j_1 m_1, j_2 m_2 | j m \rangle \langle j_1 m_1, j_2 m_2 | j' m' \rangle = \delta_{jj'} \delta_{mm'}$$

$$\sum_{j, m} \langle j_1 m_1, j_2 m_2 | j m \rangle \langle j_1 m_1', j_2 m_2' | j m \rangle = \delta_{m_1 m_1'} \delta_{m_2 m_2'}. \tag{B.21}$$

One can, however, make use of the Wigner $3j$-symbol, related to the Clebsch–Gordan coefficient,

$$\begin{pmatrix} j_1 \, j_2 \, j_3 \\ m_1 m_2 m_3 \end{pmatrix} = \frac{(-1)^{j_1 - j_2 - m_3}}{\sqrt{2 j_3 + 1}} \langle j_1 m_1, j_2 m_2 | j_3 - m_3 \rangle, \tag{B.22}$$

to define a number of relations amongst permuting various columns in the Wigner $3j$-symbol. An even permutation introduces a factor $+1$, an odd permutation a phase factor $(-1)^{j_1 + j_2 + j_3}$ and a change of all $m_i \rightarrow -m_i$ gives this same phase factor change.

Extensive sets of tables of Wigner $3j$-symbols exist (Rotenberg *et al* 1959) but explicit calculations are easily performed using the closed expression (de Shalit and Talmi (1963); Heyde (1991) for a FORTRAN program).

B.3 Racah recoupling coefficients—Wigner $6j$-symbols

When describing a system of three independent angular momentum operators $\hat{j}_1, \hat{j}_2, \hat{j}_3$, one can, as under (B.15), form the six commuting operators

$$\{\hat{j}_1^2, \hat{j}_{1,z}, \hat{j}_2^2, \hat{j}_{2,z}, \hat{j}_3^2, \hat{j}_{3,z}\}, \tag{B.23}$$

with the common eigenvectors

$$|j_1 m_1\rangle|j_2 m_2\rangle|j_3 m_3\rangle. \tag{B.24}$$

Forming the total angular momentum operator

$$\hat{j} = \hat{j}_1 + \hat{j}_2 + \hat{j}_3, \tag{B.25}$$

we can then form three independent sets of commuting operators

$$\{\hat{j}^2, \hat{j}_z, \hat{j}_1^2, \hat{j}_2^2, \hat{j}_3^2, \hat{j}_{12}^2\}$$
$$\{\hat{j}^2, \hat{j}_z, \hat{j}_1^2, \hat{j}_2^2, \hat{j}_3^2, \hat{j}_{23}^2\}$$
$$\{\hat{j}^2, \hat{j}_z, \hat{j}_1^2, \hat{j}_2^2, \hat{j}_3^2, \hat{j}_{13}^2\}, \tag{B.26}$$

with the sets of eigenvectors

$$|(j_1 j_2) j_{12}, j_3; jm\rangle$$
$$|j_1 (j_2 j_3) j_{23}; jm\rangle$$
$$|(j_1 j_3) j_{23}, j_2; jm\rangle, \tag{B.27}$$

respectively.

Between the three equivalent sets of eigenvectors (B.27), transformations that change from one basis to the other can be constructed. Formally, such a transformation can be written as

$$|j_1 (j_2 j_3) j_{23}, jm\rangle = \sum_{j_{12}} \langle (j_1 j_2) j_{12}, j_3; j | j_1 (j_2 j_3) j_{23}; j \rangle |(j_1 j_2) j_{12}, j_3; jm\rangle. \tag{B.28}$$

By explicitly carrying out the recoupling from the basis configurations as used in (B.28), one derives the detailed expression for the recoupling or transformation coefficients which no longer depend on the magnetic quantum number m. In this situation, a full sum over all magnetic quantum numbers of products of four Wigner 3 j -symbols results. The latter, become an angular momentum invariant quantity, the Wigner 6 j -symbol, and leads to the following result (Wigner 1959, Brussaard 1967)

$$\langle j_1 (j_2 j_3) j_{23}; j | (j_1 j_2) j_{12}, j_3; j \rangle$$
$$= (-1)^{j_1 + j_2 + j_3 + j} \sqrt{(2 j_{12} + 1)(2 j_{23} + 1)} \begin{Bmatrix} j_1 & j_2 & j_{12} \\ j_3 & j & j_{23} \end{Bmatrix}, \tag{B.29}$$

with the Wigner 6 j -symbol {...} defined as

$$\begin{Bmatrix} j_1 & j_2 & j_3 \\ l_1 & l_2 & l_3 \end{Bmatrix} = \sum_{\text{all } m_i, m_i'} (-1)^{\Sigma j_i + \Sigma l_i + \Sigma m_i + \Sigma m_i'} \begin{pmatrix} j_1 & j_2 & j_3 \\ m_1 & m_2 & m_3 \end{pmatrix}$$
$$\times \begin{pmatrix} j_1 & l_2 & l_3 \\ -m_1 & m_2' & -m_3' \end{pmatrix} \begin{pmatrix} l_1 & j_2 & l_3 \\ -m_1' & -m_2 & m_3' \end{pmatrix}$$
$$\times \begin{pmatrix} l_1 & l_2 & j_3 \\ m_1' & -m_2' & -m_3 \end{pmatrix}, \tag{B.30}$$

(here one sums over all projection quantum numbers m_i, m_i' both of which occur in different $3j$-symbols with opposite sign).

These Wigner $6j$-symbols exhibit a large class of symmetry properties under the interchange of various rows and/or columns, which are discussed in various texts on angular momentum algebra (de-Shalit and Talmi 1963, Rose and Brink 1967, Brussaard 1967, Brink and Satchler 1962).

B.4   Spherical tensor and rotation matrix

In quantum mechanics, the rotation operator associated with a general angular momentum operator $\hat{\jmath}$ and describing the rotation properties of the eigenvectors $|jm\rangle$ for an (active) rotation about an angle (α) around an axis defined by a unit vector $\mathbb{1}_n$, becomes

$$U_R = \mathrm{e}^{(-\mathrm{i}/\hbar)\hat{\alpha}\cdot\hat{J}}. \tag{B.31}$$

Representations of this general rotation operator are formed by the set of square $n \times n$ matrices that follow the same group rules as the rotation operator U_R itself.

The rotated state vector, obtained by acting with U_R on $|jm\rangle$ vectors, called $|jm\rangle'$, becomes

$$|jm\rangle' = U_R|jm\rangle$$
$$|jm\rangle' = \sum_{m'}\langle jm'|U_R|jm\rangle|jm'\rangle. \tag{B.32}$$

The $(2j+1) \times (2j+1)$ matrices

$$\langle jm'|\mathrm{e}^{(-\mathrm{i}/\hbar)\hat{\alpha}\cdot\hat{J}}|jm\rangle,$$

are called the representation matrices of the rotation operator U_R and are denoted by $D_{m',m}^{(j)}(R)$, the Wigner D-matrices. Various other ways to define the D matrices have been used in angular momentum algebra. We use the active standpoint. A rotation of the coordinate axes in a way defined in figure B.1 can be achieved by the rotation operator

$$\hat{R}(\alpha, \beta, \gamma) = \mathrm{e}^{-\mathrm{i}\gamma J_{z'}}\mathrm{e}^{-\mathrm{i}\beta J_{y_1}}\mathrm{e}^{-\mathrm{i}\alpha J_z}, \tag{B.33}$$

or, equivalently, in terms of rotations around fixed axes

$$\hat{R}(\alpha, \beta, \gamma) = \mathrm{e}^{-\mathrm{i}\alpha J_z}\mathrm{e}^{-\mathrm{i}\beta J_y}\mathrm{e}^{-\mathrm{i}\gamma J_z}. \tag{B.34}$$

The general D-matrix then becomes

$$D_{m',m}^{(j)}(\alpha, \beta, \gamma) = \langle jm'|\hat{R}(\alpha, \beta, \gamma)|jm\rangle. \tag{B.35}$$

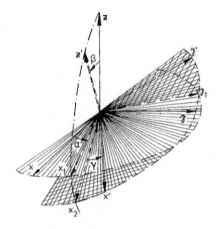

Figure B.1. Rotation of the system (x, y, z) into the (x', y', z') position, using the Euler angles (α, β, γ). A rotation around the z-axis (α) brings (x, y, z) into the position (x_1, y_1, z). A second rotation around the new y_1-axis (β) brings (x_1, y_1, z') into the position (x_2, y_1, z'). A final, third rotation around the z'-axis brings (x_2, y_1, z') into the position (x', y', z').

The following important properties are often used:

$$\sum_{m'} D^{(j)}_{m,m'}(\alpha\beta\gamma)(D^{(j)}_{n,m'}(\alpha, \beta, \gamma))^* = \delta_{m,n}, \tag{B.36}$$

$$\int D^{(j')}_{m',m}(\alpha, \beta, \gamma)D^{(j)}_{n',n}(\alpha, \beta, \gamma) \sin\beta \, d\beta \, d\alpha \, d\gamma = \frac{8\pi^2}{2j+1}\delta_{jj'}\delta_{mn}\delta_{m'n'}. \tag{B.37}$$

The $2k + 1$ components $T^{(k)}_\kappa$ represent the components of a spherical tensor operator of rank k if the components transform under rotation according to

$$T^{(k)'}_\kappa = \sum_{\kappa'} D^{(k)}_{\kappa',\kappa}(R)T^{(k)}_{\kappa'}. \tag{B.38}$$

The conjugate $(T^{(k)}_\kappa)^+$ of $T^{(k)}_\kappa$ is defined through the relation between the Hermitian conjugate matrix elements

$$\langle jm|(T^{(k)}_\kappa)^+|j'm'\rangle = \langle j'm'|T^{(k)}_\kappa|jm\rangle^*. \tag{B.39}$$

It is quite easy to prove that the $(T^{(k)}_\kappa)^+$ do not form again a spherical tensor operator; however, the tensor

$$\tilde{T}^{(k)}_\kappa \equiv (-1)^{k-\kappa}(T^{(k)}_{-\kappa})^+, \tag{B.40}$$

does transform according to the properties of a spherical tensor operator. The phase factor $(-1)^\kappa$ is unique; the factor $(-1)^k$ is somewhat arbitrary and other conventions have been used in the existing literature. The use of spherical tensor operators is advantageous, in particular when evaluating matrix elements between eigenvectors that are characterized by a good angular momentum and projection (j, m). In that context, an interesting theorem exists which allows for a separation between the magnetic projection quantum numbers and the remaining characteristics of operators and eigenvectors.

B.5   Wigner–Eckart theorem

Matrix elements $\langle jm|T_\kappa^{(k)}|j'm'\rangle$ may be separated into two parts, one of them containing information on the magnetic quantum numbers, and the other part containing more the characteristics of the eigenvectors and operator. The Wigner–Eckart theorem separates this dependence in a specific way by the result

$$\langle jm|T_\kappa^{(k)}|j'm'\rangle = (-1)^{j-m} \begin{pmatrix} j & k & j' \\ -m & \kappa & m' \end{pmatrix} \langle j\|T^{(k)}\|j'\rangle. \tag{B.41}$$

The double-barred matrix element is called the reduced matrix element whereas the angular momentum dependence is expressed through the Wigner $3j$-symbol. All the physical content is contained within the reduced matrix element. Expressed via the Clebsch–Gordan coefficient, (B.41) this becomes

$$\langle jm|T_\kappa^{(k)}|j'm'\rangle = (-1)^{2k} \frac{\langle j'm', k\kappa|jm\rangle}{\sqrt{2j+1}} \langle j\|T^{(k)}\|j'\rangle. \tag{B.42}$$

Appendix C

Second quantization—an introduction

In a quantum mechanical treatment of the matter field, a possibility exists of handling the quantum mechanical time-dependent Schrödinger (or Dirac) equation as a field equation. Using methods similar to quantizing the radiation field, a method of 'second quantization' can be set up in order to describe the quantum mechanical many-body system.

Starting from the time-dependent Schrödinger equation

$$-\frac{\hbar^2}{2m}\Delta\Psi(\vec{r},t) + U(\vec{r})\Psi(\vec{r},t) = \hat{H}\Psi(\vec{r},t) = i\hbar\frac{\partial\Psi(\vec{r},t)}{\partial t}, \qquad (C.1)$$

one can expand the solution in a basis, spanned by the solutions, to the time-independent Schrödinger equation. So, one can make the expansion

$$\Psi(\vec{r},t) = \sum_n b_n(t)\psi_n(\vec{r}). \qquad (C.2)$$

The $b_n(t)$ can be considered to be the normal coordinates for the dynamical system; the $\psi_n(\vec{r})$ then describe the normal modes of the system. Substituting (C.2) into (C.1), one obtains the equation of motion for the $b_n(t)$ as

$$\frac{db_n(t)}{dt} = -\frac{i}{\hbar}E_n b_n. \qquad (C.3)$$

We now construct the Hamilton expectation value which is a number that will only depend on the b_n coefficients starting from the expectation value

$$H = \int d\vec{r}\,\Psi^*(\vec{r},t)\left(-\frac{\hbar^2}{2m}\Delta + U(\vec{r})\right)\Psi(\vec{r},t) = \int d\vec{r}\,\mathcal{H}(\vec{r}), \qquad (C.4)$$

with $\mathcal{H}(\vec{r})$ the energy-density. Using again the expansion of (C.2) we obtain the following expression for H as

$$H = \sum_n E_n b_n^* b_n. \qquad (C.5)$$

This expression very much resembles a Hamilton function arising from an infinite number of oscillations with energy E_n and frequency $\omega_n = E_n/\hbar$. Here, one can go to a quantization of the matter field $\Psi(\vec{r}, t)$ by interpreting

$$b_n^* \rightarrow \hat{b}_n^+, \qquad b_n \rightarrow \hat{b}_n, \tag{C.6}$$

as creation and annihilation operators, respectively.

There now exists the possibility of obtaining the relevant equation of motion for the $\hat{b}_n$ (and $\hat{b}_n^+$) operators, starting from the Heisenberg equation of motion

$$i\hbar \frac{d\hat{b}_n}{dt} = [\hat{b}_n, \hat{H}]_-, \tag{C.7}$$

with

$$\hat{H} = \sum_n E_n \hat{b}_n^+ \hat{b}_n. \tag{C.8}$$

Using commutation properties for the operators, i.e.

$$[\hat{b}_n, \hat{b}_n']_- = 0, \qquad [\hat{b}_n^+, \hat{b}_{n'}^+]_- = 0, \qquad [\hat{b}_n, \hat{b}_{n'}^+] = \delta_{nn'}, \tag{C.9}$$

the equation (C.7) results in the equation

$$i\hbar \frac{d\hat{b}_n}{dt} = E_n \hat{b}_n, \tag{C.10}$$

which is formally identical to equation (C.3).

With the commutation relations (C.9), we describe the properties of Bose–Einstein particles and it is possible to form eigenstates of the number operator $\hat{N}_n \equiv \hat{b}_n^+ \hat{b}_n$. The normalized Bose–Einstein eigenvectors are constructed as

$$|n_m\rangle = \frac{1}{\sqrt{n_m!}} (\hat{b}_m^+)^{n_m} |0\rangle. \tag{C.11}$$

One can easily prove that the following properties hold

$$\hat{b}_m^+ |n_m\rangle = \sqrt{n_m + 1} |n_m + 1\rangle$$
$$\hat{b}_m |n_m\rangle = \sqrt{n_m} |n_m - 1\rangle$$
$$\hat{N}_m |n_m\rangle = n_m |n_m\rangle. \tag{C.12}$$

A product state where $n_1, n_2, \ldots, n_m, n_k, \ldots$ bosons are present corresponding to the elementary creation operators $\hat{b}_1^+, \hat{b}_2^+, \ldots, \hat{b}_m^+, \hat{b}_k^+ \ldots$ can then be constructed as the direct product state

$$|n_1, \ldots, n_m, n_k, \ldots\rangle \equiv |n_1\rangle \ldots |n_m\rangle |n_k\rangle \ldots. \tag{C.13}$$

There also exists the possibility of starting from a set of operators $\hat{b}_n^+, \hat{b}_n$ which fulfil anti-commutation relations, i.e.

$$[\hat{b}_{n'}, \hat{b}_{n'}]_+ = 0, \qquad [\hat{b}_{n'}^+, \hat{b}_{n'}^+]_+ = 0, \qquad [\hat{b}_n^+, \hat{b}_{n'}]_+ = \delta_{nn'}. \tag{C.14}$$

In evaluating the equation of motion again, but now using anti-commutation properties for the operators, we can show that

$$i\hbar \frac{d\hat{b}_n}{dt} = [\hat{b}_n, \hat{H}]_- = E_n \hat{b}_n, \tag{C.15}$$

which is again equivalent to (C.10) (and the classical equation (C.3)).

We now indicate that the anti-commutation relations imply the Pauli principle and correspond to Fermi–Dirac particles. To start with, one can easily derive that for the number operator $\hat{N}_n$ one has $\hat{N}_n^2 = \hat{N}_n$. Then, the eigenvalue equation for $\hat{N}_n$ leads to

$$\hat{N}_n|\lambda\rangle = \lambda|\lambda\rangle. \tag{C.16}$$

Since $\hat{N}_n^2 = \hat{N}_n$, one derives that

$$\hat{N}_n^2|\lambda\rangle = \lambda^2|\lambda\rangle = \lambda|\lambda\rangle \quad \text{and} \quad \lambda(\lambda - 1) = 0, \tag{C.17}$$

with $\lambda = 0, \lambda > 1$ as solutions.

So, the number operator can have as eigenvalue 0, or 1 implying the Pauli principle for this particular kind of anti-commuting operators. Starting from the eigenvalue equation

$$\hat{N}_n|n_n\rangle = n_n|n_n\rangle, \tag{C.18}$$

we then determine the properties of the states

$$\hat{b}_n^+|n_n\rangle \quad \text{and} \quad \hat{b}_n|n_n\rangle. \tag{C.19}$$

This can be done most easily by evaluating

$$\hat{N}_n \hat{b}_n^+|n_n\rangle = (1 - n_n)\hat{b}_n^+|n_n\rangle. \tag{C.20}$$

This result indicates that the state $\hat{b}_n^+|n_n\rangle$ is an eigenstate of the number operator with eigenvalue $(1 - n_n)$. We can describe this as

$$\hat{b}_n^+|n_n\rangle = c_n|1 - n_n\rangle. \tag{C.21}$$

The number c_n is derived through evaluating the norm on both sides of (C.21) and making use of the anti-commutating operator properties, i.e.

$$\langle n_n|\hat{b}_n \hat{b}_n^+|n_n\rangle = c_n^2\langle 1 - n_n|1 - n_n\rangle = c_n^2. \tag{C.22}$$

The latter expression (C.22) results in

$$c_n = \theta_n\sqrt{1 - n_n}, \tag{C.23}$$

with θ_n a phase factor. In a similar way, we finally derive that

$$\hat{b}_n|n_n\rangle = d_n|1 - n_n\rangle$$
$$d_n = \theta_n'\sqrt{n_n}. \tag{C.24}$$

In contrasting the boson and fermion statistics cases, we recall the following results (for a single mode only)

$$\hat{b}_n|n_n\rangle = \sqrt{n_n}|n_n - 1\rangle \qquad \hat{b}_n|n_n\rangle = \theta'_n\sqrt{n_n}|1 - n_n\rangle$$
$$\hat{b}_n^+|n_n\rangle = \sqrt{n_n + 1}|n_n + 1\rangle, \qquad \hat{b}_n^+|n_n\rangle = \theta_n\sqrt{1 - n_n}|1 - n_n\rangle. \quad \text{(C.25)}$$

In both casses $\hat{b}_n^+(\hat{b}_n)$ acts as a creation (annihilation) operator for bosons (fermions), respectively.

In constructing the many-fermion product rate

$$|n_1, n_2, \ldots, n_m, n_k, \ldots\rangle \equiv |n_1\rangle|n_2\rangle \ldots |n_m\rangle|n_k\rangle \ldots, \quad \text{(C.26)}$$

according to equation (C.13) but now with anti-commutating operator properties, one obtains the general results

$$\hat{b}_k|n_1, \ldots, n_k, \ldots\rangle = (-1)^{\sum_{v=1}^{k-1} n_v}\sqrt{n_k}|n_1, \ldots, 1 - n_k, \ldots\rangle$$
$$\hat{b}_k^+|n_1, \ldots, n_k, \ldots\rangle = (-1)^{\sum_{v=1}^{k-1} n_v}\sqrt{1 - n_k}|n_1, \ldots, 1 - n_k, \ldots\rangle, \quad \text{(C.27)}$$

where the phase factor expresses the fact that the $\hat{b}_k(\hat{b}_k^+)$ operator has to be put at the correct position and therefore has to be permuted with n_1 operators $\hat{b}_1^+$, n_2 operators $\hat{b}_2^+, \ldots, n_{k-1}$ operators $\hat{b}_{k-1}^+$; explaining this permutation factor of (C.27).

It was Pauli (Pauli 1940) who showed that the two statistics discussed here correspond to particles with integer spin (Bose statistics) and particles with half-integer spin (Fermi statistics). This spin-statistics theorem is an important accomplishment of quantum-field theory.

The above properties can also be used to define field operators via the expansion

$$\hat{\Psi}(\vec{r}, t) = \sum_n \hat{b}_n \psi_n(\vec{r})$$
$$\hat{\Psi}^+(\vec{r}, t) = \sum_n \hat{b}_n^+ \psi_n^*(\vec{r}), \quad \text{(C.28)}$$

field operators which obey the famous field (anti)commutation relations

$$[\hat{\Psi}(\vec{r}, t), \hat{\Psi}(\vec{r}', t)]_\mp = 0$$
$$[\hat{\Psi}^+(\vec{r}, t), \hat{\Psi}^+(\vec{r}', t)]_\mp = 0$$
$$[\hat{\Psi}^+(\vec{r}, t), \hat{\Psi}(\vec{r}', t)]_\mp = \delta(\vec{r} - \vec{r}'). \quad \text{(C.29)}$$

The formulation of quantum mechanics using creation and annihilation operators is totally equivalent to the more standard discussion of quantum mechanics using wavefunctions $\Psi(\vec{r}, t)$, solutions of the time-dependent Schrödinger (Dirac, Gordan–Klein, etc) equation. The present formalism though is quite interesting, since in many cases it can describe the quantum dynamics of many-body systems.

References

Nuclear Physics, Intermediate Energy Physics, Introductory Particle Physics

Bhadhuri R K 1988 *Models of the Nucleon* (Reading, MA: Addison-Wesley)

Blatt J M and Weisskopf V F 1952 *Theoretical Nuclear Physics* (New York: Wiley)

Bodenstedt E 1979 *Experimente der Kernphysik und Deuting* Teil 1, 2 and 3 (Bibliographisches Institut Mannheim/Wien/Zürich: Wisschenschaftsverlag)

Boehm F and Vogel P 1987 *Physics of Massive Neutrons* (New York: Cambridge University Press)

Bohr A and Mottelson B 1969 *Nuclear Structure* vol 1 (New York: Benjamin)

——1975 *Nuclear Structure* vol 2 (New York: Benjamin)

Bopp F W 1989 *Kerne, Hadronen und Elementarteilchen* (Stuttgart: Teubner Studienbucher Physik)

Brink D M and Satchler G R 1962 *Angular Momentum* (London: Oxford University Press)

Brussaard P J and Glaudemans P W M 1977 *Shell-Model Applications in Nuclear Spectroscopy* (Amsterdam: North-Holland)

Bucka H and de Gruyter W 1973 *Atomkerne und Elementarteilchen* (Berlin: Walter de Gruyter)

Cahn R N and Goldhaber G 1989 *The Experimental Foundations of Particle Physics* (Cambridge: Cambridge University Press)

Casten R F, Lipas P O, Warner D D, Otsuka T, Heyde K and Draayer J P 1993 *Algebraic Approaches to Nuclear Structure: Interacting Boson and Fermion Models* (New York: Harwood Academic)

Clayton D D 1968 *Principles of Stellar Evolution* (New York: McGraw-Hill)

Close F, Martin M and Sutton C 1987 *The Particle Explosion* (Oxford: Oxford University Press)

Condon E U and Odishaw H 1967 *Handbook of Physics* (New York: McGraw-Hill)

Covello A (ed) 1996 *Building Blocks of Nuclear Structure* (Singapore: World Scientific) and earlier conference proceedings

Creutz M 1983 *Quarks, Gluons and Lattices* (Cambridge: Cambridge University Press)

de-Shalit A and Fesbach H 1974 *Theoretical Nuclear Physics* (New York: Wiley)

de-Shalit A and Talmi I 1963 *Nuclear Shell Theory* (New York: Academic)

Eisenberg J M and Greiner W 1976 *Microscopic Theory of the Nucleus* vol 3 (Amsterdam: North-Holland)

——1987 *Nuclear Models* vol 1, 3rd edn (Amsterdam: North-Holland)

611

——1987 *Excitation Mechanisms of the Nucleus* vol 2, 3rd edn (Amsterdam: North-Holland)

Evans R D 1955 *The Atomic Nucleus* (New York: McGraw-Hill)

Fermi E 1950 *Course on Nuclear Physics* (Chicago, IL: University of Chicago Press)

Frauenfelder H and Henley E M 1991 *Subatomic Physics* 2nd edn (New York: Prentice-Hall)

Friedlander G and Kennedy J W 1949 *Introduction to Radio Chemistry* (New York: Wiley)

Green A E S 1955 *Nuclear Physics* (New York: McGraw-Hill)

Grotz K and Klapdor H V 1990 *The Weak Interaction in Nuclear, Particle and Astrophysics* (Bristol: Adam Hilger)

Heyde K 1991 *The Nuclear Shell Model, Springer Series in Nuclear and Particle Physics* (Berlin: Springer)

——1998 *From Nucleons to the Atomic Nucleus: Perspectives in Nuclear Physics* (Berlin: Springer)

——1999 *Basic Ideas and Concepts in Nuclear Physics* 2nd edn (Bristol: Institute of Physics Publishing)

Holstein B 1989 *Weak Interactions in Nuclei* (Princeton, NJ: Princeton University Press)

Hornyack W F 1975 *Nuclear Structure* (New York: Academic)

Iachello F and Arima A 1988 *The Interacting Boson Model* (Cambridge: Cambridge University Press)

Iachello F and Van Isacker P 1991 *The Interacting Boson–Fermion Model* (Cambridge: Cambridge University Press)

Jelley N A 1990 *Fundamentals of Nuclear Physics* (Cambridge: Cambridge University Press)

Kaplan I 1963 *Nuclear Physics* 2nd edn (Reading, MA: Addison-Wesley)

Krane K S 1987 *Introductory Nuclear Physics* (New York: Wiley)

Kumar K 1984 *Nuclear Models and the Search for Unity in Nuclear Physics* (Oslo: Universitetsforlanger)

Marmier P and Sheldon E 1969 *Physics of Nuclei and Particles* vol 1 (New York: Academic)

Mayer M G and Jensen H D 1955 *Elementary Theory of Nuclear Shell Structure* (New York: Wiley)

Mayer-Kuckuk T 1979 *Kernphysik* 2nd edn (Stuttgart: Teubner Studienbucher Physik)

Meyerhof W 1967 *Elements of Nuclear Physics* (New York: McGraw-Hill)

Morita M 1973 *Beta Decay and Muon Capture* (New York: Benjamin)

Perkins D H 1987 *Introduction to High Energy Physics* 3rd edn (Reading, MA: Addison-Wesley)

Phillips A C 1994 *The Physics of Stars* (New York: Wiley)

Preston M A and Bhaduri R K 1975 *Structure of the Nucleus* (Reading, MA: Addison-Wesley)

Ring P and Schuck P 1980 *The Nuclear Many-Body Problem* (Berlin: Springer)

Rolfs C E and Rodney W S 1988 *Cauldrons in the Cosmos* (Chicago: University of Chicago Press)

Rose M E 1965 α, β and γ *Spectroscopy* ed K Siegbahn (Amsterdam: North-Holland) vol 10

Rotenberg M, Bivins R, Metropolis N and Wooten J K Jr 1959 *The 3-j and 6-j Symbols* (Cambridge, MA: MIT Technology)

Rowe D J 1970 *Nuclear Collective Motion* (New York: Methuen)

Satchler G R 1990 *Introduction to Nuclear Reactions* 2nd edn (London: MacMillan)
Segré E 1982 *Nuclei and Particles* 3rd edn (New York: Benjamin)
Siemens P J and Jensen A S 1987 *Elements of Nuclei: Many-Body Physics with the Strong Interactions* (Reading, MA: Addison-Wesley)
Shapiro S L and Teukolsky S A 1983 *Black Holes, White Dwarfs and Neutron Stars* (New York: Wiley)
Valentin L 1981 *Subatomic Physics: Nuclei and Particles* vols 1 and 2 (Amsterdam: North-Holland)
Wapstra A H, Nygh G J and Van Lieshout R 1959 *Nuclear Spectroscopy Tables* (Amsterdam: North-Holland)
Williams W S C 1991 *Nuclear and Particle Physics* (Oxford: Oxford University Press)
Winter K (ed) 1991 *Neutrino Physics* (Cambridge: Cambridge University Press)
Wong S S M 1990 *Introductory Nuclear Physics* (Prentice-Hall International Editions)
Wu C S and Moszkowski S A 1966 *Beta-Decay* (New York: Wiley Interscience)
Wyss R (ed) 1995 *Proc. Int. Symp. on Nuclear Struct. Phenomena in The Vicinity of Closed Shells Phys. Scr.* T56

General textbooks discussing related topics

Alonso M and Finn E 1971 *Fundamental University Physics* vol II and vol III (Reading, MA: Addison-Wesley)
Feynmann R P, Leighton R B and Sands M 1965 *The Feynmann Lectures on Physics* vol III (Reading, MA: Addison-Wesley) chs 8–11
Lorrain P and Corson D R 1978 *Electromagnetism: Principles and Applications* (New York: Freeman)
Orear J 1979 *Physics* (New York: Collier MacMillan International Editions)
Panofsky W K H and Phillips M 1962 *Classical Electricity and Magnetism* 2nd edn (Reading, MA: Addison-Wesley)

Quantum mechanics texts

Flügge S 1974 *Practical Quantum Mechanics* (Berlin: Springer)
Greiner W 1980 *Theoretische Physik Band 4A: Quantumtheorie Spezielle Kapitel* (Frankfurt: Harri Deutsch-Thun)
Merzbacher E 1970 *Quantum Mechanics* 2nd edn (New York: Wiley)
Sakurai J J 1973 *Advanced Quantum Mechanics* (Reading, MA: Addison-Wesley)
Schiff L I 1968 *Quantum Mechanics* (New York: McGraw-Hill)

Mathematical references

Abramowitz M and Stegun J A 1964 *Handbook of Mathematical Functions* (New York: Dover)
Arfken G 1985 *Mathematical Methods for Physicists* 3rd edn (New York: Academic)
Wilkinson J H 1965 *The Algebraic Eigenvalue Problem* (Oxford: Clarendon)

OTHER REFERENCES

Åberg S, Flocard H and Nazarewicz W 1990 *Ann. Rev. Nucl. Part. Sci.* **40** 439
Abreu M C *et al* 1997 *Phys. Lett.* B **410** 337
Abul-Magd A Y, Harney H L, Simbel M H and Weidenmüller H A 2004 *Phys. Lett.* B **579** 278
Adams J *et al* 2003 *Phys. Rev. Lett.* **91** 072304
Adcox K *et al* 2002 *Phys. Rev. Lett.* **88** 022301
Adler S S *et al* 2003 *Phys. Rev. Lett.* **91** 072303
Ahmad Q R *et al* 2001 *Phys. Rev. Lett.* **87** 071301
——2002 *Phys. Rev. Lett.* **89** 011301
——2002 *Phys. Rev. Lett.* **89** 011302
Ahn M A *et al* 2003 *Phys. Rev. Lett.* **90** 041801
Ajzenberg-Selove F 1975 *Nucl. Phys.* A **248** 1
Alhassid Y, Bertsch G F, Liu S and Nakada H 2000 *Phys. Rev. Lett.* **84** 4313
Alhassid Y, Dean D J, Koonin S E, Lang G H and Ormand W E 1994 *Phys. Rev. Lett.* **72** 613
Alhassid Y, Liu S and Nakada H 1999 *Phys. Rev. Lett.* **83** 4265
Alt C *et al* 2003 http://arxiv.org.abs/hep-ex/0310014
Alt H *et al* 1995 *Phys. Rev. Lett.* **74** 62
Altarelli G 1994 *The Development of Perturbative QCD* (Singapore: World Scientific)
Andreyev A N *et al* 2000 *Nature* **405** 430
Arima A 2002 *Nucl. Phys.* A **704** 1c
Arima A and Iachello F 1975 *Phys. Rev. Lett.* **35** 1069
Arnold E *et al* 1987 *Phys. Lett.* B **197** 311
——1976 *Ann. Phys., NY* **99** 253
——1978 *Ann. Phys., NY* **111** 201
——1979 *Ann. Phys., NY* **123** 468
——1984 *Advances in Nuclear Physics* vol 13 (New York: Plenum)
Armbruster P and Münzenberg G 1989 *Sci. Amer.* (May) 36
Arnison G *et al* 1983 *Phys. Lett.* **122B** 103
Arrington J 2003 *Phys. Rev.* C **68** 034325
Arsene I *et al* 2003 *Phys. Rev. Lett.* **91** 072305
Athanassopoulos C *et al* 1995 *Phys. Rev. Lett.* **75** 2650
Aubert J J *et al* 1983 *Phys. Lett.* **123B** 275
Audi G, Bersillon D, Blachot J and Wapstra A H 1997 *Nucl. Phys.* A **624** 1 Database http://csn.www.in2p3.fr/amdc
Audi G and Wapstra A H 1995 *Nucl. Phys.* A **595** 409
Auerbach N 1987 *Proc. Workshop on Pion-Nucleus Physics: Future Directions and New Facilities at LAMPF* (AIP) **163** 34
——1990 *Ann. Phys., NY* **197** 376
Auger G *et al* 1994 *Nucl. Instrum. Methods Phys. Res. Sect.* A **350** 235
Austin S M and Bertsch G F 1995 *Sci. Am.* (June) 62
Avignone III F T and Brodzinski R L 1989 *Progr. Part. Nucl. Phys.* **21** 99
Bahcall J N 1990 *Sci. Amer.* (May) 26
Bahcall J N, Bahcall N and Shaviv G 1968 *Phys. Rev. Lett.* **20** 1209
Bahcall J N and Bethe H A 1990 *Phys. Rev. Lett.* **65** 2233
Baranger E U 1973 *Phys. Today* (June) 34

Barmin V V *et al* 2003 http://arxiv.org.abs/hep-ex/0304040
Barranco F and Broglia R A 1985 *Phys. Lett.* **151B** 90
Barranco M, Navarro J and Poves A 1997 *Phys. Rev. Lett.* **78** 4729
Barrett B R *et al* 2002 *Phys. Rev.* C **66** 024314
Barrett B R and Kirson M W 1973 *Adv. Nucl. Phys.* **6** 219
Barrett B R, Mihaila B, Pieper S C and Wiringa R B 2003 *Nucl. Phys. News* **13** 17
Barth J *et al* 2003 http://arxiv.org.abs/hep-ex/0307083
Bazin D *et al* 1992 *Phys. Rev.* C **45** 69
Beausang C W and Simpson J 1996 *J. Phys. G: Nucl. Phys.* **22** 527
Beck B *et al* 2000 *Eur. Phys. J.* A **8** 307
Bedaque P F and van Kolck U 2002 *Ann. Rev. Nucl. Part. Sci.* **52** 339
Bednarczyk P *et al* 1997 *Phys. Lett.* B **393** 285
Beiner M, Flocard H, Van Giai N and Quentin Ph 1975 *Nucl. Phys.* A **238** 29
Bemporad C, Gratta G and Vogl P 2002 *Rev. Mod. Phys.* **74** 297
Bergkvist K E 1972 *Nucl. Phys.* B **39** 317
Bertsch G F 1983 *Sci. Amer.* (May) 40
Bertsch G, Brown B A and Sagawa H 1989 *Phys. Rev.* C **39** 1154
Bertsch G and Esbensen H 1991 *Ann. Phys., New York* **209** 327
Bertsch G, Esbensen H and Sustich A 1990 *Phys. Rev.* C **42** 758
Bertsch G and Foxwell J 1990a *Phys. Rev.* C **41** 1300
Bertsch G and Foxwell J 1990b *Phys. Rev.* C **42** 1159
Bethe H 1936 *Phys. Rev.* **50** 332
Bethe H A and Bacher R F 1936 *Rev. Mod. Phys.* **8** 193
Bethe H and Peirles R 1934 *Nature* **133** 689
Bijker R and Frank A 2000 *Phys. Rev.* C **62** 014303
Bijker R, Giannini M M and Santopinto E 2003 http://arxiv.org.abs/hep-ph/0310281
Binder K (ed) 1979 *Monte-Carlo Methods in Statistical Physics* (Berlin: Springer)
Björnholm S and Lynn J E 1980 *Rev. Mod. Phys.* **52** 725
Blachot J and Marguier G 1990 *Nucl. Data Sheets* **60** 139
Blair J S, Farwell G W and McDaniels D K 1960 *Nucl. Phys.* **17** 641
Blomqvist J and Wahlborn S 1960 *Arkiv. Fys.* **16** 545
Blomqvist K I *et al* 1995 *Phys. Lett.* B **344** 85
Boehm F and Vogel P 1984 *Ann. Rev. Nucl. Part. Sci.* **34** 125
Boezio M *et al* 1999 *Phys. Rev. Lett.* **82** 4757
Bohigas O, Giannoni M J and Schmidt C 1984 *Phys. Rev. Lett.* **52** 1
Bohigas O and Weidenmüller H A 1988 *Ann. Rev. Nucl. Part. Sci.* **38** 421
Bohle D *et al* 1984 *Phys. Lett.* **137B** 27
Bohm A, Gadella M and Bruce Mainland G 1989 *Am. J. Phys.* **57** 1103
Bohr A 1951 *Phys. Rev.* **81** 134
——1952 *Mat. Fys. Medd. Dan. Vid. Selsk.* **26** no14
——1976 *Rev. Mod. Phys.* **48** 365
Bohr A and Mottelson B 1953 *Mat. Fys. Medd. Dan. Vid. Selsk.* **27** 16
——1979 *Phys. Today* (June)
Bohr N and Kalckar F 1937 *Mater. Fys. Med. Dan. Vid. Selsk.* **14** issue 10
Bollen G *et al* 1996 *Nucl. Instrum. Methods* A **368** 675
Börner H G, Jolie J, Hoyler F, Robinson S, Dewey M S, Greene G, Kessler E and Delattes
      R D 1988 *Phys. Lett.* B **215** 45
Bosetti P *et al* 1996 *Phys. Rev. Lett.* **76** 1204

Boyd R N 1994 *Mod. Phys.* E Suppl. 3 249

Bracco A, Alasia F, Leoni S, Vigezzi E and Broglia R A 1996 *Contemp. Phys.* **37** 183

Brandow B H 1967 *Rev. Mod. Phys.* **39** 771

Breuker H, Drevermann H, Grab Ch, Rademakers A A and Stone H 1991 *Sci. Amer.* (August) 42

Brianti G and Gabathuder E 1983 *Europhys. News* **14** 2

Brink D 1957 *Nucl. Phys.* **4** 215

Bromley D A 1979 *Phys. Scr.* **19** 204

Bromley D A 2002 *A Century of Physics* (New York: Springer)

Brown B A and Wildenthal B H 1988 *Ann. Rev. Nucl. Part. Sci.* **38** 29

Brown G E and Green A E S 1966a *Nucl. Phys.* **75** 401

——1966b *Nucl. Phys.* **85** 87

Brown G E and Kuo T T S 1967 *Nucl. Phys.* A **92** 481

Brown G E and Rho M 1983 *Phys. Today* (February) 24

Brown L and Rechenberg H 1988 *Am. J. Phys.* **56** 982

Brussaard P J 1967 *Ned. T. Nat.* **33** 202

Brussaard P J and Tolhoek H A 1958 *Physica* **24** 263

Burbridge F M, Burbridge G R, Fowler W A and Hoyle F 1957 *Rev. Mod. Phys.* **29** 547

Burleigh R, Leese M N and Tite M S 1986 *Radio Carbon* **28** 571

Cameron J A *et al* 1996 *Phys. Lett.* B **387** 266

Camp D C and Langer L M 1963 *Phys. Rev.* **129** 1782

Cardman L S 1996 *Nucl. Phys. News* **6** 25

Carlson J A 1990 *Energy Sci. Supercom.* 38

Carlson J A *et al* 1983 *Nucl. Phys.* A **401** 59

——1986 *Nucl. Phys.* A **449** 219

——1987 *Phys. Rev.* C **36** 2026

——1988 *Phys. Rev.* C **38** 1879

——1990 *Nucl. Phys.* A **508** 141c

Carlson J and Sciavilla R 1998 *Rev. Mod. Phys.* **70** 743

Casten R F 1981 *Interacting Bose–Fermi Systems in Nuclei* ed F Iachello (New York: Plenum) p 1

Casten R F and Warner D D 1988 *Rev. Mod. Phys.* **60** 389

Caurier E 1989 ANTOINE CRN Strasbourg

Caurier E, Gomez J M G, Manfredi V R and Salasnich L 1996 *Phys. Rev. Lett.* B **365** 7

Caurier E, Nowacki F and Poves A 2002 *Eur. Phys. J.* A **15** 145

Caurier E, Nowacki F, Poves A and Retamosa J 1997 *Phys. Rev.* C **58** 2033

Caurier E and Zuker A P 1994 *Phys. Rev.* C **50** 225

Caurier E *et al* 1995 *Phys. Rev. Lett.* **75** 2466

Caurier E, Nowacki F, Poves A and Retamosa J 1996 *Phys. Rev. Lett.* **77** 1954

Caurier E, Zuker A P, Poves A and Martinez-Pinedo G 1994 *Phys. Rev.* C **50** 225

Cavedon J M *et al* 1982 *Phys. Rev. Lett.* **49** 978

——1987 *Phys. Rev. Lett.* **58** 195

CEBAF 1986 *CEBAF Briefing for the Georgia Inst. of Technology* April

Celik T, Engels J and Satz H 1985 *Nucl. Phys.* B **256** 670

CERN 1982 *CERN/DOC 82-2* (January) 13, 19

——1983 *CERN Courier* **23** no 9

——1985 *CERN/DOC* (March) 11

——1990 *CERN Courier* (December) 1

CERN Courier 1994 March 3
CERN Courier 1995 December 2
Chamouard P A and Durand J M 1990 *Clefs CEA* **17** 36
Chartier M *et al* 1996 *Phys. Rev. Lett.* **77** 2400
Chen M, Imel D A, Radcliffe T J, Henrikson H and Boehm F 1992 *Phys. Rev. Lett.* **69** 3151
Chirovsky L M *et al* 1980 *Phys. Lett.* **94B** 127
Chodos A *et al* 1974 *Phys. Rev.* D **10** 2599
Christenson J H, Cronin J W, Fitch V L and Turlay R 1964 *Phys. Rev. Lett.* **13** 138
Cohen B L 1970 *Am. J. Phys.* **38** 766
——1987 *Am. J. Phys.* **55** 1076
Cohen S and Kurath D 1965 *Nucl. Phys.* A **73** 1
Conceptual Design of the Relativistic Heavy-Ion Collider *RHIC Report BNL-52195* 1989
    (May)
Cornell J (ed) *The Eurisol Report: a Feasibility Study for a European Isotope-Separator-
    On-Line Radioactive Ion Beam Facility* (GANIL, BP 55027, 14076 Caen-Cedex 5,
    France) December
Cowan C L and Reines F 1953 *Phys. Rev.* **92** 830
Crane H R 1968 *Sci. Amer.* (January) 72
Creutz M 2001 *Rev. Mod. Phys.* **73** 110
Cronin J W and Greenwood M S 1982 *Phys. Today* (July) 38
Csikay J and Szalay A 1957 *Nuovo Cim. Suppl.* Padova Conference
Damon P E *et al* 1989 *Nature* **337** 611
Darquennes D *et al* 1990 *Phys. Rev.* C **42** R804
Davidson J P 1965 *Rev. Mod. Phys.* **37** 105
Davies C and Collins S 2000 *Phys. World* **August** 35
Davies R Jr 2003 *Rev. Mod. Phys.* **75** 985
Davies R Jr, Harmer D S and Hoffmann K C 1968 *Phys. Rev. Lett.* **20** 1205
Davis R 1955 *Phys. Rev.* **97** 766
de Angelis G, Bracco A and Curien D 2003 *Europhys. News* **34–35** 181
Decamp D *et al* 1990 *Phys. Lett.* **235B** 399
Decrock P *et al* 1991 *Phys. Rev. Lett.* **67** 808
De Grand T *et al* 1975 *Phys. Rev.* D **12** 2060
Deleplanque M A *et al* 1999 *Nucl. Instrum. Methods* A **430** 292
Deleze M, Drissi S, Jolie J, Kern J, Vorlet J P 1993 *Nucl. Phys.* A **554** 1
Dembowski C *et al* 1990 *Phys. Rev. Lett.* **90** 014102
De Rujula A 1981 *Phys. Today* (July) 17
De Rujula A, Georgi H and Glashow S L 1975 *Phys. Rev.* D **12** 147
De Vries C, de Jager C W, deWitt-Huberts P K A 1983 *Fom Jaarboek* 167
Demarais D and Duggen J L 1970 *Am. J. Phys.* **58** 1079
Detraz C 1995 *Nucl. Phys.* A **583** 3c
Diakonov D *et al* 1997 *Z. Phys.* A **359** 305
Diamond R M and Stephens F S 1980 *Ann. Rev. Nucl. Part. Sci.* **30** 85
Dieperink A E L and de Witt-Huberts P K A 1990 *Ann. Rev. Nucl. Sci.* **40** 239
Dobaczewski J and Nazarewicz W 1995 *Phys. Rev.* C **51** R1070
Dobaczewski J, Flocard H and Treiner J 1984 *Nucl. Phys.* A **422** 103
Dobaczewski J, Hamamoto I, Nazarewicz W and Sheikh J A 1994 *Phys. Rev. Lett.* **72** 981
Dobaczewski J, Nazarewicz W and Werner T R 1996a *Z. Phys.* A **354** 27

Dobaczewski J, Nazarewicz W, Werner T R, Berger J F, Chinn C R and Dechargé J 1996b *Phys. Rev.* C **53** 2809

DOE Nucl. Science Adv. Committee 1983 A long-range plan for Nuclear Science *NSF Report*

DOE/NSF Nuclear Science Advisory Committee 1996 *Nuclear Science: a Long-Range Plan*

DOE/NSF Nuclear Science Advisory Committee 2002 *A Long-Range Plan for the Next Decade* Overview of opportunities in nuclear science

Doi *et al* 1983 *Progr. Theor. Phys.* **69** 602

Donné T private communication

Dufour M and Zuker A P 1996 *Phys. Rev.* C **54** 1641

Dullman Ch E *et al* 2002 *Nature* **418** 859

Eberth J *et al* 1997 *Prog. Part. Nucl. Phys.* **38** 29

Efimov V N 1970 *Phys. Lett.* B **33** 563

Efimov V N 1990 *Commun. Nucl. Part. Phys.* **19** 271

Eguchi K *et al* 2003 *Phys. Rev. Lett.* **90** 021802

Elliott J B *et al* 2002 *Phys. Rev. Lett.* **88** 042701

Elliott J P 1958 *Proc. R. Soc.* A **245** 128; 562

Elliott J P and Harvey M 1963 *Proc. R. Soc.* A **272** 557

Elliott S R, Hahn A A and Moe M K 1987 *Phys. Rev. Lett.* **59** 1649

Ellis C D and Wooster W A 1927 *Proc. R. Soc.* A **117** 109

Esbensen H 1991 *Phys. Rev.* C **44** 440

EURISOL 2003 *The EURISOL Report: a Feasibility Study for a European Isotope-Separator-On-Line Radioactive Ion Beam Facility* ed J Cornell (Caen: GANIL)

Fabergé J 1966 *CERN Courier* **6** no 10, 193

Falk D 2001 *Nature* **411** 12

Fedorov D V, Jensens A S and Riisager K 1994 *Phys. Rev.* C **50** 2372

Feenberg E and Trigg G 1950 *Rev. Mod. Phys.* **22** 399

Fermi E 1934 *Z. Phys.* **88** 161

Fesbach H 1995 *Nucl. Phys.* A **583** 871c

Firestone R B and Singh B 1994 *Table of Superdeformed Nuclear Bands and Fission Isomers* LBL-35916 UC-413, Lawrence Berkeley Laboratory

Fisher P, Kayser B and McFarland K S 1999 *Ann. Rev. Nucl. Part. Sci.* **49** 481

Focardi S and Ricci R A 1983 *Rivista del Nuovo Cim.* **6** 1

FOM 2002 De polarisatie van up-, down- and strange-quarks *FOM Jaarboek* 33

Fowler W A 1984 *Rev. Mod. Phys.* **56** 149

Friedman J I 1991 *Rev. Mod. Phys.* **63** 615

Friese J and Sümmerer K 1994 *Nucl. Phys. News* **4** 21

Frois B 1987 *Electron Scattering and Nuclear Structure, Rapport DPh-N/Saday, no 2432/03*

Frois B *et al* 1983 *Nucl. Phys.* A **396** 409c

Fuda M G 1984 *Am. J. Phys.* **52** 838

Fukuda S *et al* 1999 *Phys. Rev. Lett.* **82** 1810

——2001 *Phys. Rev. Lett.* **86** 5651

——2001 *Phys. Rev. Lett.* **86** 5656

Furry W 1939 *Phys. Rev.* **56** 1184

Futagami T *et al* 1999 *Phys. Rev. Lett.* **82** 5194

Galindo-Uribari A *et al* 1993 *Phys. Rev. Lett.* **71** 231

Gamow G 1928 *Z. Phys.* **51** 204

Garrett J D 1986 *Nature* **323**

——1987 *Shape and Pair Correlations in Rotating Nuclei NBI-TAL-87-1*

——1988 *The Response of Nuclei under Extreme Conditions* ed R A Broglia and G F Bertsch (New York: Plenum) 1

Garrett J D and Olsen K D (ed) 1991 *A Proposal for Physics with Exotic Beams at the HHIRF*

Garrett J D, Hagemann G B and Herskind B 1984 *Commun. Nucl. Part. Phys.* **13** 1

Gauthier N 1990 *Am. J. Phys.* **58** 375

Geiger H and Nutall J M 1911 *Phil. Mag.* **22** 613

——1912 *Phil. Mag.* **23** 439

Geissl H, Münzenberg G and Riisager K 1995 *Ann. Rev. Nucl. Part. Sci.* **45** 163

Geissl H *et al* 1992 *Phys. Rev. Lett.* **68** 3412

General Electric 1984 *Chart of Nuclides* 13th edn

Gerard A 1990 *Clefs CEA* **16** 27

Gerl J *et al* 1998 *VEGA-Proposal, GSI Report*

Gillet V, Green A M and Sanderson E A 1966 *Nucl. Phys.* **88** 321

Giovanozzi J *et al* 2002 *Phys. Rev. Lett.* **89** 102501

Glashow S L 1961 *Nucl. Phys.* **22** 579

——1980 *Rev. Mod. Phys.* **52** 539

——1991 *Phys. Lett.* **256B** 255

Glasmacher T *et al* 1997 *Phys. Lett.* B **395** 163

Goeke K, Polyakov M V and Vanderhaeghen M 2001 *Prog. Part. Nucl. Phys.* **47** 401

Goeke K and Speth J 1982 *Ann. Rev. Nucl. Part. Sci.* **32** 65

Goldhaber J 1986 *New Scient.* (November 13) 40

——1991 *LBL Research Review* (Spring) 22

Goldhaber M, Grodzins L and Sunyar A 1958 *Phys. Rev.* **109** 1015

Gomez del Campo J *et al* 2001 *Phys. Rev. Lett.* **86** 43

Gonin M *et al* 1996 *Nucl. Phys.* A **610** 404c

Gonzales-Garcia M C and Nir Y 2003 *Rev. Mod. Phys.* **75** 345

Gräf H-D *et al* 1992 *Phys. Rev. Lett.* **69** 1296

Greene S J *et al* 1979 *Phys. Lett.* **88B** 62

Greiner W and Stöcker H 1985 *Sci. Amer.* (January) 58

Gribbin J and Rees M 1990 *New Scient.* (13 January) 51

Gribov V and Pontecorvo B 1969 *Phys. Lett.* B **28** 493

Gröger J *et al* 2000 *Phys. Rev.* C **62** 064304

Gross D H E 2001 *Nucl. Phys.* A **681** 366

Grümmer F, Chen B Q, Ma Z Y and Krewald S 1996 *Phys. Lett.* B**387** 673

GSI-Nachrichten 1996a **4** 2

——1996b **5** 13 (in English)

GSI 2001 *An International Accelerator Facility for Beams of Ions and Antiprotons: A Conceptual Design Report* GSI November

Guardida R and Navarro J 2000 *Phys. Rev. Lett.* **84** 1144

Guhr T, Müller-Groeling A and Weidenmüller H A 1998 *Phys. Rev.* **299** 189

Guhr T and Weidenmüller H A 1990 *Ann. Phys.* **199** 412

Gurney R W and Condon E U 1928 *Nature* **122** 439

——1929 *Phys. Rev.* **33** 127

Gustafson C, Lamm I L, Nilsson B and Nilsson S G 1967 *Ark. Fys.* **36** 613

Gutbrod H and Stöcker H 1991 *Sci. Amer.* (November) 60
Guttormsen M *et al* 2000 *Phys. Rev.* C **61** 0673021
Gutzwiller M C 1990 *Chaos in Classical and Quantum Mechanics* (New York: Springer)
Gutzwiller M C 1992 *Scientific American* vol 26 (New York: Springer)
Hall N 1989 *New Scient.* (March, 31) 40
Hamada T and Johnston I D 1962 *Nucl. Phys.* B **4** 382
Hansen P G 1987 *Nature* **328** 476
Hansen P G 1991 *Nucl. Phys. News* **1** 21
——1993a *Nature* **361** 501
——1993b *Nucl. Phys.* A **553** 89c
——1996 *Nature* **384** 413
Hansen P G and Jonson B 1987 *Europhys. Lett.* **4** 409
Hansen P G, Jensen A S and Jonson B 1995 *Ann. Rev. Nucl. Part.Sci.* **45** 591
Haq U, Pandey A and Bohigas O 1982 *Phys. Rev. Lett.* **48** 1086
Harris S M 1965 *Phys. Rev.* B **138** 509
Harrison M, Peggs S and Roser T 2002 *Ann. Rev. Nucl. Part. Sci.* **52** 425
Hatakeyama S *et al* 1998 *Phys. Rev. Lett.* **81** 2016
Haxel O, Jensen J H D and Suess H E 1949 *Phys. Rev.* **75** 1766
——1950 *Z. Phys.* **128** 295
Haxton W C and Holstein B R 2000 *Am. J. Phys.* **68** 15
Haxton W C and Johnson C 1990 *Phys. Rev. Lett.* **65** 1325
Haxton W C and Stephenson Jr G J 1984 *Progr. Part. Nucl. Phys.* **12** 409
Haxton W C, Stephenson G J and Strottman D 1981 *Phys. Rev. Lett.* **47** 153
——1982 *Phys. Rev.* D **25** 2360
Hayward E 1965 *Nuclear Structure and Electromagnetic Interactions* ed N MacDonald
     (Edinburgh and London: Oliver and Boyd) p 141
Heeger K S 2001 *Europhys. News* **32/5** 80
Heisenberg W 1932 *Z. Phys.* **77** 1
Heiss W D, Nazmitdinov R G and Radu S 1995 *Phys. Rev.* C **52** 3032
Heyde K 1989 *J. Mod. Phys.* A **4** 2063
——1991 *Algebraic Approaches to Nuclear Structure (Ecole Joliot Curie de Physique
     Nucléaire IN2P3)* 153
Heyde K 1994 *The Nuclear Shell Model, Study Edition* (Berlin: Springer)
Heyde K and Meyer R A 1988 *Phys. Rev.* C **37** 2170
Heyde K, Van Isacker P, Waroquier M, Wood J L and Meyer R A 1983 *Phys. Rep.* **102** 291
Hime A 1993 *Phys. Lett.* **299B** 165
Hime A and Jelley N A 1991 *Phys. Lett.* **257B** 441
Hime A and Simpson J J 1989 *Phys. Rev.* D **39** 1837
Hirata K S *et al* 1990 *Phys. Rev. Lett.* **65** 1297
Hofmann S 1989 *Particle Emissions from Nuclei* vol II ed D N Penarer and M S Ivascu
     (Boca Raton, FL: CRC Press) p 25
Hofmann S 1998 *Rep. Prog. Phys.* **69** 827
Hofmann S and Münzenberg G 2000 *Rev. Mod. Phys.* **72** 733
——1996 *Nucl. Phys. News* **6** 26
Honma M, Mizusaki T and Otsuka T 1995 *Phys. Rev. Lett.* **75** 1284
——1996 *Phys. Rev. Lett.* **77** 3315
Honma M, Mizusaki T and Otsuka T 1996 *Phys. Rev. Lett.* **57** 3315
Horoi M, Brown B A and Zelevinsky V 2001 *Phys. Rev. Lett.* **87** 062501

Horoi M, Zel V and Brown B A 1995 *Phys. Rev. Lett.* **74** 5194
Hubbard J 1959 *Phys. Rev. Lett.* **3** 77
Hussonnois M *et al* 1991 *Phys. Rev.* C **43** 2599
Huyse M 1991 *Hoger Aggregaatsthesis* University of Leuven, unpublished
Iachello F 1981 *Chem. Phys. Lett.* **78** 581
Iachello F 1985 *Physica D* **15** 85
——2000 *Nucl. Phys. News.* **10** 12
——2000 *Phys. Rev. Lett.* **85** 3580
——2001 *Phys. Rev. Lett.* **87** 052502
——2003 *Phys. Rev. Lett.* **91** 132502
Iachello F and Levine R D 1995 *Algebraic Theory of Molecules* (Oxford: Oxford University Press)
Iachello F and Talmi I 1987 *Rev. Mod. Phys.* **59** 339
Ieki *et al* 1993 *Phys. Rev. Lett.* **70** 730
Ikeda K 1992 *Nucl. Phys.* A **538** 355c
Inglis D R 1954 *Phys. Rev.* **96** 1059
——1956 *Phys. Rev.* **103** 1786
ISL Newsletter information obtained at rick@riviera.physics.yale.edu
ISL Report 1991 *The Isospin Laboratory: Research Opportunities with Radioactive Nuclear Beams* LALP 91-51
Jackson K P 1970 *et al Phys. Lett.* B **33** 33
Jaffe R and Wilzeck F 2003 *Phys. Rev. Lett.* **91** 232003
Janssens R V F and Khoo T L 1991 *Ann. Rev. Nucl. Part. Sci.* **41** 321
Janssens R and Stephens F 1996 *Nucl. Phys. News* **6** 9
Johannsen L, Jensen A S and Hansen P G 1990 *Phys. Lett.* B **244** 357
Johnson A and Szymanski Z 1973 *Phys. Rep.* **7** 181
Johnson A, Ryde H and Sztarkier J 1971 *Phys. Lett.* **34B** 605
Johnson C W, Bertsch G F and Dean D J 1998 *Phys. Rev. Lett.* **80** 2749
Johnson C W, Koonin S E, Lang G H and Ormand W E 1992 *Phys. Rev. Lett.* **69** 3157
Jokinen A *et al* CERN-INTC-2003-018, approved proposal
Jolie J 2002 *Sci. Am.* **287** 54
Jonson B 1995 *Nucl. Phys.* A **583** 733
Jonson B 2004 *Phys. Rep.* **389** 1
Jonson B, Ravn H L and Walter G 1993 *Nucl. Phys. News* **3** 5
Julin R, Helariutta K and Muikku M 2001 *J. Phys. G: Nucl. Part. Phys.* **27** R109
Jumper E J *et al* 1984 *Arch. Chem.-III* ed J B Lambert (Washington: Am. Chem. Soc.) p 447
Kajita T and Totsuka Y 2001 *Rev. Mod. Phys.* **73** 85
Kar K, Sarkar S, Gomez J M G, Manfredi V R and Salasnich L 1997 *Phys. Rev.* C **55** 1260
Kelly J J 2002 *Phys. Rev.* C **66** 065203
Kendall H W 1991 *Rev. Mod. Phys.* **63** 597
Khoo T L *et al* 1996 *Phys. Rev. Lett.* **76** 1583
Kirsten T A 1999 *Rev. Mod. Phys.* **71** 1213
Klapdor-Kleingrothaus H V, Dietz A, Harney H L and Krivosheina I V 2001 *Mod. Phys. Lett.* A **16** 2409
Kobayashi T 1992 *Nucl. Phys.* A **538** 343C
Kobayashi T *et al* 1988 *Phys. Rev. Lett.* **60** 2599
——1989 *Phys. Lett.* B **232** 51

Koonin S E 1994 *Nucl. Phys.* A **574** 1c

Koonin S E, Dean D J and Langanke K 1997a *Phys. Rep* **278** 1

——1997b *Ann. Rev. Nucl. Part.Sci.* **47** 463

Kortelahti M O *et al* 1991 *Phys. Rev.* C **43** 484

Koshiba M 2003 *Rev. Mod. Phys.* **75** 1011

Kruppa A T, Bender M, Nazarewicz W, Reinhard P G, Vertse T and Ćwiok C 2000 *Phys. Rev.* **61** 03413

Kruppa A T, Pal K F and Rowley N 1995 *Phys. Rev.* C **52** 1818

Kündig W, Fritschi, Holzschuh E, Petersen J W, Pixley R E and Strüssi H 1986 *Weak and Electromagnetic Interactions in Nuclei* ed H V Klapdor (Berlin: Springer)

Kuo T T S 1974 *Ann. Rev. Nucl. Sci.* **24** 101

Kuo T T S and Brown G E 1966 *Nucl. Phys.* **85** 40

Kuyucak S and Li S C 1995 *Phys. Lett.* B **349** 253

La S Sindone-Ricerche e studi della Commissione di Esperti Nominato dall' Arcivescovo di Torino, Cardinal M. Pellegrino, nel 1969, Suppl. Rivista Dioscesana Torinese 1976

Lang G H, Johnson C W, Koonin S E and Ormand W E 1993 *Phys. Rev.* C **48** 1518

Langanke K 1998 *Phys. Lett.* B **438** 235

Langanke K and Poves A 2000 *Nucl. Phys. News* **10** 16

Lederer C M, Hollander J M and Perlman I 1967 *Tables of Isotopes* 6th edn (New York: Wiley)

Lederman L M 1963 *Sci. Amer.* (March) 60

Lee T D and Yang C N 1956 *Phys. Rev.* **104** 254

Lenzi S *et al* 1996 *Z. Phys.* A **354** 117

Leuscher M *et al* 1994 *Phys. Rev.* C **49** 955

Lewitowicz M *et al* 1993 *Nouvelles de GANIL* **48** 7

——1994 *Phys. Lett.* B **332** 20

——1995a *Nucl. Phys.* A **583** 857

——1995b *Nucl. Phys.* A **588** 197c

Li S C and Kuyucak S 1996 *Nucl. Phys.* A **604** 305

Lieb K P, Broglia R and Twin P 1994 *Nucl. Phys. News* **4** 21

Lieder R M and Ryde H 1978 *Adv. Nucl. Phys.* **10** 1

Lipas P 1984 *Int. Rev. Nucl. Phys.* **2** 33

——1990 private communication

Livio M, Hollowel D, Weiss A and Truran J W 1989 *Nature* **340** 281

Lo Iudice N and Palumbo F 1978 *Phys. Rev. Lett.* **41** 1532

——1979 *Nucl. Phys.* A **326** 193

Los Alamos Meson Physics Facility 1991 *Research Proposal*

Los Alamos Working Group on the use of RIB for Nuclear Astrophysics 1990

Lubimov V A *et al* 1980 *Phys. Lett.* **94B** 266

Ludlam T and McLerran L 2003 *Phys. Today* **October** 48

Machleidt R 1985 *Relat. Dyn. and Quark-Nuclear Physics* ed M B Johnson and A Picklesimer (New York: Wiley) p 71

Mackintosh R, Al-Khalili J, Jonson B and Pena T 2001 *Nucleus: a trip to the Heart of the Matter* (Bristol: Canopus Publishing Ltd)

Maddox J 1984 *Nature* **307** 207

——1988 *Nature* **336** 303

Madey R *et al* 2003 *Phys. Rev. Lett.* **91** 122002

Mahaux C, Bortignon P F, Broglia R A and Dasso C H 1985 *Phys. Rep.* **120** 1

Mahaux C and Sartor R 1991 *Adv. Nucl. Phys.* **20** 1

Majorana E 1937 *Nuovo Cim.* **5** 171

Mang H J 1960 *Phys. Rev.* **119** 1069

——1964 *Ann. Rev. Nucl. Sci.* **14** 1

Mariscotti M A J, Scharff-Goldhaber G and Buck B 1969 *Phys. Rev.* **178** 1864

Marques Moreno F M 2002 *Preprint* LPCC 03-02 and 'Halos, molecules and multineutrons' *Proc. Ecole-Joliot-Curie de Physique Nucléaire* September 2002

Martinez-Pinedo G *et al* 1996 *Phys. Rev.* C **54** R2150

Martinez-Pinedo G, Zuker A P, Poves A and Caurier E 1997 *Phys. Rev.* C **55** 187

Mattauch H E, Thiele W and Wapstra A H 1965 *Nucl. Phys.* **67** 1

Mayer M G 1949 *Phys. Rev.* **75** 1969

——1950 *Phys. Rev.* **78** 22

McHarris W and Rasmussen J O 1984 *Sci. Amer.* (January) 44

McRae W D, Etchegoyen A and Brown A B 1988 OXBASH MSU Report 524

Melby E *et al* 1999 *Phys. Rev. Lett.* **83** 3150

Meng J and Ring P 1996 *Phys. Rev. Lett.* **77** 3963

Metropolis N, Rosenbluth A W, Rosenbluth M N, Teller A H and Teller E 1953 *J. Chem. Phys.* **21** 1087

Metz A *et al* 1999 *Phys. Rev. Lett.* **83** 1542

——2000 *Phys. Rev.* C **61** 064313

Meyer-ter-Vehn J 1975 *Nucl. Phys.* A **249** 141

Michel, Nazarewicz W, Ploszajczak M and Bennaceur K 2002 *Phys. Rev. Lett.* **89** 042502

Mikheyev S P and Smirnov A Y 1985 *Sov. J. Nucl. Phys.* **42** 913

Millener D J *et al* 1983 *Phys. Rev.* C **28** 497

Mittig W *et al* 1987 *Phys. Rev. Lett.* **59** 1889

Mizusaki T, Honma M and Otsuka T 1996 *Phys. Rev.* C **53** 2786

Mizusaki T, Otsuka T, Honma M and Brown B A 2002 *Nucl. Phys.* A **704** 190c

Mizusaki T, Otsuka T, Utsuno Y, Honma M and Sebe T 1999 *Phys. Rev.* C **59** R1846

Moe M K and Rosen S P 1989 *Sci. Amer.* (November) 30

Molina R A, Gomez J M G and Retamosa J 2000 *Phys. Rev.* C **63** 014311

Moore C E 1949 *Atomic Energy Levels, Circular 467, vol 1, pXL NBS* (Washington DC)

Mordechai S and Moore C F 1991 *Nature* **352** 393

Morinaga H 1956 *Phys. Rev.* **101** 254

Morinaga H and Gugelot P C 1963 *Nucl. Phys.* **46** 210

Morita J L *et al* 1943 *Phys, Rev, Lett.* **70** 394

Mosel U 1999 *Fields, Symmetries and Quarks* (Berlin: Springer)

Motobayashi T *et al* 1991 *Phys. Lett.* B **264** 259

Mott N 1949 *Proc. Phys. Soc.* A **62** 416

——1968 *Rev. Mod. Phys.* **40** 677

Mottelson B R 1976 *Rev. Mod. Phys.* **48** 375

Mottelson B R and Nilsson S G 1959 *Mat. Fys. Skr. Dan. Vid. Selsk.* **1** no 8

Mottelson B R and Valatin J G 1960 *Phys. Rev. Lett.* **5** 511

Myers W D and Swiatecki W J 1966 *Nucl. Phys.* **81** 1

Nakada H and Alhassid Y 1997 *Phys. Rev. Lett.* **79** 2939

——1998 *Phys. Lett.* B **436** 231

Nakada H and Otsuka T 1997 *Phys. Rev.* C **55** 748, 2418

Nakada H and Sebe T 1996 *J. Phys. G: Nucl. Phys.* **22** 1349

Nakada K, Sebe T and Otsuka T 1994 *Nucl. Phys.* A **571** 467

Nakamura T *et al* 1994 *Phys. Lett.* B **331** 296

Nakano T *et al* 2003 *Phys. Rev. Lett.* **91** 012002

NAP 1999 *Nuclear physics: The Core of the Matter, The Fuel of the Stars* (Washington, DC: National Academy Press)

Navratil P and Ormand W E 2003 *Phys. Rev.* C **68** 034305

Nazarewicz W 1993 *Phys. Lett.* B **305** 196

——1999 *Nucl. Phys.* A **654** 195c

Nazarewicz W, Sherrill B, Tanihata I and Van Duppen P 1996 *Nucl. Phys. News* **6** 17

Nazarewicz W, Werner T R and Dobaczewski J 1994 *Phys. Rev.* C **50** 2860

Nazarewicz W 1998 private communication

Negele J 1985 *Phys. Today* (April) 24

Newton J O 1989 *Contemp. Phys.* **30** 277

Newton J O *et al* 1970 *Nucl. Phys.* A **141** 631

Nilsson S G 1955 *K. Dan. Vidensk. Selsk. Mat. Fys. Medd.* **29** no 16

Ninov V *et al* 1999 *Phys. Rev. Lett.* **83** 1104

——2002 *Phys. Rev. Lett.* **89** 039901

Nitsche M 1989 *New Sci.* (February 25) 55

Nolan P J and Twin P J 1988 *Ann. Rev. Nucl. Part. Sci.* **38** 533

Noyaux Atomique et Radioactivité 1996 *Dossier pour la Science (ed Française de Scientific American)* Dossier Hors Serie-October (in French)

Nuclear Structure and Nuclear Astrophysics 1992 *A Proposal* University of Tennessee Dept. Phys. UT-Th9201

NUPECC Report 1991 *Nuclear Physics in Europe, Opportunities and Perspectives*

——1993 *European Radioactive Beam Facilities*

——1997 *Nuclear Physics in Europe: Highlights and Opportunities*

——2000 *Computational Nuclear Physics*

——2002 *Radioactive Nuclear Beam Facilities*

——2004 *NUPECC Long-Range Plan*

Oganesian Yu Ts *et al* 2000 *Phys. Rev.* C **63** R011301

Oganesian Yu Ts *et al* 2004 *Phys. Rev.* C **69** 021601

OHIO Report 1997 *Scientific Opportunities with an Advanced ISOL Facilty, Columbus, Ohio*

Okolowicz J, Ploszajczak M and Rotter I 2003 *Phys. Rev.* **374** 271

Orear J 1973 *Am. J. Phys.* **41** 1131

Ormand W E, Dean D J, Johnson C W, Lang G H and Koonin S E 1994 *Phys. Rev.* C **49** 1422

Orr N A *et al* 1992 *Phys. Rev. Lett.* **69** 2050

Otsuka T 1985 Computer Package NPBOS JAERI-M85-09 unpubished

Otsuka T 2002 *Eur. Phys. J.* A **13** 69

——2002 *Nucl. Phys.* A **704** 21c

Otsuka T, Arima A and Iachello F 1978 *Nucl. Phys.* A **309** 1

Otuska T and Fukunishi N 1996 *Phys. Rep.* **264** 297

Otsuka T, Honma M and Mizusaki T 1998 *Phys. Rev. Lett.* **81** 1588

Ozawa A *et al* 1994 *Phys. Lett.* B **334** 18

Pandharipande V R, Sick I and deWitt Huberts P K A 1997 *Rev. Mod. Phys.* **69** 981

Pato M P *et al* 1994 *Phys. Rev.* C **49** 2919

Pfützner M *et al* 2002 *Eur. Phys. J.* A **14** 279

Philipp K 1926 *Naturwiss* **14** 1203

Phillips W R 1993 *Nature* **366** 14
PIAFE Collaboration 1994 *Overview of the PIAFE Project* ed H Nifenecker
Pieper S C 2002 *Eur. Phys. J.* A **13** 75
——2003 *Phys. Rev. Lett.* **90** 252501
Pieper S C, Pandharipande V R, Wiringa R B and Carlson J 2001 *Phys. Rev.* C **64** 014001
Pieper S C, Varga K and Wiringa R B 2002 *Phys. Rev.* C **66** 044310
Poves A 1997 Le modele en couches *Lectures Given at the Ecole Internationale Joliot-Curie, 'Structure nucleaire: Un Nouvel Horizon'*
——1999 *J. Phys. G: Nucl. Part. Phys.* **25** 589
——2003 Private communication
Povh B, Rith K, Scholz C and Zetsche F 2002 *Particles and Nuclei: An Introduction to the Physics Concepts* (Berlin: Springer)
Price P B 1989 *Ann. Rev. Nucl. Part. Sci.* **39** 19
Quint E N M 1988 *PhD Thesis* University of Amsterdam, unpublished
Racah G 1943 *Phys. Rev.* **63** 347
Radon T *et al* 1997 *Phys. Rev. Lett.* **78** 4701
Ragnarsson I and Sheline R K 1984 *Phys. Scr.* **29** 385
Rainwater J 1976 *Rev. Mod. Phys.* **48** 385
Raman S, Walkiewicz T A and Behrens H 1975 *At. Data Nucl. Data Tables* **16** 451
Ravn H L 1979 *Phys. Rep.* **54** 241
Ravn H L and Allardyce B W 1989 *Treatise on Heavy-Ion Science* vol 8, ed D A Bromley (New-York: Plenum) p 363
Ravn H L *et al* 1994 *Nucl. Instrum. Meth.* B **88** 441
Redi O 1981 *Phys. Today* (February) 26
Rego R A 1991 *Phys. Rev.* C **44** 1944
Reid R V 1968 *Ann. Phys., NY* **50** 411
Reines F and Cowan C L 1959 *Phys. Rev.* **113** 237
Retamosa J, Caurier E, Nowacki F and Poves A 1997 *Phys. Rev.* C **55** 1266
Richter A 1991 Private communication and invitation card at the inauguration of the S-Dalinac 18 October 1991
Richter A 1999 *Emerging Applications of Number Theory, The IMA Volumes in Mathematics and its Applications* vol 109, ed D A Hejhah *et al* (New York: Springer) p 479
——2002 Spectral properties of chaotic quantum systems: Billiards and Nuclei *Int. Symp. 'The Nuclear Many-Body System: exploring the limits'* Gent transparencies
——1993 *Nucl. Phys.* A **553** 417c
Riisager K *et al* 1990 *Phys. Lett.* B **235** 30
RNB 2002 Radioactive nuclear beams *Nucl. Phys.* A **701**
Robertson R G H and Knapp D A 1988 *Ann. Rev. Nucl. Part. Sci.* **38** 185
Rombouts S 1997 A Monte-Carlo method for fermionic many-body problems *PhD Thesis* University of Gent unpublished
Rombouts S and Heyde K 1998 *Phys. Rev. Lett.* **80** 885
Rose H J and Brink D M 1967 *Rev. Mod. Phys.* **39** 106
Rose H J and Jones G A 1984 *Nature* **307** 245
Rowley N 1985 *Nature* **313** 633
Ryckebusch J 1988 *PhD Thesis* University of Gent, unpublished
Ryckebusch J *et al* 1988 *Nucl. Phys.* A **476** 237

Rykaczewski K 1993 *Proc. 6th. Int. Conf. on Nuclei far from Stability (Inst. Phys. Conf. Ser. 132)* (Bristol: IOP Publishing) p 517
——1994 *Nucl. Phys. News* **4** 27
Rykaczewski K *et al* 1995 *Phys. Rev.* C **52** R2310
Saint-Laurent M G *et al* 1989 *Z. Phys.* A **332** 457
Saint-Laurent M G 1997 *Clefs CEA* **35** Printemps 3 (in French)
Salam A 1980 *Rev. Mod. Phys.* **52** 525
Salpeter E E 1957 *Phys. Rev.* C **107** 516
Satchler G R 1978 *Rev. Mod. Phys.* **50** 1
Satz H 1985 *Ann. Rev. Nucl. Sci.* **35** 245
——1986 *Nature* **324** 116
Schaefer T 2003 *Physics World* **June** 31
Scharff-Goldhaber G, Dover C B and Goodman A L 1976 *Ann. Rev. Nucl. Sci.* **26** 239
Scheit H *et al* 1996 *Phys. Rev. Lett.* **77** 3967
Schmid K W and Grümmer F 1987 *Rep. Prog. Phys.* **50** 731
Schneider R *et al* 1994 *Z. Phys.* A **348** 241
——1995 *Nucl. Phys.* A **588** 191c
Scholten O 1980 *PhD Thesis* University of Groningen unpublished
Schucan T H and Weidenmüller H A 1973 *Ann. Phys., NY* **76** 483
Schukraft J 1992 *Nucl. Phys. News* **2** 14
Schwarzschild B 1991 *Phys. Today* (August) 17
——1991 *Phys. Today* (May) 17
——1993 *Phys. Today* (April) 17
Serot B D and Walecka J D 1991 Relativistic Nuclear Many-Body theory *CEBAF-PR-91-037*
Sheline R K and Ragnarsson I 1991 *Phys. Rev.* C **43** 1476
Shimizu N, Otsuka T, Mizusaki T and Honma M 2002 *Nucl. Phys.* A **704** 244c
Shurtleff R and Derringh E 1989 *Am. J. Phys.* **57** 552
Siemssen R 1995 *Nucl. Phys.* A **583** 371c
Simpson J 1997 *Z. Phys.* A **358** 139
Simpson J and Hime A 1989 *Phys. Rev.* D **39** 1825
Simpson J *et al* 1984 *Phys. Rev. Lett.* **53** 648
Simpson J *et al* 2000 *Heavy Ion Phys.* **11** 159
Skyrme T H R 1956 *Phil. Mag.* **1** 1043
Smolańczuk R and Dobaczewski J 1993 *Phys. Rev.* C**48** R2166
Sorensen R A 1973 *Rev. Mod. Phys.* **45** 353
Sorge H, Shuryak E and Zahed I 1997 *Phys. Rev. Lett.* **79** 2775
Souza J A 1983 *Am. J. Phys.* **51** 545
Speth J and Van der Woude A 1981 *Rep. Prog. Phys.* **44** 46
Sprung W *et al* 1983 *Phys. Rev. Lett.* **51** 1522
Stachel J, Van Isacker P and Heyde K 1982 *Phys. Rev.* C **25** 650
Stelson P H and Grodzins L 1965 *Nucl. Data* A **1** 21
Stepanyan S *et al* 2003 http://arxiv.org.abs/hep-ex/0307018
Stephens F S 1960 *Nuclear Spectroscopy* ed F Ajzenberg-Selove (New York: Academic)
——1975 *Rev. Mod. Phys.* **47** 43
Stephens F S and Simon R S 1972 *Nucl. Phys.* A **183** 257
Stiegler U 1996 *Phys. Rep.* **277** 1

Stöckmann H-J 1990 *Quantum Chaos: An Introduction* (Cambridge: Cambridge University Press)

Stolzenberg H et al 1990 *Phys. Rev. Lett.* **65** 3104

Stratonovich R L 1957 *Dokl. Akad. Nauk.* **115** 1097 (Engl. Transl. 1958 Sov. Phys.–Dokl. **2** 416)

Strutinski V M 1967 *Nucl. Phys.* A **95** 420

——1968 *Nucl. Phys.* A **122** 1

Sur B et al 1991 *Phys. Rev. Lett.* **66** 2444

Talmi I 1984 *Nucl. Phys.* A **423** 189

Talmi I 1993 *Simple Models of Complex Nuclei* (Cambridge, MA: Harwood)

Tanihata I 1988 *Nucl. Phys.* A **488** 113c

——1996 *J. Phys. G: Nucl. Phys.* **22** 157

Tanihata I et al 1985 *Phys. Rev. Lett.* **55** 2676

——1988 *Phys. Lett.* B **206** 592

——1992 *Phys. Lett.* B **287** 307

Taubes G 1982 *Discover* (December) 98

Taylor R E 1991 *Rev. Mod. Phys.* **63** 573

The NuPECC Working Group on Radioactive Nuclear Beam Facilities, NuPECC Report, April 2000

The SPIRAL Radioactive Ion Beam Facility 1994 Report GANIL R 94 02

The UNIRIB Consortium 1995 *A White Paper* Oak Ridge

Thieberger P 1973 *Phys. Lett.* **45B** 417

Thompson I J et al 1993 *Phys. Rev.* C **47** 1317

Thompson R 1987 *New Sci.* (September) 3 56

't Hooft G 1980 *Sci. Amer.* (June) 90

Toth K et al 1984 *Phys. Rev. Lett.* **53** 1623

TRIUMF Canada's National Laboratory for Accelerator-Base Research 1993 Report

Uyttenhove J 1986 *RUG Internal Report* (June)

van der Woude A 1995 *Nucl. Phys.* A **583** 51c

Van Isacker P and Warner D D 1994 *J. Phys. G: Nucl. Part. Phys.* **20** 853

Van Neck D, Waroquier M and Ryckebusch J 1990 *Phys. Lett.* **249B** 157

——1991 *Nucl. Phys.* A **530** 347

Vautherin D and Brink D M 1972 *Phys. Rev.* C **5** 626

Velazques V, Hirsch J G, Frank A and Zuker A 2003 *Phys. Rev.* C **67** 034311

Villari A C C et al 1995 *Nucl. Phys.* A **588** 267C

Volya A, Zelevinsky V and Brown B A 2002 *Phys. Rev. Lett.* **65** 054312

von der Linden W 1992 *Phys. Rep.* **220** 53

von Weizsäcker C F 1935 *Z. Phys.* **96** 431

Walcher Th 1994 *Nucl. Phys. News* **4** 5

Wapstra A H 1958 *Handbuch der Physik* vol 1

Wapstra A H and Gove N B 1971 *Nucl. Data Tables* **9** 267

Waroquier M 1982 *Hoger Aggregaatsthesis* University of Gent, unpublished

Waroquier M, Heyde K and Wenes G 1983 *Nucl. Phys.* A **404** 269; 298

Waroquier M, Ryckebusch J, Moreau J, Heyde K, Blasi N, Van der Werf S Y and Wenes G 1987 *Phys. Rep.* **148** 249

Waroquier M, Sau J, Heyde K, Van Isacker P and Vincx H 1979 *Phys. Rev.* C **19** 1983

Weidenmüller H A 1986 *Comm. Nucl. Part. Phys.* **16** 199

Weinberg S 1974 *Rev. Mod. Phys.* **46** 255

——1980 *Rev. Mod. Phys.* **52** 515

Weisskopf V F 1961 *Phys. Today* **14** 18

——1994 *AAPPS Bull.* **4** 22

Wetherhill G W 1975 *Ann. Rev. Nucl. Sci.* **25** 283

Wietfeldt F E *et al* 1993 *Phys. Rev. Lett.* **70** 1759

Wigner E P 1937 *Phys. Rev.* **51** 106

Wilson K 1974 *Phys. Rev.* D **10** 2445

Wiringa R B and Pieper S C 2002 *Phys. Rev. Lett.* **89** 182501

Witherell M S 1988 *Nature* **331** 392

Wolfenstein L 1978 *Phys. Rev.* D **17** 2369

——1999 *Rev. Mod. Phys.* **71** S140

Wolfs F 2001 *Beam Line, Stanford Linear Accelerator Center* **31** 2

Wood J L 1992 private communication

Wood J L, Heyde K, Nazarewicz W, Huyse M and Van Duppen P 1992 *Phys. Rep.* **215** 101

Woods P J and Davids C N 1997 *Ann. Rev. Nucl. Part. Sci.* **47** 541

Wu C S, Ambler E, Hayward R W, Hoppes D D and Hudson R P 1957 *Phys. Rev.* **105** 1413

Yokoyama A 1997 *Phys. Rev.* C **55** 1282

Zdesenko Y 2002 *Rev. Mod. Phys.* **74** 663

Zeldes N *et al* 1967 *Mat. Fys. Skr. Dan. Vid. Selsk.* **3** no 5

Zelevinsky V, Brown A E, Frazier N and Horoi M 1996 *Phys. Rep.* **276** 885

Zhao V M and Arima A 2001 *Phys. Rev.* C **64** 041301(R)

Zheng X *et al* 2003 *Phys. Rev. Lett.* at press

Zhukov M V *et al* 1993 *Phys. Rep.* **231** 151

Zuber K 1998 *Phys. Rev.* **305** 295

Zuker A P, Retamosa J, Poves A and Caurier E 1995 *Phys. Rev.* C **52** R1741

Index

CPSIA information can be obtained
at www.ICGtesting.com
Printed in the USA
BVHW040528161118
533256BV00008B/37/P